Power Quality in Electrified Transportation Systems

Power Quality in Electrified Transportation Systems

Editors

Andrea Mariscotti
Leonardo Sandrolini

MDPI • Basel • Beijing • Wuhan • Barcelona • Belgrade • Manchester • Tokyo • Cluj • Tianjin

Editors
Andrea Mariscotti
University of Genova
Italy

Leonardo Sandrolini
University of Bologna
Italy

Editorial Office
MDPI
St. Alban-Anlage 66
4052 Basel, Switzerland

This is a reprint of articles from the Special Issue published online in the open access journal *Energies* (ISSN 1996-1073) (available at: https://www.mdpi.com/journal/energies/special_issues/electrified_transportation).

For citation purposes, cite each article independently as indicated on the article page online and as indicated below:

LastName, A.A.; LastName, B.B.; LastName, C.C. Article Title. *Journal Name* **Year**, *Volume Number*, Page Range.

ISBN 978-3-0365-2760-4 (Hbk)
ISBN 978-3-0365-2761-1 (PDF)

Contents

About the Editors . . . . **vii**

Preface to "Power Quality in Electrified Transportation Systems" . . . . **ix**

Hamed Jafari Kaleybar, Morris Brenna, Federica Foiadelli, Seyed Saeed Fazel and Dario Zaninelli
Power Quality Phenomena in Electric Railway Power Supply Systems: An Exhaustive Framework and Classification
Reprinted from: *Energies* **2020**, *13*, 6662, doi:10.3390/en13246662 . . . . **1**

Vítor A. Morais, João L. Afonso, Adriano S. Carvalho and António P. Martins
New Reactive Power Compensation Strategies for Railway Infrastructure Capacity Increasing
Reprinted from: *Energies* **2020**, *13*, 4379, doi:10.3390/en13174379 . . . . **37**

Si Wu, Mingli Wu, Yi Wang
A Novel Co-Phase Power-Supply System Based on Modular Multilevel Converter for High-Speed Railway AT Traction Power-Supply System
Reprinted from: *Energies* **2021**, *14*, 253, doi:10.3390/en14010253 . . . . **63**

Shaofeng Xie, Yiming Zhang and Hui Wang
A Novel Co-Phase Power Supply System for Electrified Railway Based on V Type Connection Traction Transformer
Reprinted from: *Energies* **2021**, *14*, 1214, doi:10.3390/en14041214 . . . . **81**

Chakrit Panpean, Kongpol Areerak, Phonsit Santiprapan, Kongpan Areerak and Seang Shen Yeoh
Harmonic Mitigation in Electric Railway Systems Using Improved Model Predictive Control
Reprinted from: *Energies* **2021**, *14*, 2012, doi:10.3390/en14072012 . . . . **103**

Marcin Steczek, Piotr Chudzik and Adam Szeląg
Application of a Non-carrier-Based Modulation for Current Harmonics Spectrum Control during Regenerative Braking of the Electric Vehicle
Reprinted from: *Energies* **2020**, *13*, 6686, doi:10.3390/en13246686 . . . . **119**

Mohamed Tanta, Jose Cunha, Luis A. M. Barros, Vitor Monteiro, José Gabriel Oliveira Pinto, Antonio P. Martins and Joao L. Afonso
Experimental Validation of a Reduced-Scale Rail Power Conditioner Based on Modular Multilevel Converter for AC Railway Power Grids
Reprinted from: *Energies* **2021**, *14*, 484, doi:10.3390/en14020484 . . . . **141**

Vaclav Kus, Bohumil Skala and Pavel Drabek
Complex Design Method of Filtration Station Considering Harmonic Components
Reprinted from: *Energies* **2021**, *14*, 5872, doi:10.3390/en14185872 . . . . **169**

Qiujiang Liu, Binghan Sun, Qinyao Yang, Mingli Wu and Tingting He
Harmonic Overvoltage Analysis of Electric Railways in a Wide Frequency Range Based on Relative Frequency Relationships of the Vehicle–Grid Coupling System
Reprinted from: *Energies* **2020**, *13*, 6672, doi:10.3390/en13246672 . . . . **187**

Andrea Mariscotti and Leonardo Sandrolini
Detection of Harmonic Overvoltage and Resonance in AC Railways Using Measured Pantograph Electrical Quantities
Reprinted from: *Energies* **2021**, *14*, 5645, doi:10.3390/en14185645 . . . . **203**

Yljon Seferi, Steven M. Blair, Christian Mester and Brian G. Stewart
A Novel Arc Detection Method for DC Railway Systems
Reprinted from: *Energies* **2021**, *14*, 444, doi:10.3390/en14020444 . . . . **225**

Lan Ma, Yuhua Du, Leilei Zhu, Fan Yang, Shibiao Xiang and Zeliang Shu
Decentralized Control Strategy for an AC Co-Phase Traction Microgrid
Reprinted from: *Energies* **2021**, *14*, 7, doi:10.3390/en14010007 . . . . **247**

Tingting He, Dylan Dah-Chuan Lu and Mingli Wu
Four-Quadrant Operations of Bidirectional Chargers for Electric Vehicles in Smart Car Parks: G2V, V2G, and V4G
Reprinted from: *Energies* **2021**, *14*, 181, doi:10.3390/en14010181 . . . . **263**

Szymon Haładyn
The Problem of Train Scheduling in the Context of the Load on the Power Supply Infrastructure. A Case Study
Reprinted from: *Energies* **2021**, *14*, 4781, doi:10.3390/en14164781 . . . . **281**

Mykola Kostin, Anatolii Nikitenko, Tetiana Mishchenko and Lyudmila Shumikhina
Electrodynamics of Reactive Power in the Space of Inter-Substation Zones of AC Electrified Railway Line
Reprinted from: *Energies* **2021**, *14*, 3510, doi:10.3390/en14123510 . . . . **301**

Janusz Mindykowski, Tomasz Tarasiuk and Piotr Gnaciński
Review of Legal Aspects of Electrical Power Quality in Ship Systems in the Wake of the Novelisation and Implementation of IACS Rules and Requirement
Reprinted from: *Energies* **2021**, *14*, 3151, doi:10.3390/en14113151 . . . . **321**

About the Editors

Andrea Mariscotti received his Ph.D. in Electrical Engineering in 1997 from the University of Genoa. After some years as a tenure track researcher, he joined the former Electrical Engineering Department of the University of Genoa as Assistant Professor in 2005, teaching Analog and Digital Electronics, Advanced Measurements and Signal Processing, and, recently, Electrical Measurements. His main research interests are EMC applied to industrial, military and transportation systems, including system assurance and functional safety aspects; power quality and power system modeling and analysis; electrical and electromagnetic measurement, including the design and construction of measurement setups and instrumentation; earthing, stray current and lightning protection design; and testing for electrified transports.

Leonardo Sandrolini received his Ph.D. degree in Electrical Engineering in 2000 from the University of Bologna, Bologna, Italy. Since 2001, he has been with the Department of Electrical, Electronic, and Information Engineering "Guglielmo Marconi", University of Bologna, where he is currently an Associate Professor teaching Electrical Engineering and Electromagnetic Compatibility. His research interests are in the areas of electromagnetic field theory, electromagnetic compatibility, electrical characterization of renewable energy sources and near-field wireless power transfer.

Preface to "Power Quality in Electrified Transportation Systems"

Due to the wide range of signaling, control, diagnostic and metering functions implemented nowadays with higher standards of immunity, availability and reliability, and to the ever increasing integration with renewables and smart transportation modes, Power Quality for modern railways and electrified transportation systems has widened its scope and refined its definitions. The extension of the frequency range, the inclusion of resonant behavior and system stability, the coverage of electric arcs and other transients and interactions with the high-voltage distribution and transmission upstream traction power stations are just the most significant examples.

A comprehensive presentation of the aforementioned phenomena is offered in "Power Quality Phenomena in Electric Railway Power Supply Systems: An Exhaustive Framework and Classification" by Kaleybar et al.

Yet, the most relevant impact on the grid upstream remains the more "classical" reactive power demand and low-frequency harmonic distortion, as confirmed by the significant amount of contributions discussing problems of Power Quality compensation both at the traction supply infrastructure ("New Reactive Power Compensation Strategies for Railway Infrastructure Capacity Increasing" by Morais et al.; "A Novel Co-Phase Power-Supply System Based on Modular Multilevel Converter for High-Speed Railway AT Traction Power-Supply System" by Si Wu, Mingli Wu and Yi Wang,;"A Novel Co-Phase Power Supply System for Electrified Railway Based on V Type Connection Traction Transformer" by Shaofeng Xie, Yiming Zhang, Hui Wang; "Harmonic Mitigation in Electric Railway Systems Using Improved Model Predictive Control" by Panpean et al.) and onboard ("Application of a Non-carrier-Based Modulation for Current Harmonics Spectrum Control during Regenerative Braking of the Electric Vehicle" by Steczek, Chudzik and Szeląg). With their work "Experimental Validation of a Reduced-Scale Rail Power Conditioner Based on Modular Multilevel Converter for AC Railway Power Grids", Tanta et al. have offered a significant amount of experimental results on a scaled prototype of a power conditioner. Kus, Skala and Drabek instead in "Complex Design Method of Filtration Station Considering Harmonic Components" consider the problem from a classic standpoint, optimizing the design of passive harmonic control measures (filters).

Consequential to the harmonic excitation of AC railway lines, there is the problem of resonances causing overvoltages (parallel resonances) and undue stress in components and the possible triggering of protection relays: "Harmonic Overvoltage Analysis of Electric Railways in a Wide Frequency Range Based on Relative Frequency Relationships of the Vehicle–Grid Coupling System" by Qiujiang Liu et al. investigates the impedance behavior of traction power stations over a wide frequency range, proposing a method to go beyond the limitation imposed by the switching frequency of compensating converters; "Detection of Harmonic Overvoltage and Resonance in AC Railways Using Measured Pantograph Electrical Quantities" by Mariscotti and Sandrolini discusses a straightforward approach to the analysis and interpretation of rolling stock input quantities to detect in real time incipient resonances and in general to identify the frequency ranges of abnormally large or small line impedance values.

Moving away from the core problem of the reactive power impact, harmonics and resonances of railways, other contributions have considered Power Quality problems of electrified transports in a wider context, dealing with transient phenomena (electric arcs typical of DC railways and metros in

"A Novel Arc Detection Method for DC Railway Systems" by Seferi et al.), integration with renewable energies ("Decentralized Control Strategy for an AC Co-Phase Traction Microgrid" by Lan Ma et al.) and electric vehicles ("Four-Quadrant Operations of Bidirectional Chargers for Electric Vehicles in Smart Car Parks: G2V, V2G, and V4G" by Tingting He et al.), or simply considering the impact on the infrastructure and train scheduling as one of the degrees of freedom for its optimization ("The Problem of Train Scheduling in the Context of the Load on the Power Supply Infrastructure. A Case Study" by Haładyn).

The paper "Electrodynamics of Reactive Power in the Space of Inter-Substation Zones of AC Electrified Railway Line" by Kostin et al. has proposed a different framework for the interpretation of harmonics propagation in railways, starting from fundamental electromagnetic theory.

Finally, "Review of Legal Aspects of Electrical Power Quality in Ship Systems in the Wake of the Novelisation and Implementation of IACS Rules and Requirement" has departed from land transportation systems focusing on maritime transportation, and in particular onboard electric distribution, with the ever increasing relevance of all-electric ships and the recent evolution of international regulations for harmonic distortion and power quality. In the light of the recent developments in Power Quality as a horizontal discipline across the various forms of electrified transportation systems, the papers appearing in this volume may be a useful reference and inspiration for further research.

Andrea Mariscotti, Leonardo Sandrolini
Editors

Review

Power Quality Phenomena in Electric Railway Power Supply Systems: An Exhaustive Framework and Classification

Hamed Jafari Kaleybar [1,*], Morris Brenna [1], Federica Foiadelli [1], Seyed Saeed Fazel [2] and Dario Zaninelli [1]

1 Department of Energy, Politecnico di Milano, 20156 Milan, Italy; morris.brenna@polimi.it (M.B.); federica.foiadelli@polimi.it (F.F.); dario.zaninelli@polimi.it (D.Z.)

2 School of Railway Engineering, Iran University of Science and Technology, Tehran 16844, Iran; fazel@iust.ac.ir

* Correspondence: hamed.jafari@polimi.it

Received: 14 November 2020; Accepted: 14 December 2020; Published: 17 December 2020

Abstract: Electric railway power systems (ERPS) as one of the most critical and high-power end-user loads of utility grids are characterized by outlandish power quality (PQ) problems all over the world. The extension and evolution of different supply topologies for these systems has resulted in significant and various forms of distortions in network voltage and current in all ERPS, the connected power system, and adjacent consumers. During the last years, numerous studies have been offered to investigate various aspects of PQs in a specific supplying topology. Variation in the supply structure of the ERPS and different types of locomotives has propelled the observation of different PQ phenomena. This versatility and development have led to confront considerable types of two-way interactive interfaces as well as reliability and PQ problems in ERPS. In addition, the lack of standards explicitly dedicated to ERPS has added to the ambiguity and complexity of this issue. In this paper, an extensive review of PQ distortions and phenomena in different configurations of ERPS is proposed and a systematic classification is presented. More than 140 scientific papers and publications are studied and categorized which can provide a fast review and a perfect perspective on the status of PQ indexes for researchers and experts.

Keywords: power quality; electric railway system; harmonics; unbalance; resonance; voltage distortions; reactive power; EMI

1. Introduction

The outstanding characteristics of electric railway power systems (ERPS) such as safe transport, high-power/capacity, high-speed, great reliability and resilience, and environment-friendly manner make them one of the popular and promising transport systems [1]. These systems have experienced considerable improvement and evolution during the last decades. Historical, geographical, and economical reasons have led to the establishment of different ERPSs structures in various countries [2]. This development and diversity have resulted in facing different forms of distortions and power quality (PQ) issues not only in ERPSs, but also in the supplier utility grid and other adjacent loads. Generally, PQ issues in ERPSs have been investigated individually by dividing the systems into three areas of DC, 16.67/25 Hz AC and 50/60 Hz AC. The DC type ERPSs introduced as the primary technology with lower requirements and capacity in transportation electrification. Urban railway systems including subways, trams, and light-rails are the most popular types of DC systems. Voltage/current harmonics, waveform transients, system imbalance, and low power factor are the foremost phenomena which are reported in the so-far published researches related to these systems [3–5]. The hardness and complexity of AC motors to supply the high-power traction loads with the industrial frequency caused to establishment of 16.67/25 Hz system taking advantage of rotary frequency converters. The frequency transient, stability problems driven by electromechanical transients, and amplification of negative sequence current or imbalance in synchronous generators are the main PQ issues of these systems [6–8]. With the subsequent developments of technologies and high demands for transportation, industrial frequency-based ERPSs have become promoted and more popular. Primary systems in this context are established based on 1 × 25 kV ERPSs, which can be found in different forms of without return conductor, with return feeder, and with booster transformer. The most important PQ phenomena reported in these systems include imbalance, harmonics, low-frequency oscillations, low power factor, voltage drops, and electromagnetic interference (EMI) based issues due to the return currents and arcing [9–11]. The next-generation AC ERPSs were adopted based on autotransformer based 2 × 25 kV systems concentrating on mitigation of EMI issues, voltage drops, and arcing problems [12–14]. The PQ problems related to AC ERPSs are so crucial that different configurations of passive and active compensation technologies [15–18] are proposed to mitigate them in the literature. Meanwhile, extensive studies are underway regarding replacing existing systems with modern power electronics (PE)-based ERPSs as co-phase, advanced co-phase, and different types of VSC and MMCs [19–23]. These ERPSs are recognized as green types because the VSCs can control their output voltage and currents matching with the desired PQ. During the last years, multiple studies have been addressed some of PQ indexes in a specific supplying ERPS. In [24,25] PQ analysis in high-speed railways (HSR) and 1 × 25 kV ERPSs have been carried out. Mariscotti has analyzed PQ issues in 16.67 Hz and DC ERPSs specifically in terms of conditioning and measurement in [26,27]. The analysis in 2 × 25 kV ERPSs emphasizing EMI phenomena has been carried out in [28]. In [29] an inclusive study of harmonic problems in ERPSs has been studied. However, there is still a lack of a comprehensive study and framework that addresses all PQ phenomena and classifies them based on ERPS types. In addition, unlike the power systems, for which multiple standards and resources have been developed, the ERPS suffers from a deficiency of a comprehensive and all-encompassing resource. This has been reinforced the ambiguity and complexity of PQ analysis in ERPSs. Meanwhile, selecting a suitable method to relieve PQ issues needs a complete knowledge and identification of main sources, features, influencing factors, and occurrence environment. Motivated by the above-mentioned shortcomings, this paper presents an exhaustive definition and classification of PQ indexes and distortions together with a brief review of various ERPSs configuration and the classified reported PQ phenomena in literature based on each ERPS type. The rest parts of the paper are structured as follows: Section 2 describes and portrays all the PQ phenomena and influencing factors in ERPS. In Section 3 the investigation of phenomena based on the classified ERPS type is presented. In Section 4 the classification of reported PQ phenomena in literature is discussed. Finally, Section 5 concludes the paper.

2. Power Quality Phenomena in ERPS

PQ issues, disturbances, or phenomena are terms used to characterize voltage or current deviations from its ideal waveform. Different forms of PQ phenomena have been occurred all around the world according to the designed and operating structure of ERPS. In this section, the reported PQ indexes in EPRSs together with originated and influencing factors are described and categorized based on their type.

2.1. System Unbalance

One of the most important and common PQ phenomena in EPRSs is system unbalance which emanated from single-phase power supplying AC EPRS [30]. In normal conditions, when the network is balanced, the three-phase voltages/currents are identical in magnitude and the phase differences are 120°. When the system is gotten imbalanced, the three-phase voltages/currents are not the same in magnitude and the phases are unsymmetrical.

So far, different definitions of unbalance have been proposed in a three-phase system. The issue of unbalance can be categorized into two subjects: voltage unbalance and current unbalance. These voltage and current unbalances may cause extra losses, communication interference, motor overheating, and malfunction of relays [30–37]. The degree of these unbalances depends on the train's movement, the tractive profile of electric locomotives, the type of traction substation transformer (TST) [35], and the power-supply scheme. Numerous problems due to imbalance motivated experts to look at it with a more specialized view and to propose standards for measuring the imbalance. The three key definitions of imbalance are provided by the IEEE [38], IEC [39], and NEMA [40] over the years.

- IEEE std 112-1991

In 1991, the IEEE introduced its second definition of voltage imbalance under standard 112 to define an indicator for voltage imbalance, in which the maximum deviation of the effective value of the phase voltage from the mean effective value of the phase voltage, relative to the average value of the phase voltage according to Equation (1) [38]. Where PVUR denotes phase voltage unbalance rate.

$$\%PVUR = \frac{\max\{|V_a - V_{avg}|, |V_b - V_{avg}|, |V_c - V_{avg}|\}}{V_{avg}} \times 100 \tag{1}$$

$$V_{avg} = \frac{V_a + V_b + V_c}{3} \tag{2}$$

- IEC 60034-26

The IEC standard defines voltage imbalance as the ratio of the negative or zero sequence component to the positive sequence component. In simple words, it is a voltage variation in a power system in which the voltage amplitudes or the phase angle differences between them are not equal. The negative and positive sequence of line voltage can be calculated using the Fortescue matrix as Equation (4) [39]. Where ε_V denotes voltage imbalance ratio.

$$\%\varepsilon_V = \frac{V^-}{V^+} \times 100 = \frac{V_{ab}^-}{V_{ab}^+} \times 100 \tag{3}$$

$$\begin{pmatrix} V_{ab}{}^{\circ} \\ V_{ab}{}^{+} \\ V_{ab}{}^{-} \end{pmatrix} = \frac{1}{3} \begin{pmatrix} 1 & 1 & 1 \\ 1 & a & a^2 \\ 1 & a^2 & a \end{pmatrix} \begin{pmatrix} V_{ab} \\ V_{bc} \\ V_{ca} \end{pmatrix} \tag{4}$$

$$\begin{cases} V_{ab} = V_a - V_b \\ V_{bc} = V_b - V_c \\ V_{ca} = V_c - V_a \end{cases} \tag{5}$$

- ANSI/NEMA Standard MG1-1993

In 1993, NEMA introduced a standard for unbalance that included only the effective magnitude of line-by-line voltages. In this definition, the maximum deviation of the effective value of line voltages from the average effective value of line voltages, compared to the average value of line voltages as follows [40]. Where LVUR denotes line voltage unbalance rate.

$$\%LVUR = \frac{\max\{|V_{ab} - V_{avg}|, |V_{bc} - V_{avg}|, |V_{ca} - V_{avg}|\}}{V_{avg}} \times 100 \tag{6}$$

In general, the voltage/currents unbalance factor (*VUF/CUF*) is defined to measure the intensity of the system unbalance as given in Equations (7) and (8).

$$\%VUF = \frac{V^-}{V^+} \times 100 \tag{7}$$

$$\%CUF = \frac{I^-}{I^+} \times 100 \tag{8}$$

where V^- and I^- are the negative sequence voltage and currents, and V^+ and I^+ present the positive sequence voltage and currents. A higher amount of factors declare a high value of negative-sequence current (NSC) injected to the ERPS. Depending on the adopted TST type in ERPS, the factors are changed. Figure 1 illustrates the measured real imbalance situation of Wuhan–Guangzhou high-speed electrified railway [41].

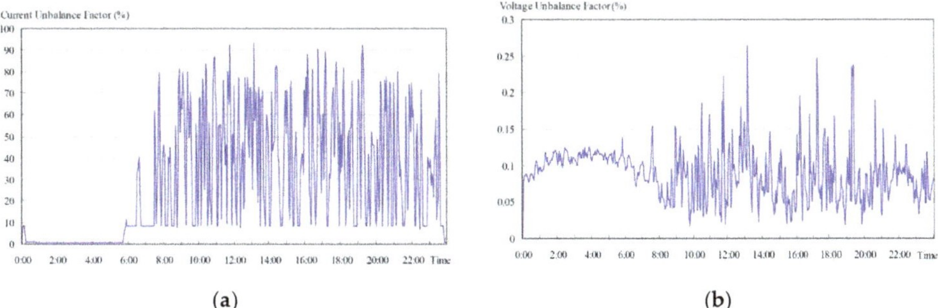

Figure 1. The measured unbalance ratio in Wuhan–Guangzhou high-speed electrified railway system: (**a**) Current unbalance factor; (**b**) Voltage unbalance factor [41].

2.2. Harmonics

The ERPS is one of the main harmonic provenances in the public power grid. Depending on the supplying topology type, different forms of harmonic phenomena can emerge.

2.2.1. Low Order Harmonics

Low order harmonics (LOH) known also as main harmonics are the most critical kind of harmonics in ERPS. They have been noticed as the motivation of numerous functioning obstacles both to the power grid and ERPS. The amplitude of the LOH is greater than the other harmonic spectrum ingredients. Accordingly, the negative efficacy of LOH can decline the fundamental signal amplitude impressively. Intensive shaking and noises in motors and generators, exceeding of heat and loss in transformers and transmission lines, harmful impacts and destructions in relays and other protection systems, and instability of power network are some of their main adversely effects [42]. In general, the main LOH in ERPS can be classified into three categories as follows.

- Background Harmonics

This category also known as internal harmonics of ERPS is generated by the power supplying system in the absence of operational trains. In addition, they can be turned out by adjacent contiguous nonlinear loads linked to the joint busbar as a point of common coupling. The harmonic spectrum ranges for background harmonics are pretty much odd inherent harmonics like 3rd, 5th, 7th, … , 19th [43,44].

- Train Internal Harmonics

In AC ERPS, trains and their PE converter based interior driving systems are assumed in the guise of basic harmonic origination. Most functioning trains in the world even now contain thyristor/diode-based PE converters in their configuration, which turn out current harmonics and accordingly voltage harmonic and distortion [24,45–47]. The low order ingredients as 3rd, 5th, 7th, … , 21st, and 23rd are the most highlighted ones measured in these kinds of trains. Figure 2a demonstrates the harmonic content of conventional thyristor/diode-based trains measured in the ERPS. Conversely, the modern trains equipped with four-quadrant converters (4QCs) evolve less harmonic pollution thanks to the adoption of high switching frequency (pulse-width modulation) (PWM). However, as shown in Figure 2b even the amplitude of LOH has declined significantly (the vertical axes scale has been decreased), the high order components around switching frequency are considered as a substantial problem. These components are known as the characteristic harmonics which will be explained in the next sections.

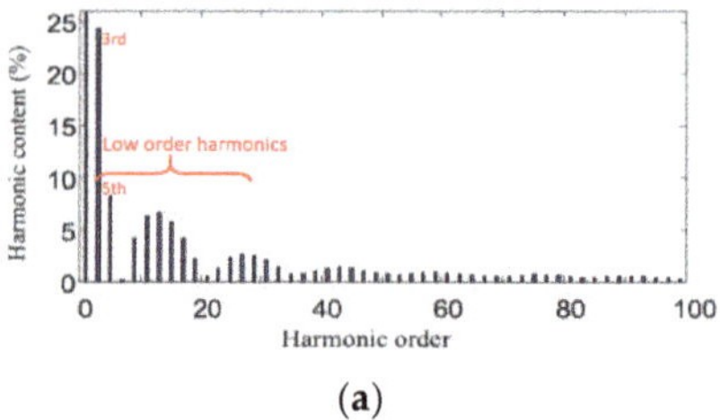

(a)

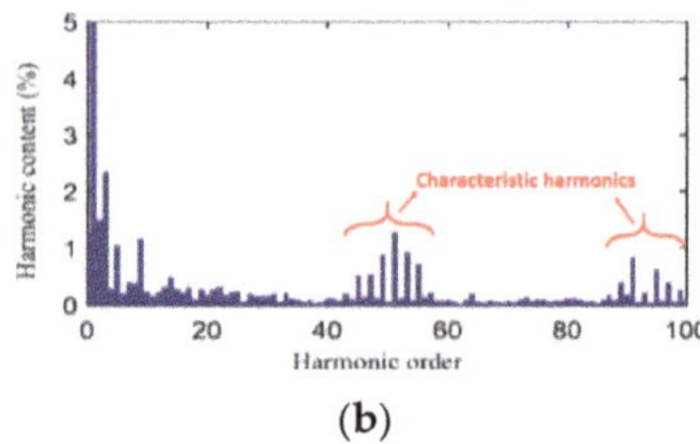

(b)

Figure 2. The harmonic spectrum of the traditional and modern trains. (**a**) Traditional SS4 locomotives. (**b**) High-speed modern CRH2 locomotives [24].

- DC Substation Harmonics

In addition to the trains, the internal structure of ERPS is considered as the foremost and significant harmonic origination in the power grid. Contingent on the AC or DC type supplying system, type of converter in traction power substation (TPSS) and voltage level, various forms of harmonic phenomena can arise. Metro (subway) EPRSs which are fed by multi-pulse conventional rectification substations are a pivotal reason for LOH distortion in the primary side of the grid [48–51]. Moreover, DC traction motor based traditional locomotives functioning in AC systems are massive harmonic contamination loads regarding the high-power utilization of nonlinear rectification converter. The harmonic ingredients in the primary-side current of AC/DC substations can be defined as the function of pulse number:

$$h = kp \pm 1 \; k \in \mathbb{N} \tag{9}$$

where p is the pulse number of converter and h expresses the harmonic order. The popular n-pulse rectifier based traditional substations in EPRSs is 6, 12, 18, and 24 pulses. The primary-side current waveforms and fast Fourier transform (FFT) investigation for these types of TPSSs are simulated and demonstrated on a small scale in Figure 3. As can be seen from the figures, by increasing the number of pulses, the total harmonic distortion (THD) percentage has been decreased and the harmonic components are shifted from low orders to high orders.

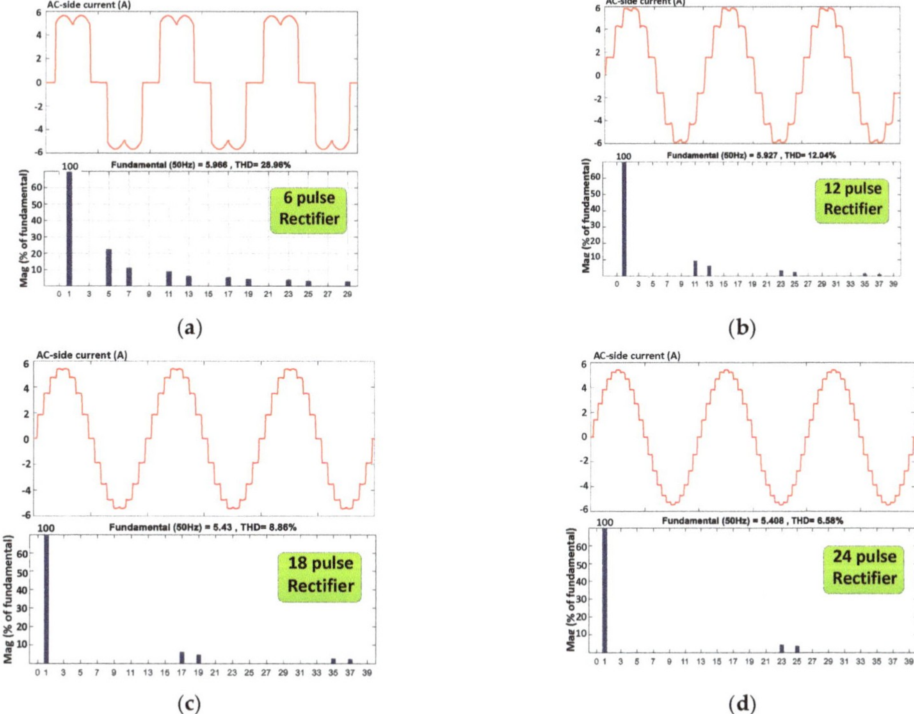

(a) (b) (c) (d)

Figure 3. The primary-side current of n-pulse AC/DC converters with FFT analysis. (**a**) 6-pulse (**b**) 12-pulse (**c**) 18-pulse (**d**) 24-pulse.

2.2.2. Inter-Harmonics

In ERPS inter-harmonics (InH) are emerged by AC motors controlled by variable frequency drives (VFDs). They are related to the fundamental main frequency (f_o) and the input AC frequency (f_i) which supplies VFD. These kinds of harmonics are realized between the typical and characterized harmonics of the VFDs. The main reason for occurrence is the compilation among the switching tasks applied to invert the DC-link voltage to the three-phase AC voltage with the DC current ripple [52–55]. The frequency of the inter-harmonics can be characterized as

$$f_{Ih} = \left| np_i f_i \pm mp_o f_o \right| \text{ n, m} \in \mathbb{Z} \tag{10}$$

where p_i and p_o denote the number of pulses in the rectification and inversion process. The inter-harmonics afford pulsating torque harmonics, stimulate a linked traction motor and torsional resonance path, wheel wear, and a critical impact on the torsional behavior of the entire train.

2.2.3. Low Frequency Oscillation

The low frequency oscillation (LFO) is an impermanent phenomenon and a vehicle-grid interlinkage issue situated by the impedance inconsistency between the ERPS and the modern trains particularly those equipped with 4QCs. Furthermore, LFO can be produced by the rotary frequency converter (RFC) used in several European ERPS [56,57]. In LFO circumstances the voltage and current demonstrate a magnitude and phase oscillation. It can lead to several significant problems, such as the conservation system breakdown, over voltage and current, damage to the onboard devices, and even divergence oscillation of voltage and current leading to railway accidents and obstruction or inconsistency issues [56–65]. Three types of LFO scenarios are defined in the literature depending on the structure of ERPS type. Figure 4. demonstrates the LFO in catenary voltage and currents.

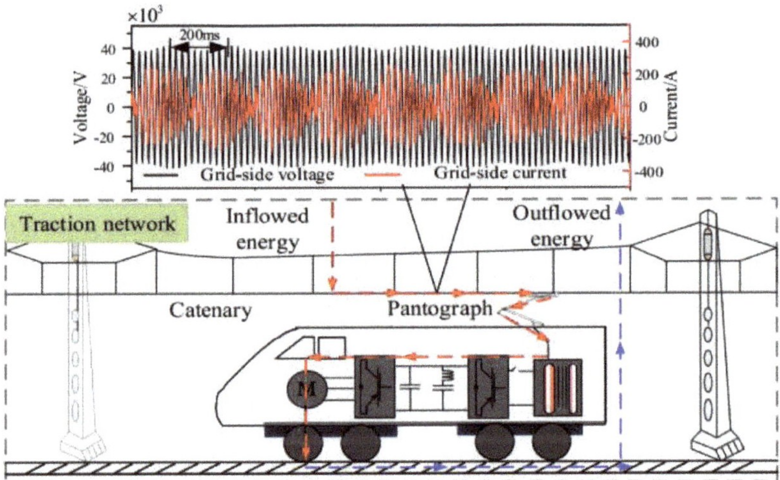

Figure 4. The overhead catenary voltage and current with low frequency oscillation (LFO) phenomenon [59,60].

- LFO in RFC based ERPS

In the age of starting railway electrification, based on the low requests, existing low-capacity power grids, and consequences regarding using of powerful motors in industrial frequency, the low-frequency systems got more attention. These kinds of ERPS are established in several countries as Austria, Germany, Switzerland, Norway, Sweden, and the USA functioning with a frequency of 16.67 or 25 Hz. The frequency below the power grid frequency demonstrates the urgency of the requirement for RFC. The LFO is primarily measured in these kinds of ERPS. The usual oscillation content in this group is reported in the range of (0.1–0.3) p.u [56–58].

- LFO in ERPS without RFC

The consistently incremental request for passenger and mass transfer as well as subsequent advancement of technologies and power networks promote the construction and development of transformer-based AC systems operating with industrial frequency and without RFCs. The conventional 1 × 25 kV and 2 × 25 kV autotransformer based ERPS are the most popular embraced structures. Notwithstanding, LFO in such a network often occurs by concurrent functioning of multiple trains. The ultra-critical quantity of operating trains which leads to the LFO occurrence is reported approximately 6–8. The usual oscillation content in this group of LFO is reported in the range of (0.01–0.12) p.u [59–63].

- Irregular LFO

The latter category of LFO is non-periodic and immethodical oscillations that can enhance current and voltage magnitude unfavorably. The irregular LFO has been reported mostly in China [64,65]. It is not possible to assign a special frequency oscillation range for this type of LFO. Figure 5 shows the measured voltage and current waveforms including various types of LFO. As shown in the figure, the most undesirable type is related to irregular LFO with higher current magnitude reinforcement.

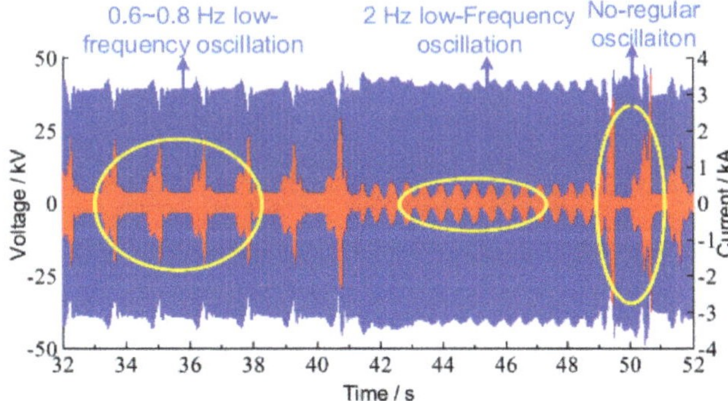

Figure 5. Measured various types of LFO phenomena [64].

2.2.4. Harmonic Resonance

The current LOH produced by trains mainly in modern ERPS can spread all around the power supply networks. Contrarily, the inductance and capacitive specifications of the overhead catenary system (OCS), may make a distributed LC circuit that can lead to parallels or series resonance at the specific frequencies. The interaction of these current harmonics and internal resonance may cause the harmonic resonance (HR) phenomenon at some characteristic frequencies in ERPS. By way of explanation, some harmonic ingredients are strengthened by the resonance. The HR can create serious issues such as drastic voltage distortion, electromagnetic interference in communication and signaling system, overheating and losses, and misdeed of protection equipment [66–79].

Figure 6 illustrates the harmonic spectrums for measured OCS currents/voltages reported in two main HSR lines in China and Italy containing HR phenomenon. This phenomenon which is reported in many countries can be divided into parallel and series scenarios. Generally, the composition of inductive and capacitive features of OCS can lead to either a series resonance (when L and C are in series) or a parallel resonance (when L and C are in parallel).

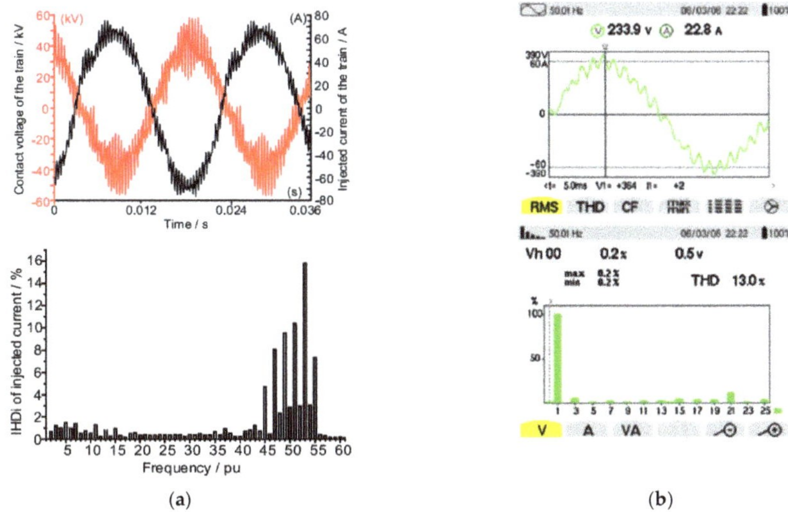

Figure 6. Distorted pantograph voltage/current under the high-frequency resonance with FFT analysis in China and Italy. (**a**) JingHu high-speed railways (HSR) [64] (**b**) Italy HSR [79].

- Parallel HR

The parallel HR (PHR) is most likely HR scenario in ERPS due to the inductive and capacitive specification of multi-conductor OCS. The specified frequency oscillation range for PHR based on measurements is between 10 and 55 p.u. The critical HR incidents correlate with the parallel resonance which is the basic concern in TPSS. The PHR has been reported in some countries which are classified based on the order in Table 1.

Table 1. Different types of reported harmonic resonance (HR) in the literature.

No.	Type of HR	Frequency Period (p.u)	Location
1	Parallel	15–20, 45–55, 35–59, 50–64	China [66–69]
2	Parallel	24–30	Korea [43]
3	Parallel	21–29,39,121,139	Italy [28,70]
4	Parallel	20–60	United Kingdom [71]
5	Parallel	49–51	Thailand [72]
6	Parallel	29–41	Japan [73]
7	Parallel	Up to 63	Zimbabwe [74]
8	Parallel	13–20	Czech Republic [75]
9	Parallel	21,81	Iran [76]
10	Parallel	<25	Germany [77]
11	Series	3–7	Spain [78]
12	Series	64, 72, 80	Italy [79]

- Series HR

The series HR (SHR) is an infrequent HR phenomenon in EPRSs. Functioning of FACT device or conditioners as static VAR compensators, STATCOM and Steinmetz theory may interact with ERPS impedance and create SHR. The specified oscillation content in this group is reported in the range of (3–9) p.u. The SHR has been reported in some countries which are mentioned in Table 1.

2.2.5. Harmonic Instabilities

The most basic factor of harmonic instabilities (HI) genesis is the higher switching frequency of PWM based 4QCs in recent and modern trains. The interplay among the multiple switching frequency and ERPS interior resonance frequency reinforce an intense oscillation with a sort of insignificant damping termed as HI [80–87]. In addition, the high-frequency features of the closed-loop control system for 4QCs including voltage and the current controller can interact with ERPS and originate HI. Accordingly, HI may amplify the voltage and current harmonics and cause the system to be instable. Compared with HR as a stable phenomenon and harmonic reinforcement, the HI is an unstable phenomenon. The specified oscillation content for this phenomenon is dependent on the resonance frequencies.

Nevertheless, according to the measurements and reports, the resonance orders can be in the range of 2nd to 100th with field test intensifications of 23rd to 24th and 47th to 55th [64]. It can lead to critical issues such as overvoltage problems, stimulate some harmonic ingredients, explosion or malfunction of protection devices together with overheating and loss problems. HI phenomenon has been stated in China [64], Switzerland [81], and Italy [82]. Figure 7 demonstrates a measured OCS voltage waveform including HI. For the purpose of having a rapid review and a suitable classification, the harmonic phenomena organization chart together with the frequency range is presented in Figure 8. This figure illustrates an epitome of the harmonic indexes with their different types in ERPS. Meanwhile, the possible per unit range for each phenomenon has been determined and demonstrated. A schematic based comparison together with a technical discussion between harmonic phenomena can be found in [64,80].

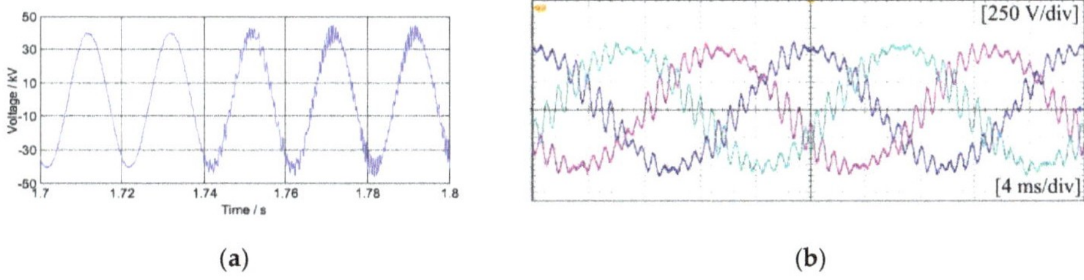

(a) (b)

Figure 7. Measured overhead catenary system (OCS) voltage and experiment waveform with harmonic instabilities (HI). (**a**) Traction power substation (TPSS) voltage in electric railway power systems (ERPS) [64]. (**b**) Phase-to-phase unstable voltage [83].

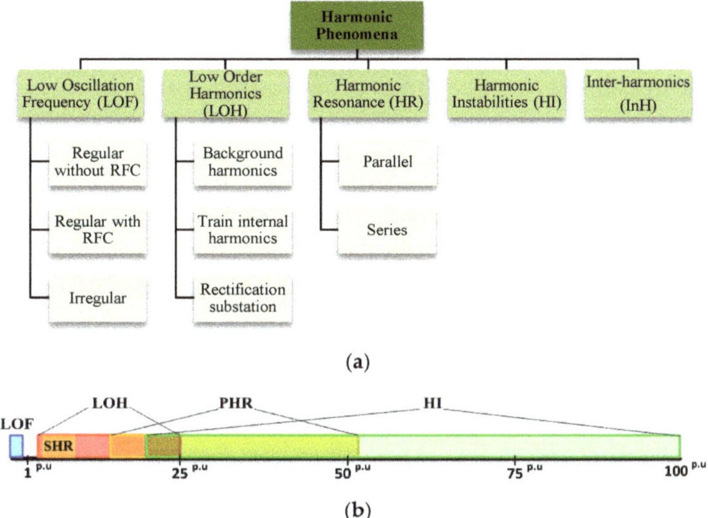

(a)

(b)

Figure 8. Classification of harmonic indexes in ERPS. (**a**) Organization chart. (**b**) Occurrence frequency range.

2.3. Reactive Power and Low Power Factor

Notwithstanding the application of advanced PWM based trains, the majority of working trains in ERPS all over the world are still based on AC/DC rectifiers using thyristor or diode. This can cause the current distortion in the primary-side of the substation and overlap commutation angle. Therefore, it can reduce the power factor (PF) of the system substantially. On the other side, considering the impacts of ERPS impedance including the inductive reactance features of the OCS, the PF will be reduced by about 1–5%. The low PF in the system can be the origination of many problems including, lower efficiency, high power losses, heating of devices, and high voltage drops across the line [45,88–95]. This is so serious for power grids that fine subscribers with a low power factor of 0.9 in PCC. However, the average amount of PF in ERPS depending on the type of supply is measured in the range of 0.70–0.84 [93]. Figure 9 demonstrates the measurements of PF for three different popular HST trains operating in the world. According to these figures, for the light-loading situation and coasting mode, the amount of PF is proportionally lower and rapidly when the train current/power increase PF reaches close to unity. Power factor is commonly specified by a particular description assuming a balanced situation without harmonics on the system. However, in a real practical system, the PF can be affected by the harmonic components and imbalance condition in the ERPS. Accordingly, the type of traction transformer and its loading characteristics will play an important role to determine the PF in AC

ERPS [95]. The performance of popular traction transformers in ERPS is based on the different PF definitions as fundamental PF (PF_1), effective PF (PF_E), vector PF (PF_V), and arithmetic PF (PF_E) [95] in IEEE Std-1459 is demonstrated in Table 2. The practical measurement in this table reveals the better performance of Y/d and Scott transformer in presence of harmonics for both balanced and imbalanced conditions.

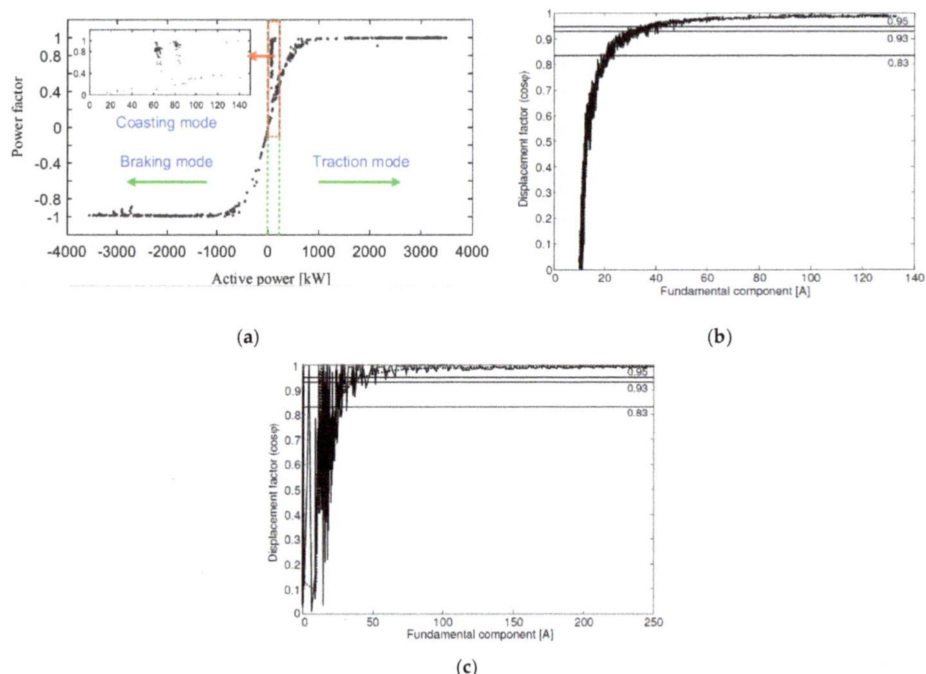

Figure 9. Measured power factor (PF) range for different kinds of modern trains. (**a**) CRH2A-China [94], (**b**) TGV-France [12], (**c**) ETR500-Italy [12].

Table 2. Performance of different traction transformers based on PF definitions in the harmonic presence.

Transformer Type	Load Condition	PF_1	PF_E	PF_V	PF_A	Overall
Single-phase	Balanced Unbalanced	high high	very low very low	medium medium	very low very low	very low
V/V	Balanced Unbalanced	low medium	low low	high high	low medium	medium
Y/d	Balanced Unbalanced	very high very high	medium medium	very high very high	medium low	very high
Scott	Balanced Unbalanced	medium low	very high high	low low	very high very high	high
Le-Blanc	Balanced Unbalanced	very low very low	high very high	very low very low	high high	low

2.4. Transient Events

PQ phenomena in ERPS includes a wide range of disturbances and different types of deviations in voltage amplitude or waveform. The PQ deviation events/disturbances can be classified based on the disposition of the waveform distortion, duration, rate of rising, and amplitude for each category of electromagnetic disturbances. A comprehensive classification of PQ electromagnetic phenomena for power systems can be found in [96]. However, due to the specific and different features, a specific classification based on the characteristics of ERPS has been carried out. The transient events are unwelcome and momentary in nature. Generally, transient events can be classified into two groups of impulsive and oscillatory which comprised the waveshape of both current and voltage transient.

2.4.1. Impulsive Transients (ImT)

The ImT is an abrupt, frequency variation from the nominal condition of voltage/current, which is usually unidirectional in polarity. It is specified by the peak value, rate of rising or decay, and duration times. The most popular reason for the occurrence of impulsive transients in ERPS are as follow:

- Lightning

Every time thunderstorms happen, ERPS has exposed to impulses transients of lightning. The strikes can be direct lightning to any conductor which is in the upper of the ground or indirect form as the induced voltage in a part of the system caused by close lightning [97–100]. The lightning energy can damage and completely destroy the equipment. Figure 10 illustrates the measured waveforms during lightning occurrence in four position distance of Swedish railway

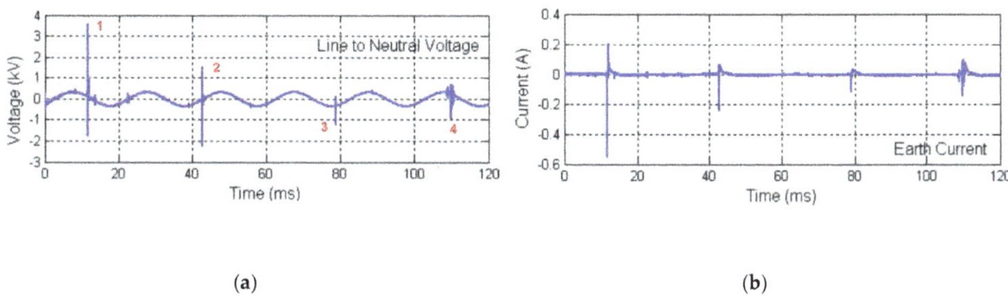

(a) (b)

Figure 10. Measured waveforms during lightning occurrence in 4 position distance of Swedish railway facility [98]. (**a**) Line-to-neutral voltage (**b**) Local ground to rail current waveforms.

- Switching of circuit breakers

Some remarkable voltage transients are measured during the operating switching of circuit breakers the ERPS. This type of transients can be classified in the impulsive group because of the sudden rise and duration of occurrence. The two possible switchings in ERPS can be defined as switching of the main high-voltage busbar for maintenance purposes and switching of converters in TPSS for daily operations [101]. The measured voltage transient for the mentioned switching of breakers in Rome subway is shown in Figure 11.

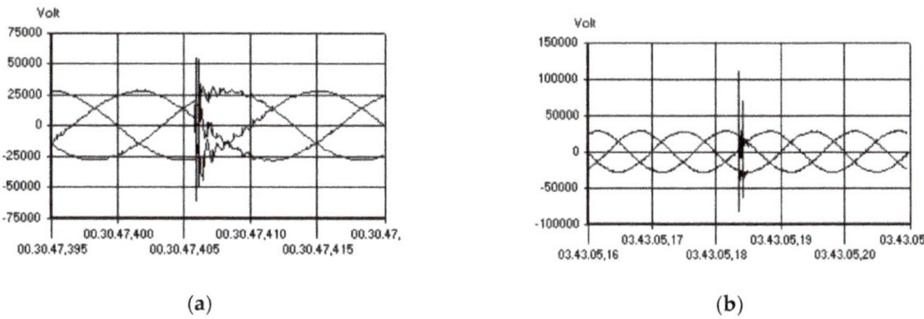

(a) (b)

Figure 11. Measured two types of transient voltage in MV busbar of Rome subway system [101]. (**a**) Recorded type 1 transient voltage. (**b**) Recorded type 2 transient voltage.

- Abnormal changes in tractive efforts

During the train movement in the route, several situations may cause sudden changes in the driving of traction motors and consequently sudden changes in voltage or currents. Some of these factors and conditions are abnormal abruptly generated tractive effort by passing neutral sections and high current absorption which can make step-change in pantograph voltage. In addition, the other type of transient can be measured during the passing neutral zone in changing over the supplying substation of sections [27,102]. The measured voltage transients for the two mentioned situations in Italian ERPS are shown in Figure 12.

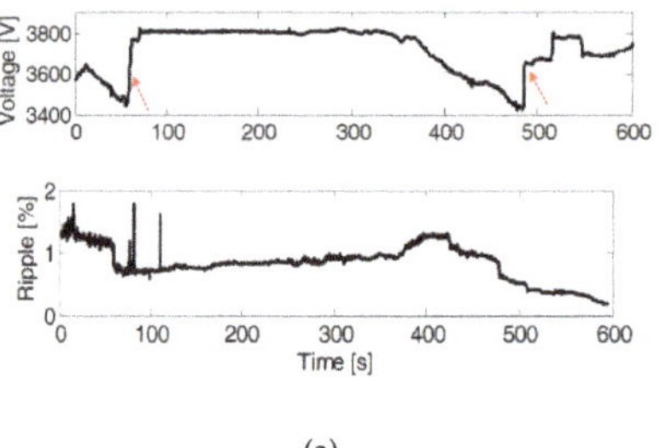

(a)

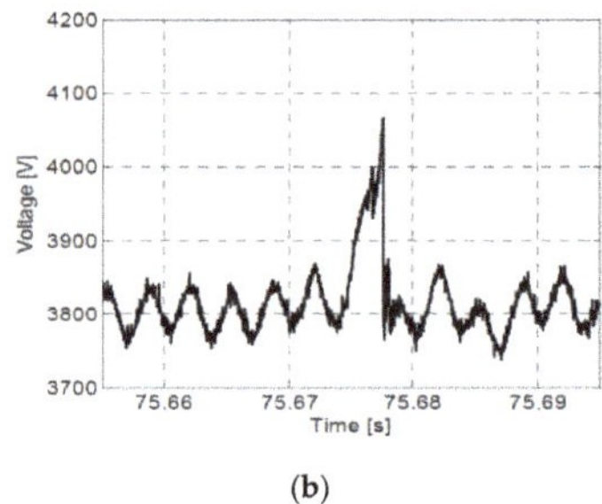

(b)

Figure 12. Measured transient voltage in 3 kV DC Italian ERPS [102]. (**a**) Caused by sudden generated tractive effort. (**b**) Caused by passing neutral section and changing supplying TPSS.

2.4.2. Oscillatory Transients (OsT)

The OsT is an abrupt, and non-power frequency variation of the steady-state situation of voltage/current, which is usually bidirectional in polarity. OsT includes a waveform with instantaneous amount changes of polarity quickly for several times and commonly declining within a fundamental-frequency period. It is specified by the amplitude, spectral content, and duration times. They can be classified into three groups of low (<5 kHz), medium (5–500 kHz), and high (0.5–5 MHz) frequency oscillations [96]. The most popular reason for the occurrence of OsT in ERPS are as follows:

- Changing in operational condition and modulation patterns

During the train movement in the route, some circumstances may lead to having changes in the operation of locomotives or modulation patterns of drive converters related to traction motors and consequently changes in voltage or currents. Some of these factors and conditions are as sudden braking, train wheel slide/slip, change in driving pattern, and extra torque. The field measurement of this phenomenon is demonstrated in Figure 13a.

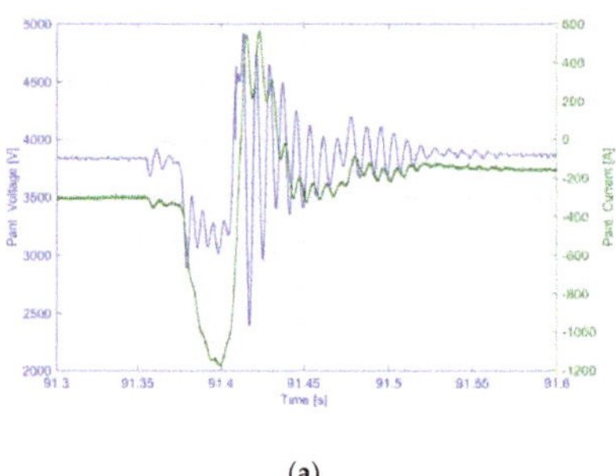

(a)

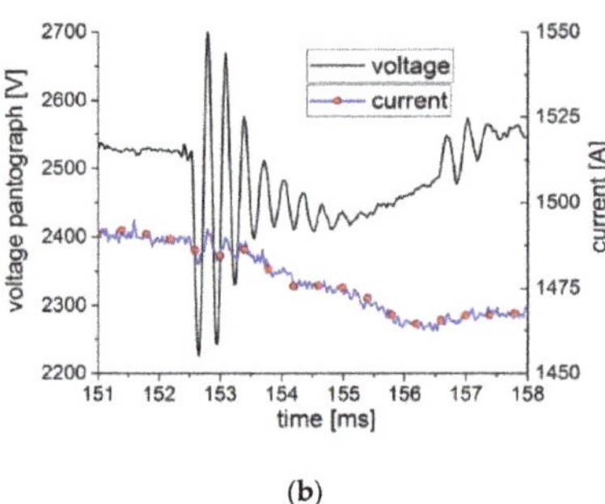

(b)

Figure 13. Measured transient voltage and currents in 3 kV DC ERPS. (**a**) Pantograph voltage and ripple index caused by changes in operational conditions [104]. (**b**) Pantograph voltage and current during oscillatory transients (OsT) caused by sliding/jump effect [103].

- Sliding contact and pantograph jump over OCS

Due to the sliding contact between the OCS and suspended pantograph on top of trains during their movement, the electromagnetic transient events which are a very common phenomenon can reduce the effective voltage and continuity of train operation. By increasing the operational speed of trains especially in HSR and high-power lines the related problems will get worse. Moreover, during this interaction, the arcing issue can appear which is detrimental for the signaling system [103]. However, based on the ERPS features we have classified this phenomenon in the radiated interference indices category which will be discussed in the next sessions. Field measurement of voltage/current transient caused by sliding contact for 3 kV DC ERPS is illustrated in Figure 13b.

- Inrush current of the locomotive transformer

The very low frequency OsT with less than hundreds Hz typically is connected with ferroresonance and energization of a power transformer. In the power system, this phenomenon happens when the system resonance leads to the amplification of low-frequency ingredients of transformer inrush current [96].

In ERPS especially AC type, this can occur when the unmagnetized transformer of the locomotive is connected to the OCS. When the onboard train's transformer is connected to the OCS, an inrush current issue arises because of the nonlinear conditions caused by a saturated transformer [105,106]. A measured inrush current transient in Switzerland is shown in Figure 14. This phenomenon can be mitigated by suitably sizing and designing transformers and filters.

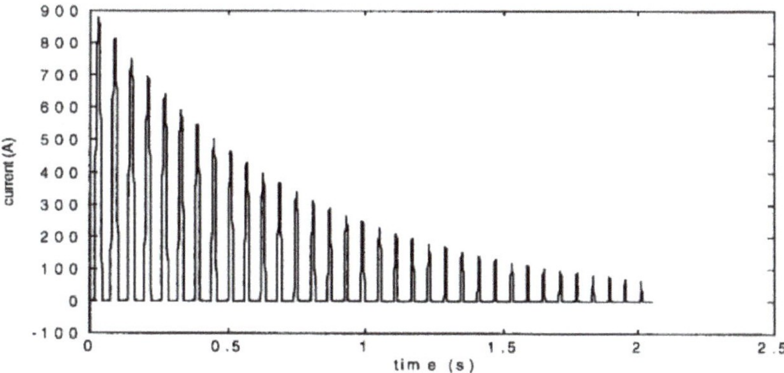

Figure 14. Measured inrush current transient in Switzerland BLS Re465 locomotive [105].

- Capacitor bank energization

The capacitor bank energization can arise both in low and medium frequency OsT. The low frequency OsT in ERPS is more common due to the energization and switching of capacitor banks [104–106]. Due to the low PF and voltage drop in railway systems, reactive power compensation using capacitors is prevalent in terms of different compensators as capacitor bank, passive filters, thyristor-switched capacitor (TSC), static VAR compensator (SVC), and railway power conditioner (RPC) [107].

2.5. Short Duration rms Variations

These kinds of variations are characterized as the variations in voltage/current for a period not exceeding one minute. According to occurrence duration, it can be in three forms of an instantaneous, momentary, or temporary phenomenon. Meanwhile, depending on the reason of appearance, the variations can be classified into three formats of voltage dips, voltage surge/rises, and interruptions [96]. However, the duration of the variations does not meet all the time the values assigned for the power grid standard due to the special features of ERPS. The measured voltage dip in Korea and Rome metro lines are represented in Figure 15.

2.5.1. Voltage Dips (Sag)

Voltage dip is a kind of phenomenon when a short duration decrement occurs in rms value of voltage. This decrement is defined in the range of 10–90% of nominal amplitude. It can be measured in TPSS and on-board. The reasons for and influencing factors of occurrence in ERPS are as follows [108–112]:

- During the fault occurrence in the line.
- The high value of starting current absorbed by traction motors.
- Sudden load changes or supplying high-power locomotives.
- TPSS transformers energization.
- The motor blocking caused by the segregation of pantograph and OCS in the vibration situations or neutral sections.
- TPSS equipment triggering such as lightning, escalator, air-conditioners, heaters, etc.

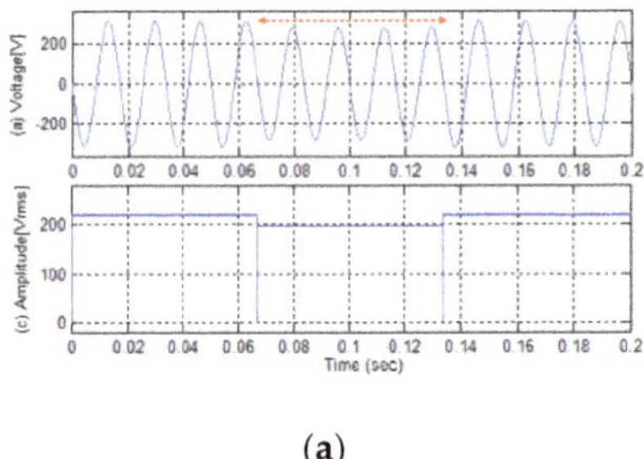

(a)

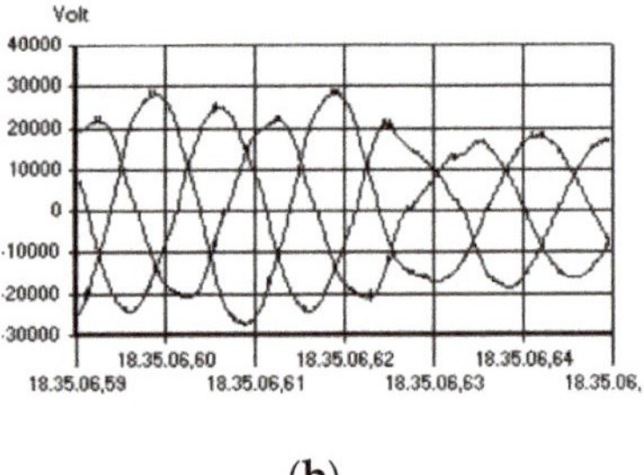

(b)

Figure 15. Measured voltage dip in Korea. (**a**) 4-cycle voltage sag signal without a noise [111]; (**b**) 63% voltage sag recorded in Rome metro line [101].

2.5.2. Voltage Rises (Swell)

The voltage rises known also as swell are characterized in IEEE 1159-2019 standard as the 10–80% increment in the rms value of voltage in a short duration of less than one minute [96]. It is infrequent in comparison to the voltage sag. This phenomenon can be caused in ERPS by different factors like:

- Sudden load changing or supplying high-power locomotives.
- TPSS transformers de-energization.
- The motor blocking caused by the segregation of pantograph and OCD in the vibration situations or neutral sections.
- TPSS equipment ceasing such as lightning, escalator, air-conditioners, heaters, etc.
- De-energization or disconnecting of high-power loads.

It is worth mentioning that the impacts of voltage swell to the ERPS in comparison to voltage sag are more hazardous. They can destruct equipment, leading to overheating and loss issues together with a malfunction on protection devices [108–112]. Figure 16 demonstrates the examples of reported voltage swell in Korean HSR.

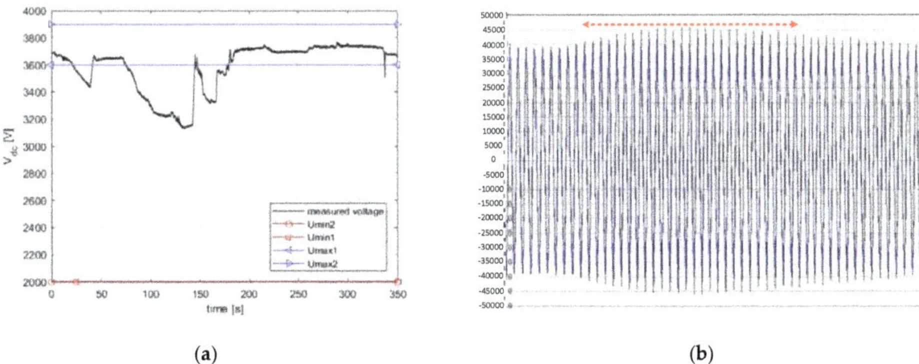

(a) (b)

Figure 16. Measured voltage swell in ERPS. (**a**) 3 kV DC line in Italy [109]. (**b**) Korea HSR (G7) [108].

2.5.3. Interruption (InR)

Interruptions have been defined by IEEE 1159-2019 as a short-duration in which the rms voltage amplitude is less than 10% of nominal voltage [96]. The impressive external determinants in interruption occurrence are the cutting of the fuse, the operating of the circuit breaker, failure and fault in the power system equipment, etc. The foremost reason for the short interruption in ERPS is the abrupt disconnection between the contact wire and pantograph. Losing of data, ruining of susceptible equipment, unwelcome tripping of protective devices, and relays and malfunction of data processing equipment are substantial ruinous impacts [3,109,112–114]. Figure 17 demonstrates the example of measured interruption in 25 kV AC ERPS line of Italy.

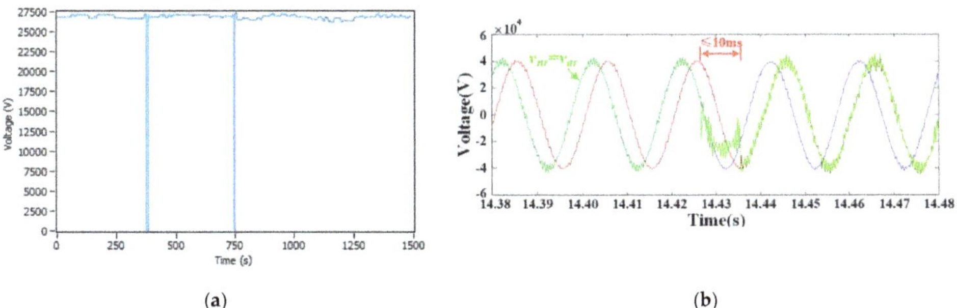

(a) (b)

Figure 17. The measured interruption in ERPS. (**a**) Uncontrolled InR in 25 kV AC ERPS line of Italy [112]; (**b**) Controlled 10 ms InR with smart electric neutral section executer [113].

2.6. Long Duration rms

These kinds of variations are characterized as the variations in voltage/current for a period longer than one minute. However, due to the instantaneous variation of loads and time-varying features of ERPS, the duration can be lower in such a system. Depending on the creating factors of the variation, this phenomenon can be classified into three formats of overvoltage, undervoltage, and sustained interruption [96].

2.6.1. Overvoltage (OvG)

When the system voltage increments and exceeds the higher limit of the designed nominal rate, it is known as an overvoltage situation. This phenomenon is in the range of a 10–20% increase in rms voltage less with a duration longer than 1 min [96]. However, it should be noted that the maximum allowed overvoltage in ERPS is 20%. Based on [99], more than 16% of malfunctions in electric power systems are originated from overvoltage. However, the time/location-varying traction load in ERPS has made it more prone to accept such a phenomenon. Figure 18a illustrates the overvoltage situation during the braking of trains in Metro de Medellin. Various influencing factors can lead to overvoltage occurrence in ERPS as [6,12,115–119]:

- Voltage increase in the OCS in the case of regenerative braking and lack of consumer trains in the network.
- The interlinkage of system harmonics and pantograph impedance and created resonances.
- System instabilities.
- Oscillations happening in the onboard controllers.
- Automatic passing of neutral zones.
- Impedance unconformity at the inverter and traction motor terminals.
- Lightening overvoltage.
- Switching or other atmospheric phenomena.
- Functioning of split-phase breakers in case of phase changing procedure.

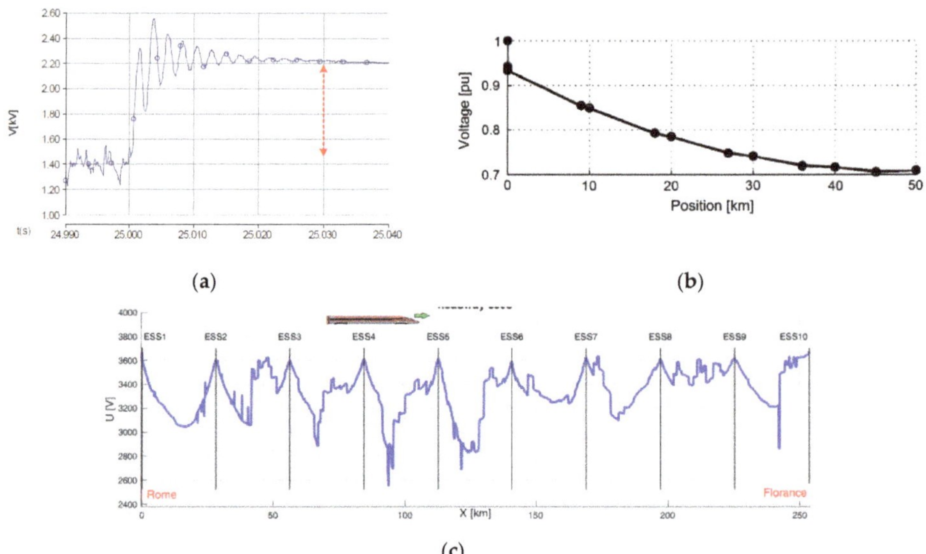

Figure 18. The measured overvoltage and under voltage in ERPS. (**a**) Overvoltage situation during braking of trains in Metro de Medellin without dissipating resistance [117] (**b**) Voltage drop/undervoltage in 25 kV ERPS [120]. (**c**) Simulated voltage drop across the Rome–Florence HSR line.

The overvoltage can damage electronic equipment, traction, and compressor motors located in the trains and the insulators. It can also cause malfunction and failure on protection and relays.

2.6.2. Undervoltage (UvG)

When the system voltage decrements and exceeds the lower limit of the designed nominal rate, it is known as undervoltage (UvG) situation. According to IEEE 1159-2019, this phenomenon is in the range of a 10% to 20% decrease in rms voltage less with a duration longer than 1 min. However, it should be noted that the minimum allowed voltage drop or undervoltage situation in ERPS is −33% [1]. Therefore, ERPS can tolerate most of these UvG situations. The analysis of voltage drops in ERPS is important for all rating, design, and planning stages to confirm that even under the worst situation, the voltage at the pantograph satisfies the related standards [120–123]. In DC ERPS, voltage drops are assigned just by the resistive component of the line, omitting the transient phenomena. On the other hand, the AC ERPS lines are affected not only by the resistive component but also by the inductive voltage drop and the reactive power included in the system. Meanwhile, the type of DC power supply including unilateral, bilateral, single-point parallel, and multi-parallel connections can influence the voltage drops [1]. Figure 18b,c show the voltage drop situation over the distance from TPSS in 25 kV AC and 3 kV DC ERPS, respectively.

2.6.3. Interruption Sustained (InRS)

The decrement of the voltage to less than 10% of the nominal value for a duration of more than one minute is characterized as sustained interruption [96]. Like short interruption, this phenomenon can be caused by some influencing factors like operating of the fuses or circuit breaker, failure, and fault in the line equipment undesirable tripping of protective devices and relays [101,109,112–114]. In Figure 19 sustained interruption with periods of about 3.5 min, which has been occurred at the substation of Rome subway is demonstrated.

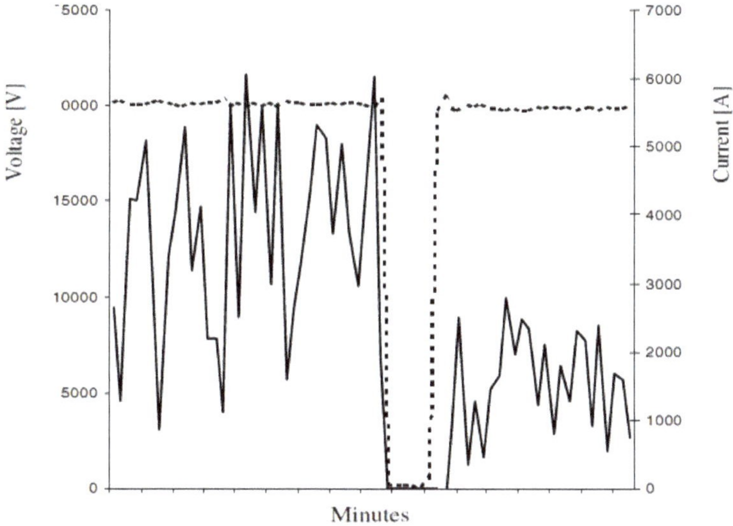

Figure 19. The measured sustained interruption with 3.5 min duration in Rome metro line [101].

2.7. *Voltage Fluctuation (Flicker)*

Voltage Fluctuations are characterized as the change in nominal voltage in the range from 0.1 to 7% with frequencies less than 25 Hz. The time-varying specification and abrupt load changes in ERPS can lead to a fast variation of the traction current results in sudden changes of OCS voltage known as voltage flicker. The voltage flicker can have resulted in light flickering, destruct the equipment, unwanted triggering of relays, and protection devices. The currents flowing in ERPS equipment may lead to pulsating forces with high amplitude and which can be potentially harmful to TPSS equipment. Static frequency converters (SFC), the operation of high-power traction motors, and arcing equipment are the main reasons for flicker occurrence [41,124–126]. Figure 20 demonstrates the examples of measured short and long duration voltage flicker in ERPS TPSS.

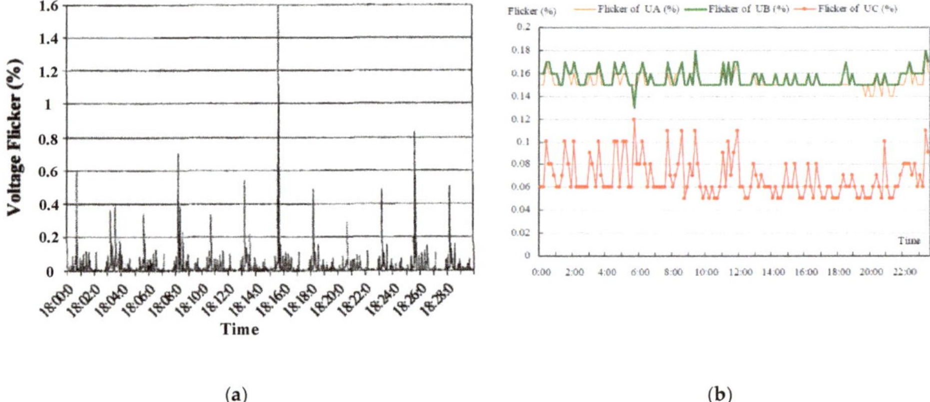

Figure 20. Measured voltage flicker in railway substation. (**a**) Short duration flicker [124]. (**b**) Long duration flicker in three-phase high-voltage busbar [41].

2.8. Waveform Distortion

Waveform distortions are characterized as a steady-state deflection from an ideal fundamental power frequency sinusoidal. According to IEEE 1159-2019, there are five different forms of waveform distortion as harmonics, inter-harmonics, DC offset, notch, and noise. Due to the importance of the harmonic issue in ERPS, we classified and discussed the first two forms separately in harmonic sections.

2.8.1. DC offset

The existence of a dc component in voltage or current in the AC side is characterized as dc offset. In ERPS, the most popular time to occurrence of this phenomenon is controlling 4QCs with indirect current control methods [127,128]. Meanwhile, in the six-step mode control of permanent magnet machines, the difference between the positive and negative half-cycle length can lead to this phenomenon on the AC side [129]. DC offset can cause DC magnetization of magnetic equipment, saturate the transformers, enhance heating problems, stressing of insulation. It can also impact the stability of 4QCs.

2.8.2. Notch

The notch is a kind of voltage disturbance which can be occurred mostly by power electronics devices. Technically, it can be generated by the synchronous transition of two semiconductors persisting on the equal dc output terminal for a short time period, when two out of three of the ac inputs are short-circuited. This can be found especially in voltage source converters controlling by SVM modulation technique which drives traction motors. Meanwhile, this phenomenon can be originated in 4QC drives [3,71,111]. Based on its specification, this phenomenon can be found in both transient and oscillation forms. In addition, due to the periodic situation and frequency oscillation features it can be classified as harmonic distortions. However, implementing high capacitors or batteries can mitigate this phenomenon.

2.8.3. Noise

According to IEEE 1159-2019, noise is characterized as an undesirable electric signal with contents lower than 200 kHz which can be imposed on the voltage/current of conductors [96]. In ERPS, it can be occurred by power electronic devices, control circuits, arcing equipment, and traction locomotives with onboard rectifiers. Meanwhile, impulsive noise and radio frequency (RF) noises have been addressed in some papers [100,130–132] as electromagnetic transients caused by the sliding contact between the

OCS and the pantograph. The common amplitude of the noise is lower than 0.01 p.u of the voltage magnitude. The communication and signaling systems in ERPS are the most vulnerable parts which are at risk and damage. Utilizing filters and isolation systems, this phenomenon can be mitigated. Due to the high potential of ERPS associated with electromagnetic noises, this phenomenon has been covered and categorized in the conductive electromagnetic interference group.

2.9. Electromagnetic Interference (EMI)

Electromagnetic interference (EMI) is characterized as a phenomenon in which an electromagnetic field (EF) interposes with another, cause to the contortion of both fields. In other words, electromagnetic coupling between a source of interference and a general sufferer system is demonstrated whenever an interlinkage happens between the EF produced by the source and the sufferer system. As a consequence, initiating a transfer of energy between them unfavorably modifies the physical features and performance of the system. EMI and electromagnetic compatibility (EMC) issues play a significant role in the generic performance of the ERPS and signaling systems. The EMI disturbances in ERPS can be manifested by formations of current/voltage, electric/magnetic field coupling and they can be classified into the four types of conduct, inducted, electrostatic, and radiated [1,131–145]. The different impacts of EMI between ERPS, infrastructure, and circumambient environment are demonstrated in Figure 21. These EMIs are known as

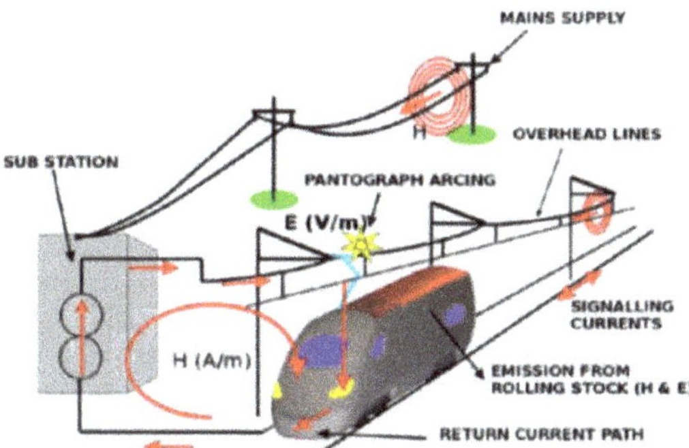

Figure 21. Electromagnetic interference (EMI) occurrence places in ERPS [134].

- The induced interference voltage because of the inductive/capacitive coupling of three-phase ac power grid transmission lines nearby to the OCS and TPSS.
- The induced interference voltage by inductive/capacitive coupling of OCS conductors.
- The conducted interference between rails and signaling systems/track circuits.
- The inducted/radiated interference originated by pantograph arcing.

2.9.1. Conducted EMI

Conducted EMI (CEMI) phenomenon occurs when a source and sufferer system share one or more conductors. CEMI can be produced by injection of distorted current or harmonics into the power lines from devices with nonlinear features, or by transient overvoltage originated by switching and atmospheric phenomena [1]. In most ERPS, the running rail conductors are utilized simultaneously as a return circuit and as the conductors for locomotive detection and signaling systems. In this situation, the trains and TPSSs which are the source of harmonic currents can interfere with the signaling system and disrupt the signal of track circuits [1]. The frequency analysis and separation for power systems and signaling systems together with designing frequency filters are the main procedures to mitigate this phenomenon. The earth current or known as the stray current is the other type of foremost conductive coupling that returns currents flow through the earth and re-enter to the rails nearby the TPSS especially in DC ERPS. The high amount of currents injected into the soil can lead to electrochemical corrosion in metal devices buried in the earth. Meanwhile, it can result in high rail to earth or touch voltage [103,131–136]. Figure 22a shows a simple diagram of stray current phenomena in DC ERPS. The other form of conducted EMI mentioned in the literature are emissions related to the PWM converters which drive the train's induction motors [137]. They can be stimulated and generate RF noises. Figure 22b shows an example of measured CEMI in PWM 4QC with power factor correction (PFC) unit.

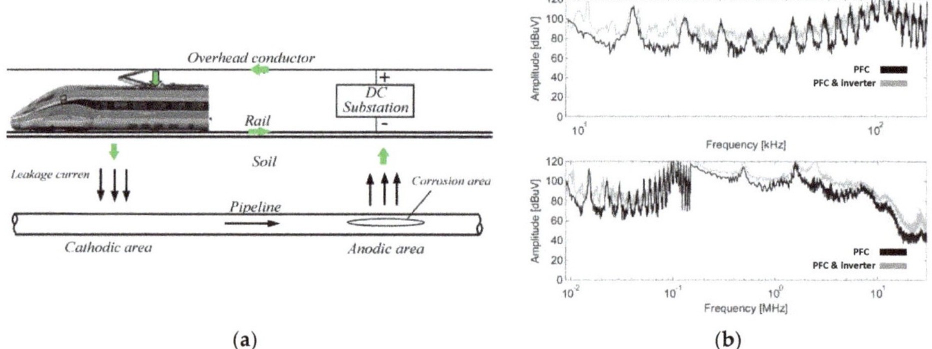

(a) (b)

Figure 22. Different forms of conducted EMI (CEMI) in ERPS. (**a**) Simple diagram of stray current phenomena in DC ERPS. (**b**) Output measured EMI in 4QC without filter (output frequency = 15 Hz) [137].

2.9.2. Inducted EMI

Inductive EMI (IEMI) happens when the magnetic flux is related to the source current interface with a secondary system. ERPS's lines especially AC types are known as a source of such EMI phenomenon since they can induce the electromotive force (emf) on the close paralleled conductors.

This longitudinal emf and the induced voltage is one of IEMI impacts. The other type is related to physical positioning, cause transverse emf to generate considerable audio frequency noise in communication circuits and power controlling cable networks [131–142]. The measured induced voltages on a conductor parallel to ERPS for both 1 × 25 kV and 2 × 25 kV are shown in Figure 23.

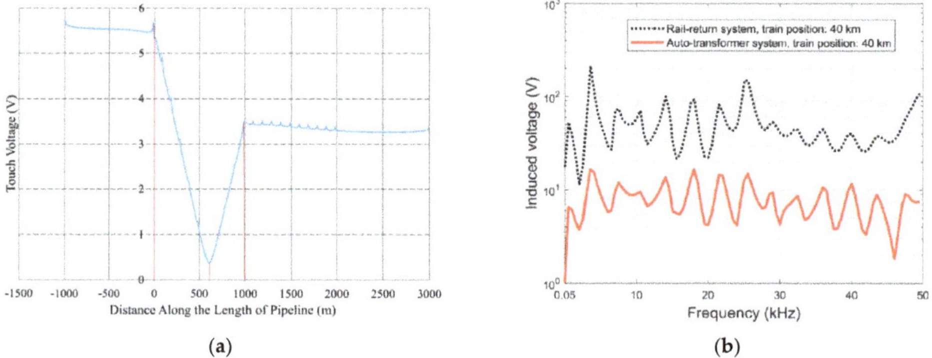

(a) (b)

Figure 23. The induced and touch voltage on a conductor parallel to ERPS track. (**a**) Induced voltage based on the distance of the conductor [138]. (**b**) Induced voltage based on frequency for both 1 × 25 kV and 2 × 25 kV ERPSs [139].

2.9.3. Electrostatic/Capacitance EMI

Electrostatic EMI (EEMI) can be caused by the production of electrostatic electric fields (EF) and happens when an EF source (primary system) has a considerable parasitic capacitance against the victim circuit (secondary system), leading to undesirable interference and induced voltage on the secondary system [1]. EEMI is executed when the railway lines are close to the interfering OCS lines (inducing EF source) and the conductors in the line (rails, signaling and telecommunication cables, metal objects) receiving interference. The noise generated by EEMI becomes more critical in the case of a high voltage supply. EEMI has been reported mostly during the regenerative braking of trains in field test measurements with frequencies ranged 8–20 kHz and the maximum transient rail current amount of 10 A [131,132]. It can be categorized in static type, where the conductors of a system with high-voltage make a potential divider toward the capacitors to earth or other metal objects, and dynamic type, where a sudden change in voltage or inherent track admittance like charging and discharging of capacitors or switching semiconductors may cause *CdV/dt* currents [131,132]. The diagrams of both static and dynamic EEMI are displayed in Figure 24 marking capacitor coupling between wires, ground, and metal subjects closed.

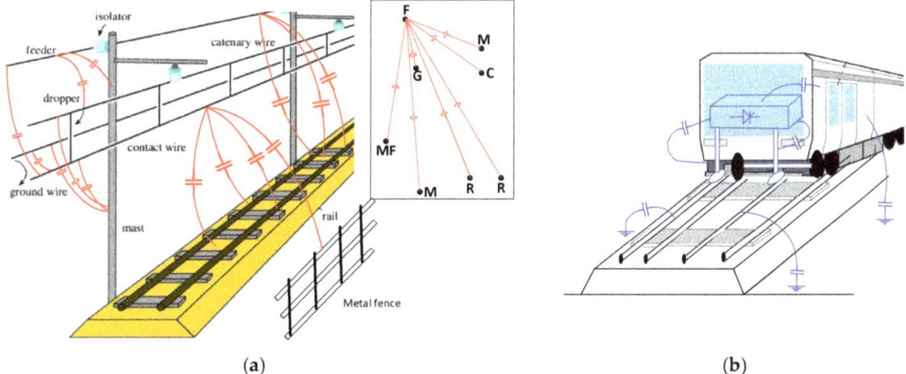

(a) (b)

Figure 24. The static and dynamic electrostatic (EEMI) in ERPS. (**a**) Static EEMI. (**b**) Dynamic EEMI.

2.9.4. Radiated EMI

One of the foremost common phenomena in ERPS especially in HSR is radiated EMI (REMI) known also as arcing. The interlinkage between the pantograph and contact wire or between the train's brushes and the third or fourth rail in ERPS and neutral intersection points are the main influencing factors and environments in creating REMI. To put it another way, loss of mechanical touch between a train and supplying wires because of excessive shaking and disturbances at ERPS tracks, the electric arc can occur. Even though the duration of this phenomenon is so short, it is highly nonlinear features lead to distorted currents with RF spectra that makes radiated emissions. The arcing issue can distort voltages and currents of ERPS and produce different kinds of transient problems [131–145]. Meanwhile, they can create DC components in AC points which may breakdown of dielectrics. However, the foremost issue caused by REMI is an intervention in the wireless or radio-based telecommunication systems, traction power, and signaling system. Various factors like train speed, absorbed current, inductance feature of OCS, network power factor, etc., can impact the arcing phenomenon. Figure 25a,b show examples of pantograph and brush-based arcing phenomenon in ERS.

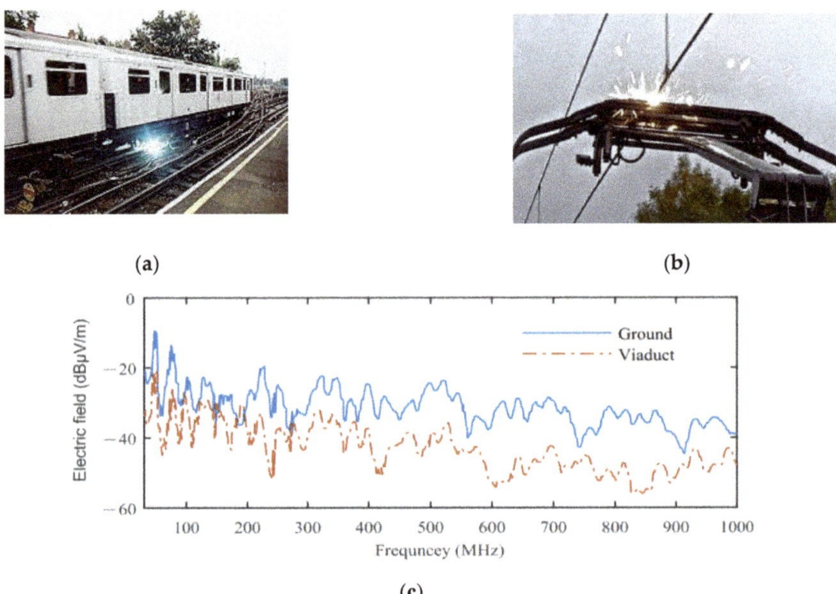

Figure 25. Arcing phenomenon and measured radiated emission in ERPS. (**a**) Pantograph arcing. (**b**) Brush arcing. (**c**) The measured electric field emission in the range frequency of 30 MHz to 1 GHz with train positions on ground and viaduct [145].

The impact of arc in ERPS can be either as radiated or inducted. The radiated EMI type which is discussed in this section is related to the high-frequency part of the pantograph current transient. The inducted type which is related to the low-frequency part of the pantograph current transient is discussed in OsT section. Figure 25c illustrates the measured radiated electric field emission by the pantograph-catenary interaction in the two train positions (viaduct and ground) at 10 m position. As it is obvious in the figure, the frequency band of REMI, in this case, is in the range of 30 MHz to 1 GHz. To have a comprehensive review and appropriate categorization of all mentioned phenomena, the PQ phenomena organized chart is presented in Figure 26. This figure illustrates the overview of the main PQ phenomena together with their different types in ERPS.

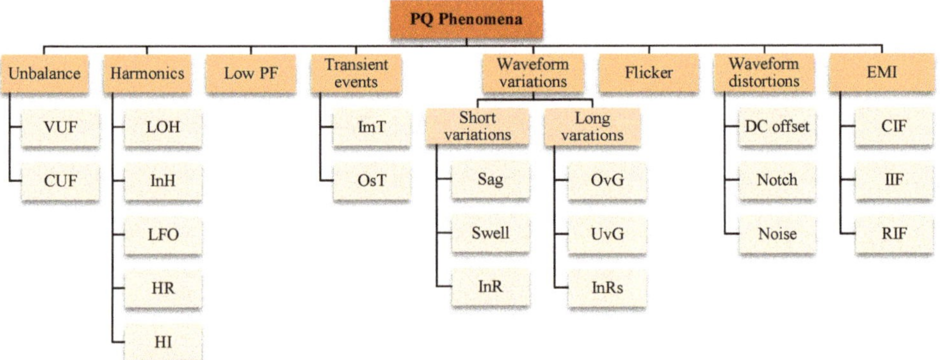

Figure 26. Classification and organization chart of power quality (PQ) phenomena in ERPS.

3. Investigation of PQ Phenomena Occurrence Based on ERPS Type

Railway electrification has experienced a significant modification and evolution process over time. A sharp increase in population growth and enhanced demand for high-power and high-speed transport together with historical, geographical, and economic policies and features of the countries, have led to different structures of ERPSs all around the world. Depending on the configuration of each ERPS, PQ characteristics and influencing parameters are different. In this section, a brief review of various ERPSs configurations together with the classified reported PQ phenomena in literature based on supplying system type are presented. As mentioned, at the beginning of the ERPS the transformer-based TPSS were prevalent and popular. They were based on the DC supplying system which feeds DC motors directly. The TPSSs in such an ERPS is resourced with the three-phase multiple pulse rectifiers to cover the locomotive's energy. They lead to a wide spectrum of harmonics. However, in addition to the harmonic problems, voltage fluctuations, voltage imbalance, voltage drops, and other transient events are reported in different DC TPSS around the world. Meanwhile, the brush-based arcs and the EMI based problems due to the stray currents are the other issues of DC ERPS. Inventing advanced PE-based DC TPSS and taking benefits of bidirectional switches, the DC voltage regulation, allowing the regenerative braking energy harvesting of trains, reduction of voltage drop and voltage fluctuation has occurred. With the growing developments of power technologies and electrical motors, the AC transformer-based TPSS were appeared and got popular. The 1 × 25 kV simple and back feeder based structures are initial configurations have been adopted. The main problem of these systems is voltage and current imbalance due to the single-phase feeding systems. Furthermore, the LPF, harmonics, transient problems, and EMI issues, and telecommunication interfaces regarding the return currents are other forms of PQ indices that have been reported in the literature. To decrease the return current of rail and mitigate EMI issues, the booster transformer (BT) based 1 × 25 kV ERPS was presented. However, the large arcing and voltage drops in BT sections could cause essential damage to the OCS. Further, the BT sections reduce the operating speed of trains. Motivated by these problems, the autotransformer based 2 × 25 kV ERPS has been employed. Reducing inductive and telecommunication interfaces, voltage drops, and arcing problems are the foremost features of these systems. Nonetheless, LFO and HI issues regarding the high-speed 4QC based locomotives and impedance interplay among the system and trains are the fundamental obstacles of these ERPS. To decrease the arcing problems and enhance the functioning speed of trains close the TPSSs the specialists provided a new ERPS topology with a lower number of neutral sections and insulating areas. They are known as co-phase system. Disregarding the converter performance in the co-phase ERPS configuration, the PQ issues are like other traditional ERPS except arcing and related problems which have been reduced. During the development of AC industrial frequency-based ERPS, several countries like Germany, Norway, Sweden, Austria, and Switzerland employed the 16.67 Hz ERPS using RFC. The main critical issues of such an ERPS are stability and transient problems owing to the synchronously coupled motor and generators. Furthermore, the modern kind of frequency converter termed the static frequency converter (SFC) comprises high-frequency harmonics and HR and HI phenomena. In addition, a wide range of perturbation among SFC and RFC manifest during their parallel functioning. However, as a consequence of the PE-based equipment, SFC can mitigate the system imbalance, control power factor, and regulate voltage.

With the rapid developments of power electronics and high-voltage switches, a new ERPS structure called advanced co-phase system is presented. Due to the three-phase to single-phase converter and their capability in controlling output voltage and currents, these types of ERPS can compensate harmonics, imbalance, and PF. Meanwhile, with the elimination of neutral zones, the arcing problems and other related voltage variation can be suppressed. In some cases, the three-phase to single-phase converters are based on the modular multilevel converter either direct (AC/AC) or indirect (AC/DC/AC). The MMC based ERPS is connected directly to the utility grid via the three-phase side of MMC. These systems symmetrically transfer active power from the three-phase grid to the single-phase OCS which leads to compensate NSC and harmonics and improve PF simultaneously. Eliminating bulky traction transformers, integrating of OCS, and removing the neutral sections which lead to obviate the arcing problems and voltage fluctuation and transients are the advantages of mentioned ERPS. To have a quick review and appropriate classification of all mentioned ERPS structures, the related organized chart is provided in Figure 27. This figure illustrates a comprehensive classification of ERPS types based on internal configuration.

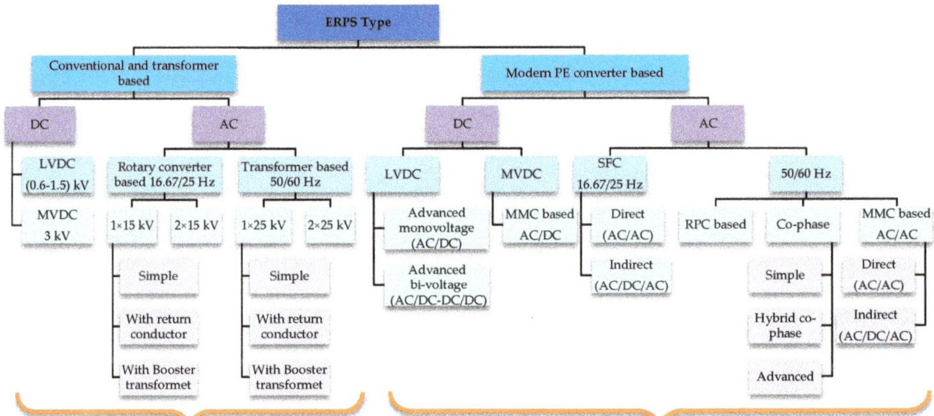

Figure 27. Classification and organization chart of different ERPS configuration.

4. Discussion and Classification

Choosing the right method to mitigate PQ issues requires comprehensive knowledge and identification of main sources, the characteristics, influencing factors, and occurrence environment. Unlike the power grid, for which numerous standards and resources have been developed, the ERPS suffers from a lack of a comprehensive and all-encompassing resource. As discussed before, by the proliferation and development of ERPSs, different forms of PQ phenomena emerge. Generally speaking, all kinds of ERPS can be categorized according to their internal configuration in two groups, transformer-based and modern power electronic converter based. The first category known as the most popular and predominant system is the preference of experts in the designing stage due to the lower expenditure costs and effortless operation. However, as evaluated in a specific framework, these systems deal with many PQ issues which even may enhance their costs remarkably. Meanwhile, as a perspective view and compatibility evaluation with future smart grid-based ERPS, power flow control in these ERPSs contain many complexities and difficulties.

On the contrary, the second group ERPSs taking advantage of the modern converter have not only admirable performance in terms of power quality but also significant potential to achieve future smart grid-based networks. However, the high cost of PE converters is still a basic weak point for this type which has led them to remain in the theoretical stage for the time being and prevents the developments and implementation of these systems. The overall classification of PQ phenomena based on the sources, ERPS system structure, and the probability of occurrence together with an overview of some main studies which have been carried out in each index are provided in Table 3.

Table 3. Reported classified PQ indexes in different ERPS.

Phenomena	Type	Causes and Sources	System Type & References
Unbalance	VUF	single-phase power supply, asymmetric faults	AC 1 × 25 kV [30], AC 2 × 25 kV [37,41]
	CUF	single-phase power supply, asymmetric faults	AC 2 × 25 kV [31,41], AC 1 × 25 kV [33], AC 15 kV-16.67 Hz [36]
Harmonics	LOH	background harmonics, train internal harmonics, DC rectifier substation	AC 1 × 25 kV [24,44–47], AC 2 × 25 kV [42,43], LVDC [29,48,50], MVDC [49,51]
	InH	AC motors controlled by variable frequency drives, onboard PWM converters	AC 1 × 25 kV [54,55], 2 × 25 kV [53], 15 kV-16.67 Hz [52], MVDC [53]
	LFO	impedance mismatch between the railway network and PWM trains, rotary converter	AC 1 × 25 kV [59–62], 2 × 25 kV [64,65] 15 kV-16.67 Hz [56–58], MVDC [63]
	HR	interaction of current harmonics and internal resonance of LC circuit of OCS	AC 1 × 25 kV [71–75], 2 × 25 kV [66–70,76,79], 15 kV-16.67 Hz [77]
	HI	interaction between high switching frequency of PWM converters in modern trains and internal resonance of OCS, high-frequency specifications of the closed-loop control system for 4QC	AC 1 × 25 kV [85,86], 2 × 25 kV [80,82,84], 15 kV-16.67 Hz [81,87], MVDC [82]
Low Power Factor	LPF	overlap commutation angel in conventional rectifier, inductive reactance features of the OCS, supplying sections with different phase	AC 1 × 25 kV [89,90,95], 2 × 25 kV [92–94], 15 kV-16.67 Hz [88], MVDC [12], LVDC [12,51]
Transient events	ImT	lightening, switching of circuit breakers, abnormal changing in tractive efforts, sliding contact between pantograph and OCS, passing neutral zone	AC 1 × 25 kV [99,100], 2 × 25 kV [97], 15 kV-16.67 Hz [98], MVDC [99,102], LVDC [101,102]
	OsT	changing in operational condition and modulation patterns, sliding contact and pantograph jump over OCS, inrush current of locomotive transformer, capacitor bank energization	AC 1 × 25 kV [106,111], 2 × 25 kV [1], 15 kV-16.67 Hz [105], MVDC [103,104], LVDC [103,104]
Short duration rms variation	Sag	fault occurrence, high current absorbed by traction motors, sudden load changes or supplying high-power locomotives,	AC 1 × 25 kV [108,111,112], 2 × 25 kV [110], 15 kV-16.67 Hz [105,109], MVDC [109]
	Swell	TPSS transformers energization/de-energization, motor blocking caused by the segregation of pantograph and OCS, neutral sections, TPSS equipment triggering	AC 1 × 25 kV [108,111,112], 2 × 25 kV [110], 15 kV-16.67 Hz [105,109], MVDC [109]
	InR	train passes NZ, pantograph bounce operating of the circuit breaker, failure, and fault, abruptly disconnection between the contact wire and pantograph	AC 1 × 25 kV [108,109,112,113], 2 × 25 kV [108,113], 15 kV-16.67 Hz [109], MVDC [109], LVDC [3,114]
Long duration rms variation	OvG	regenerative braking and lack of consumer trains, resonance, system instabilities, passing of neutral zones, lightning, switching or other atmospheric phenomena, functioning of split-phase breakers	AC 1 × 25 kV [99,115], 2 × 25 kV [12,116], 15 kV-16.67 Hz [6], MVDC [117], LVDC [117,118]
	UvG	Type of line (unilateral, bilateral, ...), high traffic of line, resonance, system instabilities, passing of neutral zones,	AC 1 × 25 kV [122], 2 × 25 kV [120,121], 15 kV-16.67 Hz [123], MVDC [1,122], LVDC [1,117,122]
	InRS	Train passes NZ, pantograph bounce operating of the circuit breaker, failure, and fault, abruptly disconnection between the contact wire and pantograph	AC 1 × 25 kV [108,109,112,113], 2 × 25 kV [108,113], 15 kV-16.67 Hz [109], MVDC [109], LVDC [3,114]
Voltage fluctuation (Flicker)	FlK	time-varying specification, abrupt load changes in ERPS, static frequency converters, operation of high-power traction motors, arcing equipment	AC 1 × 25 kV [41,108,124,125], 2 × 25 kV [108], 15 kV-16.67 Hz [124], LVDC [124]
Waveform distortion	DC offset	indirect current control method of 4QCs, inequality of the positive and negative half-cycle length in six-step mode control	AC 1 × 25 kV [127,129], 2 × 25 kV [127,129], 15 kV-16.67 Hz [127], MVDC [127], LVDC [128]
	Notch	power electronics devices, 4QC drives, abruptly disconnection between the contact wire and pantograph	AC 1 × 25 kV [45,71,111], 2 × 25 kV [18,111],
	Noise	power electronic devices, control circuits, arcing equipment and traction locomotives with onboard rectifiers, sliding contact between the OCS and the pantograph	AC 1 × 25 kV [100,131,132], 2 × 25 kV [100], 15 kV-16.67 Hz [131,132], MVDC [130], LVDC [130–132]

Table 3. *Cont.*

Phenomena	Type	Causes and Sources	System Type & References
Electromagnetic Interference	CEMI	injection of distorted current or harmonics into the power lines from devices with nonlinear features, transient overvoltage originated by switching and atmospheric phenomena, return current flowing through earth	AC 1 × 25 kV [1,134,137], 2 × 25 kV [1,134,137], 15 kV-16.67 Hz [1,137,140], MVDC [1,135,136,145], LVDC [1,135,145]
	IEMI	Interfacing of magnetic flux related to the source current with a secondary closed system, physical positioning causing transverse emf	AC 1 × 25 kV [1,134,138,139,143,144], 2 × 25 kV [1,138,139,142–144], 15 kV-16.7 Hz [1,138,140]
	EEMI	production of electrostatic electric fields, capacitive interference between closed conductors and earth, the sudden change in voltage or inherent track admittance causing charging and discharging of capacitors or switching semiconductors	AC 1 × 25 kV [1,133], 2 × 25 kV [1,133], 15 kV-16.7 Hz [1,133]
	REMI	The interlinkage between the pantograph and contact wire or between the train's brushes and the third or fourth rail, neutral intersection points	AC 1 × 25 kV [1,143], 2 × 25 kV [1,143], 15 kV-16.7 Hz [1], MVDC [103,136,145], LVDC [103,145]

5. Conclusions

Since the start of railway electrification, ERPSs have experienced substantial modifications and developments. This diversity including the types of railway power supply systems, AC or DC-based power supply, the types of converters adopted in trains and TPSS, controlling and drive systems, and the dedicated frequency has led to various power quality issues reported all around the world. Furthermore, the lack of specific standards and encompassing resources dedicated to ERPS has increased the obscurity and complexity of its PQ. In this paper, a comprehensive review of PQ phenomena in terms of characteristics, influencing factors, and occurrence sources in different configurations of ERPSs is proposed and a systematic classification is presented. Meanwhile, a detailed review of the reported PQ phenomena in the literature is classified. Unbalance of system, various types of harmonics, low power factor, different aspects of transients events, and waveform deviations, with EMI are addressed with details as the outstanding indexes. Classifying and assessing of diverse transformer-based and PE modern converter based ERPS exposed that the latter types are more efficient and have good performance regarding the PQ issues. Moreover, due to the suitable power flow capability and possible connection to the weak utility networks, which can mitigate and reduce the PQ issues significantly, the converter based ERPS are realized as an appropriate option for smart grid-based future studies of ERPS.

Author Contributions: In preparation of this paper, H.J.K. did the methodology, data curation, software-based simulations, and writing—original draft, M.B., F.F., and D.Z. did the validation, visualization, and the resources and funding acquisition; S.S.F. did the editing, and supervision; All authors have read and agreed to the published version of the manuscript.

Funding: This research received no external funding.

Conflicts of Interest: The authors declare no conflict of interest.

References

1. Brenna, M.; Foiadelli, F.; Zaninelli, D. *Electrical Railway Transportation Systems*; IEEE Press Series on Power Engineering: Hoboken, NJ, USA, 2018.
2. Serrano-Jimenez, D.; Abrahamsson, L.; Castano-Solis, S.; Sanz-Feito, J. Electrical railway power supply systems: Current situation and future trends. *Int. J. Elect. Power Energy Syst.* **2017**, *92*, 181–192. [CrossRef]
3. Magro, M.C.; Mariscotti, A.; Pinceti, P. Definition of Power Quality Indices for DC Low Voltage Distribution Networks. In Proceedings of the IEEE Instrumentation and Measurement Technology Conference Proceedings, Sorrento, Italy, 24–27 April 2006; pp. 1885–1888.
4. Popescu, M.; Bitoleanu, A. A Review of the Energy Efficiency Improvement in DC Railway Systems. *Energies* **2019**, *12*, 1092. [CrossRef]

5. Kaleybar, H.J.; Kojabadi, H.M.; Brenna, M.; Foiadelli, F.; Fazel, S.S.; Rasi, A. An Inclusive Study and Classification of Harmonic Phenomena in Electric Railway Systems. In Proceedings of the 2019 IEEE International Conference on Environment and Electrical Engineering and 2019 IEEE Industrial and Commercial Power Systems Europe (EEEIC/I&CPS Europe), Genova, Italy, 11–14 June 2019; pp. 1–6.
6. Chmielewski, T.; Oramus, P.; Koska, K. Switching transients in a 2 × 15 kV 16.7 Hz autotransformer railway system. *IET Gener. Transm. Distrib.* **2018**, *12*, 235–240. [CrossRef]
7. Mariscotti, A. Uncertainty of the Energy Measurement Function deriving from Distortion Power Terms for a 16.7 Hz Railway. *Acta Imeko* **2020**, *9*, 25–31. [CrossRef]
8. Laury, J.; Bollen, M.H.J.; Abrahamsson, L. Transient stability analysis of low frequency railway grids. *Comput. Railw. XV Railw. Eng. Des. Oper.* **2016**, *162*, 213–223.
9. Seferi, Y.; Blair, S.M.; Mester, C.; Stewart, B.G. Power Quality Measurement and Active Harmonic Power in 25 kV 50 Hz AC Railway Systems. *Energies* **2020**, *13*, 5698. [CrossRef]
10. Morrison, R.E. Power quality issues on AC traction systems. In Proceedings of the Ninth International Conference on Harmonics and Quality of Power, Proceedings (Cat. No.00EX441), Orlando, FL, USA, 1–4 October 2000; Volume 2, pp. 709–714.
11. Kaleybar, H.J.; Kojabadi, H.M.; Brenna, M.; Foiadelli, F.; Fazel, S.S. A two-phase three-wire quasi-Z-source based railway power quality compensator for AC rail networks. In Proceedings of the 2017 IEEE Int. Conf. on Environment and Electrical Engineering and 2017 IEEE Industrial and Commercial Power Systems Europe (EEEIC/I&CPS Europe), Milan, Italy, 6–9 June 2017; pp. 1–6.
12. Mariscotti, A. Results on the Power Quality of French and Italian 2 × 25 kV 50 Hz railways. In Proceedings of the 2012 IEEE International Instrumentation and Measurement Technology Conference Proceedings, Graz, Austria, 13–16 May 2012; pp. 1400–1405.
13. Brenna, M.; Foiadelli, F. Sensitivity Analysis of the Constructive Parameters for the 2 × 25 kV High-Speed Railway Lines Planning. *IEEE Trans. Power Deliv.* **2010**, *25*, 1923–1931. [CrossRef]
14. Kaleybar, H.J.; Kojabadi, H.M.; Brenna, M.; Foiadelli, F.; Fazel, S.S. An active railway power quality compensator for 2 × 25 kV high-speed railway lines. In Proceedings of the 2017 IEEE Int. Conf. on Environment and Electrical Engineering and 2017 IEEE Industrial and Commercial Power Systems Europe (EEEIC/I&CPS Europe), Milan, Italy, 6–9 June 2017; pp. 1–6.
15. Sun, Z.; Jiang, X.; Zhu, D.; Zhang, G. A novel active power quality compensator topology for electrified railway. *IEEE Trans. Power Electron.* **2004**, *19*, 1036–1042. [CrossRef]
16. Xu, Q.; Ma, F.; He, Z.; Chen, Y.; Guerrero, J.M.; Luo, A.; Li, Y.; Yue, Y. Analysis and Comparison of Modular Railway Power Conditioner for High-Speed Railway Traction System. *IEEE Trans. Power Electron.* **2017**, *32*, 6031–6048. [CrossRef]
17. Wang, H.; Liu, Y.; Yan, K.; Fu, Y.; Zhang, C. Analysis of static VAr compensators installed in different positions in electric railways. *IET Electr. Syst. Transp.* **2015**, *5*, 129–134. [CrossRef]
18. Kaleybar, H.J.; Brenna, M.; Foiadelli, F.; Fazel, S.S. Regenerative Braking Energy and Power Quality Analysis in 2 × 25 kV High-Speed Railway Lines Operating with 4QC Locomotives. In Proceedings of the 2020 11th Power Electronics, Drive Systems, and Technologies Conference (PEDSTC), Tehran, Iran, 4–6 February 2020; pp. 1–6.
19. He, X.; Shu, Z.; Peng, X.; Zhou, Q.; Zhou, Y.; Zhou, Q.; Gao, S. Advanced Cophase Traction Power Supply System Based on Three-Phase to Single-Phase Converter. *IEEE Trans. Power Electron.* **2014**, *290*, 5323–5333. [CrossRef]
20. Xie, B.; Li, Y.; Zhang, Z.; Hu, S.; Zhang, Z.; Luo, L.; Cao, Y.; Zhou, F.; Luo, R.; Long, L. A Compensation System for Cophase High-Speed Electric Railways by Reactive Power Generation of SHC&SAC. *IEEE Trans. Ind. Electron.* **2018**, *65*, 2956–2966.
21. He, X.; Peng, J.; Han, P.; Liu, Z.; Gao, S.; Wang, P. A Novel Advanced Traction Power Supply System Based on Modular Multilevel Converter. *IEEE Access* **2019**, *7*, 165018–165028. [CrossRef]
22. Ronanki, D.; Williamson, S.S. Modular Multilevel Converters for Transportation Electrification: Challenges and Opportunities. *IEEE Trans. Transp. Electrif.* **2018**, *4*, 399–407. [CrossRef]
23. Brenna, M.; Foiadelli, F.; Kaleybar, H.J. The Evolution of Railway Power Supply Systems Toward Smart Microgrids: The concept of the energy hub and integration of distributed energy resources. *IEEE Electrif. Mag.* **2020**, *8*, 12–23. [CrossRef]

24. He, Z.; Zheng, Z.; Hu, H. Power quality in high-speed railway systems. *Int. J. Rail Transp.* **2016**, *4*, 71–97. [CrossRef]
25. Župan, A.; Teklić, A.T.; Filipović-Grčić, B. Modeling of 25 kV electric railway system for power quality studies. In Proceedings of the Eurocon, Zagreb, Croatia, 1–4 July 2013; pp. 844–849.
26. Mariscotti, A. Direct Measurement of Power Quality Over Railway Networks with Results of a 16.7-Hz Network. *IEEE Trans. Instrum. Meas.* **2011**, *60*, 1604–1612. [CrossRef]
27. Mariscotti, A. Characterization of Power Quality transient phenomena of DC railway traction supply. *ACTA IMEKO* **2012**, *1*, 26–35. [CrossRef]
28. Brenna, M.; Foiadelli, F.; Zaninelli, D. Electromagnetic Model of High Speed Railway Lines for Power Quality Studies. *IEEE Trans. Power Syst.* **2010**, *25*, 1301–1308. [CrossRef]
29. Brenna, M.; Foiadelli, F.; Kaleybar, H.J.; Fazel, S.S. Power Quality Indicators in Electric Railway Systems: A Comprehensive Classification. In Proceedings of the 2019 IEEE Milan PowerTech, Milan, Italy, 23–27 June 2019; pp. 1–6.
30. Kuo, H.-Y.; Chen, T.-H. Rigorous evaluation of the voltage unbalance due to high-speed railway demands. *IEEE Trans. Veh. Technol.* **1998**, *47*, 1385–1389.
31. Chen, S.-L.; Li, R.; Hsi, P.-H. Traction system unbalance problem-analysis methodologies. *IEEE Trans. Power Deliv.* **2004**, *19*, 1877–1883. [CrossRef]
32. Kaleybar, H.J.; Farshad, S. A comprehensive control strategy of railway power quality compensator for AC traction power supply systems. *Turk. J. Electr. Eng. Comput. Sci.* **2016**, *24*, 4582–4603. [CrossRef]
33. Zhang, D.; Zhang, Z.; Wang, W.; Yang, Y. Negative Sequence Current Optimizing Control Based on Railway Static Power Conditioner in V/v Traction Power Supply System. *IEEE Trans. Power Electron.* **2016**, *31*, 200–212. [CrossRef]
34. Sutherland, P.E.; Waclawiak, M.; McGranaghan, M.F. System impacts evaluation of a single-phase traction load on a 115-kV transmission system. *IEEE Trans. Power Deliv.* **2006**, *21*, 837–844. [CrossRef]
35. Kaleybar, H.J.; Kojabadi, H.M.; Fallah, M.; Fazel, S.S.; Chang, L. Impacts of traction transformers on power rating of Railway Power Quality Compensator. In Proceedings of the IEEE 8th International Power Electronics and Motion Control Conference (IPEMC-ECCE Asia), Hefei, China, 22–26 May2016; pp. 2229–2236.
36. Gorski, M.; Heising, C.; Staudt, V.; Steimel, A. Single-phase 50-kW, 16.7-Hz railway-grid Lab Representation using a DC-excited slip-ring induction generator. In Proceedings of the Compatibility and Power Electronics, Badajoz, Spain, 20–22 May 2009; pp. 204–209.
37. Huh, J.; Shin, H.; Moon, W.; Kang, B.; Kim, J. Study on Voltage Unbalance Improvement Using SFCL in Power Feed Network with Electric Railway System. *IEEE Trans. Appl. Supercond.* **2013**, *23*, 3601004.
38. IEEE. IEEE Standard Test Procedure for Polyphase Induction Motors and Generators. In *IEEE Standard 112*; IEEE: New York, NY, USA, 2004.
39. IEC Standard Rotating Electrical Machines-Part26: Effects of Unbalanced Voltages on the Performance of Three-Phase Cage Induction Motors, IEC Standard 60034-26. 2006. Available online: https://webstore.iec.ch/publication/129 (accessed on 17 December 2020).
40. *Motors and Generators*; ANSI/NEMA Standard MG1; National Electrical Manufacturers Association: Washington, DC, USA, 2003.
41. Yu-Quan, L.; Guo-Pei, W.; Huang-Sheng, H.; Li, W. Research for the effects of high-speed electrified railway traction load on power quality. In Proceedings of the 4th International Conference on Electric Utility Deregulation and Restructuring and Power Technologies (DRPT), Weihai, China, 6–9 July 2011; pp. 569–573.
42. Wang, J.; Zhang, M.; Li, S.; Zhou, T.; Du, H. Passive filter design with considering characteristic harmonics and harmonic resonance of electrified railway. In Proceedings of the 8th International Conference on Mechanical and Intelligent Manufacturing Technologies (ICMIMT), Cape Town, South Africa, 3–6 February 2017; pp. 174–178.
43. Lee, H.; Lee, C.; Jang, G.; Kwon, S.-H. Harmonic analysis of the Korean high-speed railway using the eight-port representation model. *IEEE Trans. Power Deliv.* **2006**, *21*, 979–986. [CrossRef]
44. Gao, S.; Li, X.; Ma, X.; Hu, H.; He, Z.; Yang, J. Measurement-based compartmental modeling of harmonic sources in traction power-supply system. *IEEE Trans. Power Deliv.* **2017**, *32*, 900–909. [CrossRef]
45. Tan, P.-C.; Morrison, R.E.; Holmes, D.G. Voltage form factor control and reactive power compensation in a 25-kV electrified railway system using a shunt active filter based on voltage detection. *IEEE Trans. Ind. Appl.* **2003**, *39*, 575–581.

46. Tan, P.-C.; Loh, P.C.; Holmes, D.G. A robust multilevel hybrid compensation system for 25-kV electrified railway applications. *IEEE Trans. Power Electron.* **2004**, *19*, 1043–1052. [CrossRef]
47. Kaleybar, H.J.; Farshad, S.; Asadi, M.; Jalilian, A. Multifunctional control strategy of Half-Bridge based Railway Power Quality Conditioner for Traction System. In Proceedings of the 2013 13th International Conference on Environment and Electrical Engineering (EEEIC), Wroclaw, Poland, 1–3 November 2013; pp. 207–212.
48. Ying-Tung, H.; Lin, K.-C. Measurement and characterization of harmonics on the Taipei MRT DC system. *IEEE Trans. Ind. Appl.* **2004**, *40*, 1700–1704.
49. Skarpetowski, G.; Zajac, W.; Czuchra, W. Analytical Calculation of Supply Current Harmonics Generated by Train Unit. In Proceedings of the 12th International Power Electronics and Motion Control Conference, Portoroz, Slovenia, 30 August–1 September 2006; pp. 1378–1384.
50. Ogunsola, A.; Mariscotti, A.; Sandrolini, L. Measurement of AC side harmonics of a DC metro railway. In Proceedings of the 2012 Electrical Systems for Aircraft, Railway and Ship Propulsion, Bologna, Italy, 16–18 October 2012; pp. 1–5.
51. Terciyanli, A.; Acik, A.; Cetin, A.; Ermis, M.; Cadirci, I.; Ermis, C.; Demirci, T.; Bilgin, H.F. Power Quality Solutions for Light Rail Public Transportation Systems Fed by Medium-Voltage Underground Cables. *IEEE Trans. Ind. Appl.* **2012**, *48*, 1017–1029. [CrossRef]
52. Caramia, P.; Morrone, M.; Varilone, P.; Verde, P. Interaction between supply system and EMU loco in 15 kV-16 2/3 hz AC traction systems. In Proceedings of the Power Engineering Society Summer Meeting. Conference Proceedings (Cat. No.01CH37262), Vancouver, BC, Canada, 15–19 July 2001; Volume 1, pp. 198–203.
53. Caramia, P.; Carpinelli, G.; Varilone, P.; Verde, P.; Gallo, D.; Langella, R.; Testa, A. High speed AC locomotives: Harmonic and interharmonic analysis at a vehicle test room. In Proceedings of the Ninth International Conference on Harmonics and Quality of Power. Proceedings (Cat. No.00EX441), Orlando, FL, USA, 1–4 October 2000; Volume 1, pp. 347–353.
54. Qiujiang, L.; Mingli, W.; Junqi, Z.; Kejian, S.; Liran, W. Resonant frequency identification based on harmonic injection measuring method for traction power supply systems. *IET Power Electron.* **2018**, *11*, 585–592. [CrossRef]
55. Spangenberg, U. Variable frequency drive harmonics and interharmonics exciting axle torsional vibration resulting in railway wheel polygonization. *Veh. Syst. Dyn.* **2020**, *58*, 404–424. [CrossRef]
56. Laury, J.; Abrahamsson, L.; Bollen, M.H. A rotary frequency converter model for electromechanical transient studies of 1623 Hz railway systems. *Int. J. Electr. Power Energy Syst.* **2019**, *106*, 467–476. [CrossRef]
57. Danielsen, S. Electric Traction Power System Stability: Low-Frequency Interaction between Advanced Rail Vehicles and a Rotary Frequency Converter. Ph.D. Dissertation, Norwegian University of Science and Technology Faculty of Information Technology, Mathematics and Electrical Engineering, Trondheim, Norway, 2010.
58. Laury, J.; Abrahamsson, L.; Bollen, M.H. Transient stability of rotary frequency converter fed low frequency railway grids: The impact of different grid impedances and different converter station configurations. In Proceedings of the ASME. ASME/IEEE Joint Rail Conference, Pittsburgh, PA, USA, 18–21 April 2018.
59. Wang, H.; Mingli, W.; Sun, J. Analysis of low-frequency oscillation in electric railways based on small-signal modeling of vehicle-grid system in dq frame. *IEEE Trans. Power Electron.* **2015**, *30*, 5318–5330. [CrossRef]
60. Liu, Z.; Wang, Y.; Liu, S.; Li, Z.; Zhang, H.; Zhang, Z. An Approach to Suppress Low-Frequency Oscillation by Combining Extended State Observer with Model Predictive Control of EMUs Rectifier. *IEEE Trans. Power Electron.* **2019**, *340*, 10282–10297. [CrossRef]
61. Liu, Z.; Geng, Z.; Hu, X. An approach to suppress low frequency oscillation in the traction network of high-speed railway using passivity-based control. *IEEE Trans. Power Syst.* **2018**, *33*, 3909–3918. [CrossRef]
62. Jiang, K.; Zhang, C.; Ge, X. Low-frequency oscillation analysis of the train-grid system based on an improved forbidden-region criterion. *IEEE Trans. Ind. Appl.* **2018**, *54*, 5064–5073. [CrossRef]
63. Quesada, I.; Lazaro, A.; Martinez, C.; Barrado, A.; Sanz, M.; Fernandez, C.; Vazquez, R.; Gonzalez, I. Modulation technique for low frequency harmonic cancellation in auxiliary railway power supplies. *IEEE Trans. Ind. Electron.* **2011**, *58*, 3976–3987. [CrossRef]
64. Hu, H.; Tao, H.; Blaabjerg, F.; Wang, X.; He, Z.-Y.; Gao, S. Train–network interactions and stability evaluation in high-speed railways—Part I: Phenomena and modeling. *IEEE Trans. Power Electron.* **2018**, *33*, 4627–4642. [CrossRef]

65. Hu, H.; Zhou, Y.; Li, X.; Lei, K. Low-Frequency Oscillation in Electric Railway Depot: A Comprehensive Review. *IEEE Trans. Power Electron.* **2020**, *36*, 295–314. [CrossRef]
66. Liu, J.; Yang, Q.; Zheng, T.Q. Harmonic analysis of traction networks based on the CRH380 series EMUs accident. In Proceedings of the 2012 IEEE Transportation Electrification Conference and Expo (ITEC), Dearborn, MI, USA, 18–20 June 2012; pp. 1–6.
67. He, Z.; Hu, H.; Zhang, Y.; Gao, S. Harmonic Resonance Assessment to Traction Power-Supply System Considering Train Model in China High-Speed Railway. *IEEE Trans. Power Deliv.* **2014**, *29*, 1735–1743. [CrossRef]
68. Hu, H.; He, Z.; Gao, S. Passive Filter Design for China High-Speed Railway with Considering Harmonic Resonance and Characteristic Harmonics. *IEEE Trans. Power Deliv.* **2015**, *30*, 505–514. [CrossRef]
69. Song, W.; Jiao, S.; Li, Y.; Wang, J.; Huang, J. High-frequency harmonic resonance suppression in high-speed railway through single-phase traction converter with LCL filter. *IEEE Trans. Transp. Electrif.* **2016**, *2*, 347–356. [CrossRef]
70. Brenna, M.; Foiadelli, F.; Zaninelli, D. New Stability Analysis for Tuning PI Controller of Power Converters in Railway Application. *IEEE Trans. Ind. Electron.* **2011**, *58*, 533–543. [CrossRef]
71. Morrison, R.E.; Barlow, M.J. Continuous Overvoltages on A.C. Traction Systems. *IEEE Trans. Power Appar. Syst.* **1983**, *PAS-102*, 1211–1217. [CrossRef]
72. Lutrakulwattana, B.; Konghirun, M.; Sangswang, A. Harmonic resonance assessment of 1 × 25kV, 50 Hz traction power supply system for suvarnabhumi airport rail link. In Proceedings of the 2015 18th International Conference on Electrical Machines and Systems (ICEMS), Pattaya, Thailand, 25–28 October 2015; pp. 752–755.
73. Maeda, T.; Watanabe, T.; Mechi, A.; Shiota, T.; Iida, K. A hybrid single-phase power active filter for high order harmonics compensation in converter-fed high speed trains. In Proceedings of the Power Conversion Conference—PCC '97, Nagaoka, Japan, 6 August 1997; Volume 2, pp. 711–717.
74. Morrison, R.E.; Corcoran, J.C.W. Specification of an overvoltage damping filter for the National Railways of Zimbabwe. *IEE Proc. B Electr. Power Appl.* **1989**, *136*, 249–256. [CrossRef]
75. Kolar, V.; Palecek, J.; Kocman, S.; Vo, T.T.; Orsag, P.; Stýskala, V.; Hrbac, R. Interference between electric traction supply network and distribution power network-resonance phenomenon. In Proceedings of the 14th International Conference on Harmonics and Quality of Power—ICHQP 2010, Bergamo, Italy, 26–29 September 2010; pp. 1–4.
76. Mehdavizadeh, F.; Farshad, S.; Veysi Raygani, S.; Shahroudi, M.R. Resonance verification of Tehran-Karaj electrical railway. In Proceedings of the First Power Quality Conferance, Tehran, Iran, 14–15 September 2010; pp. 1–6.
77. Heising, C.; Bartelt, R.; Staudt, V.; Steimel, A. Single-phase 50-kW 16.7-Hz four-quadrant line-side converter for railway traction application. In Proceedings of the 13th International Power Electronics and Motion Control Conference, Poznan, Poland, 1–3 September 2008; pp. 521–527.
78. Sainz, L.; Caro, M.; Caro, E. Analytical study of the series resonance in power systems with the Steinmetz circuit. *IEEE Trans. Power Deliv.* **2009**, *24*, 2090–2098. [CrossRef]
79. Brenna, M.; Capasso, A.; Falvo, M.C.; Foiadelli, F.; LaMedica, R.; Zaninelli, D. Investigation of resonance phenomena in high speed railway supply systems: Theoretical and experimental analysis. *Electr. Power Syst. Res.* **2011**, *810*, 1915–1923. [CrossRef]
80. Hu, H.; Tao, H.; Wang, X.; Blaabjerg, F.; He, Z.; Gao, S. Train–network interactions and stability evaluation in high-speed railways—Part II: Influential factors and verifications. *IEEE Trans. Power Electron.* **2018**, *33*, 4643–4659. [CrossRef]
81. Mollerstedt, E.; Bernhardsson, B. Out of control because of harmonics an analysis of the harmonic response of an inverter locomotive. *IEEE Control Syst.* **2000**, *20*, 70–81.
82. Brenna, M.; Foiadelli, F. Analysis of the filters installed in the interconnection points between different railway supply systems. *IEEE Trans. Smart Grid* **2012**, *3*, 551–558. [CrossRef]
83. Wang, Y.; Wang, X.; Blaabjerg, F.; Chen, Z. Harmonic Instability Assessment Using State-Space Modeling and Participation Analysis in Inverter-Fed Power Systems. *IEEE Trans. Ind. Electron.* **2017**, *64*, 806–816. [CrossRef]
84. Pan, P.; Hu, H.; Yang, X.; Blaabjerg, F.; Wang, X.; He, Z. Impedance Measurement of Traction Network and Electric Train for Stability Analysis in High-Speed Railways. *IEEE Trans. Power Electron.* **2018**, *332*, 10086–10100. [CrossRef]

85. Tao, H.; Hu, H.; Zhu, X.; Zhou, Y.; He, Z. Harmonic instability analysis and suppression method based on αβ-frame impedance for trains and network interaction system. *IEEE Trans. Energy Convers.* **2019**, *34*, 1124–1134. [CrossRef]
86. Zhang, X.; Wang, L.; Dunford, W.; Chen, J.; Liu, Z. Integrated Full-Frequency Impedance Modeling and Stability Analysis of the Train-Network Power Supply System for High-Speed Railways. *Energies* **2018**, *11*, 1714. [CrossRef]
87. Mariscotti, A.; Slepicka, D. Analysis of frequency stability of 16.7 Hz railways. In Proceedings of the IEEE International Instrumentation and Measurement Technology Conference, Binjiang, China, 10–12 May 2011; pp. 1–5.
88. Mariscotti, A. Behaviour of Spectral Active Power Terms for the Swiss 15 kV 16.7 Hz Railway System. In Proceedings of the IEEE 10th International Workshop on Applied Measurements for Power Systems (AMPS), Aachen, Germany, 25–27 September 2019; pp. 1–6.
89. Raygani, S.V.; Tahavorgar, A.; Fazel, S.S.; Moaveni, B. Load flow analysis and future development study for an AC electric railway. *IET Electr. Syst. Transp.* **2012**, *2*, 139–147. [CrossRef]
90. Morais, V.A.; Afonso, J.L.; Carvalho, A.S.; Martins, A.P. New Reactive Power Compensation Strategies for Railway Infrastructure Capacity Increasing. *Energies* **2020**, *13*, 4379. [CrossRef]
91. Hafezi, H.; Faranda, R. Open UPQC series and shunt units cooperation within Smart LV Grid. In Proceedings of the 6th International Conference on Clean Electrical Power (ICCEP), Santa Margherita Ligure, Italy, 27–29 June 2017; pp. 304–310.
92. Hafezi, H.; Faranda, R. Power quality and custom power: Seeking for a common solution in LV distribution network. In Proceedings of the IEEE International Conference on Environment and Electrical Engineering and IEEE Industrial and Commercial Power Systems Europe (EEEIC/I&CPS Europe), Milan, Italy, 6–9 June 2017; pp. 1–6.
93. Hu, S.; Xie, B.; Li, Y.; Gao, X.; Zhang, Z.; Luo, L.; Krause, O.; Cao, Y. A Power Factor-Oriented Railway Power Flow Controller for Power Quality Improvement in Electrical Railway Power System. *IEEE Trans. Ind. Electron.* **2017**, *64*, 1167–1177. [CrossRef]
94. Wang, K.; Hu, H.; Zheng, Z.; He, Z.; Chen, L. Study on Power Factor Behavior in High-Speed Railways Considering Train Timetable. *IEEE Trans. Transp. Electrif.* **2018**, *4*, 220–231. [CrossRef]
95. Huang, C.-P.; Wu, C.-J.; Peng, S.-K.; Yen, J.-L.; Han, M.-H. Loading characteristics analysis of specially connected transformers using various power factor definitions. *IEEE Trans. Power Deliv.* **2006**, *21*, 1406–1413. [CrossRef]
96. IEEE. *IEEE Std 1159-2019. IEEE Draft Recommended Practice for Monitoring Electric Power Quality*; IEEE: New York, NY, USA, 2009.
97. Bigharaz, M.H.; Hosseinian, S.H.; Afshar, A.; Suratgar, A.A.; Dehcheshmeh, M.A. A comprehensive simulator of AC autotransformer electrified traction system. *Int. J. Power Energy Convers.* **2019**, *10*, 129–147. [CrossRef]
98. Theethayi, N.; Thottappillil, R.; Yirdaw, T.; Liu, Y.; Gotschl, T.; Montano, R. Experimental Investigation of Lightning Transients Entering a Swedish Railway Facility. *IEEE Trans. Power Deliv.* **2007**, *22*, 354–363. [CrossRef]
99. Asmontas, I.; Gudzius, S.; Markevicius, L.A.; Morkvenas, A.; Ticka, V. The investigation of overvoltage transient processes in railway electric power feeding systems. In Proceedings of the Electric Power Quality and Supply Reliability, Tartu, Estonia, 11–13 June 2012; pp. 1–6.
100. Kharbech, S.; Dayoub, I.; Zwingelstein-Colin, M.; Simon, E.P. Blind Digital Modulation Identification for MIMO Systems in Railway Environments with High-Speed Channels and Impulsive Noise. *IEEE Trans. Veh. Technol.* **2018**, *67*, 7370–7379. [CrossRef]
101. Lamedica, R.; Maranzano, G.; Marzinotto, M.; Prudenzi, A. Power quality disturbances in power supply system of the subway of Rome. In Proceedings of the IEEE Power Engineering Society General Meeting, Denver, CO, USA, 6–10 June 2004; Volume 1, pp. 924–929.
102. Mariscotti, A. DC railway line voltage ripple for periodic and aperiodic phenomena. In Proceedings of the International Measurement Conference Imeko, Natal, RN, Brazil, 27–30 September 2011.
103. Crotti, G.; Delle Femine, A.; Gallo, D.; Giordano, D.; Landi, C.; Luiso, M.; Mariscotti, A.; Roccato, P.E. Pantograph-to-OHL Arc: Conducted Effects in DC Railway Supply System. *IEEE Trans. Instrum. Meas.* **2019**, *680*, 3861–3870. [CrossRef]

104. Mariscotti, A.; Giordano, D. Experimental Characterization of Pantograph Arcs and Transient Conducted Phenomena in DC Railways. *Acta Imeko* **2020**, *9*, 10–17. [CrossRef]
105. Terwiesch, P.; Menth, S.; Schmidt, S. Analysis of transients in electrical railway networks using wavelets. *IEEE Trans. Ind. Electron.* **1998**, *45*, 955–959. [CrossRef]
106. Xie, C.; Tennakoon, S.; Langella, R.; Gallo, D.; Testa, A.; Wixon, A. Harmonic impedance measurement of 25 kV single phase AC supply systems. In Proceedings of the Ninth International Conference on Harmonics and Quality of Power. Proceedings (Cat. No.00EX441), Orlando, FL, USA, 1–4 October 2000; Volume 1, pp. 214–219.
107. Kaleybar, H.J.; Brenna, M.; Foiadelli, F. Dual-loop generalized predictive control method for two-phase three-wire railway active power quality controller. *Trans. Inst. Meas. Control* **2020**, in press. [CrossRef]
108. Im, Y.C.; Lee, T.H.; Han, I.S.; Park, C.S.; Kim, K.H.; Shin, M.C. Analysis about the tendencies of power quality indices according to the running states of the electric train. In Proceedings of the Transmission & Distribution Conference & Exposition: Asia and Pacific, Seoul, Korea, 26–30 October 2009; pp. 1–6.
109. Femine, A.D.; Gallo, D.; Landi, C.; Luiso, M. Discussion on DC and AC Power Quality Assessment in Railway Traction Supply Systems. In Proceedings of the IEEE International Instrumentation and Measurement Technology Conference (I2MTC), Auckland, New Zealand, 20–23 May 2019; pp. 1–6.
110. Zhang, S.; He, Z.; Lee, W.; Mai, R. Voltage-Sag-Profiles-Based Fault Location in High-Speed Railway Distribution System. *IEEE Trans. Ind. Appl.* **2017**, *53*, 5229–5238. [CrossRef]
111. Yoo, J.H.; Shin, S.K.; Park, J.Y.; Cho, S.H. Advanced railway power quality detecting algorithm using a combined TEO and STFT method. *J. Electr. Eng. Technol.* **2015**, *10*, 2442–2447. [CrossRef]
112. Seferi, Y.; Clarkson, P.; Blair, S.M.; Mariscotti, A.; Stewart, B.G. Power Quality Event Analysis in 25 kV 50 Hz AC Railway System Networks. In Proceedings of the IEEE 10th International Workshop on Applied Measurements for Power Systems (AMPS), Aachen, Germany, 25–27 September 2019; pp. 1–6.
113. Zhang, Z.; Zheng, T.Q.; Li, K.; Hao, R.; You, X.; Yang, J. Smart Electric Neutral Section Executer Embedded with Automatic Pantograph Location Technique Based on Voltage and Current Signals. *IEEE Trans. Transp. Electrif.* **2020**, *6*, 1355–1367. [CrossRef]
114. Agustoni, A.; Borioli, E.; Brenna, M.; Simioli, G.; Tironi, E.; Ubezio, G. LV DC Networks for Distributed Energy Resources. In Proceedings of the CIGRE Symposium on Power Systems with Dispersed Generation, Athens, Greece, 13–16 April 2005.
115. Zhezhelenko, I.V.; Sayenko, Y.L.; Gorpinich, A.V.; Nesterovych, V.V.; Baranenko, T.K. Analysis of resonant modes in the single-phase industrial AC electrified railway systems. In Proceedings of the 10th International Conference on Electrical Power Quality and Utilisation, Lodz, Poland, 15–17 September 2009; pp. 1–4.
116. Wang, Q.; Lu, J.; Wang, Q.; Duan, J. Transient overvoltage study of auto-passing neutral section in high-speed railway. In Proceedings of the IEEE Transportation Electrification Conference and Expo, Asia-Pacific (ITEC Asia-Pacific), Harbin, China, 7–10 August 2017; pp. 1–5.
117. Suárez, M.A.; González, J.W.; Celis, I. Transient overvoltages in a railway system during braking. In Proceedings of the IEEE/PES Transmission and Distribution Conference and Exposition: Latin America (T&D-LA), Sao Paulo, Brazil, 8–10 November 2010; pp. 204–211.
118. Delfino, F.; Procopio, R.; Rossi, M. Overvoltage protection of light railway transportation systems. In Proceedings of the IEEE Bologna Power Tech Conference Proceedings, Bologna, Italy, 23–26 June 2003; Volume 4, p. 8.
119. Pons, E.; Colella, P.; Rizzoli, R. Overvoltages in DC Urban Light Railway Systems: Statistical Analysis and Possible Causes. In Proceedings of the IEEE International Conference on Environment and Electrical Engineering and 2018 IEEE Industrial and Commercial Power Systems Europe (EEEIC/I&CPS Europe), Palermo, Italy, 12–15 June 2018; pp. 1–6.
120. Pilo, E.; Rouco, L.; Fernandez, A.; Abrahamsson, L. A Monovoltage Equivalent Model of Bi-Voltage Autotransformer-Based Electrical Systems in Railways. *IEEE Trans. Power Deliv.* **2012**, *27*, 699–708. [CrossRef]
121. Celli, G.; Pilo, F.; Tennakoon, S.B. Voltage regulation on 25 kV AC railway systems by using thyristor switched capacitor. In Proceedings of the Ninth International Conference on Harmonics and Quality of Power. Proceedings (Cat. No.00EX441), Orlando, FL, USA, 1–4 October 2000; Volume 2, pp. 633–638.

122. Tomita, M.; Fukumoto, Y.; Ishihara, A.; Suzuki, K.; Akasaka, T.; Caron, H.; Kobayashi, Y.; Onji, T.; Herve, C. Train Running Test Transmitted by Superconducting Feeder Cable and Study as an Example of Line in Japan and France. *IEEE Trans. Appl. Supercond.* **2020**, *30*, 1–7. [CrossRef]
123. Heising, C.; Oettmeier, M.; Gorski, M.; Staudt, V.; Steimel, A. Implications of resonant circuit adjustment errors to the DC-link voltage in single-phase 16.7-Hz-railway applications. In Proceedings of the Compatibility and Power Electronics, Badajoz, Spain, 20–22 May 2009; pp. 210–216.
124. Kneschke, T. Voltage flicker calculations for single-phase AC railroad electrification systems. In Proceedings of the 2003 IEEE/ASME Joint Railroad Conference, Chicago, IL, USA, 24 April 2003; pp. 161–170.
125. Matta, V.; Kumar, G. Unbalance and voltage fluctuation study on AC traction system. In Proceedings of the 2014 Electric Power Quality and Supply Reliability Conference (PQ), Rakvere, Estonia, 11–13 June 2014; pp. 315–320.
126. Guide, Power Quality Application. In *Voltage Disturbances*; Standard EN 50160; 2004; Available online: https://www.google.co.jp/url?sa=t&rct=j&q=&esrc=s&source=web&cd=&ved=2ahUKEwiz5YLz-dPtAhUqwYsBHaFjBE4QFjAAegQIAxAC&url=https%3A%2F%2Fcopperalliance.org.uk%2Fuploads%2F2018%2F03%2F542-standard-en-50160-voltage-characteristics-in.pdf&usg=AOvVaw031QxJ_AWz-qyKIaM3lLO- (accessed on 17 December 2020).
127. He, L.; Xiong, J.; Ouyang, H.; Zhang, P.; Zhang, K. High-Performance Indirect Current Control Scheme for Railway Traction Four-Quadrant Converters. *IEEE Trans. Ind. Electron.* **2014**, *612*, 6645–6654. [CrossRef]
128. Choudhury, A.; Pillay, P.; Williamson, S.S. DC-Bus Voltage Balancing Algorithm for Three-Level Neutral-Point-Clamped (NPC) Traction Inverter Drive with Modified Virtual Space Vector. *IEEE Trans. Ind. Appl.* **2016**, *52*, 3958–3967. [CrossRef]
129. Liu, J.; Zhang, W.; Xiao, F.; Lian, C.; Gao, S. Six-Step Mode Control of IPMSM for Railway Vehicle Traction Eliminating the DC Offset in Input Current. *IEEE Trans. Power Electron.* **2019**, *34*, 8981–8993. [CrossRef]
130. Kawasaki, K. Method to Calculate Fluctuations in the Strength of Radio Noise Emitted from Electric Railway Systems. *Q. Rep. RTRI* **2009**, *50*, 158–161. [CrossRef]
131. Hill, R.J. Electric railway traction. Part 7: Electromagnetic interference in traction systems. *Power Eng. J.* **1997**, *11*, 259–266. [CrossRef]
132. Hill, R.J. Electric railway traction. Part 6: Electromagnetic compatibility disturbance-sources and equipment susceptibility. *Power Eng. J.* **1997**, *11*, 31–39. [CrossRef]
133. Ogunsola, A.; Mariscotti, A. Electromagnetic compatibility in railways. In *Analysis and Management*; Springer: Berlin/Heidelberg, Germany, 2013.
134. Morant, A.; Wisten, Å.; Galar, D.; Kumar, U.; Niska, S. Railway EMI impact on train operation and environment. In Proceedings of the International Symposium on Electromagnetic Compatibility-EMC EUROPE, Rome, Italy, 17–21 September 2012; pp. 1–7.
135. Ogunsola, A.; Mariscotti, A.; Sandrolini, L. Estimation of Stray Current From a DC-Electrified Railway and Impressed Potential on a Buried Pipe. *IEEE Trans. Power Deliv.* **2012**, *27*, 2238–2246. [CrossRef]
136. Tellini, B.; Macucci, M.; Giannetti, R.; Antonacci, G.A. Conducted and radiated interference measurements in the line-pantograph system. *IEEE Trans. Instrum. Meas.* **2001**, *50*, 1661–1664. [CrossRef]
137. Mariscotti, A. Design of an EMI filter for static converters with off-line simplified measurements. In Proceedings of the IEEE EUROCON 2009, St.-Petersburg, Russia, 18–23 May 2009; pp. 1493–1497.
138. Charalambous, C.; Charalambous, A.; Demetriou, A.; Lazari, A.L.; Nikolaidis, A.I. Effects of Electromagnetic Interference on Underground Pipelines Caused by the Operation of High Voltage AC Traction Systems: The Impact of Harmonics. *IEEE Trans. Power Deliv.* **2018**, *33*, 2664–2672. [CrossRef]
139. Konefal, Z.; Fei, T.; Armstrong, R. AC railway electrification systems—An EMC perspective. *IEEE Electromagn. Compat. Mag.* **2019**, *8*, 62–69.
140. Niska, S. Electromagnetic interference: A major source of faults in Swedish railway. *Int. J. Perform. Eng.* **2009**, *5*, 187–196.
141. Lu, H.; Zhu, F.; Liu, Q.; Li, X.; Tang, Y.; Qiu, R. Suppression of Cable Overvoltage in a High-Speed Electric Multiple Units System. *IEEE Trans. Electromagn. Compat.* **2019**, *61*, 361–371. [CrossRef]
142. Mariscotti, A. Induced Voltage Calculation in Electric Traction Systems: Simplified Methods, Screening Factors, and Accuracy. *IEEE Trans. Intell. Transp. Syst.* **2011**, *12*, 201–210. [CrossRef]

143. Midya, S.; Bormann, D.; Schutte, T.; Thottappillil, R. Pantograph Arcing in Electrified Railways—Mechanism and Influence of Various Parameters—Part II: With AC Traction Power Supply. *IEEE Trans. Power Deliv.* **2009**, *24*, 1940–1950. [CrossRef]
144. Liu, Y.; Chang, G.W.; Huang, H.M. Mayr's Equation-Based Model for Pantograph Arc of High-Speed Railway Traction System. *IEEE Trans. Power Deliv.* **2010**, *25*, 2025–2027. [CrossRef]
145. Li, M.; Wen, Y.; Sun, X.; Wang, G. Analysis of Propagation Characteristics of Electromagnetic Disturbance from the Off-Line of Pantograph-Catenary in High-Speed Railway Viaducts. *Chin. J. Electron.* **2020**, *29*, 966–972. [CrossRef]

Publisher's Note: MDPI stays neutral with regard to jurisdictional claims in published maps and institutional affiliations.

Article

New Reactive Power Compensation Strategies for Railway Infrastructure Capacity Increasing

Vítor A. Morais [1,*], João L. Afonso [2], Adriano S. Carvalho [1] and António P. Martins [1]

[1] Department of Electrical and Computers Engineering, University of Porto, 4200-465 Porto, Portugal; asc@fe.up.pt (A.S.C.); ajm@fe.up.pt (A.P.M.)

[2] Centro ALGORITMI, University of Minho, 4800-058 Guimarães, Portugal; jla@dei.uminho.pt

* Correspondence: v.morais@fe.up.pt

Received: 31 July 2020; Accepted: 23 August 2020; Published: 25 August 2020

Abstract: In AC railway electrification systems, the impact of reactive power flow in the feeding voltage magnitude is one aspect contributing to the quality of supply degradation. Specifically, this issue results in limitations in the infrastructure capacity, either in the maximum number of trains and in maximum train power. In this paper, two reactive power compensation strategies are presented and compared, in terms of the theoretical railway infrastructure capacity. The first strategy considers a static VAR compensator, located in the neutral zone and compensating the substation reactive power, achieving a maximum capacity increase up to 50% without depending on each train active power. The second strategy adapts each train reactive power, achieving also a capacity increase around 50%, only with an increase of the train apparent power below 10%. With a smart metering infrastructure, the implementation of such compensation strategy is viable, satisfying the requirements of real-time knowledge of the railway electrification system state. Specifically, the usage of droop curves to adapt in real time the compensation scheme can bring the operation closer to optimality. Thus, the quality of supply and the infrastructure capacity can be increased with a mobile reactive power compensation scheme, based on a smart metering framework.

Keywords: electric traction systems; mobile reactive power compensation; power quality; railway power systems; railway infrastructure capacity; smart railways

1. Introduction

The Railway transportation system has huge power requirements, leading the railway operators to be focused on the increase of the energy efficiency in order to reduce the energy consumption bill. According to [1], the railway sector has a 9% market share in transportation of passengers and goods in the European Union, with an increase of 8.9% between 2005 and 2015. In addition, this market share is only achieved with a final energy consumption of 2%, in comparison with other sectors.

With the mission of "Moving European Railway Forward", the Shift2Rail European program [2] targets the reduction of costs, increases the capacity, reliability, and punctuality. In particular, this program contributes to doubling the railway capacity [3].

The increase of railway infrastructure capacity is an extensive research area where the evaluation is made with the application of definitions, metrics methodologies, and tools [4]. Despite there being no standard definition of railway capacity, it can be defined as the number of trains that can safely pass over a segment of line, within a selected time period.

Regarding the electrification aspects, generally the railway infrastructure capacity is directly affected by the current collection quality of the electric train, which is normally determined by both the mechanical and electrical parts. The mechanical part concerns the train–infrastructure interactions, like pantograph-catenary [5,6], and the wheel-rail [7], which determines the stability of electric transmission and is a source of electrical issues. Due to the inductive characteristic of railway transmission line, this will directly affect the quality of supply of electric trains, being also a source of electrical issues.

Specifically, due to the electrical characteristics of the railway electrification, the increase of railway infrastructure capacity leads to an increase in line voltage drop. According to the IEC60850 [8], in AC 1×25 kV electrification scheme, the lowest non-permanent voltage is 0.7 p.u. (17.5 kV), whereas the lowest permanent voltage is 0.76 p.u. (19 kV). Specifically, if the voltage is higher than the lowest permanent voltage, the train can consume all its demand power; if the lowest non-permanent voltage is reached, a limit of operation is achieved and the train power must be clipped to zero.

It is well known that the higher the reactive power flow, the higher the voltage drop across the catenary line [9]. The study of reactive power, and power flow depends on a model for the catenary.

The catenary line is usually modeled as a multi-conductor line model, for each of the conductors (rails, buried cables, aerial protection cables, feeder cables, and contact line, as example). This results in a matrix of self and mutual impedances, where the main diagonal (self impedance) depends on the analysis of the ring formed by the respective conductor and the earth return, and the remaining elements of the matrix corresponds to the mutual impedances [10]. These elements can be obtained from the application of the Carson formulas, as proposed in [11] and as demonstrated its railway application in [12].

For power flow calculations and for 1×25 kV electrification scheme, the multi-conductor line model can be simplified to consider the same voltage level in the conductors. First, the earth conductance is not considered, as well as the rail–earth conductance and the capacitance matrix. Therefore, this can be simplified to a lumped-parameters line model. As illustrated in [12], one approach to perform this simplification is in adding all the admittance matrix elements corresponding to the same voltage level. For 1×25 kV, this results in a 2×2 impedance matrix that can be further simplified to a PI model.

This simplified model allows a proper usage of common power flow solvers in the railway electrification analysis, namely reactive power flow control. This paper proposes an adaptation on the railway reactive power control, towards an increase in the railway infrastructure capacity. This adaptation will directly result in a reduction of the losses, as a first objective, bringing clear advantages to the railway infrastructure.

This paper is structured in seven sections. The following Section 2 presents a literature review. Section 3 presents the materials and models, starting with a basic model and covering the used simulation framework. Section 4 presents the methodology for reactive power compensation, with an illustration for a particular scenario. Section 5 presents the rest of the methodology and the results for the increase in the railway infrastructure capacity with the adaptation of the reactive power in the catenary. Section 6 discusses a conceptual architecture to implement this reactive power compensation strategy, with the usage of a smart metering framework. Finally, the conclusions of this work are presented in Section 7.

2. Literature Review

Several works have been addressed in the literature regarding strategies for power quality improvement in railways. In AC electrification, two main types of devices based on power electronics are usually implemented: voltage stabilization devices, or voltage boosters, and line current balancers [13].

The main objective of the high voltage boosters is to inject reactive power into the line, with a level adapted in real time. Usually, this is achieved with a static VAR compensator (SVC) and tuned LC filters for specific harmonics [13].

The purpose of line current balancers is to minimize the unbalance caused in the transmission/distribution network by the railway electrification.

These two types of devices are necessary to comply with the increase for power demand. Usually, infrastructure managers adopt systems in the Traction Power Substation (TPS) site that can either balance the line currents and inject reactive power (to boost the catenary voltage).

In [14,15], comparative studies on several Railway Power Conditioners (RPC) topologies are presented, to be employed in the TPS. Traditional RPC comprises two back-to-back converters and two isolation/coupling transformers [16,17]. From this topology, several have been derived, such as the active power quality compensator (AQPC) which comprises a three phase converter [18], or the hybrid power quality compensator (HPQC) in which the APQC is combined with a Static VAR compensator [19,20]; In addition, modular multilevel converter (MMC) topologies have been researched in current years [21].

Despite the inability to control the line current unbalances, the inclusion of voltage stabilization devices in the opposite site of the TPS (in the end of a traction feeder section, the neutral zone) will strongly support this desired voltage boost. Thus, this compensation strategy allows more powerful trains without violating the standards (e.g., IEC60850 [8]). In [22,23], strategies are presented to include compensation systems at the end of a traction section feeder. In [24], the 3 kV increase in the minimum voltage in the catenary is highlighted, with further details of this project in [25]. However, this improvement is achieved with a system occupying a very large area. The compensation scheme for Static VAR Compensator is also studied in [26], with the usage of a neural network for online operation.

In the PhD thesis of [27] and later in [28], an alternative to the inclusion of bulky SVCs is proposed, with the adoption of mobile reactive power compensators. This is achieved with the reactive power injection within each train. This compensation strategy is further extended with the work of [29], where the compensation scheme is based in a genetic algorithm heuristic. Later in [30], the usage of modern locomotives as mobile reactive power compensators is evaluated and compared, where they can be more efficient than SVC. However, the limitations related to the control of leading power factor as well as the need for very fast algorithms are considered that do not justify the usage of a power factor different than the unitary.

From the knowledge of the authors, the possibility of operating the modern trains with variable power factor has not been actively researched in recent years. However, from the authors' point of view, the reason for this is not in the advantages, but in the difficulty to implement such a strategy, since it requires real-time knowledge of the state of the railway electrification.

Regarding the infrastructure capacity increase, most works are focused on the operational and logistics. In [31], the main concepts and methods to perform capacity analysis are reviewed. In theory, the capacity is defined as the number of trains running over a line section, during a time interval, with trains running at minimum headway. This capacity mostly depends on infrastructure constraints (signaling system, power traction constraints, single/double tracks, speed limits, etc.), on traffic parameters (timetables, priorities, type of trains, etc.) and on operation parameters (track interruptions, train stop time, etc.). In [3], increasing the railway infrastructure capacity by increasing the speed of freight trains is proposed. Specifically, in the case of a delay, these trains are allowed to have higher maximum speed.

In this work, the infrastructure capacity increase with the adoption of reactive power compensation strategies is reviewed. In the following section, the models and the used framework to demonstrate the infrastructure capacity improvement are presented.

3. Materials and Methods

This section covers the models and simulation framework required for power flow analysis and for reactive power control. The combination of these models, the simulation framework, and the compensation strategy will be the scientific contribution for a new approach to reactive power control in the railway system.

3.1. Basic Model

The basic model considers the Traction Power Substation (TPS), the catenary line, and the electric train. The electric train is simplified from the dynamic model as follows:

$$\frac{dv}{dt} = \frac{F_T(v) - w(v) - g(x)}{M(t)} \tag{1}$$

where $F_T(v)$ is the traction force, $w(v)$ is the aerodynamic resistive force, and $g_(x)$ represents the track gradient and curvature forces. The electric active power is directly related to the train dynamic movement and the reactive power is the one to control. At higher speeds, the maximum traction force is limited by the maximum available power [32]. Therefore, the train can be simplified to a decoupled active and reactive power load.

In this work, the considered line model is represented as the PI line model, Z_L. Then, the railway power flow can be analyzed in Figure 1, where the train, V_t, is supplied through the catenary line, Z_l, from a 25 kV TPS, V_s.

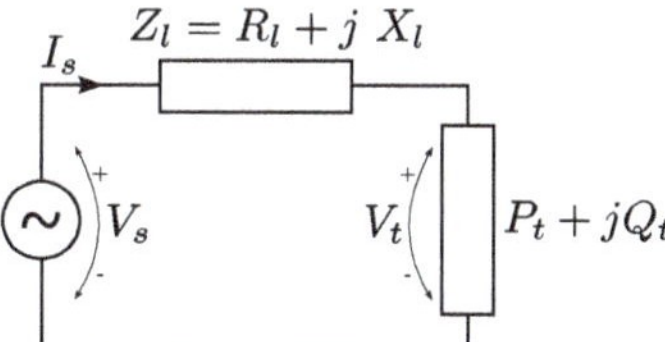

Figure 1. Steady state equivalent circuit.

The apparent, active, and reactive power flows (S_s, P_s and Q_s, respectively) in the TPS can be obtained from the following expressions:

$$\begin{aligned} S_s &= V_s I_s^* \\ &= P_s + jQ_s \end{aligned} \tag{2}$$

The current in the branch can be expressed as:

$$I_s = \frac{\hat{V}_s - \hat{V}_t \cos(\delta) - j\,\hat{V}_t \sin(\delta)}{R_l + j\,X_l} \tag{3}$$

where $V_s = \hat{V}_s \angle 0$ and $V_t = \hat{V}_t \angle \delta$. Thus, replacing the expression in (2),

$$\begin{aligned} P_s &= \frac{\hat{V}_s^{\,2} R_l - \hat{V}_s\,\hat{V}_t\,R_l\,\cos(\delta) - \hat{V}_s\,\hat{V}_t\,X_l\,\sin(\delta)}{R_l^2 + X_l^2} \\ Q_s &= \frac{\hat{V}_s^{\,2} X_l + \hat{V}_s\,\hat{V}_t\,R_l\,\sin(\delta) - \hat{V}_s\,\hat{V}_t\,X_l\,\cos(\delta)}{R_l^2 + X_l^2} \end{aligned} \tag{4}$$

It can be seen that the active and reactive power flow in the TPS is dependent on the magnitude and phase of the train voltage, as well as on the line characteristics.

Supported by the expression in (4), let's consider a variation of the train voltage magnitude (from 15 kV to 30 kV) and the train voltage phase (between $-\pi/4$ to $\pi/4$). Assuming a line distance of 30 km and $X/R = 3$, Figure 2 shows the TPS active and reactive power as a function of the train voltage and phase.

The lines in Figure 2 show the isobaric lines where the power flow at TPS is the same. In particular, in each figure, the lines having, respectively, 0 MW for Figure 2a and 0 MVAr for Figure 2b is highlighted.

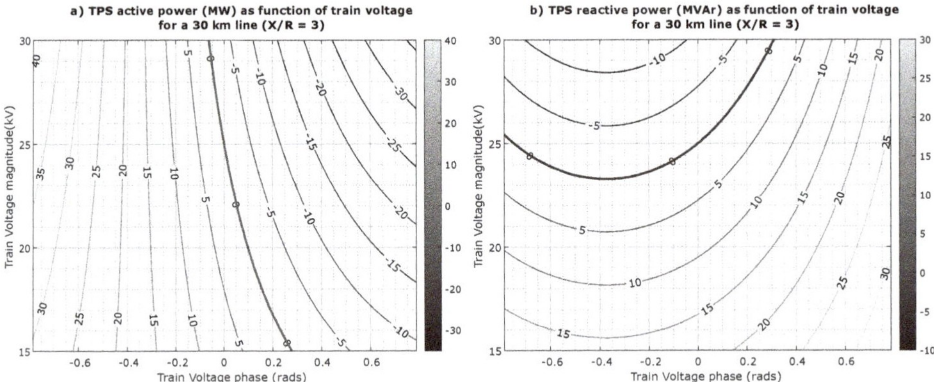

Figure 2. Sensibility analysis for different conditions of train voltage and phase: (**a**) Active power flow at TPS; (**b**) Reactive power flow at TPS.

From the analysis of Figure 2a, it is possible to view that a variation on the phase of the train will considerably affect the active power flow in the TPS (the variation of the train voltage is barely related to a variation on active power flow at TPS).

Regarding Figure 2b, it is visible that, for a train voltage phase angle of, around, −0.4 radians, the reactive power only depends on the train voltage.

If the train active and reactive power is obtained from the application of previous sensibility analysis, it is possible to perform a correlation of TPS power flow and train power flow. Specifically, the train power is given by:

$$P_t + jQ_t = V_t I_s^* \tag{5}$$

Replacing the I_s^* by expression in (3), it is obtained for P_t and Q_t the following:

$$\begin{aligned} P_t &= \frac{\hat{V}_t\,\hat{V}_s\,R_l\,\cos(\delta) + \hat{V}_t\,\hat{V}_s\,X_l\,\sin(\delta) - \hat{V}_t^{\,2}\,R_l\,\cos(2\delta) - \hat{V}_t^{\,2}\,X_l\,\sin(2\delta)}{R_l^2 + X_l^2} \\ Q_t &= \frac{\hat{V}_t\,\hat{V}_s\,R_l\,\sin(\delta) - \hat{V}_t^{\,2}\,R_l\,\sin(2\delta) + \hat{V}_t^{\,2}\,X_l\,\cos(2\delta) - \hat{V}_t\,\hat{V}_s\,X_l\,\cos(\delta)}{R_l^2 + X_l^2} \end{aligned} \tag{6}$$

The TPS active and reactive power can be related to the train power, through a variable change. However, considering the expressions in (4) and (6), there is no simple mathematical solution that results in the TPS power as function of the train power.

A simple procedure (to conduct a variable change of the train voltage and phase, in order to obtain the TPS power, from expression (4), and the train power, from expression (6)), towards an evaluation of the dependence of TPS power from the train voltage is considered. The result of this evaluation can be analyzed graphically, in Figure 3.

Figure 3a presents isobaric curves of active power at TPS, as a function on the train active and reactive power. In particular, in this result, it is visible that TPS active power is more dependent on train active power. In Figure 3b, the dependence of TPS reactive power is more dependent on the train reactive power.

To conclude, in previous analysis, a dependence on TPS reactive power and train voltage magnitude are visible. Specifically, by having an inductive reactive power flow in the TPS, the train voltage magnitude will reduce, as visible in Figure 2a. In addition, by changing the train reactive power value, this results in an adaptation of the reactive power in the TPS.

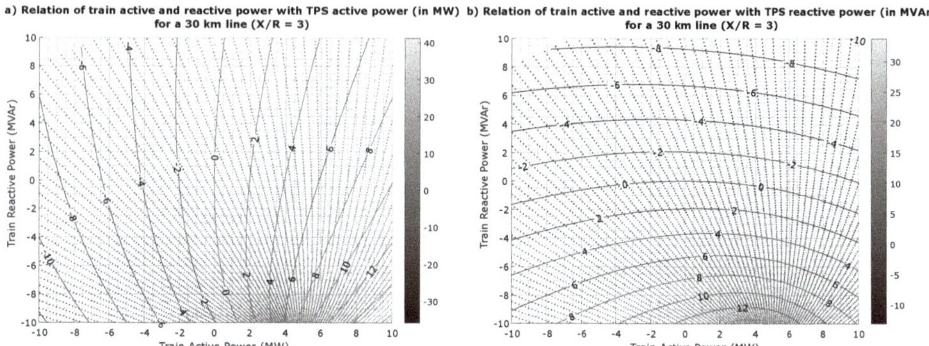

Figure 3. Sensibility analysis for different conditions of train power: (**a**) active power flow at TPS; (**b**) reactive power flow at TPS.

Thus, the adaptation of the reactive power by changing the train reactive power to a capacitive power factor results in a reduction of the TPS reactive power and an increase of the train voltage level.

3.2. Simulation Framework

The simulation framework of this work is now presented in Figure 4, where the TPS, a railway line, and a single train are illustrated.

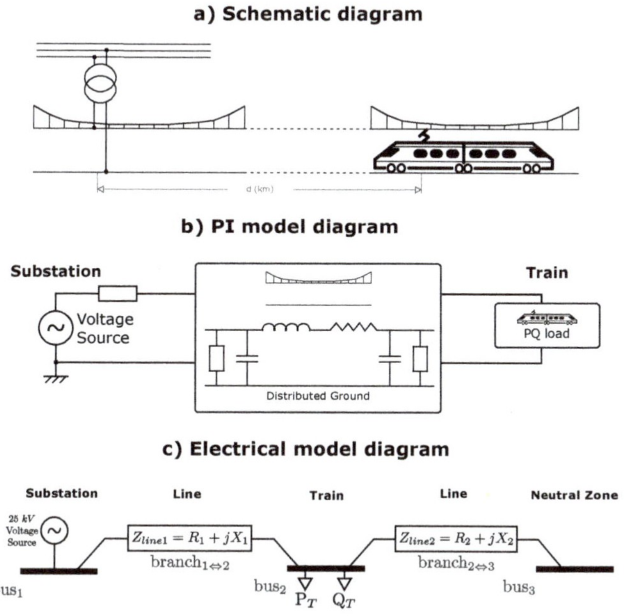

Figure 4. Framework of the 1 × 25 kV models: (**a**) illustration of physical representation; (**b**) PI model diagram; (**c**) considered bus-branch model for MatPower (Note: the traction transformer is not considered in this work).

The power flow problem considered in this simple model is a nonlinear problem, as previously discussed. The usage of MatPower [33], and, in particular, the Newton–Raphson solver, allows this problem to be solved. Therefore, for a fixed supply voltage and specific branch parameters, the train power consumption can be fixed regardless of the voltage drop in the line. With this, the voltages in the nodes can be calculated, as well as the injected supply power in the TPS.

Considering the simple model in Figure 4c, this has four variable parameters: (i) a variable line distance, $\boldsymbol{d_L}$; (ii) a variable train power, $\boldsymbol{P_T}$; (iii) a variable train power factor, $\boldsymbol{PF_T}$; and (iv) a variable line X/R ratio, $\boldsymbol{X/R_L}$.

To better evaluate the model, these parameters can be spawned across a surface of possible parameters $\boldsymbol{S}(\boldsymbol{L_d}, \boldsymbol{X/R_L}, \boldsymbol{P_T}, \boldsymbol{PF_T}) \in \mathbb{R}^4$, where $\boldsymbol{d_L} \in \ ]0 \ \ d_{max}]$, $\boldsymbol{X/R_L} \in \ [X/R_{min} \ \ X/R_{max}]$, $\boldsymbol{P_T} \in [P_{min} \ \ P_{max}]$ and $\boldsymbol{PF_T} \in [0 \ \ 1]$. The surface of possible parameters has infinite possible solutions.

One element of the surface of parameters can be parametrized as set of parameters $\boldsymbol{SP}(\boldsymbol{L_d}, \boldsymbol{X/R_L}, \boldsymbol{P_T}, \boldsymbol{PF_T}) \in \boldsymbol{S}$, where

$$\text{Set of parameters } \boldsymbol{SP} = \begin{cases} \boldsymbol{L_d} & (in\ km) \\ \boldsymbol{X/R_L} & \\ \boldsymbol{P_T} & (in\ MW) \\ \boldsymbol{PF_T} & \\ Voltage\ @\ bus_k & k = \{1,\ 2\} \\ Power\ bus_k & k = \{1,\ 2\} \\ Line\ losses & P\ and\ Q \end{cases} \tag{7}$$

To study the behavior of this model, it is possible to generate several $\boldsymbol{SP}$ elements, using distribution probability function for each of the four parameters, inside the defined interval of values. Assuming that, for each parameter, N random possibilities are generated. Then, the surface of solutions depends on testing N^4 different elements, resulting in a huge computational power required for this model and a hard task to evaluate the model. A more direct analysis, in particular, a sensibility analysis, will better illustrate and validate the behavior of the model.

3.3. Sensibility Analysis

A sensibility analysis is an adequate tool to evaluate the variation of certain input variables, in particular, the variation of the parameters. In this analysis, three parameters can be defined as variable and the fourth can be fixed, as better explained in the following results.

Considering a testing surface, given by the expression:

$$\boldsymbol{S} = \begin{cases} \boldsymbol{L_d} & [0.1\ :\ 2\ :\ 30.1] & (in\ \text{km}) \\ \boldsymbol{X/R_L} & \{2\ :\ 1\ :\ 5\} & \\ \boldsymbol{P_T} & [-20\ :\ 2\ :\ 20] & (in\ \text{MW}) \\ \boldsymbol{PF_T} & \{0.8\ 0.9\ 0.95\ 1\} & \text{ind.} \end{cases} \tag{8}$$

By fixing the $\boldsymbol{X/R_L}$ ratio, then this leaves room to variate the other three variables. Thus, the sensibility analysis can be seen within the train voltage value, as visible in Figure 5.

Complementarily, the second sensibility analysis is made by fixing the $\boldsymbol{PF_T}$ value. The remaining parameters will be variate towards an evaluation of the resultant voltage value, as visible in Figure 6.

From the evaluation of the parameters of the model, it is clear that the $\boldsymbol{X/R_L}$ ratio will affect the train voltage. Specifically, a higher $\boldsymbol{X/R_L}$ ratio results in higher voltage drops in the line, which is a characteristic of high inductive lines. Nevertheless, the characteristics of the line depend on several aspects related to the design of the electrification, and, therefore, in this work, a fixed value for the $\boldsymbol{X/R_L}$ ratio (specifically, $R_L = 0.15\ \Omega/\text{km}$ and $X_L = 0.45\ \Omega/\text{km}$) is considered.

The only aspect that can be manipulated is the train power factor. By having higher power factor values, this reduces the voltage drop. Therefore, in the following section, the reactive power compensation is detailed.

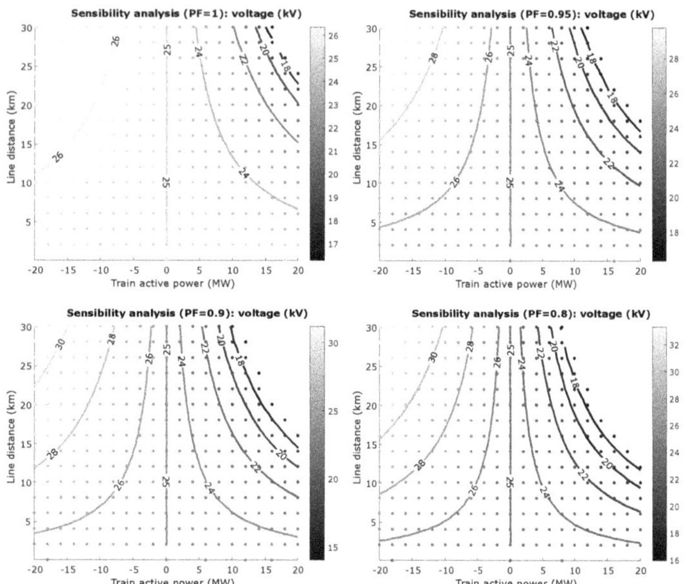

Figure 5. Sensibility analysis for $X/R_L = 3$: voltage levels for different train power factor values, spawned across different train active power values and line distances.

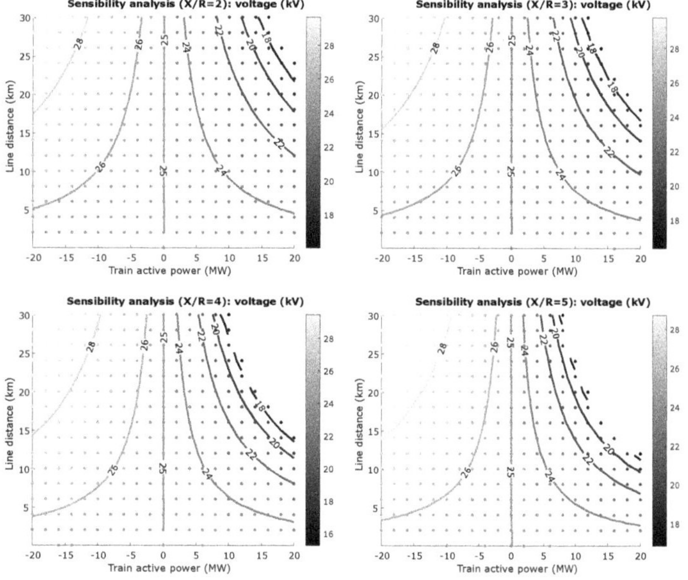

Figure 6. Sensibility analysis for $PF_T = 0.95$: voltage levels for different X/R_L ratios.

4. Reactive Power Compensation

As previously illustrated, there is a clear advantage in controlling the reactive power in the railway electrification. This section covers the used methodology to improve the traction power supply with the adaptation of the train power factor, using references coming from measurements from TPS. In this section, a simple reactive power compensation algorithm is presented first, and is demonstrated in the performance of such algorithm.

4.1. Algorithm for Reactive Power Compensation

The adopted optimization strategy was based on the compensation strategy from [27], where the train reactive power is iteratively adapted, based on solving the power flow problem. This simple algorithm is presented in the following Algorithm1, based on fixed steepest descent method, where Q_{MIN} is a tolerance value for the reactive power and the λ is the step size for the iterative process.

Algorithm 1: Reactive Power Compensation Using Fixed Steepest Descent Method

```
begin
    SET Q_k = 0Mvar;
    SET λ = λ_0
    while NOT termination criteria do
        LAUNCH (Power Flow Algorithm)
        GET Q_SST
        if |Q_SST| ≥ Q_min then
            SET Q_{k+1} = Q_k − λ Q_SST
        else
            BREAK
        end
    end
end
```

Considering the following set of parameters **SP**, where

$$\boldsymbol{SP} = \begin{cases} \boldsymbol{L_d} &= 30\text{ km} \\ X/R_L &\approx 3 \qquad\qquad R = 0.155\ \Omega/\text{km} \\ \boldsymbol{P_T} &= 10\text{ MW} \\ \boldsymbol{PF_T} &= 0.90\text{ ind.} \quad (\text{initial value}) \end{cases} \tag{9}$$

Figure 7 presents the results for the proposed algorithm (with fixed $\lambda_0 = 0.25$), for the optimization of **SP**.

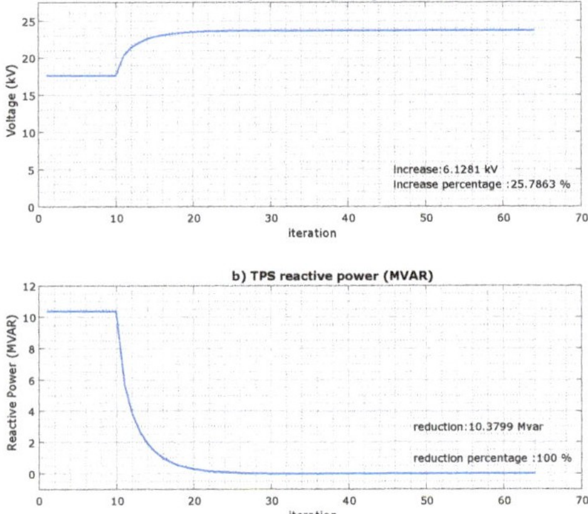

Figure 7. Illustration of algorithm evolution for reactive power compensation: (**a**) evolution of train voltage; (**b**) evolution of SST reactive power. Note, for illustration purposes, that the reactive compensation procedure is only enabled at iteration $it = 10$.

This result clearly illustrates the major advantages of this reactive power compensation. The train voltage increases 25.8%, and the resultant reactive power in traction substation is zero.

Regarding the train operation, as illustrated in Figure 8, the train active power consumption is unaffected, as expected; the reactive power is considerably changed, from an inductive power factor to a capacitive one; the big advantage is on the reduction of the apparent power consumption which is directly related to a reduction in train power losses (in train transformer and power converters).

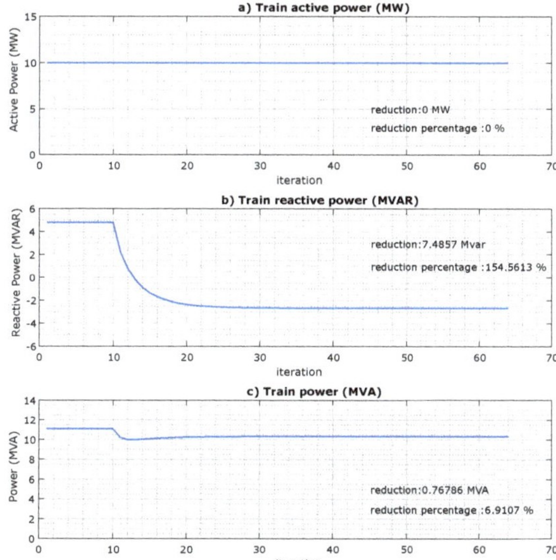

Figure 8. Illustration of algorithm evolution for reactive power compensation: (**a**) evolution of train active power; (**b**) evolution of train reactive power; (**c**) evolution of train apparent power.

Finally, the major advantage of this strategy is visible in Figure 9, where the power losses in catenary can be reduced by half.

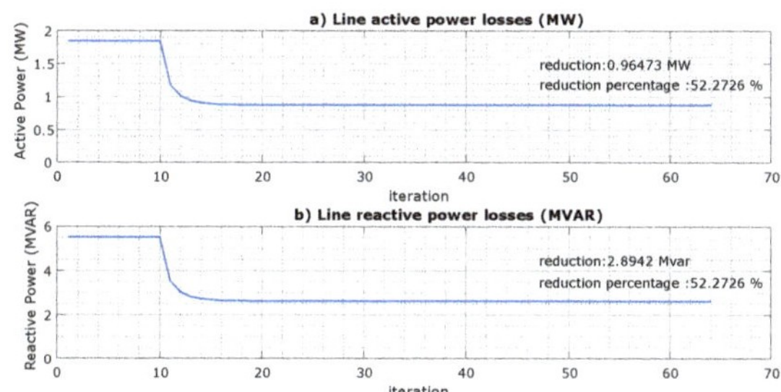

Figure 9. Illustration of algorithm evolution for reactive power compensation: (**a**) evolution of active power losses in catenary; (**b**) evolution of reactive power losses in catenary.

However, these results must be taken with caution, regarding the predefined parameters and specifically the initial inductive train power factor of 0.9. In the following, a sensibility analysis will be made to better evaluate the potential improvement from initial different power factors.

4.2. Sensibility Analysis

Figure 9 presents the visible active and reactive power losses in the catenary, for **SP** in (9). A sensibility analysis will be performed for a fixed $X/R_L \approx 3$, with the results in Figure 10.

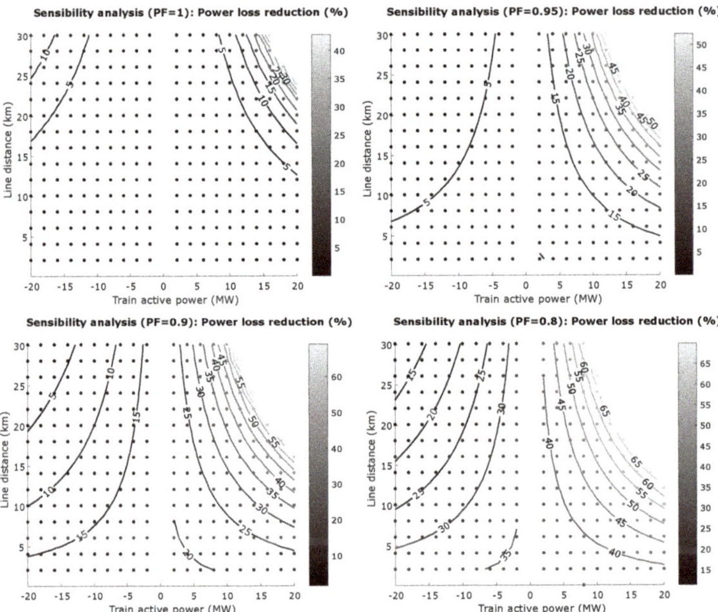

Figure 10. Sensibility analysis for reactive power compensation: line power loss reduction, in percentage, for different power factors. Note that the values near 0 MW or near 0 km are not relevant for this demonstration.

As expected, for lower values for $\mathbf{PF}_T$, this results in higher reduction in the line losses. This value is expected since trains will have more margin to adapt the power factor value to a capacitive one.

In the following section, this algorithm will be included to compensate the reactive power in two situations: (i) compensation in NZ through a PWM controlled SVC, using measurements from TPS; and (ii) compensation made by each train.

5. Increase of the Railway Infrastructure Capacity

This section proposes to increase the infrastructure capacity of a railway line (increase of the number of trains), with the adoption of a reactive power compensation strategy.

The railway infrastructure capacity will be considered, in this section, as the maximum number of trains that can exist in a railway line, all of them separated with the same distance among each others, and that the voltage levels on the line are according to IEC60850 [8]. The security issues such as minimum distance that a train must be apart from each other will not be considered.

Consider a railway line branch, with fixed length separating the TPS and a neutral zone, and having N trains. In order to better evaluate the infrastructure capacity, in the implemented model illustrated in Figure 11, the distances $d_1, d_2, \cdots, d_{n+1}$ are all the same, as well as all the train powers, P_T and Q_T.

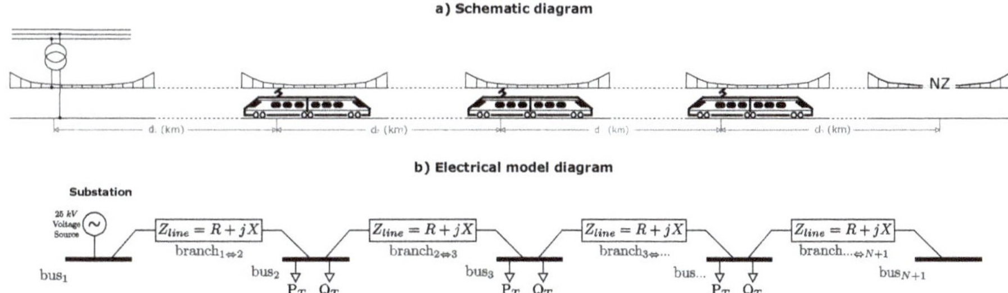

Figure 11. Framework for the increase of railway capacity: (**a**) illustration of physical representation; (**b**) considered bus-branch model for MatPower.

With this consideration, three different case studies can be listed:

- Study of railway capacity without compensation (Baseline);
- Installation of a Static VAR Compensator located at neutral zone;
- On-board compensation in all trains;

5.1. Baseline: Railway Capacity without Compensation

This section takes into account a railway line having a fixed length $D_{TPS\Leftrightarrow NZ} = 29.2$ km (corresponding to the maximum distance of a track section of a real 250 km railway line, from the knowledge of the authors).

The procedure to evaluate the railway capacity is illustrated in Figure 12, where the addition of trains to the railway line is iteratively tested.

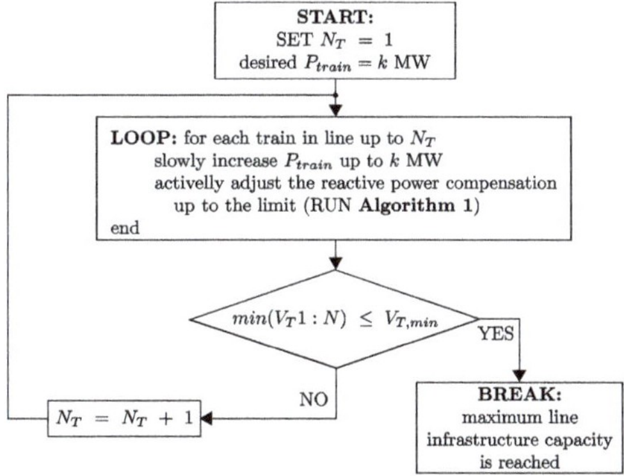

Figure 12. Flowchart to test the increasing of capacity procedure.

Specifically, considering that the line has N trains, each train will be separated among them by a fixed distance:

$$d_T = \frac{D_{TPS\Leftrightarrow NZ}}{N+1} \tag{10}$$

It is noteworthy that the usage of variable train distances will result in opening a degree of freedom that, only with an extensive probabilistic analysis (for different distances), is it possible to obtain reasonable results. Nevertheless, similar results are expected.

In this baseline case study, the procedure in Figure 12 will not consider the active adjustment of reactive power. The maximum infrastructure capacity is then achieved when the voltage in the line is lower than the IEC60850 minimum non-permanent voltage (17.5 kV [8]). The decision for choosing the non-permanent voltage was arbitrary between the two minimum IEC60850 voltage limits (both voltage level values are valid for the following analysis, expecting similar results; the window between these two levels must not be considered as a steady state train operation).

Figure 13 illustrates the results, where a relation of the number of trains and the minimum voltage level is illustrated for different cases.

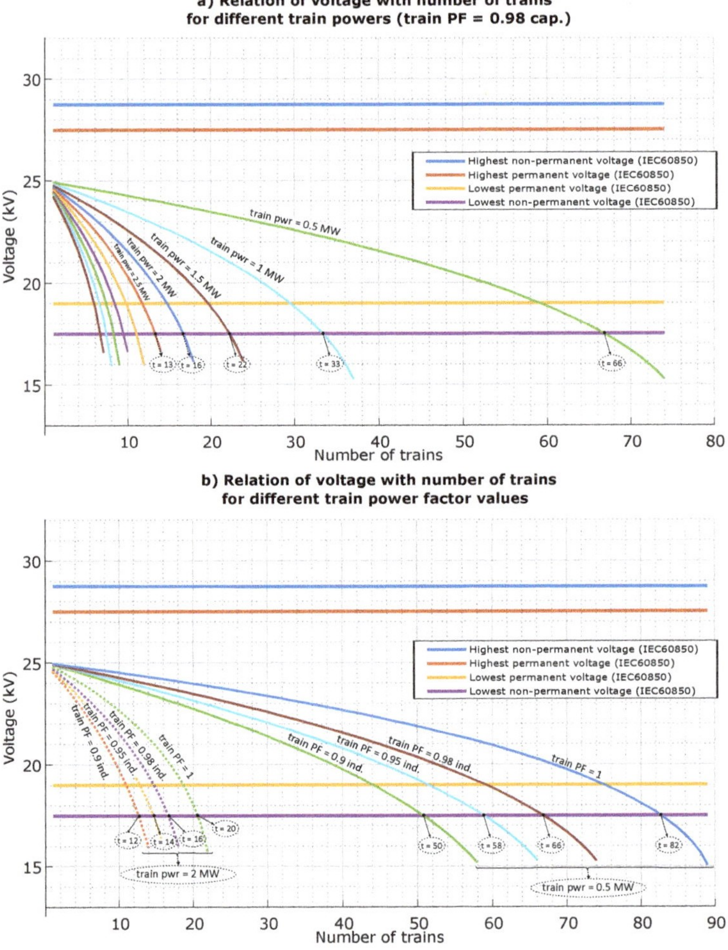

Figure 13. Relation of number of trains with voltage in neutral zone: (**a**) variation of active power in each train, for fixed $PF = 0.98$ ind.; (**b**) variation of power factor in each train, for fixed active power of 0.5 MW and 2 MW.

Specifically, Figure 13a presents the dependence of minimum voltage level and the number of trains, for different power values in each train. In theory, it is possible to have 16 trains in the line, all consuming 2 MW with $PF = 0.98$ ind., without achieving the lower value of non-permanent voltage (17.5 kV according to IEC60850).

Figure 13b shows the voltage for different train power factors, where the increase to unitary power factor from $PF = 0.9$ ind. results in an increase of the railway capacity around 67% (for $P_t = 2$ MW, increase from 12 to 20 trains).

Table 1 extends the evaluation of railway capacity for different train power consumptions and for different train power factors.

Table 1. Maximum number of trains, for different train power consumptions and for different train power factors. Note: the percentage reduction from unitary power factor is presented in parentheses. As an example, the baseline for $P_T = 2$ MW and $PF = 1$ is 20 trains; then, for $PF = 0.9$ ind., it is only possible to have 12 trains (8 less than the unitary power factor, corresponding −40% less than baseline).

	Train Power Factor			
Train Active Power	**0.9 ind.**	**0.95 ind.**	**0.98 ind.**	**1**
0.5 MW	50 (−39.0%)	58 (−29.3%)	66 (−19.5%)	82 (0%)
1 MW	25 (−39.0%)	29 (−29.3%)	33 (−19.5%)	41 (0%)
2 MW	12 (−40.0%)	14 (−30.0%)	16 (−20.0%)	20 (0%)
3 MW	8 (−38.5%)	9 (−30.8%)	11 (−15.4%)	13 (0%)
4 MW	6 (−40.0%)	7 (−30.0%)	8 (−20.0%)	10 0%)
5 MW	5 (−37.5%)	5 (−37.5%)	6 (−25.0%)	8 (0%)

From this baseline, two possible strategies for the railway reactive compensation will be covered, considering each train starting with $PF = 0.98$ ind.

5.2. Reactive Power Compensation in the Neutral Zone

In the second case study, the reactive power compensation will be made with a SVC in the neutral zone, having the objective of minimizing the SST reactive power. Figure 14 presents the lower voltage in line, as a function of the number of trains in the line.

Table 2 extends the evaluation of railway capacity for different power consumptions and for different compensation limits at the neutral zone.

Table 2. Maximum number of trains, for different train power consumptions and for different Neutral Zone power limits. Note: the percentage improvement from baseline is presented in parentheses (where 0 MVAr means no compensation). As an example, the baseline for $P_T = 2$ MW is 16 trains; then, for $Q_{NZ} = 20$ MVAr, it is possible to have nine more trains (+56% more than baseline).

	Neutral Zone Power Limit (for Compensation)				
Train Active Power	**0 MVAr**	**5 MVAr**	**10 MVAr**	**20 MVAr**	**30 MVAr**
0.5 MW	66 (0%)	81 (+22.73%)	90 (+36.36%)	100 (+51.52%)	105 (+59.09%)
1 MW	33 (0%)	40 (+21.21%)	45 (+36.36%)	50 (+51.52%)	52 (+57.58%)
2 MW	16 (0%)	20 (+25%)	22 (+37.50%)	25 (+56.25%)	26 (+62.50%)
3 MW	11 (0%)	13 (+18.18%)	15 (+36.36%)	16 (+45.45%)	17 (+54.55%)
4 MW	8 (0%)	10 (+25%)	11 (+37.50%)	12 (+50%)	12 (+50%)
5 MW	6 (0%)	8 (+33.33%)	8 (+33.33%)	9 (+50%)	10 (+66.67%)

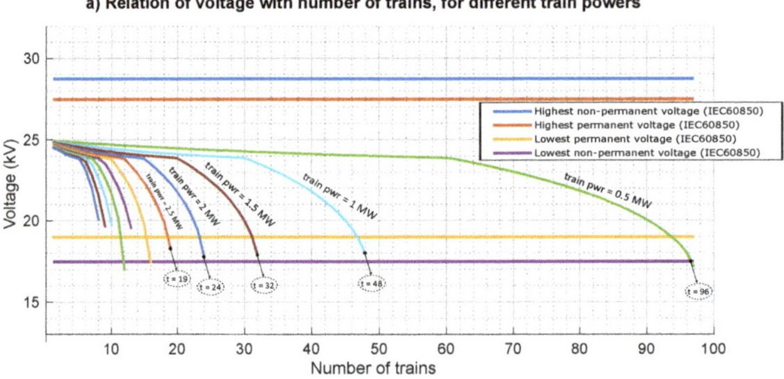

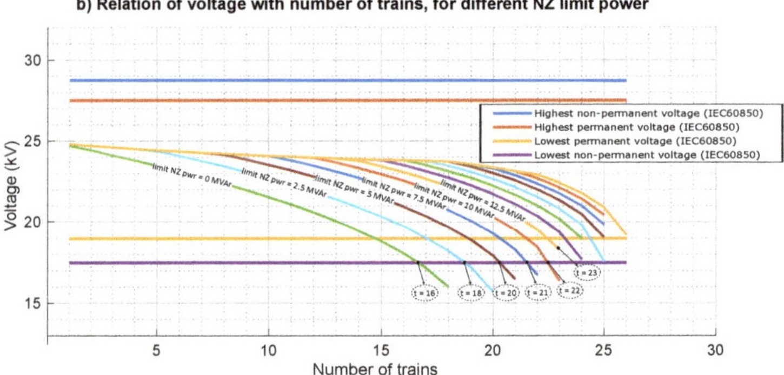

Figure 14. Relation of number of trains ($PF = 0.98$ ind.) with lower voltage in line: (**a**) variation of active power in each train, for fixed NZ limit power (15 MVA maximum power, as example, visible after 30 trains for 1 MW train power); (**b**) variation of NZ limit power, for fixed train active power of 2 MW.

From the results of Table 2, it is possible to identify a relation of the maximum number of trains, $N_{trains,max}$, in the line and the power of each train, $P_{train}[MW]$, following (11):

$$N_{trains,max} = K_{NZ,lim} * \frac{1}{P_{train}} \tag{11}$$

In addition, it is possible to estimate the $K_{NZ,lim}$ parameter, since it follows a polynomial function and is dependent on the limit power of the SVC of NZ, P_{NZ}. For the obtained results, this can be extrapolated to the expression in (12):

$$N_{trains,max}(P_{NZ}, P_{train}) = \frac{K_2(P_{NZ})^2 + K_1 P_{NZ} + K_0}{P_{train}} \tag{12}$$

where $K_0 = 33.4$, $K_1 = 1.35$ and $K_2 = -0.0252$, for this case study.

The evaluation of the percentage improvement of the railway infrastructure capacity, in Table 2, shows that this improvement is mostly dependent on the NZ reactive power. Specifically, it follows a polynomial trendline, as illustrated in Figure 15.

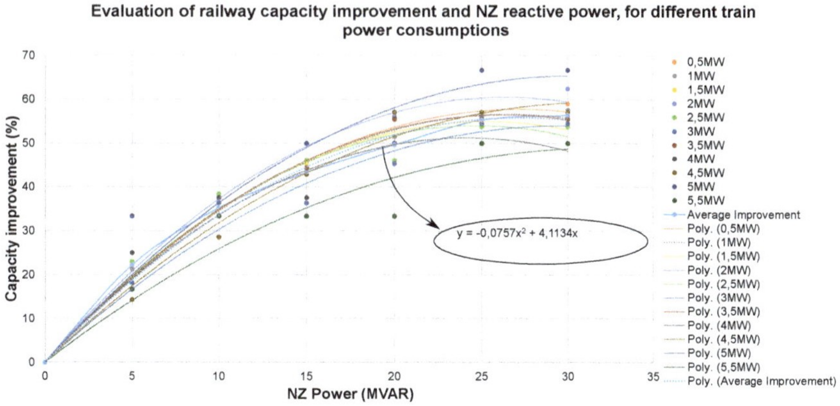

Evaluation of railway capacity improvement and NZ reactive power, for different train power consumptions

Capacity improvement (%)

NZ Power (MVAR)

y = -0,0757x² + 4,1134x

0,5MW; 1MW; 1,5MW; 2MW; 2,5MW; 3MW; 3,5MW; 4MW; 4,5MW; 5MW; 5,5MW; Average Improvement; Poly. (0,5MW); Poly. (1MW); Poly. (1,5MW); Poly. (2MW); Poly. (2,5MW); Poly. (3MW); Poly. (3,5MW); Poly. (4MW); Poly. (4,5MW); Poly. (5MW); Poly. (5,5MW); Poly. (Average Improvement)

Figure 15. Relation of capacity improvement with the increase of the NZ reactive power. The dots display the Table 2 percentage improvement values; the smaller dot lines present polynomial regression curves from each of those points. The blue dash-dot line (Average improvement) presents a polynomial regression curve from the average of all the improvement values.

The analysis of the trendline shows that, for the considered railway line, the maximum train capacity improvement is around 50% to 60%. However, for higher NZ installed power (above 25 MVAr), it is expected that the capacity improvement will flatten, mostly due to the voltage limitation in the NZ (according to the railway standards, the SVC can not impose a voltage higher than 1.1 p.u).

5.3. Mobile Reactive Power Compensation

Previously, it was considered that the reactive power compensation is performed in NZ. It was visible that the capacity improvement is mostly dependent on the NZ power capability and not in the train power demand.

In this section, the reactive power compensation is performed in each train, where it will be limited by the maximum compensation, which means minimum capacitive power factor. By considering a train operating in any power factor, then the apparent power is given by the following expression:

$$|S| = \frac{P}{PF}, \quad P \in \Re_0^+, \ PF \in \ ]0\ 1] \tag{13}$$

The variation of the apparent power in relation with the unitary power factor condition is given by:

$$|\Delta S| = \frac{P}{PF} - P \tag{14}$$

If expressed in percentage of P, this variation is only dependent on the power factor:

$$\Delta S[\%] = \left(\frac{1 - PF}{PF}\right) * 100 \tag{15}$$

Figure 16 illustrates the variation of apparent power, $\Delta S[\%]$, for different train power factors.

Similarly to the NZ reactive power compensation algorithm, the mobile reactive power compensation algorithm will adapt the reactive power in one bus (e.g., bus k), in order to minimize the reactive power in another bus (specifically, bus $k-1$). However, the difference here is that the same algorithm is replicated to all of the trains in the line.

To better illustrate the expected behavior of the compensation scheme, if the train in bus_3 in Figure 11 is considered, this train compensation objective is to have a zero value of reactive power feeding the $branch_{2\Leftrightarrow3}$, at bus_2. This is achieved with the adjustment of power factor, as exemplified in Figure 17, where, from the baseline train reactive power Q_T, this is reduced to Q^*.

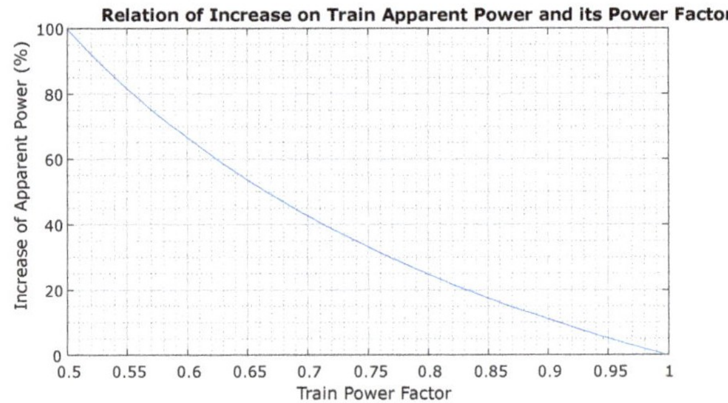

Figure 16. Relation of changing power factor, from unitary one, with an increase of train apparent power.

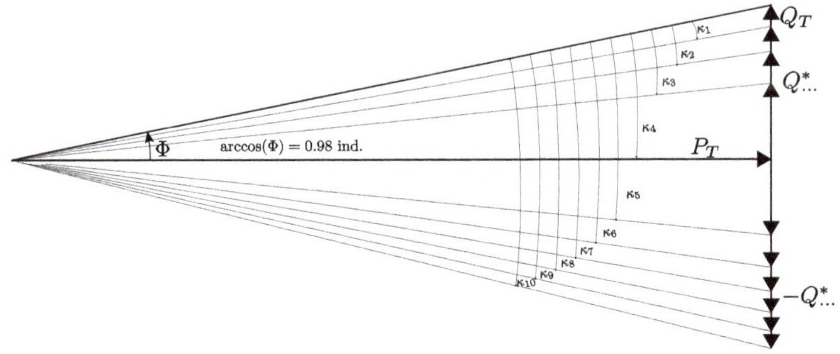

Figure 17. Illustration of the adjustment of power factor for mobile reactive power compensation.

In Figure 17 and, in the following results, the train will start with an inductive power factor (0.98) as a baseline. Furthermore, fixed increments of κ_x will be considered, where $\arccos(\kappa_1) = 0.005$.

Figure 18 presents the results of the mobile reactive power compensation, where the maximum train limit PF is 0.98 cap., in Figure 18a, and follows the different κ_x, in Figure 18b.

In Figure 18b, it is clear that simple adaptation of reactive power in each train will result in an increase in the number of trains. Specifically, changing the PF from 0.98 ind. to 0.94 cap. will result in an increasing in 50% in the number of trains. This result is obtained with the increase of the train apparent power in 6.4% (by using Equation (15) for PF 0.94 cap.). All the results are presented in Table 3.

Table 3. Maximum number of trains, for different train power consumptions and for different power factor limit. Note: the percentage improvement from baseline (train PF = 0.98 ind.) is presented in parentheses. As an example, the baseline for $P_T = 2$ MW is 16 trains; then for PF = 0.92 cap., it is possible to have eight more trains (+50% more than baseline).

	Train Active Power					
Train Power Factor	**0.5 MW**	**1 MW**	**2 MW**	**3 MW**	**4 MW**	**5 MW**
0.98 ind.	66 (+0%)	33 (+0%)	16 (+0%)	11 (+0%)	8 (+0%)	6 (+0%)
0.985 ind.	74 (+12%)	37 (+12%)	18 (+13%)	12 (+9%)	9 (+13%)	7 (+17%)
0.99 ind.	77 (+17%)	38 (+15%)	19 (+19%)	12 (+9%)	9 (+13%)	7 (+17%)
0.995 ind.	80 (+21%)	40 (+21%)	20 (+25%)	13 (+18%)	10 (+25%)	7 (+17%)

Table 3. *Cont.*

Train Power Factor	Train Active Power					
	0.5 MW	1 MW	2 MW	3 MW	4 MW	5 MW
1.00	82 (+24%)	41 (+24%)	20 (+25%)	13 (+18%)	10 (+25%)	8 (+33%)
0.99 cap.	86 (+30%)	43 (+30%)	21 (+31%)	14 (+27%)	10 (+25%)	8 (+33%)
0.98 cap.	88 (+33%)	44 (+33%)	22 (+38%)	14 (+27%)	11 (+38%)	8 (+33%)
0.97 cap.	90 (+36%)	45 (+36%)	22 (+38%)	15 (+36%)	11 (+38%)	9 (+50%)
0.96 cap.	92 (+39%)	46 (+39%)	23 (+44%)	15 (+36%)	11 (+38%)	9 (+50%)
0.94 cap.	95 (+44%)	47 (+42%)	24 (+50%)	16 (+45%)	12 (+50%)	9 (+50%)
0.92 cap.	98 (+48%)	49 (+48%)	24 (+50%)	16 (+45%)	12 (+50%)	10 (+67%)

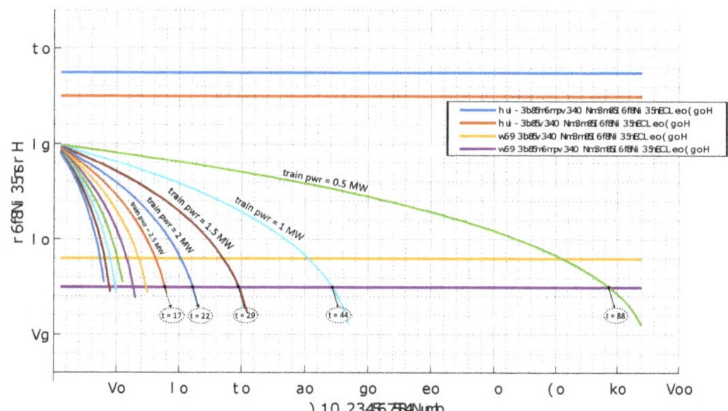

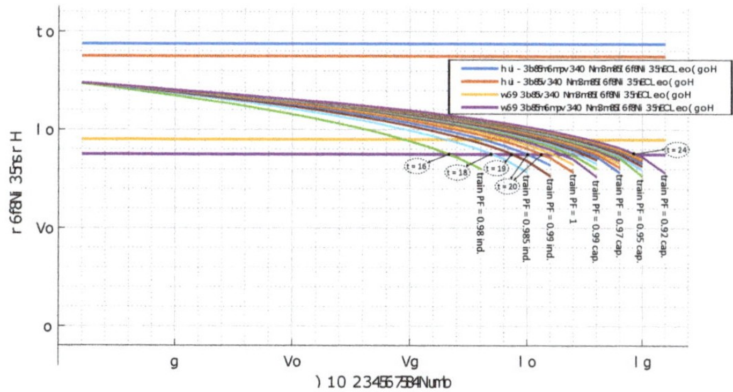

Figure 18. Relation of number of trains (PF = 0.98 ind.) with lower voltage in line: (**a**) variation of active power in each train, for fixed train PF limit (0.98 cap.); (**b**) variation of compensation, for fixed train active power of 2 MW.

It is possible to identify a correlation between the capacity improvement, in percentage, and the limit for train power factor. These results are presented in Table 4.

Table 4. Comparison of power factor with apparent power increase and average improvement of railway infrastructure capacity.

Train Power Factor	Power Increase	Capacity Improvement
0.98 ind.	−2.0%	0.0%
0.985 ind.	−1.5%	10.7%
0.99 ind.	−1.0%	15.5%
0.995 ind.	−0.5%	20.2%
1.00	0.0%	23.9%
0.99 cap.	1.0%	28.6%
0.98 cap.	2.0%	32.7%
0.97 cap.	3.1%	39.1%
0.96 cap.	4.2%	40.2%
0.95 cap.	5.3%	41.1%
0.94 cap.	6.4%	44.9%
0.93 cap.	7.5%	46.5%
0.92 cap.	8.7%	51.2%

It can be easily concluded that, if the trains operate with capacitive power factors, the number of trains in the line can be increased. However, the implementation of this strategy should be in compliance with the standard EN 50388-1 [34], by ensuring that the capacitive reactive power compensation made by each train is clipped to zero, once the catenary nominal voltage is reached. In addition, in regenerative mode as stated in the same standard, the voltage is likely to increase and, then, capacitive reactive power compensation must be avoided.

A smart railway framework will take advantage of these results, in order to justify the advantages for either the railway operators and the infrastructure managers. This framework will be covered in the following section.

6. Smart Railway Framework

The advantages of controlling the reactive power flow in the railway electrification were presented in previous sections, both in the reduction of the line losses and in the increase of the railway infrastructure capacity. In this section, the practical implementation of the proposal of this work is discussed. Specifically, a conceptual architecture to implement a reactive power compensation strategy is presented, with the usage of a smart metering framework. This framework targets advantages both for the infrastructure managers and the train operators.

6.1. The Problem of Mobile Reactive Power Compensation

One limitation of the railway infrastructure is in the impossibility of each train to measure the power flowing along the catenary (it is impossible to have current sensors measuring the downstream and upstream currents). The on-board energy measurement and data transmission to ground stations [35] have been implemented in the past few years by railway operators, and the details on this measurement have been actively researched [36].

A reactive power compensation strategy must have accurate measurements on the power flow in the catenary. Specifically, as illustrated in Figure 19, the train T_N must have the knowledge of the power flow at upstream, in the train T_{N-1} node.

This knowledge of the catenary power flow is not an easy task due to several issues:

- It requires the implementation of on-board energy meters in all trains (or in the majority) and in the TPS;
- Requires data reporting to a central station (data gathering);
- It is needed to calculate the power flow in each node and dynamically adapt this calculation mechanism to consider all trains in the traction section (power flow calculation);

- In the case of reactive power compensation strategy, the generated setpoints must be sent to each train (setpoints updating)
- All of these procedures must be made within real-time constraints.

The optimal operation for reactive power compensation will be achieved without delay in the *Data Gathering ⟶ Power flow calculation ⟶ Generation of reactive power references ⟶ Setpoints updating* procedure flow.

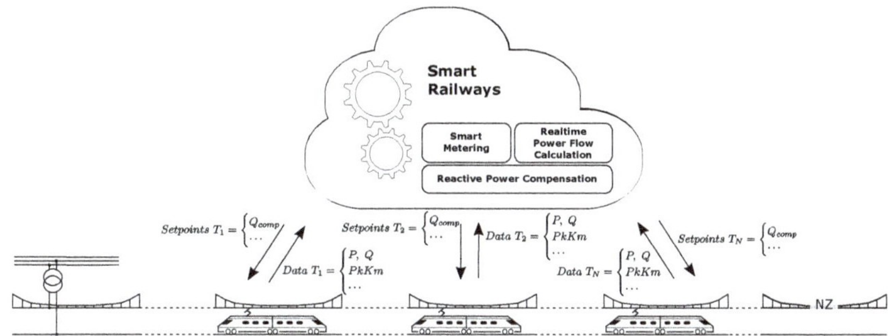

Figure 19. Integration of reactive power compensation in a smart railway framework.

However, in a practical implementation of a reactive power compensation strategy, strategies to avoid non-optimal operation should be adopted. It is clear that the power flow prediction algorithms (that estimate future consumptions) are necessary.

Considering a power flow prediction algorithm that depends on real-time data and can predict a near future, the reactive power compensation algorithm can be more successful to operate in optimal conditions.

Assuming that the smart railways framework is able to predict a time window (as example, 10 s) with a specific accuracy (as example 20% in error) then, in the worst case, the predictor algorithm generates a train reactive power setpoint with a 20% error.

6.2. A Solution for Mobile Reactive Power Compensation

It is necessary to the train to have not only the setpoint of the reactive power, but also the resultant catenary voltage after the compensation. Naturally, the reactive power injected by a train will affect the voltage of the same train; this is a feedback process. The expected and correct voltage value can be used to improve the algorithm in the following way:

- If the train voltage is above the expected voltage, then it means that the amount of reactive power injected is above the optimal value;
    - Then, the on-board reactive power compensation system (viewed as an algorithm that adapts the power factor depending on the desired reactive power value) will reduce the value of reactive power.
- If the train voltage is below the expected, then the on-board reactive power compensation system will increase the injection of reactive power.

A droop-like approach can be used to help stabilize the reactive power control.

This on-board adaptation will follow a droop characteristic curve, as illustrated in Figure 20: the higher the deviation of the voltage and the closer the train is to the NZ, the higher the correction to Q_{comp}.

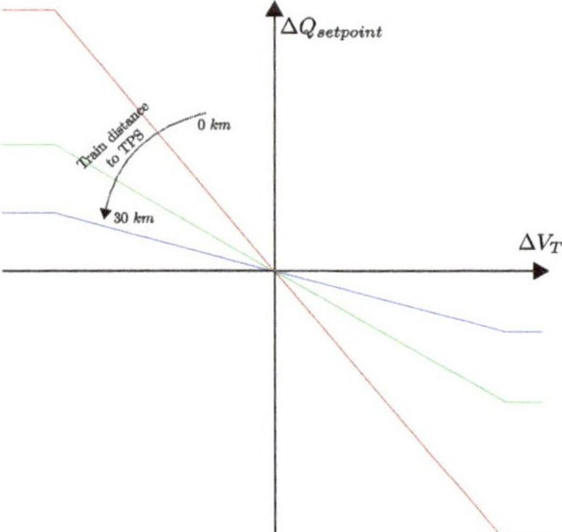

Figure 20. Droop strategy for on-board adjustment. The ΔV_T is the difference between the real train voltage and the expected train voltage. The $\Delta Q_{setpoint}$ is the output of the droop curve, where this value is added to the predicted Q_{comp}.

The slope of the droop characteristic curve might be difficult to obtain. As an example, if all trains have the same droop characteristic curve, it is possible to have a resonance behavior. This train–network interactions and resonance is a well known issue [37], where the converter control loops must be immune to low-frequency oscillation, harmonic resonance, and harmonic instability phenomena (specifically with the tuning of current and voltage control loops, as well as the estimation of the phase angle of the incoming voltage). Therefore, the proposed droop approach most likely reduces a possible low frequency oscillation in the reactive power.

Therefore, in this conceptual implementation, the droop characteristic must be dependent on the distance between the train and the TPS, as well as the number of trains separating a compensation one and the TPS. It is expected that the trains closer to the TPS are requested for a higher contribution of the reactive power, in comparison to the trains closer to the neutral zone.

Considering the example in Figure 21, the usage of the real measured voltage for on-board adaptation of the reactive power will contribute to having an operation closer to the optimal (in comparison with only following the Q_{comp} setpoints).

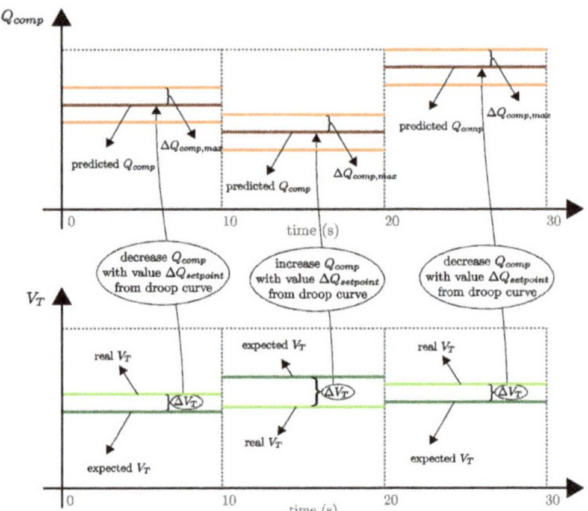

Figure 21. Illustration of the on-board adaptation, supported by the predicted Q_{comp} and $V_{T,estim.}$ from the smart railways framework.

6.3. The Path to Reactive Power Compensation

The SVC compensation strategy is based on the inclusion of power electronic devices within the proximity of the NZ. Currently, the compensation strategy is local and does not take into consideration the measurements of the reactive power in the TPS. In this work, a communication channel is considered, where the reactive power is measured in the TPS and used in the NZ to inject reactive power.

The proposed mobile reactive power compensation scheme requires that each train is able to adapt its own power factor. This requirement leaves some of the currently used trains, since the technology used does not allow this level of adaptation. The infrastructure managers can promote modernization of train fleet, with cost reduction of operation for trains able to adapt the reactive power.

As one outcome from UIC International Railway Solutions (IRS) 90930 stakeholders workshop [38], from July 2020 onwards, the European rail sector should implement a new standard for energy metering, which includes the installation of on-board energy metering systems as well as data exchanges on ground [39]. Therefore, the need for smart metering is now closer to being achieved.

Finally, as a preliminary incentive, the billing should accommodate the injection of the reactive power, similarly to the situation happening in certain countries, where the regenerated energy is billed in favor of the railway operator. The clear advantages for the infrastructure managers should allow, in theory, the elimination of the reactive (capacitive) power billing. Further billing strategies should accommodate this effort that the railway operators might take, in order to improve the power quality. The injection of reactive power can be seen as a service that the railway operators can provide.

7. Discussion and Conclusions

In this section, a final discussion and conclusions on the outcomes of this work are presented. This work results from the recent lack of coverage of railway reactive power control in the literature. Specifically, the reactive power control towards increase of railway infrastructure capacity is not an active research topic, from the knowledge of the authors.

From the presented results in this work and as already highlighted in the literature, the increase of railway energy efficiency is clear, not only with the reduction of the reactive power consumption from the transmission/distribution system operator, but also with the reduction of line power losses. The example of Figure 9 illustrates the potential of reduction of catenary power losses up to 50%.

The approach to study the infrastructure capacity improvement is the new hypothesis covered with this work. Specifically, two reactive power compensation strategies are compared, regarding the railway infrastructure capacity. The integration of a static VAR compensation in the neutral zone can increase the railway capacity up to 50%, without its compensation factor depending on the train active power. The second (proposed) compensation strategy considers the integration of the reactive power compensation within each train. With this strategy, it is possible to increase the railway capacity up to 50%, only with an increase of train apparent power below 10%.

The findings in this work were obtained with certain open degrees of freedom, such as each train power consumption, the NZ SVC power limitations, and the maximum capacitive power factor for each train. Certain degrees of freedom were closed with justifications made throughout the article. As an example, the X/R_L ratio was fixed since it depends on the characteristics of the electrification. The line distance was also fixed. It is expected that these two parameters do not affect the railway capacity (in percentage). The other fixed parameter is the distance between each train. It is clear that this corresponds to an ideal situation (in a realistic situation, the train dynamic constraints, journey timetables, signaling, among others, will affect each train position, and the distance between trains will not be all the same). However, only with an extensive statistical analysis is it possible to evaluate if variable train distances result in different results. It is expected, as an example, that, if the majority of trains are more concentrated near the TPS, the resultant minimum voltage will be higher than the fixed distance presented in this work.

Later, in this work, a smart railway framework is proposed, focused on solving the issues regarding this reactive power compensation strategy. A solution is then presented based on a droop controller and a smart metering strategy, which enables the trains to be closer to an optimal point of operation. With this work, the quality of supply in the railway network can be increased, with the adoption of a mobile reactive power compensation strategy based on a smart metering framework.

Author Contributions: For this paper, the investigation as well as the methodology, software development and validation was conducted by V.A.M. under supervision of J.L.A., A.S.C., and A.P.M. The original draft preparation was made by V.A.M. and A.P.M. All authors have read and agreed to the published version of the manuscript.

Funding: This research was funded by FCT (Fundação Ciência e Tecnologia) under grant PD/BD/128051/2016. This research is also associated with the Shift2Rail In2Stempo project (grant 777515).

Conflicts of Interest: The authors declare no conflict of interest.

Abbreviations

The following abbreviations are used in this manuscript:

APQC	Active Power Quality Compensator
HPQC	Hybrid Power Quality Compensator
IRS	International Railway Solutions
MMC	Modular Multilevel Converter
MRPC	Mobile Reactive Power Compensation
NZ	Neutral Zone
PF	Power Factor
PWM	Pulse Width Modulation
RPC	Railway Power Conditioner
SVC	Static VAR Compensator
TPS	Traction Power Substation
UIC	Union Internationale des Chemins de fer—International Union of Railways

References

1. IEA. Energy Consumption and CO_2 Emissions Focus on Passenger Rail Services. International Energy Agency (IEA) and International Union of Railways (UIC). Available online: https://uic.org/IMG/pdf/handbook_iea-uic_2017_web3.pdf (accessed on 24 August 2020).
2. Shift2Rail Joint Undertaking. *Shift2Rail Multi-Annual Action Plan, November 2015*; Shift2Rail Joint Undertaking: Brussels, Belgium, 2015.
3. Geischberger, J.; Moesters, M. Impact of faster freight trains on railway capacity and operational quality. *Int. J. Transp. Dev. Integr.* **2020**, *4*, 274–285. [CrossRef]
4. Pouryousef, H.; Lautala, P. Hybrid simulation approach for improving railway capacity and train schedules. *J. Rail Transp. Plan. Manag.* **2015**, *5*, 211–224. [CrossRef]
5. Bruni, S.; Bucca, G.; Collina, A.; Facchinetti, A. *Dynamics of the Pantograph Catenary System*; CRC Press: Boca Raton, FL, USA, 2019; pp. 613–647. [CrossRef]
6. Song, Y.; Liu, Z.; Rxnnquist, A.; Navik, P.; Liu, Z. Contact Wire Irregularity Stochastics and Effect on High-speed Railway Pantograph-Catenary Interactions. *IEEE Trans. Instrum. Meas.* **2020**. [CrossRef]
7. Wang, Z.; Cheng, Y.; Mei, G.; Zhang, W.; Huang, G.; Yin, Z. Torsional vibration analysis of the gear transmission system of high-speed trains with wheel defects. *Proc. Inst. Mech. Eng. Part F J. Rail Rapid Transit* **2019**, *234*, 123–133. [CrossRef]
8. IEC 60850. *Railway Applications—Supply Voltages of Traction Systems*; International Electrotechnical Commission (IEC): Geneva, Switzerland, 2007.
9. Sauer, P.W. Reactive Power and Voltage Control Issues in Electric Power Systems. In *Power Electronics and Power Systems*; Kluwer Academic Publishers: Boston, MA, USA, 2005; pp. 11–24. [CrossRef]
10. Brenna, M.; Foiadelli, F.; Zaninelli, D. *Electrical Railway Transportation Systems*; IEEE Press Series on Power Engineering: Hoboken, NJ, USA, 2018. [CrossRef]

11. Carson, J.R. Wave Propagation in Overhead Wires with Ground Return. *Bell Syst. Tech. J.* **1926**, *5*, 539–554. [CrossRef]
12. Pilo, E. Diseño óptimo de la electrificación de Ferrocarriles de alta Velocidad. Ph.D. Thesis, Universidad Pontificia de Comillas, Madrid, Spain, 2003.
13. Aeberhard, M.; Courtois, C.; Ladoux, P. Railway traction power supply from the state of the art to future trends. In Proceedings of the SPEEDAM 2010, Pisa, Italy, 14–16 June 2010; IEEE: New York City, NY, USA, 2010. [CrossRef]
14. Langerudy, A.T.; Mariscotti, A.; Abolhassani, M.A. Power Quality Conditioning in Railway Electrification: A Comparative Study. *IEEE Trans. Veh. Technol.* **2017**, *66*, 6653–6662. [CrossRef]
15. Tanta, M.; Pinto, J.G.; Monteiro, V.; Martins, A.P.; Carvalho, A.S.; Afonso, J.L. Topologies and Operation Modes of Rail Power Conditioners in AC Traction Grids: Review and Comprehensive Comparison. *Energies* **2020**, *13*, 2151. [CrossRef]
16. Luo, A.; Wu, C.; Shen, J.; Shuai, Z.; Ma, F. Railway static power conditioners for high-speed train traction power supply systems using three-phase V/V transformers. *IEEE Trans. Power Electron.* **2011**, *26*, 2844–2856. [CrossRef]
17. Dai, N.; Wong, M.; Lao, K.; Wong, C. Modelling and control of a railway power conditioner in co-phase traction power system under partial compensation. *IET Power Electron.* **2014**, *7*, 1044–1054. [CrossRef]
18. Sun, Z.; Jiang, X.; Zhu, D.; Zhang, G. A novel active power quality compensator topology for electrified railway. *IEEE Trans. Power Electron.* **2004**, *19*, 1036–1042. [CrossRef]
19. Lao, K.; Dai, N.; Liu, W.; Wong, M. Hybrid Power Quality Compensator With Minimum DC Operation Voltage Design for High-Speed Traction Power Systems. *IEEE Trans. Power Electron.* **2013**, *28*, 2024–2036. [CrossRef]
20. Lao, K.; Wong, M.; Dai, N.Y.; Wong, C.; Lam, C. Analysis of DC-Link Operation Voltage of a Hybrid Railway Power Quality Conditioner and Its PQ Compensation Capability in High-Speed Cophase Traction Power Supply. *IEEE Trans. Power Electron.* **2016**, *31*, 1643–1656. [CrossRef]
21. Tanta, M.; Pinto, J.G.; Monteiro, V.; Martins, A.P.; Carvalho, A.S.; Afonso, J.L. Deadbeat Predictive Current Control for Circulating Currents Reduction in a Modular Multilevel Converter Based Rail Power Conditioner. *Appl. Sci.* **2020**, *10*, 1849. [CrossRef]
22. Celli, G.; Pilo, F.; Tennakoon, S.B. Voltage regulation on 25 kV AC railway systems by using thyristor switched capacitor. In Proceedings of the Ninth International Conference on Harmonics and Quality of Power Proceedings (Cat. No.00EX441), Orlando, FL, USA, 1–4 October 2000; Volume 2, pp. 633–638. [CrossRef]
23. Tan, P.C.; Loh, P.C.; Holmes, D.G. A Robust Multilevel Hybrid Compensation System for 25-kV Electrified Railway Applications. *IEEE Trans. Power Electron.* **2004**, *19*, 1043–1052. [CrossRef]
24. Fracchia, M.; Courtois, C.; Talibart, A.; Bordignon, P.; Consani, T.; Merli, E.; Bacha, S.; Stuart, M.; Zoeter, K.; Burrarini, G.G.; et al. High Voltage Booster for Railway Applications. World Congress on Railway Research. 2001. Available online: http://www.railway-research.org/IMG/pdf/278.pdf (accessed on 24 August 2020).
25. Zanotto, L.; Piovan, R.; Toigo, V.; Gaio, E.; Bordignon, P.; Consani, T.; Fracchia, M. Filter Design for Harmonic Reduction in High-Voltage Booster for Railway Applications. *IEEE Trans. Power Deliv.* **2005**, *20*, 258–263. [CrossRef]
26. Raygani, S.V.; Moaveni, B.; Fazel, S.S.; Tahavorgar, A. SVC implementation using neural networks for an AC electrical railway. *WSEAS Trans. Circuits Syst.* **2011**, *10*, 173–183.
27. Kulworawanichpong, T. Optimizing AC Electric Railway Power Flows with Power Electronic Control. Ph.D. Thesis, University of Birmingham, Birmingham, UK, 2003.
28. Kulworawanichpong, T.; Goodman, C.J. Optimal area control of AC railway systems via PWM traction drives. *IEE Proc. Electr. Power Appl.* **2005**, *152*, 33–70. [CrossRef]
29. Tahavorgar, A.; Raygani, S.V.; Fazel, S. Power quality improvement at the point of common coupling using on-board PWM drives in an electrical railway network. In Proceedings of the 2010 1st Power Electronic & Drive Systems & Technologies Conference (PEDSTC), Tehran, Iran, 17–18 February 2010; IEEE: New York, NY, USA, 2010; pp. 412–417. [CrossRef]
30. Raygani, S.V.; Tahavorgar, A.; Fazel, S.S.; Moaveni, B. Load flow analysis and future development study for an AC electric railway. *IET Electr. Syst. Transp.* **2012**, *2*, 139–147. [CrossRef]

31. Abril, M.; Barber, F.; Ingolotti, L.; Salido, M.; Tormos, P.; Lova, A. An assessment of railway capacity. *Transp. Res. Part E Logist. Transp. Rev.* **2008**, *44*, 774–806. [CrossRef]
32. Morais, V.A.; Afonso, J.L.; Martins, A.P. Modeling and Validation of the Dynamics and Energy Consumption for Train Simulation. In Proceedings of the 2018 International Conference on Intelligent Systems (IS), Funchal-Madeira, Portugal, 25–27 September 2018; IEEE: New York, NY, USA, 2018; pp. 288–295. [CrossRef]
33. Zimmerman, R.D.; Murillo-Sanchez, C.E.; Thomas, R.J. MATPOWER: Steady-State Operations, Planning, and Analysis Tools for Power Systems Research and Education. *IEEE Trans. Power Syst.* **2011**, *26*, 12–19. [CrossRef]
34. CENELEC EN 50388-1. *Railway Applications—Fixed Installations and Rolling Stock—Technical Criteria for the Coordination between Traction Power Supply and Rolling Stock to Achieve Interoperability—Part 1: General*; CENELEC: Brussels, Belgium, 2017.
35. CENELEC EN 50463-2. *Railway Applications—Energy Measurement on Board Trains*; CENELEC: Brussels, Belgium, 2017.
36. Mariscotti, A. Impact of Harmonic Power Terms on the Energy Measurement in AC Railways. *IEEE Trans. Instrum. Meas.* **2020**, 6731–6738. [CrossRef]
37. Hu, H.; Tao, H.; Wang, X.; Blaabjerg, F.; He, Z.; Gao, S. Train–Network Interactions and Stability Evaluation in High-Speed Railways—Part II: Influential Factors and Verifications. *IEEE Trans. Power Electron.* **2018**, *33*, 4643–4659. [CrossRef]
38. UIC. IRS 90930 (Traction Energy Settlement and Data Exchange). Information Published on 9 October 2018 in the UIC Electronic Newsletter "UIC eNews" Nr 617. Available online: https://bit.ly/30Urpsw (accessed on 24 August 2020).
39. Zasiadko, M. New Stage of Energy Measuring for European Rail Sector. 2019 Available online: https://www.railtech.com/policy/2019/10/28/new-stage-of-energy-measuring-for-european-rail-sector/ (accessed on 24 August 2020).

Article

A Novel Co-Phase Power-Supply System Based on Modular Multilevel Converter for High-Speed Railway AT Traction Power-Supply System

Si Wu, Mingli Wu * and Yi Wang

Institute of Traction Power Supply, School of Electrical Engineering, Beijing Jiaotong University, Beijing 100044, China; 16117406@bjtu.edu.cn (S.W); ywang5@bjtu.edu.cn (Y.W.)

* Correspondence: mlwu@bjtu.edu.cn

Abstract: The existing problems of the traction power-supply system (i.e., the existence of the neutral section and the power quality problems) limit the development of railways, especially high-speed railways, which are developing rapidly worldwide. The existence of the neutral section leads to the speed loss and traction loss as well as mechanical failures, all of which threaten the fast and safe operation of the train and the system. Meanwhile, the power quality problems (e.g., the negative sequence current, the reactive power, and the harmonic) can bring a series of problems that cannot be ignored on the three-phase grid side. In response, many researchers have proposed co-phase power-supply schemes to solve these two problems simultaneously. Given that the auto-transformer (AT) power-supply mode has become the main power-supply mode for the high-speed railway traction power-supply system, it has a bright future following the rapid development of the high-speed railway. In addition, there is no co-phase power-supply scheme designed for AT power-supply mode in the existing schemes. Therefore, the main contribution of this paper is to propose a specifically designed power-supply mode more suitable for the AT, as well as to establish the control systems for the rectifier side and the inverter side. In addition, for the proposed scheme, the operation principle is analyzed, the mathematical model is built, and the control system is created, and its functionality is verified by simulation, and its advantages are compared and summarized finally. The result proves that it can meet functional requirements. At the same time, compared with the existing co-phase power-supply scheme, it saves an auto-transformer in terms of topology, reduces the current stress by 10.9% in terms of the current stress of the switching device, and reduces the power loss by 0.25% in terms of the entire system power loss, which will result in a larger amount of electricity being saved. All of this makes it a more suitable co-phase power-supply scheme for the AT power-supply mode.

Keywords: co-phase power-supply system; traction power supply; modular multilevel converter (MMC); AT power-supply mode; power quality; neutral section

Citation: Wu, S.; Wu, M.; Wang, Y. A Novel Co-Phase Power-Supply System Based on Modular Multilevel Converter for High-Speed Railway AT Traction Power-Supply System. *Energies* **2021**, *14*, 253. https://doi.org/10.3390/en14010253

Received: 21 November 2020
Accepted: 25 December 2020
Published: 5 January 2021

1. Introduction

There are still two significant problems in the current railway traction power-supply system:

(1) Neutral Section:

Due to the single-phase, non-linear, random fluctuating load characteristics of the locomotive, the railway traction power-supply system adopts the rotating phase sequence method to supply power to the locomotive (i.e., the phase of the traction network will change every 20 km–25 km). Therefore, the areas of different phase sequences need to be isolated, which leads to the existence of a non-electrical zone, which is the neutral section [1].

The neutral section results in the loss of the speed and the traction of the train as well as the mechanical wear and tear in traction power-supply systems and locomotives.

In addition, there already have some cases of train ramp accidents and substation tripping accidents due to the neutral section. Thus, it really brings major safety hazards to the fast and safe operation of the train and system. In particular, high-speed railways will pass through the neutral section more frequently because of their high speed. For the development of railways, especially high-speed railways, it is necessary to solve the problem of the neutral section [2].

In response to the problems brought by the neutral section, researchers have proposed many schemes. Recently, more attention has been paid to the continuous power passing through neutral section scheme. e.g., mechanical switches ground auto-passing neutral section scheme [3], electronic switches ground auto-passing neutral section scheme [4,5], flexible ground auto-passing neutral section scheme [6]. Although these schemes can solve the power problems brought by the neutral section to some extent, such as reduce the time of power loss and reduce the over-voltage and over-current in the process of passing through the neutral section, they do not fundamentally eliminate the neutral section area on the traction network, so the mechanical weakness is still there, which still cause mechanical losses and economic costs.

(2) Power Quality Problem:

The locomotive loads with non-linear, single-phase, and random fluctuation characteristics bring the power quality problems to the power grid, e.g., the negative sequence current, harmonics, and reactive power.

The power quality problems will pose a hazard to the grid and equipment. Negative sequence currents can be harmful to generators, asynchronous motors, transformers, transmission lines, and communication systems. Harmonics, especially the high-order harmonics brought by the "AC-DC-AC" electric locomotives that have been widely used in high-speed railways and heavy-haul railways in recent years, are prone to cause high-frequency resonance, which can easily cause damage to electrical equipment and accidents. Reactive power can increase the loss of lines and equipment, increase the voltage drop of lines and transformers, and bring adverse effects to the power grid and equipment [7–9].

In response to the power quality problems, researchers have proposed many schemes. A passive filter in [10] was used to solve the harmonics and reactive power problems. The Static Var Compensator (SVC) in [11,12] was used to address the negative sequence current and reactive power, while it is quite limited in its ability to suppress harmonics. The Static Synchronous Compensator (STATCOM) in [13] has better harmonic characteristics and can use for negative sequence current, reactive power, and harmonics compensation. The Active Power Filter (APF) in [14] can solve power quality problems effectively, but the cost is relatively high. However, these solutions are only proposed to address power quality issues, but do not attempt to address the problem of the neutral section.

Recently, some researchers have proposed co-phase power-supply schemes to simultaneously solve the neutral section and power quality problems. Among them, one category is the compensation co-phase power-supply scheme, and the other category is the through co-phase power-supply scheme. The compensation co-phase power-supply schemes can solve the power quality problems and the neutral section at the same time. However, it can only eliminate the neutral section at the substation but cannot eliminate the neutral section in the entire traction network. This kind of co-phase power-supply schemes includes China's first co-phase power-supply unit, which is put into operation at the Mershan substation in 2010 [10] and the first single three-phase combined co-phase power-supply unit at the Shangyu substation in 2014 [15], as well as the Active Power Compensator (APC) [16], the Railway Power Conditioner (PRC) [17,18], the Hybrid Power Quality Conditioner (HPQC) [19–22], and the Hybrid Electrical Magnetic Power Quality Compensator (HEMPQC) [23]. The through co-phase power-supply schemes can simultaneously and completely solve the power quality problems and eliminate all neutral section of the whole traction network, making the traction network whole line through, e.g., the schemes based on the Cascaded H-Bridge (CHB) converter in [24,25] and the schemes based on the Modular Multilevel Converter (MMC) in [26].

However, these co-phase power-supply schemes are not specifically detailed in terms of the specific power-supply mode. As is well known, high-speed railways are developing rapidly all over the world since they attract great attention from many countries due to their high speed, large transportation capacity, low average energy consumption, light environmental impact, and good economic benefits. To improve the power-supply capacity of the traction network, reduce the number of traction substation and the electromagnetic interference to the adjacent metal conductors, Japan first use the AT power-supply mode in high-speed railways [27].

As for the two prominent problems existing in the railway system. The existence of the neutral section is a common problem in existing electrified railways, and it exists worldwide. At the present stage, there are problems of the neutral section in electrified railways in China, Japan, France, and the United Kingdom. Only Germany built a power grid on the railway with a frequency different from that of the public grid to isolate it from the public grid, achieve the same phase across the entire line, cancel the neutral section, and realize the co-phase power-supply system. Power quality is also a common problem in electrified railways worldwide. The Japanese Shinkansen used STATCOM [28] and RPC [29,30] to solve the power quality problem in 1993 and 2002, respectively. With the rapid development of high-speed railways, the AT power-supply mode has been vigorously promoted and has become the main power-supply mode for high-speed railways [31]. In view of the wide application of AT power-supply mode in high-speed railways and its bright prospects, scheme to these two prominent problems in AT power-supply system are valuable to study.

Therefore, the main contribution and innovation of this paper is that under the background that there is no co-phase power-supply scheme designed for AT power-supply mode, a co-phase power-supply scheme designed for AT power-supply mode and more suitable for it is proposed, and the control systems for the rectifier side and the inverter side is created, and the control section that can make the newly added bridge arm voltage balance is added to the control system of the inverter side. In addition, for this scheme, its operation principle is analyzed, its mathematical model is established, its control system is created, its functionality is verified by simulation, and finally, its superiority over other schemes when applied to the AT power-supply mode is analyzed and summarized. The results prove that the topology proposed in this paper is superior for AT power-supply system in the following aspects, in terms of topology, it eliminates an auto-transformer, in terms of current stress of the insulated gate bipolar transistor (IGBT) of the bridge arm of the inverter, it reduces the current stress by 10.9%, in terms of power loss, it reduces the power loss of the system by 0.25%, which means that for a 40 MW traction substation, 100 kWh of power can be saved in one hour, and the long-term power savings of many substations will be significant.

This paper is organized as follows. In Section 2, the operating principle and mathematical model of the proposed scheme are described. In Section 3, the control system is illustrated. In Section 4, the simulation verification and advantages analysis of the proposed topology are elaborated. In Section 6, the conclusion is stated.

2. Operation Principle of the Proposed Co-Phase Power-Supply Scheme

2.1. The Topology of the Proposed Scheme

In the existing railway power-supply system, the Auto-Transformer (AT) power-supply mode can not only reduce the interference of the electrified railway to the adjacent communication lines but also has good technical and economic indexes for the traction power-supply system, which make it more suitable to the operation of high-speed and high-power electric locomotives. In terms of structure, as shown in Figure 1, AT power-supply mode is composed of traction transformer, auto-transformer, catenary, steel rail, and positive feeder. The wiring form of traction transformer in AT power-supply mode can be divided into the three-phase–two-phase balanced transformer shown in Figure 1a (taking the most widely used Scott wiring three-phase-two-phase balanced transformer [32] as an example, and adding the required auto-transformer in the substation to the secondary side) , the three-phase crossing wiring traction transformer shown in Figure 1b [33], the V/X wiring traction transformer shown in Figure 1c [34], and the secondary midpoint withdrawable single-phase wiring traction transformer shown in Figure 1d [35]. The auto-transformer in the AT power-supply mode is connected in parallel between the contact suspension and the positive feeder, eliminating the segmentation caused by the addition of the transformer to the contact network. The distance between two auto-transformers is generally 8–15 km.

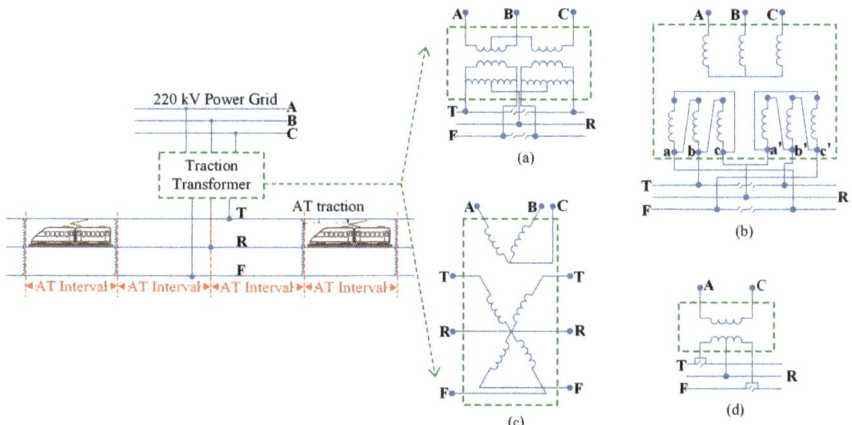

Figure 1. Traditional AT traction power-supply system components: (**a**) three-phase-two-phase balanced wiring transformer; (**b**) three-phase cross wiring transformer; (**c**) V/X wiring transformer; (**d**) secondary midpoint withdrawable single-phase wiring transformer.

Researchers have proposed several schemes to the existing problems of neutral section and power quality in the system. Among them, the schemes to neutral section can be divided into power-off neutral section passing scheme and live-line neutral section passing scheme according to whether the main circuit breaker is closed or not. The live-line neutral section passing schemes can be divided into column switch neutral section passing scheme, mechanical switch ground automatic neutral section passing scheme, electronic switch ground automatic neutral section passing scheme and flexible ground automatic neutral section passing scheme [36]. The performance of each scheme in terms of functionality, operation, cost and maintenance has been listed in Table 1 below.

It can be seen from the table that in terms of functionality, power loss time and over-voltage, the three ground automatic neutral section passing schemes have a shorter power-loss time, which brings less speed loss, and at the same time, less over-voltage impact. Among them, the best is the flexible ground automatic neutral section passing scheme, which realizes the complete uninterrupted power supply and no over-voltage shock. The other solutions have relatively long power-loss time, relatively large speed loss, and over-voltage impact. Manual power-off neutral section passing scheme is the worst, since the power-loss time is long, and the speed loss is large, and it is easy to cause over-voltage impact. In operation, according to the fatigue degree of the train crew, except for the manual power-off neutral section passing scheme, there is no need for the train crew to operate, which will not cause the passing through the neutral section with electricity caused by human operation errors. In terms of cost and maintenance, the three ground automatic neutral section passing schemes are more expensive to invest in, with the flexible above-ground automatic over-phase being the most expensive. In terms of switch lifetime, the three ground automatic neutral section passing scheme involve switches. The mechanical switch has a short lifetime and high maintenance costs. The switches involved in the electronic switch ground automatic neutral section passing scheme and the flexible ground automatic neutral section passing scheme have a long lifetime and the maintenance cost is not high [4,6,36–38].

Table 1. Comparative analysis of neutral section passing schemes in terms of functionality, operation, cost and maintenance.

Comparative Aspects	Specific Comparison Content	Power-Off Neutral Section Passing Scheme		Live-Line Neutral Section Passing Scheme			
		Manual Power-off Neutral Section Passing Scheme	Vehicle-Mounted Automatic Neutral Section Passing Scheme	Column Switch Neutral Section Passing Scheme	Mechanical Switch Ground Automatic Neutral Section Passing Scheme	Electronic Switch Ground Automatic Neutral Section Passing Scheme	Flexible Ground Automatic Neutral Section Passing Scheme
Functionality	power-loss time (speed loss)	Long power-loss time, big speed loss	Long power-loss time, big speed loss	Relatively long power-loss time, relatively big speed loss	Relatively short power-losstime, relatively small speed loss	Short power-loss time, small speed loss	Completely realize uninterrupted power neutral section passing
	over-voltage	Easy to cause over-voltage shock	Over-voltage shock	Over-voltage shock	Mechanical switch brings operating over-voltage	the operating over-voltage caused by mechanical switches solved	No over-voltage shock
Operation	the fatigue degree of the train crew	Easy to cause train crew fatigue	No train crew operation required	No train crew operation required	No train crew operation required	No train crew operation required	No train crew operation required
Cost and maintenance	Investment	Small	Relatively small	Relatively small	Relatively big	Relatively big	Big
	Switch lifetime (if included in the scheme)	None	None	None	Short switch lifetime	Long switch lifetime	Long switch lifetime

Those schemes to power quality problems include passive filter, Static Var Compensator (SVC), Static Synchronous Compensator (STATCOM), Active Power Filter (APF), Railway Power Conditioner (RPC). As shown in Table 2, in comparison, passive filters are used to control harmonics and reactive power, but they can only eliminate specific harmonics. In addition, it is easy to resonate with the system impedance and cause harmonic amplification. Since the device cannot follow the dynamic changes in time, the fixed capacitor value in each passive filter cannot totally compensate the frequently change reactive power, then cannot reach a unitary power factor correction [39]. SVC is used for negative sequence and reactive power compensation, but its suppression effect on harmonics is limited, and it will bring harmonic problems by itself. In addition, may produce series and parallel resonance. Compared to the SVC, STATCOM behaves as a bidirectional reactive power compensator, with a faster response time, larger load capability, lower harmonic contents and more compact design structure [39]. APF can eliminate harmonics well, but when APF is used to compensate negative sequence and reactive power, its cost is high, so it is usually used in combination with passive compensation. PRC is a relatively successful power management program, which effectively realizes the comprehensive compensation of reactive power, negative sequence and harmonic current [39–41].

Table 2. Comparative analysis of the performance of the schemes for the power quality problems.

Name of Each Scheme	Negative Sequence	Harmonics	Reactive Power	Others
Passive filter	None	Govern harmonics, but only eliminate specific harmonics. And it is easy to resonate with the system impedance and cause harmonic amplification	Govern reactive power, but with poor continuity, which cannot be compensated well due to the frequent fluctuation of the traction load.	Require large space for installation and high implementation cost. the change in the filter parameters affected by the heat or lifetime, which gradually caused misoperations in the filter functionality
SVC	Compensation for negative sequence	Limited harmonic suppression and it will bring harmonic problems by itself	Compensation for reactive power	May produce series-parallel resonance
STATCOM	Compensation for negative sequence	Low harmonic content	Fast response for reactive power compensation	Small Land area
APF	Compensation for negative sequence	Harmonic suppression	Compensation for reactive power	The cost is high when used to compensate negative sequence and reactive power, so it is usually combined with passive compensation.
RPC	Effective realization of reactive power compensation	Effective implementation of negative sequence compensation	Effective harmonic compensation	None

Recently, the co-phase power-supply scheme proposed by researchers can simultaneously and well solve the problems of neutral section and power quality. These co-phase power-supply schemes can be divided into compensative co-phase power-supply schemes and continuous co-phase power-supply schemes. Compensative co-phase power-supply schemes include combined co-phase power-supply schemes, APC, RPC, HPQC, etc. Continuous co-phase power-supply schemes include continuous co-phase power-supply schemes based on cascaded H-bridge structure and continuous co-phase power-supply schemes based on modular multilevel converter (MMC). The compensative co-phase power-supply scheme is to combine with the transformer to supply power to the traction network. The negative sequence reactive harmonics are compensated by the installed power electronic devices on the secondary side of the traction transformer, but it can only eliminate the neutral section in the substation. The neutral sections where the districts are located are still there, and the full line continuous cannot be achieved. The continuous co-phase power-supply scheme uses a back-to-back power electronic structure to directly hang to the traction network. The amplitude and phase of the output voltage of this structure can be controlled. It can achieve the same phase of the traction network across the entire line, thereby achieving a full line continuous. At the same time, it can also compensate the negative sequence current, reactive power and harmonics well.

The continuous co-phase power-supply scheme replaces the traction transformer in the traditional traction power-supply system, which connects the three-phase power grid and the traction network with a co-phase power-supply device, which is generally a back-to-back structure consisting of switching devices. By controlling the rectifier side of the back-to-back structure, the power quality problems (i.e., negative sequence current, reactive power, and harmonics) that may be brought to the power grid side can be addressed. At the same time, by controlling the inverter side in the back-to-back structure, the amplitude, phase, and frequency of the output voltage supplied to the traction network are controllable to achieve the same phase of the output voltage of the substations, thereby achieving the same phase of the entire traction network, thereby eliminating the neutral section. Therefore, the through co-phase power supply can wholly and simultaneously solve the problems of the neutral section and power quality problems.

If the co-phase power-supply scheme that has been proposed in the most paper is applied to the AT power-supply system, as shown in Figure 2a, an auto-transformer is required between the back-to-back structure of the topology and the traction network. The structure proposed in this paper, as shown in Figure 2b, uses a bridge leg to replace the role of the auto-transformer. The newly added bridge leg should be connected to the rail (i.e., the R line in Figure 2), and the modules on it should be controlled to ensure that the voltage between the contact grid (i.e., the T line in Figure 2) and the rail, as well as the voltage between the positive feeder (i.e., the F line in Figure 2) and the rail, are 27.5 kV single-phase AC voltages of equal amplitude and 180 degrees difference in phase, i.e., $V_{TR} = 27.5\angle 0^\circ$ and $V_{FR} = 27.5\angle 180^\circ$. The functionality and advantages of this scheme will be analyzed and illustrated in detail in Section 4.

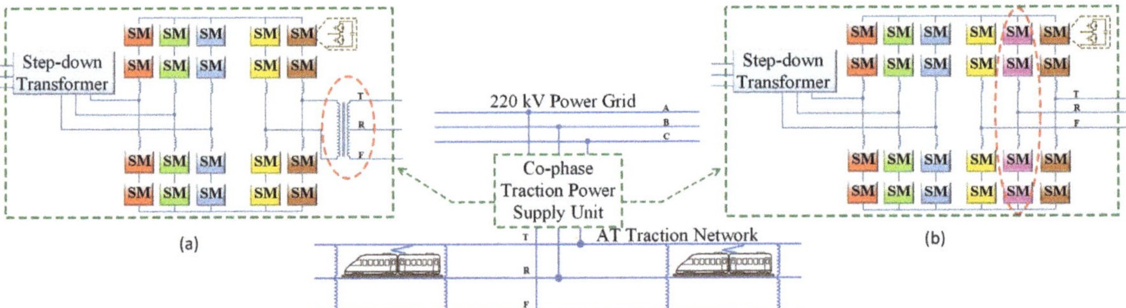

Figure 2. Co-phase power-supply system topologies: (**a**) The scheme already have been proposed but is not specific to AT supply mode; (**b**) The scheme proposed in this paper for AT supply mode.

2.2. The Mathematical Model of the Proposed Scheme

As for the mathematical model, the topology proposed in this paper is a back-to-back structure consisting of the rectifier side, which is composed of three bridge legs corresponding to the three-phase step-down transformer, and the inverter side, which is composed of three bridge legs corresponding to the two single-phase output voltages to the traction network. The mathematical model of the rectifier side and the inverter side will be illustrated separately below.

2.2.1. The Rectifier Side

According to the equivalent circuits as shown in Figure 3 and Kirchhoff's voltage law, we have Equation (1) [42]

$$\begin{cases} U_{si} = -L\frac{di_u}{dt} - U_{ui} - U_{NO} + \frac{1}{2}U_{dc}; (i = a, b, c) \\ U_{si} = L\frac{di_l}{dt} + U_{li} - U_{NO} - \frac{1}{2}U_{dc}; (i = a, b, c) \end{cases} \tag{1}$$

In Equation (1), $L = L_s + L_0$, and R_0 is so small that can be ignored. Bringing in $U_{NO} = 0$, adding and subtracting the two-equation in (1) from each other yields (2)

$$\begin{cases} U_{si} = \frac{L}{2}\frac{d(i_{li} - i_{ui})}{dt} + \frac{1}{2}(U_{li} - U_{ui}); (i = a, b, c) \\ U_{dc} = L\frac{d(i_{li} + i_{ui})}{dt} + (U_{li} + U_{ui}); (i = a, b, c) \end{cases} \tag{2}$$

As shown in Figure 3, i_s is the AC side source current, i_c is the circulating current, i_u is the upper arm current, i_l is the lower arm current, and i_{dc} is the DC bus current, the relationship between them is as follows [43].

$$\begin{cases} i_s = i_l - i_u \\ i_c = \frac{i_u + i_l}{2} \end{cases} \tag{3}$$

$$\begin{cases} i_u = -\frac{i_s}{2} + i_c \\ i_l = \frac{i_s}{2} + i_c \end{cases} \tag{4}$$

$$i_{dc} = \sum_{i=a,b,c} \frac{(i_{ui} + i_{li})}{2} \tag{5}$$

As for the instantaneous power, the rectifier side was analyzed, and the results were as follows [43]

$$\begin{cases} P_{au} = u_{au} i_{au} \\ P_{al} = u_{al} i_{al} \end{cases} \tag{6}$$

Ignoring the harmonic components yields the bridge arm's current and voltage expression [43].

$$\begin{cases} U_{au} = \frac{1}{2} u_{dc} - \sqrt{2} U_a \sin \omega t \\ U_{al} = \frac{1}{2} u_{dc} + \sqrt{2} U_a \sin \omega t \end{cases} \tag{7}$$

$$\begin{cases} I_{au} = -\frac{1}{2} \sqrt{2} I_a \sin(\omega t + \varphi) - \frac{1}{3} I_{dc} \\ I_{al} = \frac{1}{2} \sqrt{2} I_a \sin(\omega t + \varphi) - \frac{1}{3} I_{dc} \end{cases} \tag{8}$$

The expression for instantaneous power is obtained by bringing (7) and (8) into (6).

$$\begin{cases} P_{au} = A - B(\omega t) + C(2\omega t) \\ P_{al} = A + B(\omega t) + C(2\omega t) \end{cases} \tag{9}$$

$$\begin{cases} A = \frac{1}{2} U_a I_a \cos \varphi - \frac{1}{6} U_{dc} I_{dc} \\ B(\omega t) = \frac{\sqrt{2}}{4} U_{dc} I_a \sin(\omega t + \varphi) - \frac{\sqrt{2}}{3} U_a I_{dc} \sin \omega t \\ C(2\omega t) = -\frac{1}{2} U_a I_a \cos(2\omega t + \varphi) \end{cases} \tag{10}$$

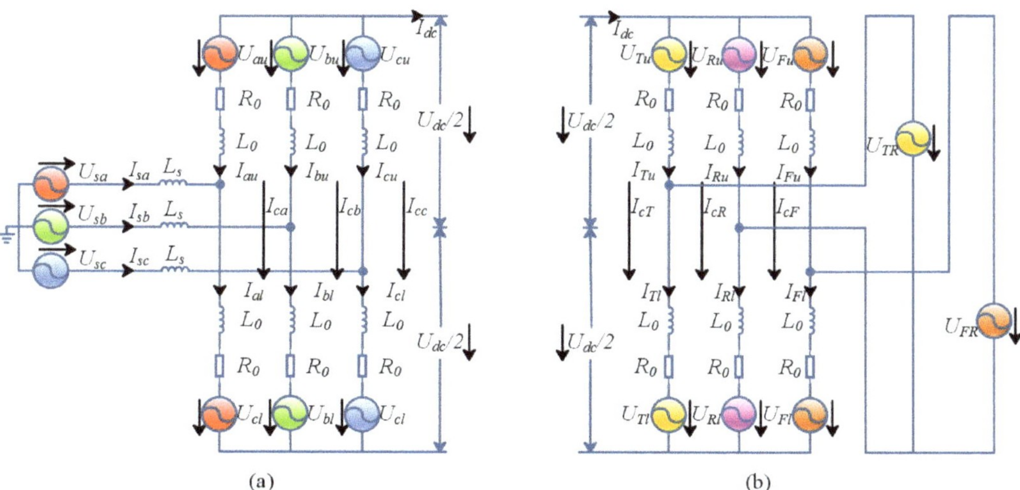

Figure 3. The equivalent circuit of the proposed topology: (**a**) Rectifier side equivalent circuit of the proposed topology; (**b**) Inverter side equivalent circuit of the proposed topology.

2.2.2. The Inverter Side

The inverter side uses a bridge leg to replace the AT self-coupling transformer. As for voltage of the inverter side, the voltage between the T line and the R line as well as the voltage between the F line and the R line are two 27.5 kV single-phase AC voltages of equal amplitude and 180° difference in phase, i.e., $V_{TR} = 27.5\angle 0^\circ$ and $V_{FR} = 27.5\angle 180^\circ$.

As for the current of the inverter side, the current in the upper and lower bridge arms is a superposition of the AC side current and the circulating current, while the circulating current is a superposition of the fundamental frequency component, the double frequency component, and the DC component, which is come from the DC bus current. In addition, regarding the output side current, When the train is within one AT interval before or after the substation, there will be current on the R line. In contrast, when the train is not within one AT interval before or after the substation, there will be no current on the R line, and the T/F line current's amplitude is a constant value.

As for the instantaneous power, there will be an equal distribution of the direct-current across the three bridge legs, ignoring the harmonic component, and the expressions for the voltages are as below [43].

$$\begin{cases} U_{Tu} = \frac{1}{2}u_{dc} - \sqrt{2}U_T \sin \omega t \\ U_{Tl} = \frac{1}{2}u_{dc} + \sqrt{2}U_T \sin \omega t \end{cases} \tag{11}$$

And the current expressions are as below [43].

$$\begin{cases} I_{Tu} = \frac{1}{2}\sqrt{2}I_T \sin(\omega t + \varphi) + \frac{1}{3}I_{dc} \\ I_{Tl} = -\frac{1}{2}\sqrt{2}I_T \sin(\omega t + \varphi) + \frac{1}{3}I_{dc} \end{cases} \tag{12}$$

And the expression for the instantaneous power in the bridge legs are as follows.

$$\begin{cases} P_{Tu} = A - B(\omega t) + c(2\omega t) \\ P_{Tl} = A + B(\omega t) + c(2\omega t) \end{cases} \tag{13}$$

$$\begin{cases} A = -\frac{1}{2}U_T I_T \cos\varphi + \frac{1}{6}U_{dc}I_{dc} \\ B(\omega t) = -\frac{\sqrt{2}}{4}U_{dc}I_T \sin(\omega t + \varphi) + \frac{\sqrt{2}}{3}U_T I_{dc} \sin \omega t \\ C(2\omega t) = \frac{1}{2}U_T I_T \cos(2\omega t + \varphi) \end{cases} \tag{14}$$

3. Control System for the Proposed Topology

According to the analysis of the operation principle above, the control objectives below need to be met as a co-phase power-supply device, For the rectifier side, the control objectives of no negative sequence current, reactive current, and harmonic component on the rectifier side should be met, and at the same time, the DC bus voltage should be stabilized near the reference value, so as to avoid power quality problems caused by the traction network and to maintain the stability of the DC side. For the inverter side, it is necessary to achieve the control objectives of controllable amplitude and phase of the output voltage to ensure that the entire traction network substations have the voltage of the same phase and amplitude. In addition, because of the newly added bridge leg, it is necessary to ensure that the voltage between the contact network and the rail as well as the voltage between the positive feeder and the rail are 27.5 kV single-phase AC voltages of equal amplitude and 180° in phase.

3.1. The Control System for the Rectifier Side

As for the rectifier side, according to the mathematical model in Section 2 , add and subtract Equation (1) to each other, the following equation can be obtained [43].

$$\begin{cases} \frac{L}{2}\frac{dI_s}{dt} = V_{sa} - \frac{-V_u + V_l}{2} - \frac{R}{2}I_s \\ L\frac{dI_c}{dt} = \frac{V_d}{2} - \frac{V_u + V_l}{2} - RI_c \end{cases} \tag{15}$$

Defining $\frac{-V_u+V_l}{2}$ as V_C and $\frac{V_u+V_l}{2}$ as V_S, we can get the following equation [43].

$$\begin{cases} V_C = V_{sa} - \frac{L}{2}\frac{dI_s}{dt} - \frac{R}{2}I_s \\ V_S = \frac{V_d}{2} - L\frac{dI_c}{dt} - RI_c \end{cases} \tag{16}$$

And we can get V_u and V_l if there is V_C and V_S [43].

$$\begin{cases} V_u = V_C - V_S \\ V_l = V_C + V_S \end{cases} \tag{17}$$

By using the PI controller, according to the following expression, we can get V_C and V_S.

$$\begin{cases} V_C(s) = V_{sa}(s) - \frac{R}{2}I_s(s) - [i_s^*(s) - i_s(s)](k_{p1} + \frac{k_{i1}}{s}) \\ V_S(s) = \frac{V_d(s)}{2} - RI_c(s) - [i_c^*(s) - i_c(s)](k_{p1} + \frac{k_{i1}}{s}) \end{cases} \tag{18}$$

From this, the control system on the rectifier side can be obtained, as shown in Figure 4. And the modulation method is selected to be CPS-SPWM [44].

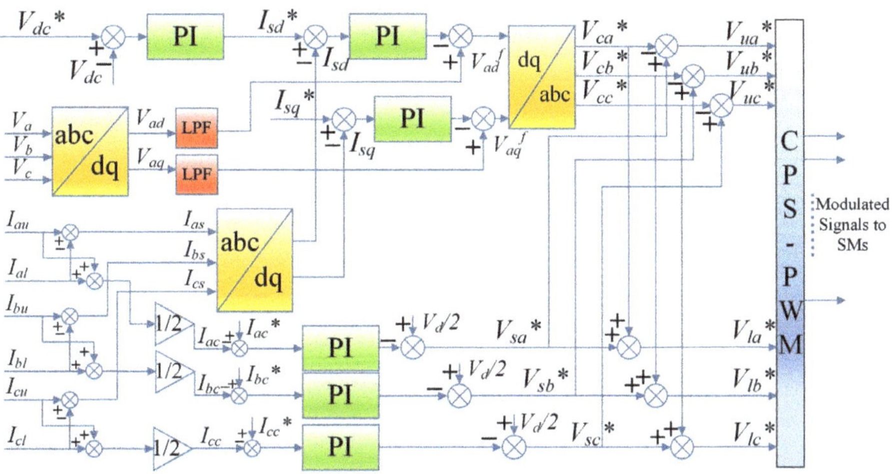

Figure 4. The control system for the rectifier side of the proposed topology.

3.2. The Control System for the Inverter Side

As for the inverter side, based on the mathematical model built for the inverter side, the following expression can be obtained [42].

$$\begin{cases} V_{Tu} = \frac{v_{dc}}{2} - L\frac{di_{Tu}}{dt} - U_T \\ V_{Tl} = \frac{v_{dc}}{2} - L\frac{di_{Tl}}{dt} + U_T \end{cases} \tag{19}$$

Since $i_{Tu} = \frac{i_s}{2} + i_c$ [43] and $i_{Tl} = -\frac{i_s}{2} + i_c$ [43], the equation can be transformed into:

$$\begin{cases} V_{Tu} = \frac{v_{dc}}{2} - U_T - L\frac{di_c}{dt} - \frac{L}{2}\frac{di_s}{dt} \\ V_{Tl} = \frac{v_{dc}}{2} + U_T - L\frac{di_c}{dt} + \frac{L}{2}\frac{di_s}{dt} \end{cases} \tag{20}$$

Taking the T line as an example, with the PI controller, we can get the upper and lower bridge arm reference voltages by the following expression.

$$\begin{cases} V_{Tu}(s) = \frac{v_{dc}(s)}{2} - U_T(s) - \frac{L}{2}\frac{di_s(s)}{dt} - [i_c^*(s) - i_c(s)](k_{p1} + \frac{k_{i1}}{s}) \\ V_{Tl}(s) = \frac{v_{dc}(s)}{2} + U_T(s) + \frac{L}{2}\frac{di_s(s)}{dt} - [i_c^*(s) - i_c(s)](k_{p1} + \frac{k_{i1}}{s}) \end{cases} \tag{21}$$

From this, the control system on the inverter side can be get.

As shown in Figure 5, since the output voltages are two single-phase AC voltages of equal amplitude and 180° difference in phase, if we want to convert them to the dq coordinate system for control, we need to converter them to the ABC coordinate system first. By converting V_{TR} to the A-phase and V_{FR} to the sum of the B phase and C phase, we can convert them to the ABC coordinate system and then control them.

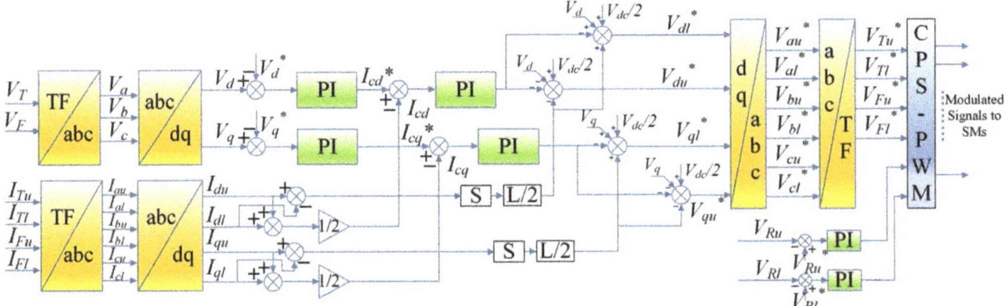

Figure 5. The control system for the inverter side of the proposed topology.

4. The Verification of the Proposed Topology by Simulation

In this section, the functionality of the proposed topology is verified by simulation. The simulation builds a realistic model of AC-DC-AC substation. The grid voltage of the system is 220 kV. Then connect the grid to a step-down transformer to transfer the voltage to the rated phase voltage of 27.5 kV on the AC grid side. Then connect the secondary side of the step-down transformer to the MMC-based back-to back structure. The rated voltage of the DC bus of the back-to-back structure is 80 kV. In addition, the rated phase voltage on the output side is 27.5 kV.

As for the functions that the device needs to meet, as analyzed above, the proposed co-phase power-supply device for the AT power-supply mode needs to meet the following functions: the rectifier side should eliminate negative sequence current, reactive current, and harmonics to ensure that it does not bring power quality problems for the power grid, and the DC bus voltage needs to be stabilized near the reference value. As for the inverter side, it is necessary to ensure that the output voltage's amplitude and phase can be controlled at the control reference value.

It can be seen from Figure 6a that the voltage and current on the rectifier side basically have no phase difference, i.e., their power factor is basically 1, which means that there is no reactive current component. The three-phase voltage and current waveforms have stable amplitude and good waveforms without negative sequence and harmonic components, meeting the functional requirements of the co-phase power-supply scheme on the rectifier side.

It can be seen from Figure 6b that the DC bus voltage can rise and stabilize at the reference value with minimal fluctuations, meeting the requirement of the DC bus voltage.

It can be seen from Figure 6c,d that the voltage's amplitude and phase on the output side can be stabilized at the reference value. And to verify that when the current changes suddenly, the voltage remains stable and basically unchanged. The load mutation is set at 1.995 s, and the current changes suddenly with it. It can be seen from Figure 6c,d that the voltage remains stable and basically unchanged. Therefore, the inverter side can meet the functional requirements of the co-phase power-supply scheme.

In summary, the simulation verified that the entire system can meet the functional requirements of the co-phase power-supply scheme.

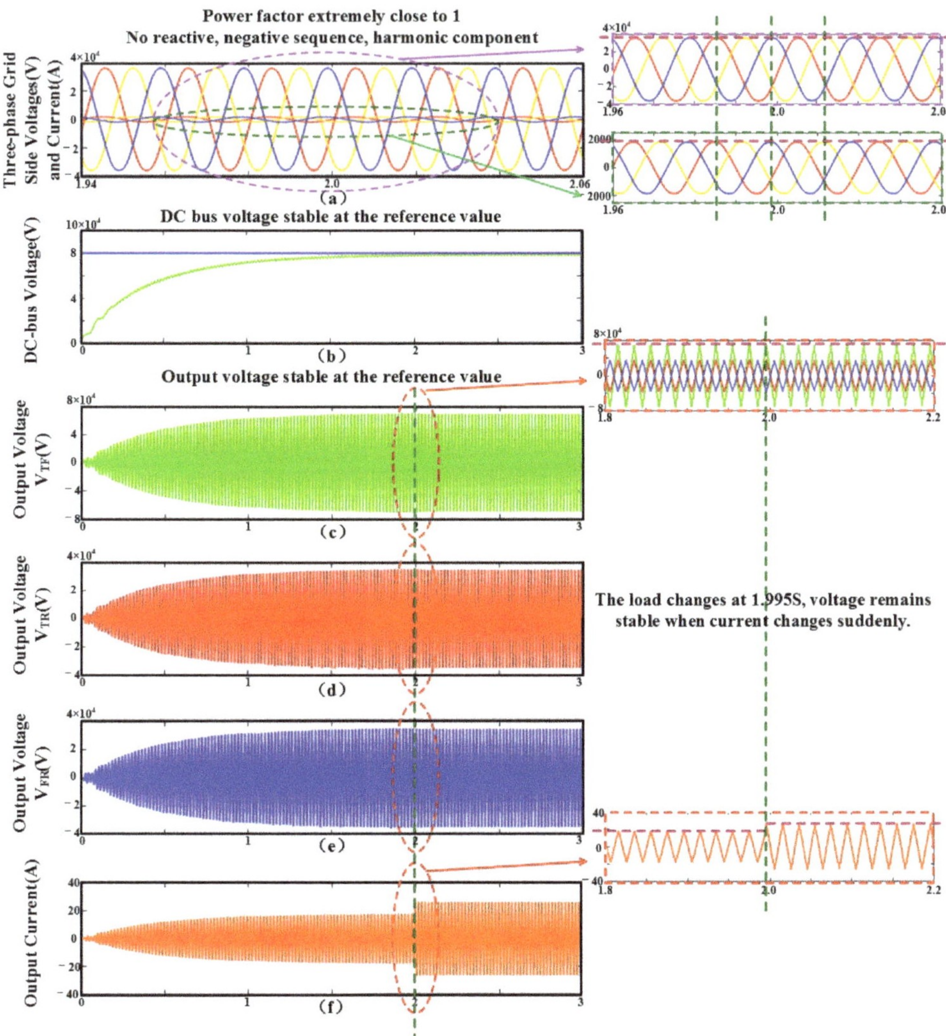

Figure 6. The simulation results of the proposed topology: (**a**) Voltage and current on the three-phase input side of the proposed topology; (**b**) DC-bus voltage of the proposed topology; (**c**) Voltage between T line and F line on the output side of the proposed topology; (**d**) Voltage between T line and R line on the output side of the proposed topology; (**e**) Voltage between F line and R line on the output side of the proposed topology; (**f**) Current on the output side of the proposed topology.

5. The Advantages Analysis of the Proposed Topology

After verifying that the proposed co-phase power-supply device topology can meet the functional requirements of the co-phase power-supply scheme, the advantages of the proposed topology needs to be analyzed and discussed.

Then the two schemes need to be compared in the same scenario. It is assumed that the scenarios of the two schemes are both AC-DC-AC substations with a rated power of 40 MW. The step-down transformer transfers the 220 kV grid voltage to the rated grid-side phase voltage of 27.5 kV, the system rated DC bus voltage is 80 kV, and the system output side rated phase voltage is 27.5 kV. As for the MMC structure, there are both three bridge legs on the rectifier side and the inverter side. Each bridge leg is divided into the upper and lower bridge arms. There are 40 sub-modules on each bridge arm. The rated capacitor voltage of the sub-module is 2 kV, and the switching device is selected to be Infineon-FZ1500R33HE3.

The following section will specifically analyze the advantages and disadvantages of the two schemes in different aspects under this scenario.

5.1. The Comparative Analysis of Current Stress of the Switching Devices

From the perspective of the current stresses borne by switching devices in the topology, qualitative analysis shows that since the inverter side adds a bridge leg, the DC component flowing from the DC bus into the inverter side bridge leg is changed from the original two equal parts to three equal parts, so that the DC component in the bridge leg decreases. As for the AC output side, when the locomotive is in one AT interval before or after the substation, there will be a current inflow on the newly added bridge leg, because the sum of the TRF line current is zero, T line current outflow, and R and F line current inflow, so the current's amplitude of the bridge leg connected to the F line is reduced. In addition, when the locomotive is not located in one AT interval before or after the substation, there is no difference in the AC current with other topology. Therefore, in a comprehensive view, when the power is constant, the current stress of the switching devices in the inverter side bridge leg of the topology proposed in this paper is smaller.

At the same time, quantitative analysis shows that the current stress value of the inverter side switching device is indeed smaller. The calculation is based on the system rated at 40 MW, and the calculation is according to the maximum current may appear in the bridge leg. For example, on the rectifier side, since $S = 3$ UI, $U_s = 27.5$ kV, the current $I_s = 484.85$ A can be obtained, and because $S = U_{dc}I_{dc}$, the reference value of U_{dc} is 80 kV, $I_{dc} = 500$ A can be obtained, thus the maximum current in the bridge arm is $\frac{\sqrt{2}}{2}I_s + \frac{I_{dc}}{3} = 509.51$ A. However, on the inverter side, the output voltages are two single-phase voltages, thus $S = 2$ UI, $U_o = 27.5$ kV, the current $I_o = 727.27$ A can be obtained, so the maximum current in the bridge arm is $\frac{\sqrt{2}}{2}I_o + \frac{I_{dc}}{3} = 680.93$ A. Therefore, in summary, the current maximum value of the switching devices on the rectifier and inverter sides is, respectively 509.51 A and 680.93 A.

It is similarly calculated that the maximum currents in the bridge arm of the rectifier-inverter side of the already proposed topology are respectively 509.51 A and 764.24 A.

It can be seen that the current stress of the inverter side switching devices of the topology proposed in this paper is significantly reduced.

5.2. The Comparative Analysis of the Power Loss of the Whole System

From the perspective of power loss, the loss of the device can be calculated theoretically. The power loss of a switching device consists of two parts: the conduction loss and the turn-on turn-off loss. The expression is as below [45,46].

$$\begin{cases} P_{tot} = P_{Ttot} + P_{Dtot} \\ P_{Ttot} = P_{Tcon} + P_{Ton} + P_{Toff} \\ P_{Dtot} = P_{Dcon} + P_{Doff} \end{cases} \tag{22}$$

The P_{Ttot} and P_{Dtot} means the total power loss of the IGBT and diode separately, and the P_{Tcon} P_{Ton} P_{Toff} are the conduction loss and turn-on turn-off losses of the IBGT in a fundamental output period, and the P_{Dcon} P_{Doff} are the conduction loss and turn-off loss of the diode. All these losses make up the total loss of a switching device. These losses can be calculated individually by these equations below [45,46].

$$\begin{cases} P_{Tcon} = \frac{\omega}{2\pi} \int_{T_s}^{T_s+2\pi/\omega} i_C(\tau) v_{CE}(i_C(\tau)) d\tau \\ P_{Dcon} = \frac{\omega}{2\pi} \int_{T_\alpha}^{T_\alpha+2\pi/\omega} i_F(\tau) v_F(i_F(\tau)) d\tau \\ P_{Ton} = \frac{\omega}{2\pi} \sum_{\alpha=1}^{N_\alpha} \{ \frac{v_{CE,off}(t_\alpha)}{v_{CE,ref}} E_{on}(i_C(t_\alpha)) \} \\ P_{Toff} = \frac{\omega}{2\pi} \sum_{\beta=1}^{N_\beta} \{ \frac{v_{CE,off}(t_\beta)}{v_{CE,ref}} E_{off}(i_C(t_\beta)) \} \\ P_{Doff} = \frac{\omega}{2\pi} \sum_{\gamma=1}^{N_\gamma} \{ \frac{v_{F,off}(t_\gamma)}{v_{CE,ref}} E_{rec}(i_F(t_\gamma)) \} \end{cases} \quad (23)$$

In these equations, $i_C(\tau)$ and $i_F(\tau)$ can be obtained based on the rated values, and V_{CE} and V_F can be obtained based on the corresponding current values and the output characteristic curves of the switching devices provided in [47]. E_{on} E_{off} E_{rec} can be obtained from the corresponding current and the device loss curves provided in [47]. $V_{CE,off}$ $V_{F,off}$ can also be obtained from the corresponding current and the output characteristic curve. Set the rated power of the system to 40 MW. From this, it can be calculated that the power loss of all switching devices in the topology proposed in this paper in one fundamental period is 613.67 kW, while the power loss of the switching devices in the topology has been commonly proposed is 502.28 kW.

At the same time, the Infineon Online Power Simulation Tool (IPOSIM) (produced by Infineon in Munich, Germany and provided on the official website) can be used to approximate the power loss of the switching device after setting the value of the DC link voltage, RMS current, frequency, and the switching frequency on it. The Infineon Online Power Simulation Tool (IPOSIM) (produced by Infineon in Munich, Germany and provided on the official website) calculation result of the power loss of the topology proposed in this paper is 625.44 kW, while the result of the topology commonly proposed is 536.80 kW. However, from the perspective of switching devices, the power loss proposed in this paper is relatively high, from the perspective of the entire system of the scheme, the scheme proposed in this paper eliminates an auto-transformer. Take an auto-transformer with a nameplate capacity of 31.5 MVA as an example. Its actual electromagnetic capacity is only 13.3 MVA, saving it can save about $30,000 and save about 0.5% of energy loss, which is 66.5 kW.

So as for the loss of the whole system, we need to add the loss of the transformer based on the loss of switching devices. Because of the high efficiency of high-voltage large-capacity transformers, we set the efficiency of high-voltage large-capacity transformers at 99.5%, and set the rated system power to 40 MW, then the efficiency of the whole system (i.e., considering the previous step-down transformer and the energy losses of the switching devices) is 97.96%, and the efficiency of the whole system of the commonly proposed scheme (i.e., considering the previous step-down transformer, an auto-transformer and the energy losses of the switching devices) is 97.71%.

Although the numerical difference in efficiency between the two systems is not massive, considering that the railway system is a large-capacity system, for example, with a rated capacity at 40 MW, a difference of 0.25% would result in an energy difference of 100 kWh per hour, so the scheme proposed in this paper can save 100 kWh per hour compared to the commonly proposed scheme, which has excellent economic benefits in the long term.

6. Conclusions

Against the background that there is no previous co-phase power-supply scheme especially designed for the AT power-supply mode, this article proposes a co-phase power-supply scheme for the AT power-supply mode traction power-supply system that has become the main power-supply mode for the high-speed railways. The feasibility and functionality of the scheme are verified by establishing mathematical models, creating control systems, and simulating. It proves that the scheme proposed in this paper can meet the needs of the co-phase power-supply system. In addition, on this basis, a comparative analysis with the system that has been commonly proposed has verified that this scheme has obvious advantages in the current stress of the switching device and the power loss of the entire system. Therefore, this solution is a more suitable co-phase power-supply scheme for the AT power-supply mode traction power-supply system.

This scheme can be used in the AT traction power-supply system of high-speed railways to be installed in substations to eliminate neutral section and at the same time eliminate power quality problems such as negative sequence, reactive power, and harmonics injected into the public grid. It ensures the safe and economical operation of the traction network and the power grid and promotes the increase in the speed of high-speed railways. The application value of this project can be popularized in the traction power-supply system of high-speed electrified railways around the world, and remove existing obstacles for the development of electrified railways worldwide.

Author Contributions: Conceptualization, investigation and data curation, S.W.; methodology, S.W. and M.W. and Y.W.; software and validation, formal analysis, S.W.; writing—original draft preparation, S.W.; writing—review and editing, M.W. and Y.W.; supervision and project administration, M.W. All authors have read and agreed to the published version of the manuscript.

Funding: This research was funded by Fundamental Research Funds for the Central Universities with grant number 2018YJS150.

Data Availability Statement: Data sharing not applicable. No new data were created or analyzed in this study. Data sharing is not applicable to this article.

Conflicts of Interest: The authors declare no conflict of interest.

Abbreviations

The following abbreviations are used in this manuscript:

AT	Auto-transformer
MMC	Modular multilevel converter
SVC	Static var compensator
STATCOM	Static synchronous compensator
APF	Active power filter
APC	Active power compensator
RPC	Railway power conditioner
HPQC	Hybrid power quality conditioner
HEMPQC	Hybrid electrical magnetic power quality compensator
CHB	Cascaded H-Bridge
V_{TR}	Voltage of the contact network to the rails (kV)
V_{FR}	Voltage of the positive feeder to the rails (kV)
U_{si}	The voltage of phase i on the AC grid side connected to the back-to-back rectifier side based on MMC (i = a, b, c) (kV)
L	The integrated equivalent inductance in the single-phase equivalent circuit of MMC.(mH)
i_u	Upper arm current of MMC back-to-back structure (A)
i_l	Lower arm current of MMC back-to-back structure (A)
U_{ui}	Sum of the voltages of the i-phase upper bridge arm sub-modules of the MMC structure (kV)
U_{li}	Sum of the voltages of the i-phase lower bridge arm sub-modules of the MMC structure (kV)
U_{dc}	Voltage of the DC bus in the MMC back-to-back structure (kV)
i_s	The source current of AC grid side connected to the back-to-back rectifier side based on MMC (A)
i_c	The circulating current of the MMC-based back-to-back structure (A)
i_{dc}	Current of the DC bus in the MMC back-to-back structure (A)
P_{au}	Instantaneous power of the A-phase upper bridge arm (kW)
LD	linear dichroism
P_{al}	Instantaneous power of the A-phase lower bridge arm(kW)
V_C	Defined as $\frac{-V_u+V_l}{2}$ (kV)
V_S	Defined as $\frac{V_u+V_l}{2}$ (kV)
P_{tot}	Total power loss of the switching device in a fundamental output period (mJ)
P_{Ttot}	Total power loss of the IGBT in a fundamental output period (mJ)
P_{Dtot}	Total power loss of the diode in a fundamental output period (mJ)
P_{Tcon}	Conduction loss of the IGBT in a fundamental output period (mJ)
P_{Ton}	Turn-on loss of the IGBT in a fundamental output period (mJ)
P_{Toff}	Turn-off loss of the IGBT in a fundamental output period (mJ)
P_{Dcon}	Conduction loss of the diode in a fundamental output period (mJ)
P_{Doff}	Turn-off loss of the diode in a fundamental output period (mJ)

References

1. Zhang, Z.; Zheng, T.Q.; Li, K.; Hao, R.; You, X.; Zhang, Z.; Yang, J. Smart Electric Neutral Section Executer Embedded With Automatic Pantograph Location Technique Based on Voltage and Current Signals. *IEEE Trans. Transp. Electrif.* **2020**, *6*, 1355–1367. [CrossRef]

2. Wang, S.; Zhang, L.; Yu, G.; Dong, E.; Zou, J.; Cong, Y.; Tian, Y. Hybrid phase-controlled circuit breaker with switch system used in the railway auto-passing neutral section with an electric load. *CSEE J. Power Energy Syst.* **2019**, *5*, 545–552.
3. Yuan, J.X.; Zhou, N.; Xiao, F.R.; Min, Y.Z. Study on the Control Method and the Uninterrupted Phase-Separation Passing System with Voltage Compensation Function. *Trans. China Electrotech. Soc.* **2020**, 1–12.
4. Qiu, W.J.; Hu, Q. Research on Technologies for electronics switch automatic passing of neutral section on heavy haul railway. *Electr. Railw.* **2019**, *6*, 17–21.
5. Canales, J.M.; Aizpuru, I.; Iraola, U.; Barrena, J.A.; Barrenetxea, M. Medium-Voltage AC Static Switch Solution to Feed Neutral Section in a High-Speed Railway System. *Energies* **2018**, *11*, 27–40. [CrossRef]
6. Han, Z.Q.; Shen, R.; Zhou, Y.Z. Simulation of Flexible Automatic Neutral Section Passing System Based on RTDS. *Electr. Drive Locomot.* **2019**, *3*, 6–11.
7. Paul, P.A.; Anju, U.D.; Anoop, M.P.; Pallan, M.R.; Roshna, N.K.; Sunny, A. Effects of two phase traction loading on a three phase power transformer. In Proceedings of the 2014 Annual International Conference on Emerging Research Areas: Magnetics, Machines and Drives (AICERA/iCMMD), Kottayam, India, 24–26 July 2014; pp. 1–5.
8. Joseph, V.P.; Thomas, J. Power quality improvement of AC railway traction using railway static power conditioner a comparative study. In Proceedings of the 2014 International Conference on Power Signals Control and Computations (EPSCICON), Thrissur, India, 6–11 January 2014; pp. 1–6.
9. Gunavardhini, N.; Chandrasekaran, M.; Sharmeela, C.; Manohar, K. A case study on Power Quality issues in the Indian Railway traction sub-station. In Proceedings of the 2013 7th International Conference on Intelligent Systems and Control (ISCO), Coimbatore, India, 4–5 January 2013; pp. 7–12.
10. Hu, H.; He, Z.; Gao, S. Passive Filter Design for China High-Speed Railway With Considering Harmonic Resonance and Characteristic Harmonics. *IEEE Trans. Power Deliv.* **2015**, *1*, 505–514. [CrossRef]
11. Hu, L.; Morrison, R.E. The stability and power flow limits of an electrified railway system with voltage controlled SVCs. In Proceedings of the 1994 Fifth International Conference on Power Electronics and Variable-Speed Drives, London, UK, 26-28 October 1994; pp. 562–567.
12. Ledwich, G.; George, T.A. Using phasors to analyze power system negative phase sequence voltages caused by unbalanced loads. *IEEE Trans. Power Syst.* **1994**, *3*, 1226–1232. [CrossRef]
13. Lee, S.H.; Bae, I.S.; Jung, C.H.; Kim, J.-O. A study on system stability improvement of distribution system with high speed electric railway using STATCOM. In Proceedings of the 2003 IEEE PES Transmission and Distribution Conference and Exposition (IEEE Cat. No.03CH37495), Dallas, TX, USA, 7–12 September 2003; Volume 1, pp. 61–67.
14. Wu, L.R.; Mingli, W. Single-phase cascaded H-bridge multi-level active power filter based on direct current control in AC electric railway application. *IET Power Electron.* **2017**, *6*, 637–645. [CrossRef]
15. Shen, W.T. *Study on Experiment and Modification of a Combined Co-Phase Supply System with a Single-Phase and Three-Phase Modular*; Southwest Jiaotong University: Chengdu, China, 2016.
16. Shu, Z.; Xie, S.; Lu, K.; Zhao, Y.; Nan, X.; Qiu, D.; Zhou, F.; Gao, S.; Li, Q. Digital Detection, Control, and Distribution System for Co-Phase Traction Power Supply Application. *IEEE Trans. Ind. Electron.* **2013**, *5*, 1831–1839. [CrossRef]
17. Dai, N.; Wong, M.; Lao, K.; Wong, C. Modeling and control of a railway power conditioner in co-phase traction power system under partial compensation. *IET Power Electron.* **2014**, *5*, 1044–1054. [CrossRef]
18. Tanta, M.; Pinto, J.G.; Monteiro, V.; Martins, A.P.; Carvalho, A.S.; Afonso, J.L. Topologies and Operation Modes of Rail Power Conditioners in AC Traction Grids: Review and Comprehensive Comparison. *Energies* **2020**, *13*, 21–51. [CrossRef]
19. Dai, N.Y.; Lao, K.W.; Wong, M.C.; Wong, C.K. Hybrid power quality conditioner for co-phase power supply system in electrified railway. *IET Power Electron.* **2012**, *7*, 1084–1094. [CrossRef]
20. Lao, K.; Wong, M.; Dai, N.Y.; Wong, C.; Lam, C. Analysis of DC-Link Operation Voltage of a Hybrid Railway Power Quality Conditioner and Its PQ Compensation Capability in High-Speed Cophase Traction Power Supply. *IEEE Trans. Power Electron.* **2016**, *2*, 1643–1656. [CrossRef]
21. Lao, K.; Wong, M.; Dai, N.; Lam, C.; Wong, C.; Lam, L.W. Analysis in the Effect of Co-phase Traction Railway HPQC Coupled Impedance on Its Compensation Capability and Impedance-Mapping Design Technique Based on Required Compensation Capability for Reduction in Operation Voltage. *IEEE Trans. Power Electron.* **2017**, *4*, 2631–2646. [CrossRef]
22. Lao, K.; Wong, M.; Dai, N.; Lam, C.; Wang, L.; Wong, C.K. Analysis of the Effects of Operation Voltage Range in Flexible DC Control on Railway HPQC Compensation Capability in High-Speed Co-phase Railway Power. *IEEE Trans. Power Electron.* **2018**, *2*, 1760–1774. [CrossRef]
23. Chen, B.; Zhang, C.; Tian, C.; Wang, J.; Yuan, J. A Hybrid Electrical Magnetic Power Quality Compensation System With Minimum Active Compensation Capacity for V/V Cophase Railway Power Supply System. *IEEE Trans. Power Electron.* **2016**, *6*, 4159–4170. [CrossRef]
24. Xiong, C.; Ma, J.; Wu, X.; Wang, S.; Feng, X. Virtual co-phase traction power supply system adopting the cascaded H-bridge multilevel converter. *Electron. Lett.* **2016**, *10*, 865–866. [CrossRef]
25. Li, L.; Wu, M.; Wu, S.; Li, J.; Song, K. A Three-Phase to Single-Phase AC-DC-AC Topology Based on Multi-Converter in AC Electric Railway Application. *IEEE Access* **2019**, *7*, 111539–111558. [CrossRef]
26. He, X.; Peng, J.; Han, P.; Liu, Z.; Gao, S.; Wang, P. A Novel Advanced Traction Power Supply System Based on Modular Multilevel Converter. *IEEE Access* **2019**, *7*, 165018–165028. [CrossRef]

27. Ma, Q.A. *Study on Some Problems of Traction Power Supply System in High-Speed Railway*; Southwest Jiaotong University: Chengdu, China, 2013.
28. Takeda, M.; Murakami, S.; Iizuka, A.; Hirakawa, M.; Kishida, M.; Hase, S.; Mochinaga, H. Development of an SVG Series for Voltage Control Over Three-phase Unbalance Caused by Railway Load. In Proceedings of the IPEC Yokohama 1995 Conference Records, Yokohama, Japan, 3–7 April 1995; pp. 603–608.
29. Ando, M.; Mochinaga, Y.; Kato, T.; Yoshizawa, J.; Gomi, T.; Miyashita, T.; Funahashi, S.; Nishitoba, M.; Oozeki, S. Development of railway static power conditioner used at substation for Shinkansen. In Proceedings of the PCC-Osaka 2002 Conference Records, Osaka, Japan, 2–5 April 2002; pp. 1108–1111.
30. Kai, M.; Ohnishi, M.; Masui, T.; Noguchi, M.; Morishima, N.; Horita, Y. Single phase STATCOM for feeding system of Tokaido Shinkansen. In Proceedings of the IPEC-Sapporo 2010 Conference Records, Sapporo, Japan, 21–24 June 2010; pp. 2165–2170.
31. Xu, Z.J. *Study on AT Network Models*; Southwest Jiaotong University: Chengdu, China, 2015.
32. Wu, M.L.; Wu, L.R. Experimental Verification of Mathematical Model and Impedance Matching of the Scott Connected Balance Transformer. *J. China Railw. Soc.* **2007**, *2*, 39–44.
33. Wu, M.L.; Fan, Y.; Zheng, T.Q. Operational Performance of the Cross-Connected Traction Transformer (1)—Mathematical Model and Theoretical Analysis. *Trans. China Electrotech. Soc.* **2004**, *9*, 11–17.
34. Zhao, Y.L. *Secondary Midpoint Withdrawable Single-Phase Traction Transformer*; Beijing Jiaotong University: Beijing, China, 2014.
35. Wu, L.R.; Guo, Y.P.; Yi, Y.J.; Huang, X.Y.; Liu, Z.Q.; Wen, W.S.; Zhang, X.S.; Fan, Z.A.; Xu, G. Development of a single-phase traction transformer for the AT power supply. *Electr. Railw.* **1998**, *4*, 3–5.
36. Hu, J.X.; Zhou, F.Y. Status and Development of Neutral Section Passing Technology for Electrified Railway Trains. *Electr. Drive Locomot.* **2019**, *3*, 1–5.
37. Gao, S.W. Study on Control Strategy for Device of Electronic Switch Type for Passing of Phase Break. *Electr. Railw.* **2020**, *3*, 16–20.
38. Wang, S.H. *Study of Uninterruptible Power Automatic Passing through Neutral Section Device for High Speed EMU*; Beijing Jiaotong University: Beijing, China, 2012.
39. Tanta, M.; Monteiro, V.; Sousa, T.J.C.; Martins, A.P.; Carvalho, A.S.; Afonso, J.L. Power quality phenomena in electrified railways: Conventional and new trends in power quality improvement toward public power systems. In Proceedings of the 2018 International Young Engineers Forum (YEF-ECE), Costa da Caparica, Portugal, 4 May 2018; pp. 25–30.
40. Wu, L.R. *Power Quality Compensation Technology in Electric Railways Based on 27.5 kV Straight Hanging Cascaded APF*; Beijing Jiaotong University: Beijing, China, 2017.
41. Gazafrudi, S.M.M.; Langerudy, A.T.; Fuchs, E.F.; Al-Haddad, K. Power Quality Issues in Railway Electrification: A Comprehensive Perspective. *IEEE Trans. Ind. Electron.* **2015**, *5*, 3081–3090. [CrossRef]
42. Li, B.B. *Research on Modular Multilevel Converter and Its Control Schemes*; Harbin Institute of Technology: Harbin, China, 2012.
43. Sharifabadi, K.; Harnefors, L.; Nee, H.P.; Norrga, S.; Teodorescu, R. *Design, Control, and Application of Modular Multilevel Converters for HVDC Transmission Systems*; John Wiley & Sons Ltd: Chichester, West Sussex, UK, 2016; pp. 34–148.
44. Yang, X.F.; Sun, H.; Zheng, T.Q. Application of Double-modulation-wave Carrier Phase-shifted SPWM in Modular Multilevel Converters. *Electr. Drive* **2011**, *10*, 15–20.
45. Rohner, S.; Bernet, S.; Hiller, M.; Sommer, R. Modulation, Losses, and Semiconductor Requirements of Modular Multilevel Converters. *IEEE Trans. Ind. Electron.* **2010**, *8*, 2633–2642. [CrossRef]
46. Rodrigues, S.; Papadopoulos, A.; Kontos, E.; Todorcevic, T.; Bauer, P. Steady-State Loss Model of Half-Bridge Modular Multilevel Converters. *IEEE Trans. Ind. Appl.* **2016**, *3*, 2415–2425. [CrossRef]
47. Wiesenthal and C. Lübke. Technical Information, EUPEC IGBT Modules FZ1500R33HE3. 2016. Available online: https://www.infineon.com/cms/en/product/power/igbt/igbt-modules/fz1500r33he3/ (accessed on 4 January 2021).

Article

A Novel Co-Phase Power Supply System for Electrified Railway Based on V Type Connection Traction Transformer

Shaofeng Xie *, Yiming Zhang and Hui Wang

School of Electrical Engineering, Southwest Jiaotong University, Chengdu 611756, China; ymzhang@my.swjtu.edu.cn (Y.Z.); huiwang@my.swjtu.edu.cn (H.W.)
* Correspondence: sfxie@swjtu.cn

Abstract: Power quality and neutral section are two technical problems that hinder the development of electrified railway to high-speed and heavy railway. The co-phase power supply technology is one of the best ways to solve these two technical problems. At present, a V type connection traction transformer is widely used in a power frequency single-phase AC traction power supply system, especially in high-speed railway. In this paper, a new type of co-phase power supply system for electrified railway based on V type connection traction transformer is proposed. One single-phase winding in the V type connection traction transformer is used as main power supply channel, and three ports are used as compensation ports. Neutral section is no longer set with traction substation, and the train is continuously powered through. The independent single-phase Static Var Generators (SVGs) are used to compensate the three-phase imbalance caused by single-phase traction load. When necessary, the power factor can be improved at the same time. The principle, structure, control strategy, and capacity configuration of the technical scheme are analyzed in this paper, and the effectiveness of the scheme is verified by using the measured data of electrified railway. The advantage of this scheme lies in the universal applicability of the V type connection traction transformer, and the flexibility of the SVG device.

Keywords: co-phase power supply technology; power quality; V type connection traction transformer; comprehensive compensation; Static Var Generator (SVG)

Citation: Xie, S.; Zhang, Y.; Wang, H. A Novel Co-Phase Power Supply System for Electrified Railway Based on V Type Connection Traction Transformer. *Energies* **2021**, *14*, 1214. https://doi.org/10.3390/en14041214

Academic Editors: Andrea Mariscotti and Leonardo Sandrolini

Received: 12 January 2021
Accepted: 19 February 2021
Published: 23 February 2021

Publisher's Note: MDPI stays neutral with regard to jurisdictional claims in published maps and institutional affiliations.

1. Introduction

As the backbone of modern comprehensive transportation system and one of the main modes of transportation, railway plays an important role in the process of social and economic development in China [1,2]. According to the railway statistics bulletin of 2019 issued by the State Railway Administration of China, by the end of 2019, business mileage has reached 139,000 km, and the railway electrification rate reached 71.9% [3].

With the development of electrified railways, power quality has always been a research hotspot, because of the characteristics of the traction power supply system [4]. A power frequency single phase AC system is widely used in China's electrified railways. The structure diagram of existing electrified railways is shown in Figure 1. As a single-phase power load essentially, traction load will cause power quality problems including mainly negative sequence in three-phase power system [5–7]. In order to reduce the influence of traction load on the three-phase unbalance at the Point of Common Coupling (PCC) of power system, the scheme of supplying power by sections in traction network and each section using three-phase power system in turn is usually adopted. The neutral section is set between two adjacent power supply sections [8–10]. However, it brings about a series of problems, such as the reduction of train speed, loss of traction force, frequent action of phase passing device, short service life, and low reliability, which seriously affect the safe and stable operation of the train. It shows that the neutral section is the weakest link in the AC traction power supply system of electric railway, and greatly restricts the traction power supply system Railway Development [11,12].

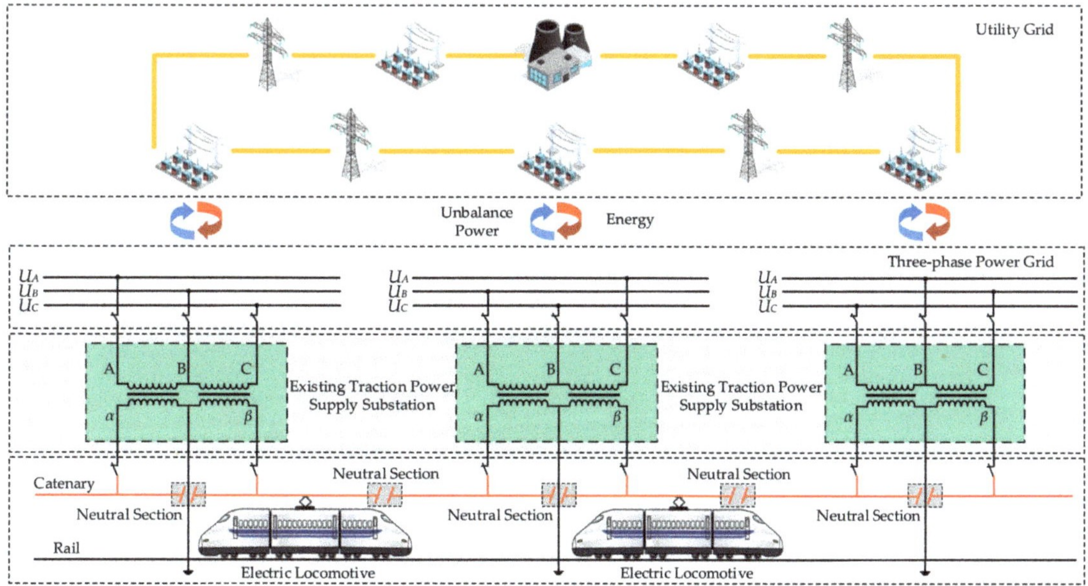

Figure 1. Structure diagram of China's existing traction power supply system.

With widespread use of AC-DC-AC electric locomotives using Pulse Width Modulation (PWM) technology, compared to the traditional AC-DC electric locomotives, the harmonic problem has been greatly improved, and the power factor is close to 1, which has overcome the shortcomings of traditional AC-DC electric locomotives [13–18]. However, at the same time, because of the huge increase in traction power of AC-DC-AC electric locomotives, the high-power single-phase traction load will cause more three-phase unbalance problems in the power system. Therefore, negative sequence will become the focus of power quality research of electrified railways.

In view of how to effectively solve the power quality problems including mainly negative sequence and neutral section, a lot of research work has been carried out, and different solutions have been put forward. The scheme to solve the negative sequence and electrical phase separation in Germany is adopting special railway power generation, transmission, and distribution system and setting a three-phase/single AC-DC-AC converter in the traction substation. At the same time, the neutral section is cancelled [19]. In Japan, Scott traction transformer and Railway Static Power Conditioner (RPC) are used in Shinkansen to compensate three-phase unbalance and voltage fluctuation so as to reduce the influence of traction load on three-phase power system [20–22]. It can solve the problem of power quality, and automatic passing neutral section technology is used to solve the problems caused by neutral section. Bilateral power supply technology is adopted in Russia to cancel neutral section between traction substations [19]. However, the traction power supply system is parallel with the power system transmission lines, and maybe there is current power in the traction power supply system, which needs to be restrained.

The idea of a co-phase power supply system to eliminate the neutral section and solve power quality problem is first proposed in [23]. The co-phase power supply refers to the power supply mode that a traction substation or multiple traction substations of a railway line supply the whole traction network with the same phase (line) voltage in the same three-phase power grid. If two adjacent traction substations adopt bilateral power supply, the co-phase power supply network formed by multiple traction substations on the line is called through co-phase power supply, referred to as through power supply [24,25].

Since the concept of co-phase power supply system was proposed, there have been some related studies on topological structure, compensation technology, and control strategy about co-phase power supply system. The co-phase power supply scheme based on three-phase-two-phase balanced connection traction transformer and Power Flow Controller (PFC) is proposed in [26]. Furthermore, more co-phase power supply schemes are proposed in [19,27,28]. Moreover, the schemes proposed in [19,26] have been applied in the Meishan traction substation of Chengdu Kunming railway, Shayu traction substation of Central South Passage of Shanxi Province, and Wenzhou City railway S1 line. The practice results show that the co-phase power supply technology can effectively solve the power quality problem with negative sequence as the main factor and cancel the neutral section at the traction substation.

In order to effectively reduce the PFC device capacity, a co-phase power supply scheme based on Hybrid Power Quality Conditioner (HPQC) was proposed in [29]. By using relatively inexpensive passive components to provide a part of reactive power, the active device capacity can be reduced. In order to obtain satisfactory compensation effects, the HPQC system often needs to consider more complex factors in design. A scheme to reduce the device capacity of RPC by using a C-type filter is proposed in [30], the C-type filter can compensate for a part of reactive power of the traction load while filtering out harmonics. Therefore, the device capacity of RPC can be reduced to a certain extent. In addition, based on the Modular Multilevel Converter (MMC) technology, the co-phase power supply schemes by using three-phase/single-phase AC-DC-AC converter is introduced in [31–33]. The outstanding advantage of these design schemes is that they can realize the balanced transformation between a three-phase system and single-phase system without generating power quality problems. However, it should be noted that in these schemes, all the power of the traction load will be transmitted through the power electronic converter, so the required device capacity will be equivalent to the capacity of a traditional traction transformer [34].

In recent years, Static Var Generator (SVG) has been widely used in various transmission and distribution systems due to its outstanding performance in compensating reactive power, etc. [35]. Compared with the traditional Static Var Compensator (SVC), SVG has the characteristics of faster dynamic response and lower harmonic content [36–38].

In view of this, a co-phase power supply scheme based on SVG for V type connection traction transformer which is widely used in high-speed railway is proposed in this paper. The feature of the scheme is that the traction load and compensation equipment share the same transformer, which can realize the comprehensive compensation of negative sequence and reactive power, effectively solve the problem of power quality, and cancel the neutral section at traction substation. Furthermore, the capacity of compensation equipment is minimized.

2. System Structure

Based on V type connection traction transformer and SVG, a novel co-phase power supply system for electrified railway is proposed in this paper. By means of unequal side V type connection traction transformer and setting single-phase SVG on multiple ports of the secondary side, the negative sequence and reactive power comprehensive compensation is carried out for three-phase unbalance and power factor, so that the power quality can meet the standard. For traction load, neutral section at the outlet of traction substation can be cancelled to realize co-phase power supply. The scheme of system is shown in Figure 2.

In Figure 2, the system mainly includes the Traction Compensation Transformer (TCT), the Comprehensive Compensation Equipment (CCE), and the Measurement and Control System (MCS). TCT is composed of unequal side V type connection traction transformer. The primary side terminal of TCT is connected with three-phase power system. The secondary side port ab of TCT is traction port, and port ac, b′c, and ad are compensation ports.

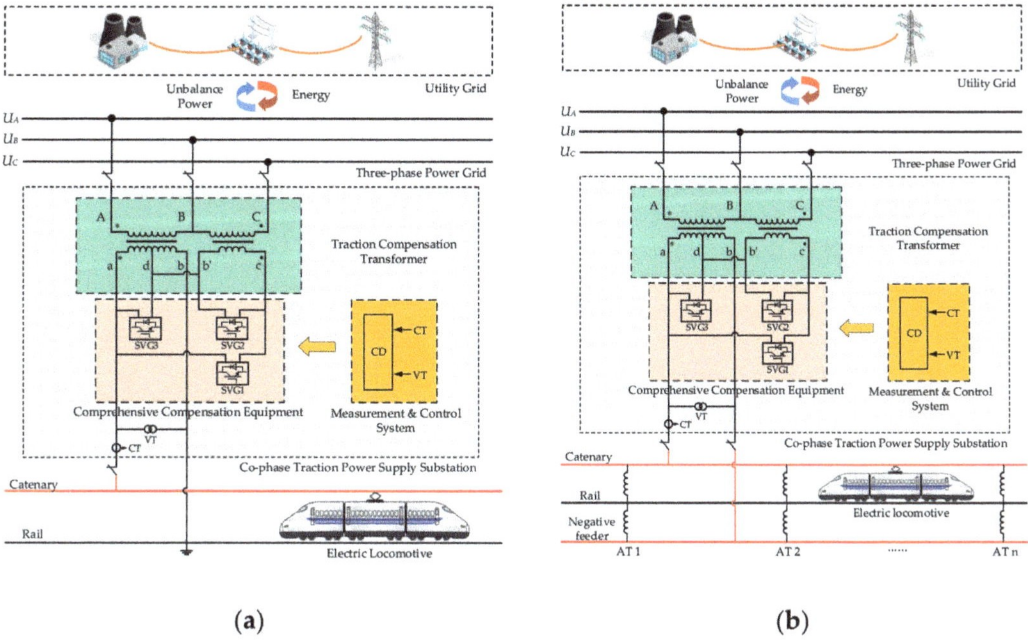

Figure 2. Structure diagram of traction substation with co-phase power supply. (**a**) Traction substation under direct power supply mode; (**b**) traction substation under Auto Transformer (AT) power supply mode.

CCE is composed of three single-phase SVGs, which are connected with compensation ports, respectively. MCS includes the Voltage Transformer (VT), the Current Transformer (CT), and the Controller Device (CD), and the signal output terminal of CD is connected with the control terminal of CCE.

Traction port ab of TCT is connected to traction network with VT and CT. If the power supply mode of traction network is direct power supply mode or direct power supply mode with return line, as shown in Figure 2a, terminal a of traction port ab is connected with traction network, terminal b is connected with rail.

If the traction network power supply mode is AT power supply mode, as shown in Figure 2b, terminal a of traction port ab of TCT is connected with traction network, and terminal b is connected with negative feeder.

3. Comprehensive Compensation Principle

The system takes the negative sequence limits of three-phase high-voltage bus and power factor as the compensation target. By controlling the reactive power generated by CCE, the reactive power and negative sequence generated by single-phase traction load are comprehensively compensated, so that the compensated power factor and negative sequence meet the requirements of the compensation target. CCE only changes the reactive power flow and does not change the active power flow.

3.1. Comprehensive Compensation Principle

According to reference [39], any traction port or compensation port λ on the secondary side of the traction transformer, the positive and negative sequence currents $\dot{I}_{\lambda}^{+}$ and $\dot{I}_{\lambda}^{-}$ generated on the primary side can be expressed as

$$\begin{cases} \dot{I}_{\lambda}^{+} = \frac{1}{\sqrt{3}} k_{\lambda} I_{\lambda} e^{-j\phi_{\lambda}} \\ \dot{I}_{\lambda}^{-} = \frac{1}{\sqrt{3}} k_{\lambda} I_{\lambda} e^{-j(2\psi_{\lambda}+\phi_{\lambda})} \end{cases}, \tag{1}$$

where I_{λ} is the effective value of the port current; ψ_{λ} is the angle at which $\dot{U}_{\lambda}$ lags $\dot{U}_{A}$; ϕ_{λ} is the power factor angle; $k_{\lambda} = U_{\lambda}/\sqrt{3}U_{A}$. U_{λ} is the effective value of the port voltage $\dot{U}_{\lambda}$.

The traction load current and power factor angle of TCT secondary traction port are set as $\dot{I}_{L}$ and ϕ_{L}, respectively. Taking the traction load in traction condition as an example, the comprehensive compensation principle of CCE is analyzed as follows.

When CCE performs comprehensive compensation, its positive and negative sequence phasor diagram is shown in Figure 3.

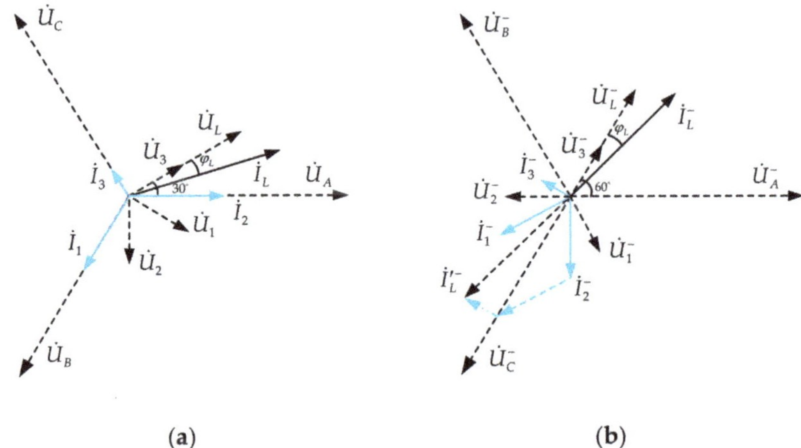

Figure 3. Phasor diagram of comprehensive compensation principle. (a) Phasor diagram of Positive sequence; (b) phasor diagram of Negative sequence.

In Figure 3, $\dot{U}_{A}$, $\dot{U}_{B}$, and $\dot{U}_{C}$ are the A, B, C three-phase voltage of the three-phase high-voltage bus, and $\dot{U}_{1}$, $\dot{U}_{2}$, and $\dot{U}_{3}$ are the compensation port voltage of TCT secondary side SVG1, SVG2, and SVG3, respectively, and $\dot{U}_{L}$ is the traction port voltage of the TCT secondary side. The corresponding negative sequence voltage are $\dot{U}_{A}^{-}$, $\dot{U}_{B}^{-}$, $\dot{U}_{C}^{-}$, $\dot{U}_{1}^{-}$, $\dot{U}_{2}^{-}$, $\dot{U}_{3}^{-}$, and $\dot{U}_{L}^{-}$. According to the topology of CCE, the relation diagram of positive sequence phasor and negative sequence phasor can be determined. Furthermore, the reactive currents generated by SVG1, SVG2, and SVG3 are $\dot{I}_{1}$, $\dot{I}_{2}$, and $\dot{I}_{3}$, respectively, where $\dot{I}_{1}$ is inductive current and $\dot{I}_{2}$ and $\dot{I}_{3}$ are capacitive current. The corresponding negative sequence currents are $\dot{I}_{1}^{-}$, $\dot{I}_{2}^{-}$, and $\dot{I}_{3}^{-}$, respectively. The resultant negative sequence current $\dot{I}'^{-}_{L}$ can be used to offset the negative sequence current $\dot{I}_{L}^{-}$ generated by the traction load. When the negative sequence is fully compensated, Equation (2) is correct:

$$\dot{I}_{L}^{-} + \dot{I}'^{-}_{L} = \dot{I}_{L}^{-} + \dot{I}_{1}^{-} + \dot{I}_{2}^{-} + \dot{I}_{3}^{-} = 0. \tag{2}$$

3.2. Comprehensive Compensation Model

According to the above analysis, it can be seen that CCE compensates the negative sequence and reactive power compensation simultaneously. Generally, the check point of negative sequence and power factor is Point of Common Coupling (PCC), so the negative sequence and reactive power generated by traction load at PCC are compensation object.

In case of CCE compensation, the active power of traction load remains unchanged at PCC, then reactive power Q_{CSS} and power factor $\cos\phi'_L$ after compensation can be expressed as follows:

$$Q_{CSS} = S_L \sin\phi_L + \sum_{k=1}^{n} S_k \sin\phi_K = (1 - K_C) S_L \sin\phi_L, \tag{3}$$

$$\cos\phi'_L = \frac{S_L \cos\phi_L}{\sqrt{(S_L \cos\phi_L)^2 + (Q_{CSS})^2}}, \tag{4}$$

where S_L and ϕ_L are the total apparent power and power factor angle of traction load. S_k and ϕ_k are the reactive power and power factor angle of SVG, respectively. n is the number of compensation ports, so n = 3 for Figure 2. K_C is the reactive power compensation degree.

According to Equation (3), K_C can be expressed as

$$K_C = -\frac{\sum_{k=1}^{n} S_k \sin\phi_k}{S_L \sin\phi_L}. \tag{5}$$

After using CCE to compensate, the negative sequence power at PCC is

$$\dot{S}^- = S_L e^{j\theta_L} + \sum\nolimits_{k=1}^{n} S_k e^{j\theta_k} = (1 - K_N) S_L e^{j\theta_L}, \tag{6}$$

where $\theta_L = 2\psi_L + \phi_L$, ψ_L, and ϕ_L are angle of $\dot{U}_L$ lagging behind $\dot{U}_A$ and traction load power factor angle; $\theta_k = 2\psi_k + \phi_k$, ψ_k, and ϕ_k are angle of compensating port k voltage lagging behind $\dot{U}_A$ and power factor angle of SVG, respectively; K_N is the negative sequence compensation degree.

According to Equation (6), K_N can be expressed as

$$K_N = -\frac{\sum_{k=1}^{n} S_k e^{j\theta_k}}{S_L e^{j\theta_L}}. \tag{7}$$

According to Equations (5) and (7), the comprehensive compensation effects of reactive power and negative sequence can be determined respectively by K_C and K_N. K_C and K_N are real numbers, and the range is $K_C, K_N \in [0, 1]$.

Equation (7) is expanded according to the real part and the imaginary part, respectively, and constitutes the simultaneous equations together with Equation (5). The comprehensive compensation model shown as Equation (8):

$$\begin{cases} K_C S_L \sin\phi_L = -\sum_{k=1}^{n} S_k sin\phi_k \\ K_N S_L \cos\theta_L = -\sum_{k=1}^{n} S_k cos\theta_k \\ K_N S_L \sin\theta_L = -\sum_{k=1}^{n} S_k sin\theta_k \end{cases}. \tag{8}$$

The reactive power generated by SVG1, SVG2, and SVG3 are S_1, S_2, and S_3, respectively, and according to Section 3.1, $\phi_1 = \pi/2$, $\phi_2 = -\pi/2$, and $\phi_3 = -\pi/2$, the comprehensive compensation model can be obtained as follows according to Equation (8):

$$\begin{cases} K_C S_L \sin\phi_L = -S_1 + S_2 + S_3 \\ K_N S_L \cos(2\psi_L + \phi_L) = S_1 \sin 2\psi_1 - S_2 \sin 2\psi_2 - S_3 \sin 2\psi_3 \\ K_N S_L \sin(2\psi_L + \phi_L) = -S_1 \cos 2\psi_1 + S_2 \cos 2\psi_2 + S_3 \cos 2\psi_3 \end{cases}, \tag{9}$$

where $\psi_L = -\pi/6$, $\psi_1 = -5\pi/6$, $\psi_2 = -\pi/2$, and $\psi_3 = -\pi/6$.

S_1, S_2, and S_3 can be solved by Equation (9):

$$\begin{cases} S_1 = \frac{1}{\sqrt{3}} K_N S_L \cos\phi_L + \frac{1}{3}(K_N - K_C) S_L \sin\phi_L \\ S_2 = \frac{1}{\sqrt{3}} K_N S_L \cos\phi_L - \frac{1}{3}(K_N - K_C) S_L \sin\phi_L \\ S_3 = \frac{1}{3}(2K_N + K_C) S_L \sin\phi_L \end{cases}, \tag{10}$$

where $S_k > 0 \ (k = 1,\ 2)$ means that SVG1 and SVG2 output inductive and capacitive reactive power, respectively; otherwise, SVG1 and SVG2 output capacitive and inductive reactive power, respectively; SVG3 always output capacitive reactive power.

If Equation (10) is divided by U_L, the reactive current generated by SVG1, SVG2, and SVG3 are shown in Equation (11):

$$\begin{cases} I_1 = \frac{k_M}{\sqrt{3}k_L}\left[K_N I_L \cos\phi_L + \frac{1}{\sqrt{3}}(K_N - K_C)I_L \sin\phi_L\right] \\ I_2 = \frac{k_M}{\sqrt{3}k_L}\left[K_N I_L \cos\phi_L - \frac{1}{\sqrt{3}}(K_N - K_C)I_L \sin\phi_L\right] \\ I_3 = \frac{k_M}{3k_L}(2K_N + K_C)I_L \sin\phi_L \end{cases}, \tag{11}$$

where $k_L = \sqrt{3}U_A/U_L, k_M = \sqrt{3}U_A/U_k \quad (k = 1,\quad 2,\quad 3)$, and satisfies $U_1 = U_2 = U_3$.

4. Comprehensive Compensation Control Strategy

As shown in Figure 4, set the instantaneous value of voltage on primary side of TCT as $U_A(t) = \sqrt{2}U_A \sin(\omega t)$. With $\dot{U}_A$ as the reference, the port voltage of traction port $\dot{U}_L$ and load current $\dot{I}_L$ are shown as follows:

$$U_L(t) = \sqrt{2}U_L \sin(\omega t - \psi_L) = \sqrt{2}U_L \sin(\omega t + \frac{\pi}{6}), \tag{12}$$

$$I_L(t) = I_{L1}(t) + I_{Lh}(t) = \sqrt{2}I_{L1} \sin(\omega t + \frac{\pi}{6} - \phi_{L1}) + I_{Lh}(t), \tag{13}$$

where $I_{L1}(t)$ is the fundamental current; $I_{Lh}(t)$ is the harmonic current; I_{L1} is the effective value of the fundamental current; ω is the angular frequency; and ϕ_{L1} is the fundamental power factor angle. The fundamental component I_{L1} of load current can be further decomposed into instantaneous active component I_{L1p} and instantaneous reactive component I_{L1q}:

$$I_{L1}(t) = \sqrt{2}I_{L1p} \sin(\omega t + \frac{\pi}{6}) - j\sqrt{2}I_{L1q} \cos(\omega t + \frac{\pi}{6}), \tag{14}$$

where $I_{L1p} = I_{L1} \cos\phi_L, I_{L1q} = I_{L1} \sin\phi_L$.

By introducing Equation (14) into Equation (13) and multiplying both sides of the equation by $\sin(\omega t + \frac{\pi}{6})$ and $\cos(\omega t + \frac{\pi}{6})$, Equations (15) and (16) can be obtained:

$$I_L(t)\sin(\omega t + \tfrac{\pi}{6}) = \left[\tfrac{\sqrt{2}}{2}I_{L1p} - \tfrac{\sqrt{2}}{2}I_{L1p}\cos(2\omega t + \tfrac{\pi}{3}) - \tfrac{\sqrt{2}}{2}I_{L1q}\sin(2\omega t + \tfrac{\pi}{3})\right] + I_{Lh}(t)\sin(\omega t + \tfrac{\pi}{6}), \tag{15}$$

$$I_L(t)\cos(\omega t + \tfrac{\pi}{6}) = \left[-\tfrac{\sqrt{2}}{2}I_{L1q} + \tfrac{\sqrt{2}}{2}I_{L1q}\cos(2\omega t + \tfrac{\pi}{3}) + \tfrac{\sqrt{2}}{2}I_{L1p}\sin(2\omega t + \tfrac{\pi}{3})\right] + I_{Lh}(t)\cos(\omega t + \tfrac{\pi}{6}). \tag{16}$$

DC components $\frac{\sqrt{2}}{2}I_{L1p}$ and $-\frac{\sqrt{2}}{2}I_{L1q}$ can be separated from Equations (15) and (16), and then I_{L1p} and I_{L1q} can be obtained.

4.1. System Control Strategy

When CCE operates, $\dot{U}_1$, $\dot{U}_2$, and $\dot{U}_3$ are, respectively, locked by phase-locked loop (PLL) to generate synchronous signals $\sin(\omega t - \frac{\pi}{6})$, $\sin(\omega t - \frac{\pi}{2})$, and $\sin(\omega t + \frac{\pi}{6})$. Under the traction condition of traction load, SVG1 outputs inductive reactive power, SVG2 and SVG3 output capacitive reactive power. Therefore, the synchronous signals of compensation current of SVG1, SVG2, and SVG3 are, respectively, $-\cos(\omega t - \frac{\pi}{6})$, $\cos(\omega t - \frac{\pi}{2})$, and $\cos(\omega t + \frac{\pi}{6})$. According to Equation (11), the expected values of compensation current of SVG1, SVG2, and SVG3, I_1^*, I_2^*, and I_3^* are

$$\begin{cases} I_1^* = -\frac{\sqrt{2}k_M}{\sqrt{3}k_L}\left[K_N I_{L1p} + \frac{1}{\sqrt{3}}(K_N - K_C)I_{L1q}\right]\cos(\omega t - \frac{\pi}{6}) \\ I_2^* = \frac{\sqrt{2}k_M}{\sqrt{3}k_L}\left[K_N I_{L1p} - \frac{1}{\sqrt{3}}(K_N - K_C)I_{L1q}\right]\cos(\omega t - \frac{\pi}{2}) \\ I_3^* = \frac{\sqrt{2}k_M}{3k_L}\left[(2K_N + K_C)I_{L1q}\right]\cos(\omega t + \frac{\pi}{6}) \end{cases}. \quad (17)$$

According to Equation (17), the block diagram of the expected value detection of compensation currents of the CCE is shown in Figure 5.

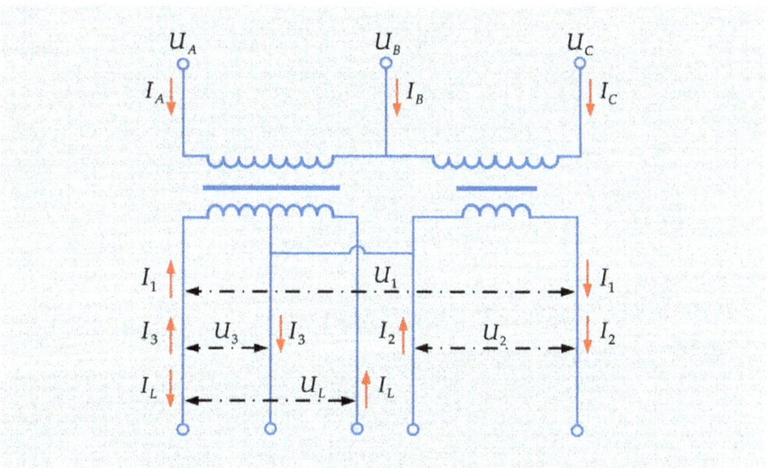

Figure 4. Simplified electrical schematic diagram of Traction Compensation Transformer (TCT).

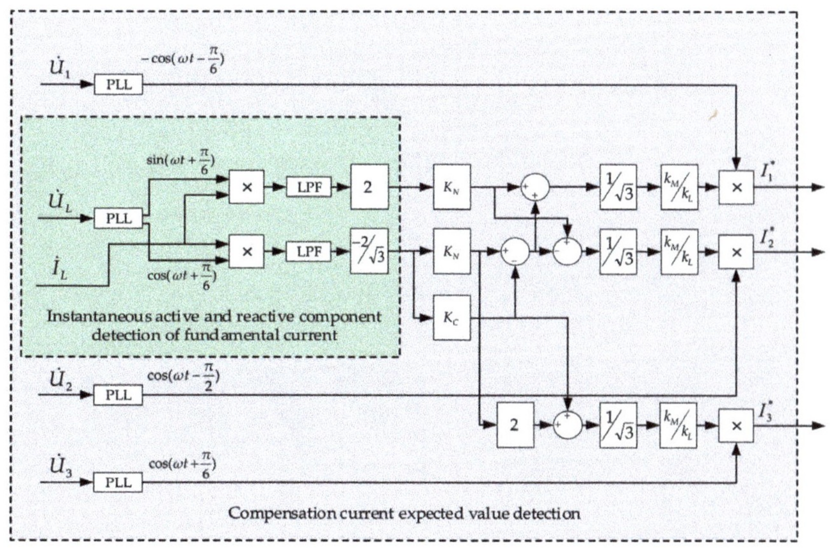

Figure 5. Block diagram of expected value detection of compensation currents.

The expected compensation currents I_1^*, I_2^*, and I_3^* are compared with the actual currents value I_1, I_2, and I_3 on the compensation port. After PI adjustment, the PWM control signal driving SVG is generated by carrier modulation technology. Therefore, the CCE can be controlled to send out the corresponding expected compensation currents. The control block diagram of CCE is shown in Figure 6.

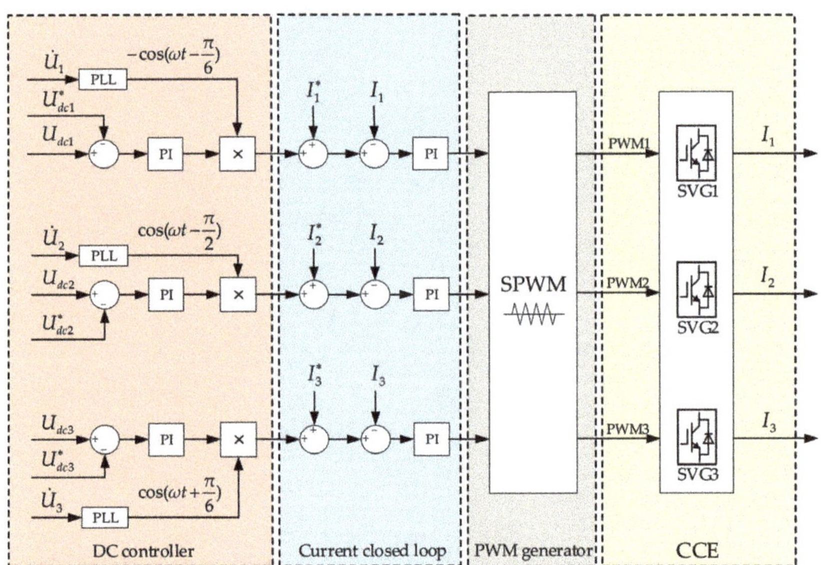

Figure 6. Control block diagram of Comprehensive Compensation Equipment (CCE).

4.2. Determination Method and Steps of K_C and K_N

In the integrated compensation control strategy, the key to achieve the compensation goal is to confirm the appropriate reactive power compensation degree K_C and negative sequence compensation degree K_N.

The relationship between the K_C and power factors can be obtained by solving Equations (3) and (4) simultaneously:

$$K_C = 1 - \sqrt{\frac{(\tan\phi'_L)^2}{(\tan\phi_L)^2}} = 1 - \sqrt{\frac{(\cos\phi'_L)^{-2} - 1}{(\cos\phi_L)^{-2} - 1}}, \tag{18}$$

where $\cos\phi_L$ and $\cos\phi'_L$ are power factors before and after compensation.

The following relationship between the limit of three-phase voltage unbalance degree at PCC ε_{U2} and the allowable negative sequence power at PCC S_ε is

$$S_\varepsilon = \varepsilon_{U2} \cdot S_d, \tag{19}$$

where S_d is the short circuit capacity at PCC.

If the residual negative sequence power caused by traction load at PCC after compensation is $\dot{S}^-$, its magnitude shall meet $\left|\dot{S}^-\right| \le S_\varepsilon$. The relationship between K_N and expected value of three-phase voltage unbalance degree ε_{U2}^* can be obtained by solving Equations (6) and (19), as shown in Equation (20):

$$K_N = \frac{S_L - \varepsilon_{U2}^* \cdot S_d}{S_L}. \tag{20}$$

At present, AC/DC/AC electric locomotive and AC/DC electric locomotive are widely used in electrified railway. According to load characteristics, the methods for determining the values of K_C and K_N can be summarized as follows, the specific steps are

(1) When the negative sequence power S_L^- generated by the traction load is greater than the allowable negative sequence power S_ε at PCC, and the power factor $\cos\phi_L$ is less than the target power factor value $\cos\phi^*$, the negative sequence and reactive power are compensated by CCE at the same time. Based on the expected target, the value of K_C and K_N can be determined according to Equations (18) and (20). Then, according to Equation (10), the S_1, S_2, and S_3 are obtained. Under the traction condition of traction load, S_1 is inductive or capacitive reactive ($S_1 > 0$ or $S_1 < 0$), S_2 is capacitive or inductive reactive ($S_2 > 0$ or $S_2 < 0$), and S_3 is capacitive reactive;

(2) When S_L^- is greater than S_ε, and $\cos\phi_L$ is greater than or equal to $\cos\phi^*$. Only the negative sequence power is compensated by CCE, and the power factor is not changed before and after compensation. Therefore, K_C can be determined by $K_C = 0$, and based on the expected target after compensation, the value of K_N can be determined according to Equation (20). Then according to Equation (10), the S_1, S_2, and S_3 are obtained. Under the traction condition of traction load, S_1 is inductive reactive, S_2 and S_3 are capacitive reactive;

(3) When S_L^- is less than or equal to S_ε, and $\cos\phi_L$ is less than $\cos\phi^*$. Only the reactive sequence power is compensated by CCE. Therefore, K_N can be determined by $K_N = 0$, and based on the expected target after compensation, the value of K_C can be determined according to Equation (18). Then, according to Equation (10), the S_1, S_2, and S_3 are obtained. Under the traction condition of traction load, S_1, S_2, and S_3 are all capacitive reactive;

(4) When S_L^- is less than or equal to S_ε, and $\cos\phi_L$ is greater than or equal to $\cos\phi^*$. The negative sequence and reactive power generated by the traction load can meet the compensation target, no additional compensation is required. Therefore, K_N and K_C can be determined by $K_N = 0$ and $K_C = 0$. CCE operates in standby.

In summary, the schematic diagram of the process of determining the values of K_C and K_N is shown in Figure 7.

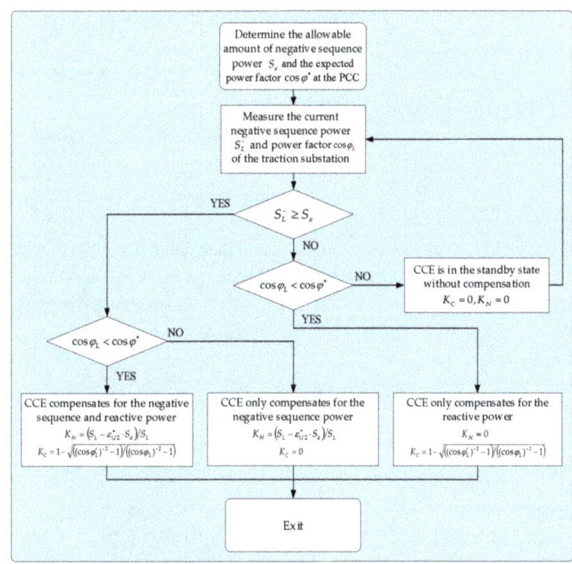

Figure 7. Schematic diagram of the process of determining the values of K_C and K_N.

5. Effectiveness Verification

5.1. Analysis and Verification of Comprehensive Compensation Scheme Based on Actual Case

Take the actual data of an electrified railway traction load as an example for illustration. The traction transformer is single-phase wiring, and its primary side is connected to the A and B phases of the 110 kV power system. The primary and secondary side transformation ratio is 110/27.5. By using the power quality test device, the amplitude and phase angle of the voltage and current of the primary and secondary sides of the traction transformer are recorded. The measurement period is 24 h. After processing the data, the diagram of 24-h load curve of the traction load is shown in Figure 8.

After further statistical analysis of the load data, it shows that the 95% probability value of the three-phase voltage unbalance degree of the traction substation is 1.2%, and the maximum is 3.6%. The maximum value has exceeded 2.6%, the limit of the three-phase voltage unbalance degree standard [40]. In addition, the daily average power factor of the traction substation is 0.79, which is lower than the economic power factor of 0.9 required by the power system. Therefore, it is necessary to set up an appropriate compensation scheme for comprehensive treatment of the negative sequence and reactive power in the traction substation. The diagram of 24-h three-phase voltage unbalance degree curve of the traction substation is shown in Figure 9.

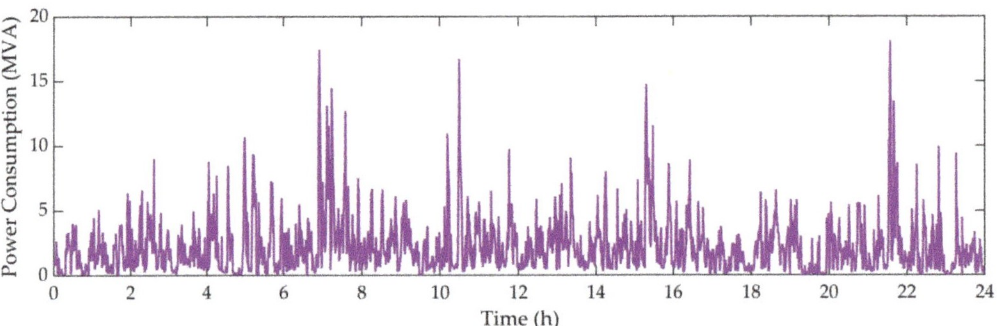

Figure 8. Diagram of 24-h load curve of the traction substation.

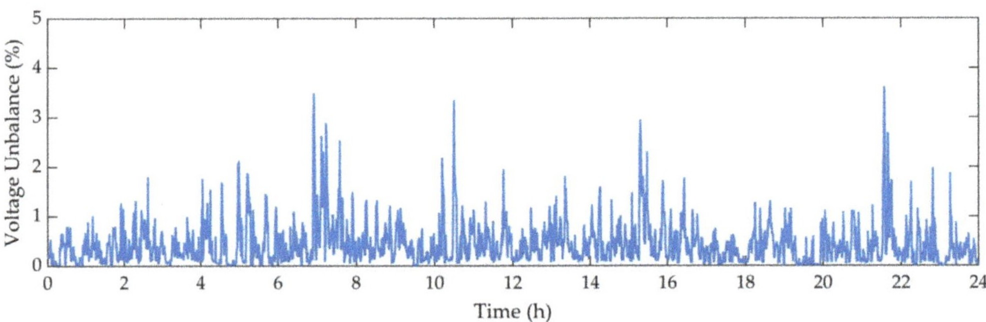

Figure 9. Diagram of 24-h three-phase voltage unbalance degree curve of the traction substation.

The design target of the compensation scheme proposed in this paper is that after compensation, the maximum value of the three-phase voltage unbalance degree of the traction substation will be reduced to a value less than the limit, for example, 2.5%, the 95% probability value will be further reduced to a value less than the limit, for example, 1%, and the daily average power factor is increased to 0.9.

In order to reduce the capacity of the compensation device as much as possible, for the condition where the three-phase voltage unbalance degree caused by the traction load is greater than 1.2%, the first negative sequence allowable amount $S_{\varepsilon 1}$ will implemented as the assessment reference value, $S_{\varepsilon 1}$ = 12.5 MVA. According to Equation (19), it can be known that the corresponding three-phase voltage unbalance degree target is 2.5%. If S_L^- is greater than $S_{\varepsilon 1}$, the expected value of the three-phase voltage unbalance degree after compensation is set to 2.5%. Otherwise, S_L^- meets the requirements of $S_{\varepsilon 1}$, so that the CCE is standby. However, for the condition where the three-phase voltage unbalance degree is less than or equal to 1.2%, the allowable second negative sequence $S_{\varepsilon 2}$ will implemented as the assessment reference value, $S_{\varepsilon 2}$ = 5 MVA. Moreover, the corresponding three-phase voltage unbalance degree target is 1.0%. If S_L^- is greater than $S_{\varepsilon 2}$, the expected value of the three-phase voltage unbalance degree after compensation is set to 1.0%. Otherwise, S_L^- meets the requirements of $S_{\varepsilon 2}$ and similarly there is also no need to compensate for the negative sequence power.

Based on the selection of the above limits, by using the determination method and steps of K_C and K_N given in Section 4.2, the values of K_C and K_N corresponding to the load at each moment of the traction substation can be calculated. As shown in Figure 10.

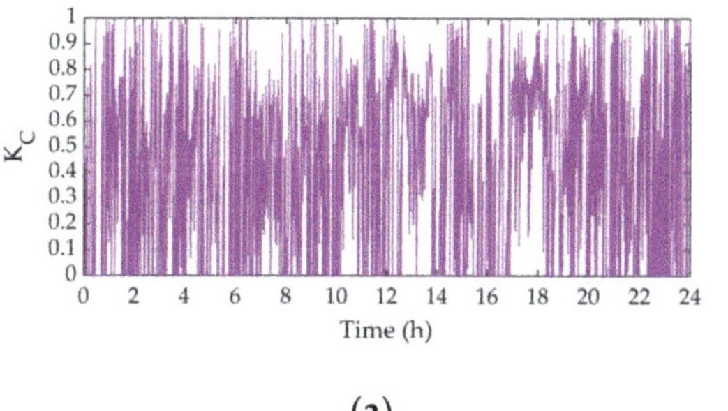

(a)

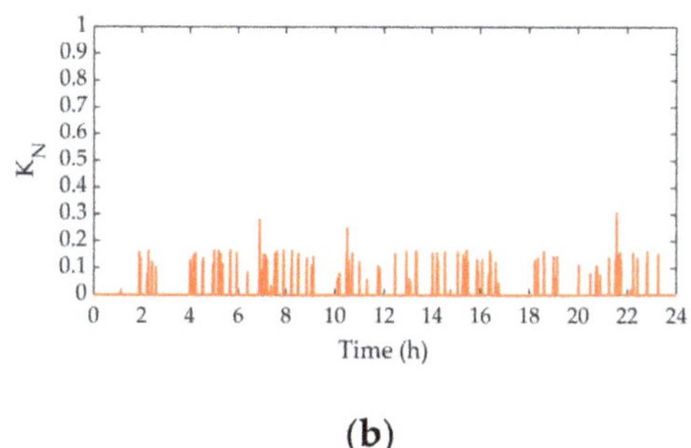

(b)

Figure 10. Diagram of 24-h K_C and K_N calculation results curve of the traction substation. (**a**) Diagram of calculation results for reactive power compensation degree K_C during a day; (**b**) diagram of calculation results for negative sequence compensation degree K_N during a day.

According to the calculation results of K_C and K_N in Figure 10, it can be further calculated by Equation (10) that the maximum reactive power compensation amount required by SVG1, SVG2, and SVG3 are 2.66 MVA, 2.66 MVA, and 3.55 MVA, respectively. Therefore, combined with actual engineering applications, a comprehensive compensation design scheme for the traction substation can be determined. The device capacity of SVG1, SVG2, and SVG3 can be selected respectively as 3 MVA, 3 MVA, and 5 MVA. Furthermore, the capacity of each device above is the actual value of required to meet the compensation target.

Based on the above compensation design scheme, after implementing the comprehensive compensation of negative sequence and reactive power for the traction substation, the diagram of 24-h device output power of the CCE is shown in Figure 11.

The diagram of 24-h three-phase voltage unbalance degree curve and power factor curve of the traction substation before and after compensation are shown in Figures 12 and 13, respectively.

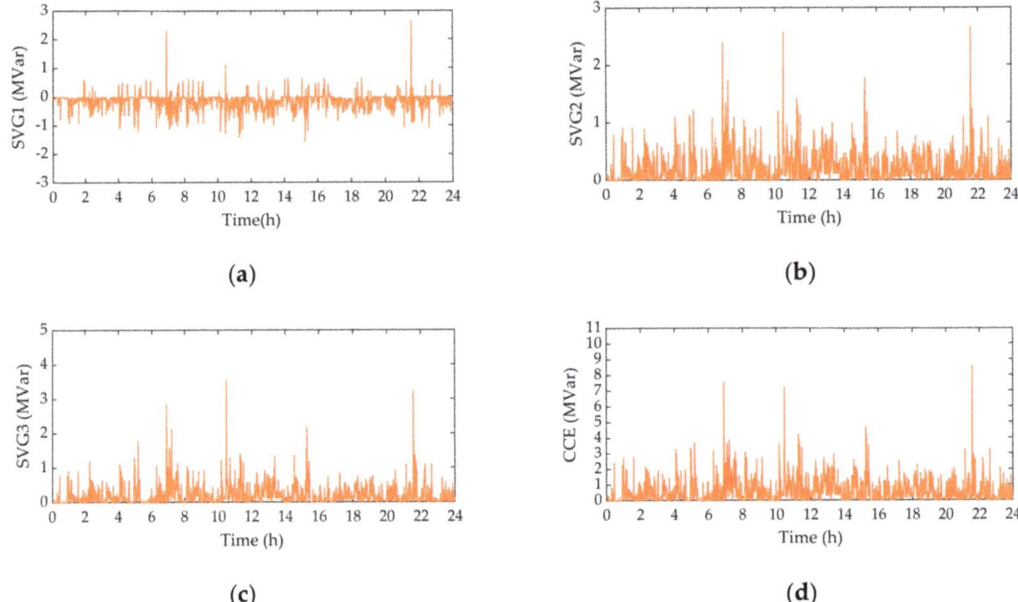

Figure 11. Diagram of device output curve of the CCE during a day. (**a**) Diagram of device output curve of the Static Var Generator (SVG)1 during a day; (**b**) diagram of device output curve of the SVG2 during a day; (**c**) diagram of device output curve of the SVG3 during a day; (**d**) diagram of total device output curve of the CCE during a day.

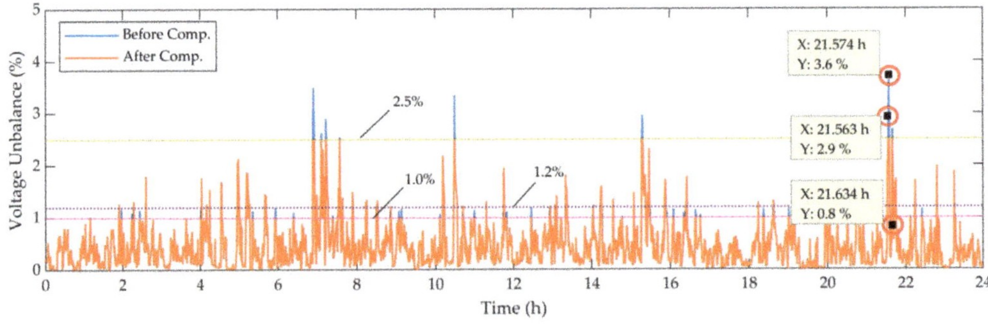

Figure 12. Diagram of 24-h three-phase voltage unbalance degree curve of the traction substation (before and after compensation).

In Figure 14, the statistical result after compensation shows that the 95% probability value of the three-phase voltage unbalance degree of the traction substation has reduced from 1.2% to 1.0%, and the maximum also has reduced from 3.6% to 2.5%. At the same time, the daily average power factor of the traction substation has increased from 0.79 to 0.91, achieving compensation design target.

5.2. Analysis and Verification of Comprehensive Compensation Control Strategy

To verify the effectiveness of the comprehensive compensation control strategy, a simulation model is established with MATLAB/SIMULINK, as shown in Figure 15. In the simulation model, the short-circuit capacity of the power system is set to 500 MVA; on the primary-side of TCT, the power system voltage is 110 kV; on the secondary side of TCT, the voltage of traction port is 27.5 kV. Furthermore, the constant power source is used for simulating the traction load.

The compensation is only needed when the negative sequence power or reactive power does not meet the requirements of the compensation target. Therefore, based on the actual traction load data within 24 h given in Section 5.1, three typical cases are selected for simulation verification in this section. Among of them, case 1 will conduct simulation analysis for the condition when negative sequence power and reactive power need to be compensated simultaneously; and case 2 will conduct simulation analysis for the condition when only the negative sequence power need to be compensated; finally, case 3 will conduct simulation analysis for the condition when only the reactive power need to be compensated.

Case 1:

In case 1, take the traction load data at 21.574 h as an example, at this time traction the load power is 18.2 MVA, three-phase voltage unbalance degree ε_{U2}^{L1} at the PCC is 3.6%, and power factor $\cos\phi_{L1}$ is 0.8. Because ε_{U2}^{L1} exceeds the limit of 2.5%, and $\cos\phi_{L1}$ is also lower than the limit of economic power factor 0.9. Therefore, according to the comprehensive compensation strategy, comprehensive compensation for negative sequence and reactive power is required. Set the compensation target of three-phase voltage unbalance degree at PCC as $\varepsilon_{U2}^{*} = 2.5\%$, and power factor compensation target as $\cos\phi^{*} = 0.9$, so that $K_N = 0.31$, $K_C = 0.35$.

CCE was put into operation at 0.3 s. The simulation results are shown in Figure 16. It can be seen that after CCE operates, the three-phase voltage unbalance degree at the PCC is rapidly reduced from 3.6% to 2.5%. At the same time, the power factor also increased rapidly from 0.8 to 0.9. The compensation target of CCE is realized for negative sequence and reactive power. Table 1 shows the results of case 1 before and after comprehensive compensation.

Case 2:

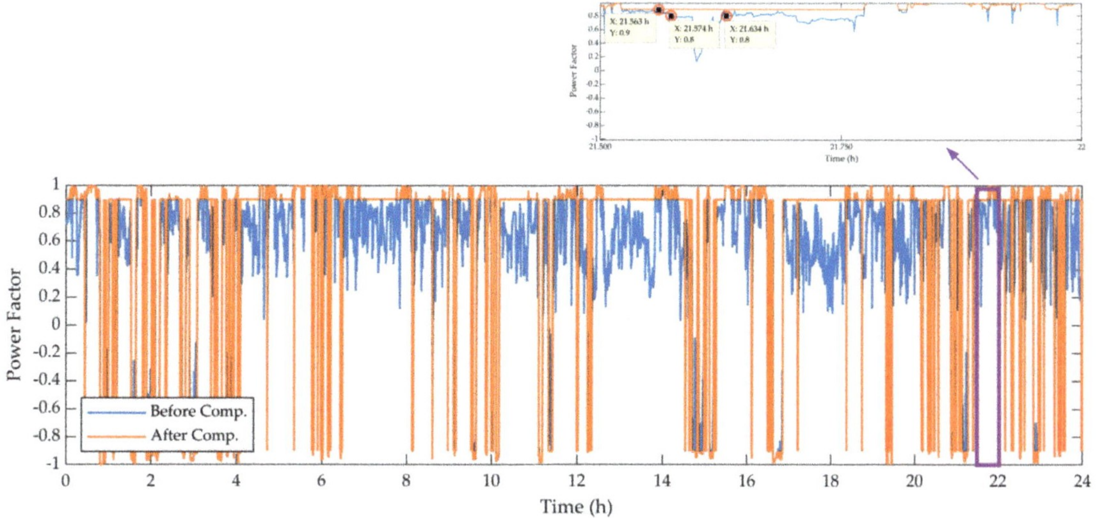

Figure 13. Diagram of 24-h power factor curve of the traction substation (before and after compensation).

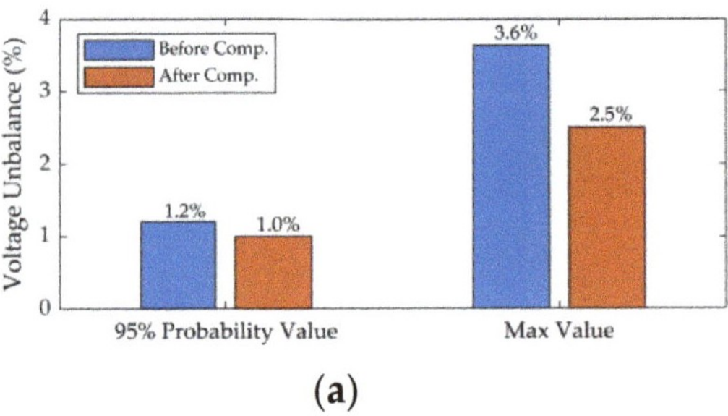

(a)

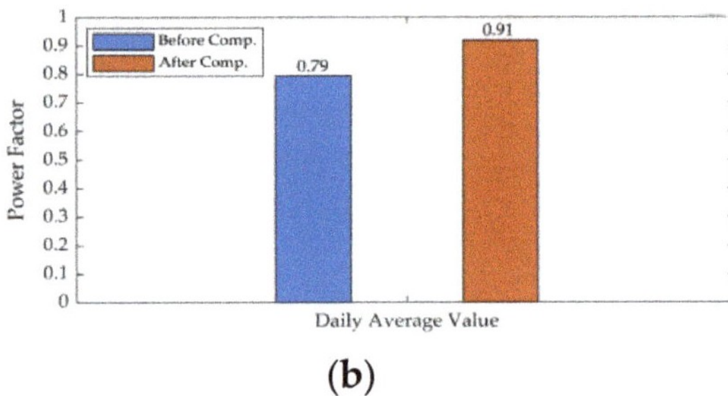

(b)

Figure 14. Diagram of statistical result of the data in Figures 12 and 13. (**a**) Diagram of statistical result of three-phase voltage unbalance degree during a day before and after compensation; (**b**) diagram of statistical result of power factor during a day before and after compensation.

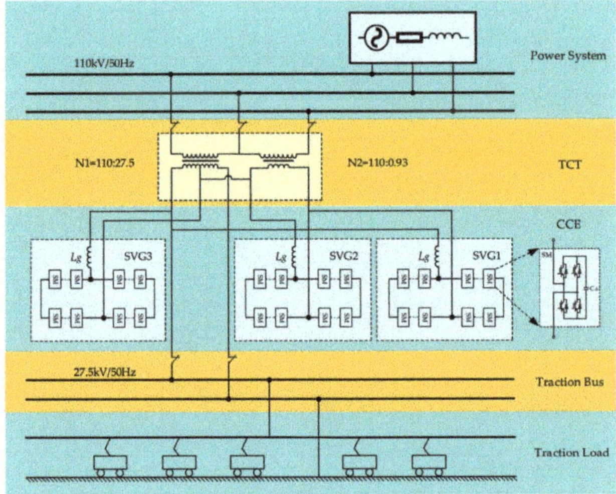

Figure 15. Diagram of comprehensive compensation control strategy simulation.

In case 2, take the traction load data at 21.563 h as an example, at this time traction load power is 14.3 MVA, three-phase voltage unbalance degree ε_{U2}^{L2} at the PCC is 2.9%, and power factor $\cos\phi_{L2}$ is 0.9. Because ε_{U2}^{L2} exceeds the limit of 2.5%, $\cos\phi_{L2}$ still meets the requirement of economic power factor. Therefore, according to the comprehensive compensation control strategy, only the negative sequence power needs to be compensated. Set the compensation target of three-phase voltage unbalance degree at PCC as $\varepsilon_{U2}^{*} = 2.5\%$, and power factor compensation target as $\cos\phi^{*} = 0.9$, so that $K_N = 0.31$, $K_C = 0$.

CCE was put into operation at 0.3 s. The simulation results are shown in Figure 17. It can be seen that after CCE operates, the three-phase voltage unbalance degree at the PCC is rapidly reduced from 2.9% to 2.5%. At the same time, because of the CCE does not inject additional reactive power into the system and the reactive power at PCC will not change, the power factor can always be stabilized at 0.9. The compensation target of CCE is realized for negative sequence power. Table 2 shows the results of case 2 before and after negative sequence power compensation.

Case 3:

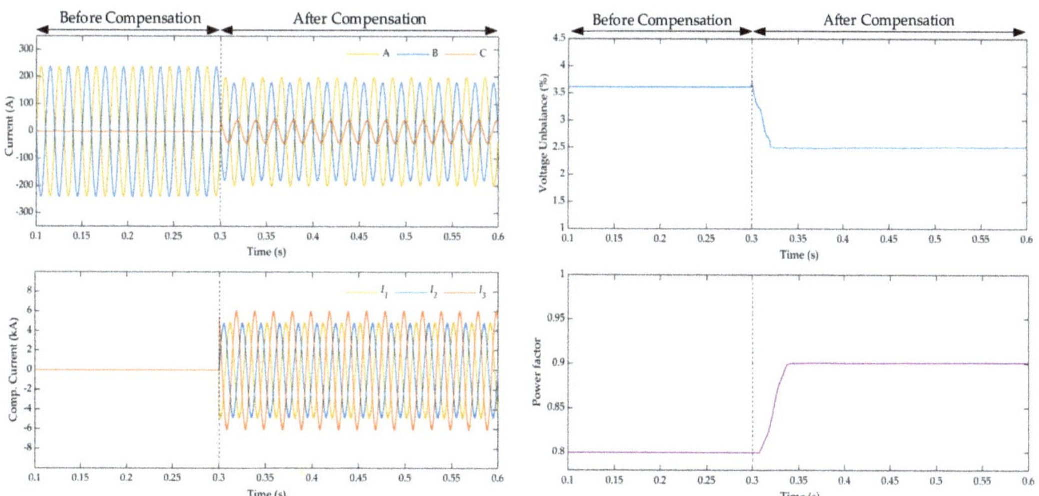

Figure 16. Simulation results at Point of Common Coupling (PCC) before and after comprehensive compensation.

Table 1. Statistical results of typical values before and after comprehensive compensation.

Parameters	Three-Phase Voltage		Three-Phase Voltage Unbalance Degree	Power Factor
	Positive Sequence	**Negative Sequence**		
Before compensation	61.89 kV	2.23 kV	3.6%	0.8
After compensation	62.28 kV	1.56 kV	2.5%	0.9

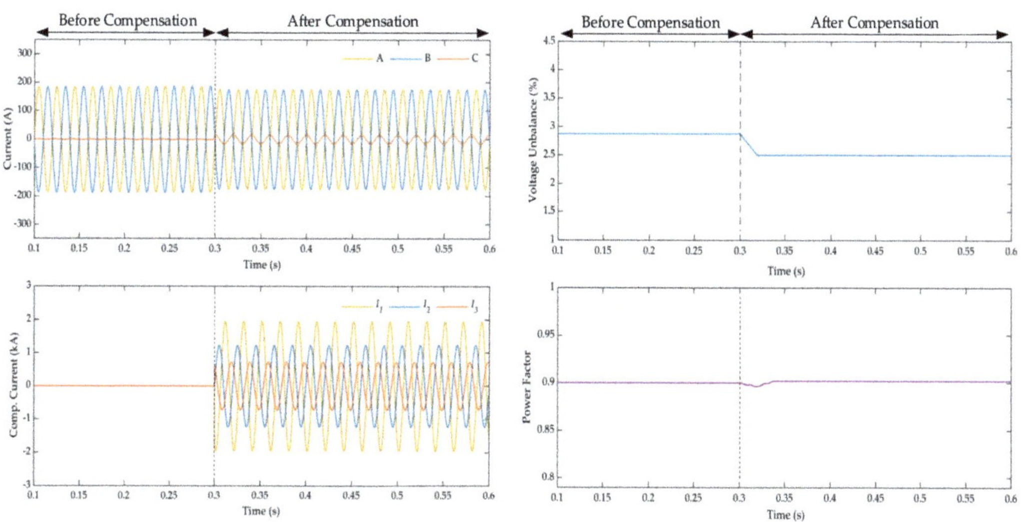

Figure 17. Simulation results at PCC before and after negative sequence power compensation.

Table 2. Statistical results of typical values before and after negative sequence power compensation.

Parameters	Three-Phase Voltage		Three-Phase Voltage Unbalance Degree	Power Factor
	Positive Sequence	Negative Sequence		
Before compensation	62.48 kV	1.80 kV	2.9%	0.9
After compensation	62.48 kV	1.56 kV	2.5%	0.9

In the case 3, take the measured load data of the traction substation at 21.634 h as an example, at this time traction load power is 4.0 MVA, three-phase voltage unbalance degree ε_{U2}^{L3} at the PCC is 0.8%, and power factor $\cos\phi_{L3}$ is 0.9. The other simulation parameters are the same as above. Because ε_{U2}^{L3} meets the 1.0% limit, $\cos\phi_{L3}$ is lower than the limit of economic power factor 0.9. Therefore, according to the comprehensive compensation control strategy, only the reactive power needs to be compensated. Set the compensation target of three-phase voltage unbalance degree at PCC as $\varepsilon_{U2}^{*} = 0.8\%$, and power factor compensation target as $\cos\phi^{*} = 0.9$, so that $K_N = 0$, $K_C = 0.35$.

CCE was put into operation at 0.3 s. The simulation results are shown in Figure 18. It can be seen that after CCE operates, the power factor at the PCC is rapidly increased from 0.8 to 0.9. At the same time, the three-phase voltage unbalance degree is always maintained at a low level of 0.8%, and no further compensation for negative sequence power is required. The compensation target of CCE is realized for reactive power. Table 3 shows the results of case 3 before and after reactive power compensation.

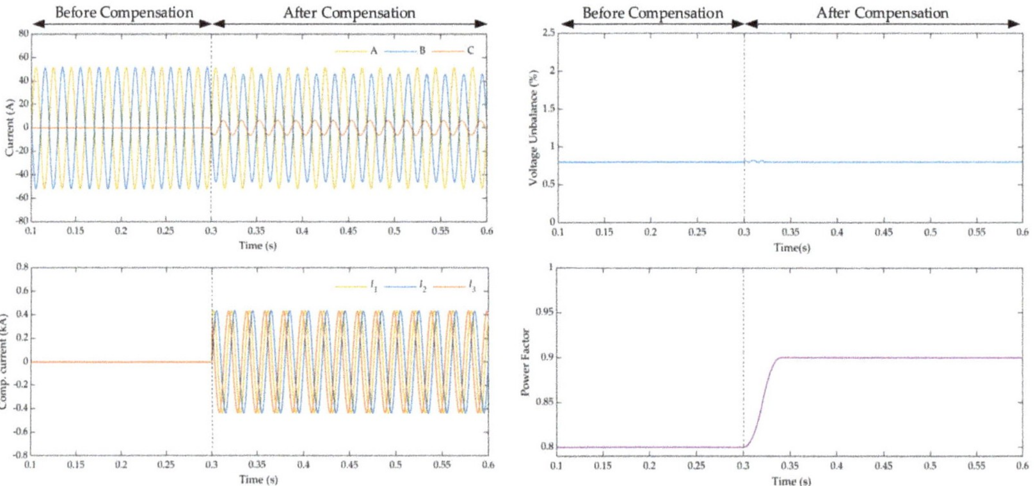

Figure 18. Simulation results at PCC before and after reactive power compensation.

Table 3. Statistical results of typical values before and after reactive power compensation.

Parameters	Three-Phase Voltage		Three-Phase Voltage Unbalance Degree	Power Factor
	Positive Sequence	Negative Sequence		
Before compensation	63.15 kV	0.50 kV	0.8%	0.8
After compensation	63.26 kV	0.51 kV	0.8%	0.9

In summary, through the above three simulation experiment results based on measured load data, the effectiveness of the comprehensive compensation control strategy proposed in this paper is fully verified. The system responds quickly and the compensation effect is better.

6. Conclusions

In this paper, a novel co-phase power supply system for electrified railway based on V type connection traction transformer was proposed to cancel neutral section at the outlet of traction substation, and compensate for the negative sequence and reactive power. It is beneficial to reduce the adverse effects caused by the train passing the neutral section and improve the safety of train operation. Moreover, it can also effectively solve the power quality problem mainly caused by the negative sequence power generated by the electrified railway.

The Traction Compensation Transformer (TCT) presented in the scheme has both traction port and compensation port, and the windings of the traction port and compensation port can be shared. It has obvious advantages such as high functional integration, effectively reducing equipment footprint and transformer manufacturing difficulty. Furthermore, the traction port of the TCT is essentially a single-phase transformer with a high capacity utilization rate, which can effectively reduce the installation capacity of the equipment.

The scheme proposed in this paper is suitable for the comprehensive treatment of negative sequence and reactive power of various AC-DC and AC-DC-AC electric locomotives, and the working conditions of the CCE are reversible. When the traction load is working under the regenerative braking conditions, it can still feed power that meets the standard to the power system.

Author Contributions: S.X. proposed the idea, completed the theoretical analysis, and developed the model. Y.Z. collected and analyzed measured data, performed the simulation, and related verification works. S.X., Y.Z. and H.W. wrote the paper. All authors have read and agreed to the published version of the manuscript.

Funding: This work was supported by the National Natural Science Foundation of China (Grant No. 51877182) and the Sichuan Science and Technology Program (Grant No. 2021YJ0028).

Institutional Review Board Statement: Not applicable.

Informed Consent Statement: Not applicable.

Data Availability Statement: Not applicable.

Conflicts of Interest: The authors declare no conflict of interest.

Abbreviations

Abbreviations	
SVG	Static Var Generator
PCC	Point of Common Coupling
RPC	Railway Static Power Conditioner
PFC	Power Flow Controller
HPQC	Hybrid Power Quality Conditioner
MMC	Modular Multilevel Converter
SVC	Static Var Compensator
TCT	Traction Compensation Transformer
CCE	Comprehensive Compensation Equipment
MCS	Measurement and Control System
VT	Voltage Transformer
CT	Current Transformer
CD	Controller Device
AT	Auto Transformer

Variables	
$\dot{U}_A$,$\dot{U}_B$,$\dot{U}_C$	Voltage of the three-phase high-voltage bus
$\dot{U}_A^-$,$\dot{U}_B^-$, $\dot{U}_C^-$	Negative sequence component of $\dot{U}_A$, $\dot{U}_B$, $\dot{U}_C$
$\dot{U}_1$, $\dot{U}_2$,$\dot{U}_3$	Compensation port voltage of TCT secondary side SVG1, SVG2, SVG3
$\dot{U}_1^-$, $\dot{U}_2^-$, $\dot{U}_3^-$	Negative sequence component of $\dot{U}_1$,$\dot{U}_2$,$\dot{U}_3$
$\dot{I}_1$,$\dot{I}_2$, $\dot{I}_3$	Compensation current generated by SVG1, SVG2, SVG3
$\dot{I}_1^-$, $\dot{I}_2^-$,$\dot{I}_3^-$	Negative sequence component of $\dot{I}_1$,$\dot{I}_2$, $\dot{I}_3$
$\dot{U}_L$	Traction port voltage of TCT secondary side
$\dot{U}_L^-$	Negative sequence component of $\dot{U}_L$
$\dot{I}_L$	Traction load current
$\dot{I}_L^-$	Negative sequence component of $\dot{I}_L$
Q_{CSS}	Reactive power of co-phase power supply traction substation after compensation
$\cos\phi_L'$	Power factor of co-phase power supply traction substation after compensation
S_L	Total apparent power of traction load
ϕ_L	Power factor angle of traction load
S_k	Reactive power generated by SVGs
ϕ_k	Power factor angle of SVGs
n	The number of compensation ports
K_C	Reactive power compensation degree
ψ_L	Angle of $\dot{U}_L$ lagging behind $\dot{U}_A$
ψ_k	Angle of compensating port k voltage lagging behind $\dot{U}_A$
K_N	Negative sequence compensation degree
S_1,S_2,S_3	Reactive power generated by SVG1, SVG2, SVG3
k_L	TCT traction port voltage and primary sideline voltage transformation ratio
k_M	TCT compensation port voltage and primary sideline voltage transformation ratio
I_{L1}	The effective value of fundamental current of traction load
ϕ_{L1}	Power factor angle of fundamental current of traction load
I_{L1p},I_{L1q}	Instantaneous active component andreactive component of I_{L1}
I_1^*,I_2^*,I_3^*	Expected values of compensation current of SVG1, SVG2, and SVG3
ε_{U2}	The limit of three-phase voltage unbalance degree at PCC
S_ε	The allowable negative sequence power at PCC
S_d	The short circuit capacity at PCC
$\dot{S}^-$	The residual negative sequence power at PCC after compensation
ε_{U2}^*	Expected value of three-phase voltage unbalance degree at PCC
$\cos\phi^*$	Expected value of power factor

References

1. Chen, M.; Chen, Y.; Wei, M. Modeling and Control of a Novel Hybrid Power Quality Compensation System for 25-kV Electrified Railway. *Energies* **2019**, *12*, 3303. [CrossRef]
2. He, X.; Ren, H.; Lin, J.; Han, P.; Wang, Y.; Peng, X.; Shu, Z. Power Flow Analysis of the Advanced Co-Phase Traction Power Supply System. *Energies* **2019**, *12*, 754. [CrossRef]
3. Railway Statistics Bulletin. 2019. Available online: http://www.gov.cn/xinwen/2020-04/30/content_5507767.htm (accessed on 29 April 2020).
4. Femine, A.D.; Gallo, D.; Giordano, D.; Landi, C.; Luiso, M.; Signorino, D. Power Quality Assessment in Railway Traction Supply Systems. *IEEE Trans. Instrum. Meas.* **2020**, *69*, 2355–2366. [CrossRef]
5. Chen, Y.; Chen, M.; Tian, Z.; Liu, Y.; Hillmansen, S. VU limit pre-assessment for high-speed railway considering a grid connection scheme. *IET Gener. Transm. Distrib.* **2019**, *13*, 1121–1131. [CrossRef]
6. Fang, L.; Xu, X.; Fang, H.; Xiao, Y. Negative-sequence current compensation of power quality compensator for high-speed electric railway. In Proceedings of the 2014 IEEE Conference and Expo Transportation Electrification Asia-Pacific (ITEC Asia-Pacific), Beijing, China, 31 August–3 September 2014; pp. 1–5.
7. Tian, X.; Li, X.; Zhou, Z. Novel Uninterruptible Phase-Separation Passing and Power Quality Compensation Scheme Based on Modular Multilevel Converter for Double-Track Electrified Railway. *Energies* **2020**, *13*, 738. [CrossRef]
8. Wu, C.; Luo, A.; Shen, J.; Ma, F.; Peng, S. A Negative Sequence Compensation Method Based on a Two-Phase Three-Wire Converter for a High-Speed Railway Traction Power Supply System. *IEEE Trans. Power Electron.* **2012**, *27*, 706–717. [CrossRef]
9. He, X.; Shu, Z.; Peng, X.; Zhou, Q.; Zhou, Y.; Zhou, Q.; Gao, S. Advanced Cophase Traction Power Supply System Based on Three-Phase to Single-Phase Converter. *IEEE Trans. Power Electron.* **2014**, *29*, 5323–5333. [CrossRef]

10. Zhou, Y.; Guo, A.; He, X. Output-transformerless traction substation based on three-phase to single-phase cascaded converter. In Proceedings of the 2014 International Power Electronics and Application Conference and Exposition, Shanghai, China, 5–8 November 2014; pp. 749–753.
11. Min, J.; Ma, F.; Xu, Q.; He, Z.; Luo, A.; Spina, A. Analysis, Design, and Implementation of Passivity-Based Control for Multilevel Railway Power Conditioner. *IEEE Trans. Ind. Inf.* **2018**, *14*, 415–425. [CrossRef]
12. Luo, A.; Ma, F.; Wu, C.; Ding, S.; Zhong, Q.; Shuai, Z. A Dual-Loop Control Strategy of Railway Static Power Regulator Under V/V Electric Traction System. *IEEE Trans. Power Electron.* **2011**, *26*, 2079–2091. [CrossRef]
13. Zhang, D.; Zhang, Z.; Wang, W.; Yang, Y. Negative Sequence Current Optimizing Control Based on Railway Static Power Conditioner in V/v Traction Power Supply System. *IEEE Trans. Power Electron.* **2016**, *31*, 200–212. [CrossRef]
14. Ma, F.; Xu, Q.; He, Z.; Tu, C.; Shuai, Z.; Lou, A.; Li, Y. A Railway Traction Power Conditioner Using Modular Multilevel Converter and Its Control Strategy for High-Speed Railway System. *IEEE Trans. Transp. Electr.* **2016**, *2*, 96–109. [CrossRef]
15. Chen, M.; Liu, R.; Xie, S.; Zhang, X.; Zhou, Y. Modeling and Simulation of Novel Railway Power Supply System Based on Power Conversion Technology. In Proceedings of the 2018 International Power Electronics Conference (IPEC-Niigata 2018-ECCE Asia), Niigata, Japan, 20–24 May 2018; pp. 2547–2551.
16. Zhu, X.; Chen, M.; Xie, S.; Luo, J. Research on new traction power system using power flow controller and Vx connection transformer. In Proceedings of the 2016 IEEE International Conference on Intelligent Rail Transportation (ICIRT), Birmingham, UK, 23–25 August 2016; pp. 111–115.
17. Kaleybar, H.J.; Brenna, M.; Foiadelli, F.; Fazel, S.S. Regenerative Braking Energy and Power Quality Analysis in 2 $\times$ 25 kV High-Speed Railway Lines Operating with 4QC Locomotives. In Proceedings of the 2020 11th Power Electronic and Drive Systems and Technologies Conference (PEDSTC), Tehran, Iran, 4–6 February 2020; pp. 1–6.
18. Mariscotti, A. Behaviour of Spectral Active Power Terms for the Swiss 15 kV 16.7 Hz Railway System. In Proceedings of the 2019 IEEE 10th International Workshop on Applied Measurements for Power Systems (AMPS), Aachen, Germany, 25–27 September 2019; pp. 1–6.
19. Li, Q. New generation traction power supply system and its key technologies for electrified railways. *J. Mod. Transport.* **2015**, *23*, 1–11. [CrossRef]
20. Mochinaga, Y.; Hisamizu, Y.; Takeda, M.; Miyashita, T.; Hasuike, K. Static power conditioner using GTO converters for AC electric railway. In Proceedings of the Power Conversion Conference, Yokohama, Japan, 19–23 April 1993; pp. 641–646.
21. Mochinaga, Y.; Uzuka, T. Development of Single Phase Feeding Power Conditioner for Shinkansen Depots. *QR RTRI* **2000**, *41*, 154–158. [CrossRef]
22. Morimoto, H.; Ando, M.; Mochinaga, Y.; Kato, T.; Yoshizawa, J.; Gomi, T.; Miyashita, T.; Funahashi, S.; Nishitoba, M.; Oozeki, S. Development of railway static power conditioner used at substation for Shinkansen. In Proceedings of the Power Conversion Conference, Osaka, Japan, 2–5 April 2002; pp. 1108–1111.
23. Li, Q.; Zhang, J.; He, W. Research on a new power supply system suitable for heavy-duty electric traction. *J. CRS* **1988**, *4*, 23–31. (In Chinese)
24. Li, Q. Unified traction power supply mode for trunk railway and urban rail transit. *Sci. Sin. Tech.* **2018**, *48*, 1179–1189. (In Chinese) [CrossRef]
25. Li, Q. Several Key Technical Issues in the Development of Traction Power Supply for High-speed Railways in China. *J. CRS* **2010**, *32*, 119–124. (In Chinese)
26. Shu, Z.; Xie, S.; Li, Q. Single-Phase Back-To-Back Converter for Active Power Balancing, Reactive Power Compensation, and Harmonic Filtering in Traction Power System. *IEEE Trans. Power Electron.* **2011**, *26*, 334–343. [CrossRef]
27. Shu, Z.; Xie, S.; Lu, K.; Zhao, Y.; Nan, X.; Qiu, D.; Zhou, F.; Gao, S.; Li, Q. Digital Detection, Control, and Distribution System for Co-Phase Traction Power Supply Application. *IEEE Trans. Ind. Electron.* **2013**, *60*, 1831–1839. [CrossRef]
28. He, X.; Guo, A.; Peng, X.; Zhou, Y.; Shi, Z.; Shu, Z. A Traction Three-Phase to Single-Phase Cascade Converter Substation in an Advanced Traction Power Supply System. *Energies* **2015**, *8*, 9915–9929. [CrossRef]
29. Lao, K.; Wong, M.; Dai, N.Y.; Wong, C.; Lam, C. Analysis of DC-Link Operation Voltage of a Hybrid Railway Power Quality Conditioner and Its PQ Compensation Capability in High-Speed Cophase Traction Power Supply. *IEEE Trans. Power Electron.* **2016**, *31*, 1643–1656. [CrossRef]
30. Habibolahzadeh, M.; Roudsari, H.M.; Jalilian, A.; Jamali, S. Improved Railway Static Power Conditioner Using C-type Filter in Scott Co-phase Traction Power Supply System. In Proceedings of the 2019 10th International Power Electronic and Drive Systems and Technologies Conference (PEDSTC), Shiraz, Iran, 12–14 February 2019; pp. 355–360.
31. Vemulapati, V.; Vijayakumar, Y.N.; Visali, N. Droop Characteristics based High Speed Traction Power Supply system using Modular Multilevel Converter. In Proceedings of the 2020 4th International Conference on Trends in Electronics and Informatics (ICOEI), Tirunelveli, India, 15–17 June 2020; pp. 111–118.
32. Vijaykumar, Y.N.; Vemulapati, V.; Visali, N.; Raju, K. Railway Power Supply System using Modular Multilevel Converter with Droop Characteristics. In Proceedings of the 2020 4th International Conference on Electronics, Communication and Aerospace Technology (ICECA), Coimbatore, India, 5–7 November 2020; pp. 12–20.
33. AlBader, M.; Enjeti, P. Three phase to Co-phase railway electrification approach using voltage synthesizing electronic phase shifter (EPS). In Proceedings of the 2016 IEEE Transportation Electrification Conference and Expo (ITEC), Dearborn, MI, USA, 27–29 June 2016; pp. 1–4.

34. Mousavi Gazafrudi, S.M.; Tabakhpour Langerudy, A.; Fuchs, E.F.; Al-Haddad, K. Power Quality Issues in Railway Electrification: A Comprehensive Perspective. *IEEE Trans. Ind. Electron.* **2015**, *62*, 3081–3090. [CrossRef]
35. Omi, M.; Kotegawa, R.; Ando, M.; Masui, T.; Horita, Y. Introduction and effectiveness of STATCOM to the independent power system of JR East. In Proceedings of the 2014 International Power Electronics Conference (IPEC-Hiroshima 2014-ECCE ASIA), Hiroshima, Japan, 18–21 May 2014; pp. 1317–1321.
36. Tamai, S. Novel power electronics application in traction power supply system in Japan. In Proceedings of the 2014 16th International Power Electronics and Motion Control Conference and Exposition, Antalya, Turkey, 21–24 September 2014; pp. 701–706.
37. Luo, L.; Chen, P.; Cui, P.; Zhang, Z.; Zhou, Y.; Xie, X.; Chen, J. A current balance compensation method for traction substation based on SVG and V/v transformer. In Proceedings of the 2017 IEEE Transportation Electrification Conference and Expo, Asia-Pacific (ITEC Asia-Pacific), Harbin, China, 7–10 August 2017; pp. 1–6.
38. Horita, Y.; Morishima, N.; Kai, M.; Onishi, M.; Masui, T.; Noguchi, M. Single-phase STATCOM for feeding system of Tokaido Shinkansen. In Proceedings of the 2010 International Power Electronics Conference-ECCE ASIA, Sapporo, Japan, 21–24 June 2010; pp. 2165–2170.
39. Li, Q.; He, J. *Analysis of Traction Power Supply System*; Southwest Jiaotong University Press: Chengdu, China, 2012.
40. *Power Quality—Three-Phase Voltage Unbalance*; National Standard of the People's Republic of China GB/T 15543-2008; Standardization Administration of the People's Republic of China: Beijing, China, 2008.

Article

Harmonic Mitigation in Electric Railway Systems Using Improved Model Predictive Control

Chakrit Panpean [1], Kongpol Areerak [1,*], Phonsit Santiprapan [2], Kongpan Areerak [1] and Seang Shen Yeoh [3]

1 School of Electrical Engineering, Suranaree University of Technology, Nakhon Ratchasima 30000, Thailand; chakritpanpean@gmail.com (C.P.); kongpan@sut.ac.th (K.A.)
2 Department of Electrical Engineering, Prince of Songkla University, Songkhla 90110, Thailand; phonsit.s@psu.ac.th
3 Power Electronics, Machines and Control Group, University of Nottingham, Nottingham NG7 2RD, UK; seang.yeoh@nottingham.ac.uk
* Correspondence: kongpol@sut.ac.th; Tel.: +66-8-1208-2873

Citation: Panpean, C.; Areerak, K.; Santiprapan, P.; Areerak, K.; Shen Yeoh, S. Harmonic Mitigation in Electric Railway Systems Using Improved Model Predictive Control. *Energies* **2021**, *14*, 2012. https://doi.org/10.3390/en14072012

Academic Editors: Andrea Mariscotti and Leonardo Sandrolini

Received: 15 March 2021
Accepted: 2 April 2021
Published: 5 April 2021

Abstract: An electric multiple unit (EMU) high-speed train is the dynamic load that degrades the power quality in an electric railway system. Therefore, a power quality improvement system using an active power filter (APF) must be considered. Due to the oscillating load current in the dynamic load condition, a fast and accurate harmonic current-tracking performance is required. As such, this paper proposes the design of a model predictive control (MPC) since the minimization of cost function in the MPC process can suitably determine APF switching states. The design technique of MPC is based on the APF mathematical model. This controller was designed to compensate the time delay in the digital control. Moreover, the synchronous detection (SD) method applied the reference current calculations, as shown in this paper. To verify the proposed MPC, the overall control of APF was implemented on a eZdsp F28335 board by using the hardware-in-the-loop technique. The testing results indicated that the proposed MPC can provide a fast and accurate harmonic current-tracking response compared with the proportional-integral controller. In the load changing condition, the MPC was still effective in providing a good result after compensation. The percentage of total harmonic distortion, the percentage current unbalance factor, and the power factor would also be kept within the IEEE Standard 519 and IEEE Standard 1459.

Keywords: harmonic mitigation; active power filter; synchronous detection; model predictive control; electric railway system

1. Introduction

Nowadays, electric multiple unit high-speed trains (EMU high-speed trains) are an important part of the transportation infrastructure. The power distribution system is the utility supply for EMU high-speed train drives. The EMU high-speed train consists of a single-phase rectifier, three-phase inverter, traction motor drive, facilities in the passenger compartment, and other loads. These components behave like a linear and non-linear load. These loads produce harmonic currents, unbalanced currents, and reactive power on the utility supply side. The degradation of power quality can lead to many disadvantages, such as power loss in transmission lines and power transformers [1,2]; the arcing [3] and irregularity [4] of the EMU's pantograph; interference in the communication systems and electric railway signaling systems [5]; electromagnetic interference of electric devices [6]; malfunction of the device [7]; and protective device failures and short-life electric devices [8]. In [9,10], the EMU-high speed train works in three typical situations. The first situation is during accelerating startup. This situation results in a fast increase in the load current. In the second situation, the EMU high-speed train moves away at a constant speed. The load current is invariable. The last situation is a braking. In this situation, the load current decreases rapidly. Thus, the load current behavior of the electric railway system is dynamically varied, which significantly affects the power quality in the transient condition. The undesirable effects on the power system, power equipment, and adjacent loads may occur. The reliability requirements of power distribution systems are necessary. Consequently, the power quality improvement in the electric railway system must be examined.

By considering the power quality conditioning in railway electrification, there are many techniques, such as static VAR compensators (SVC) [11], static synchronous compensators (STATCOM) [12], passive power filter (PPF) [13], active power filter (APF) [14], and hybrid power filter (HPF) [15]. The SVC and STATCOM are used for the imbalance current and reactive power compensations. Nevertheless, SVC and STATCOM cannot mitigate the harmonic. The PPF are widely installed in the system to mitigate the main harmonics for large loads, but the current balancing cannot be achieved using this technique. PPF parameters are designed to mitigate the harmonic at the specific frequency (resonance frequency). In the case of load changing, these parameters should be redesigned to maintain good compensation performance. It should be noted that the PPF are not suitable for dynamic load conditions, especially for EMU high-speed trains.

The APF can compensate both for imbalanced currents and reactive power, as well as perform harmonic mitigation. To solve the aforementioned problems, an APF is a suitable power filter for the electric railway system [16]. The APF would provide a faster dynamic response, higher compensation performance, and more flexibility compared with the SVC, STATCOM, PPF, and HPF. Therefore, the power quality improvement of an electric railway system using the APF is intensively studied in this paper.

The harmonic mitigation topology in an electric railway system using APF is shown in Figure 1. It is well known that the compensating current control is an important part of the APF system. Therefore, from the literature survey, it was found that there are several strategies to control the compensating current injection. In the electric railway system, the hysteresis control was applied for the harmonic mitigation in the electric railway system [17]. This control has a fast dynamic response, simplicity, and is also insensitive to the model and system parameter variations. However, the hysteresis operation provides an oscillating and high switching frequencies. It can cause the compensating current ripple. Moreover, the frequency rating of the devices (IGBTs, MOSFETs, etc.) and the heating due to switching losses must be considered for practical work. For an APF with another system, the PI controller [18] is widely used to control the compensating current. It has the proportional and integral terms. These terms offer good transient and steady state responses. However, the parameter design of the PI controller is based on a mathematical model of the system. Here, the harmonic current-tracking performance is affected during a variation of the system parameters. A proportional resonant (PR) controller is also applied for the APF control [19]. The PR controller parameters can be tuned for the specific harmonic frequency. This point has good tracking performance between the compensating current and the reference current at the selected harmonic frequency. However, the harmonic components may occur at any frequency. Thus, excellent harmonic current tracking performance cannot be completely achieved by the PR controller. In addition, in a three-phase four-wire system, a repetitive (RT) controller [20] was adopted. This controller can provide the steady-state tracking accuracy for the harmonic current control. Due to the periodic operation by the RT mechanism, the harmonic current will not be exactly compensated in the transient condition. Moreover, the stability analysis of the RT controller has become a serious problem in the design process.

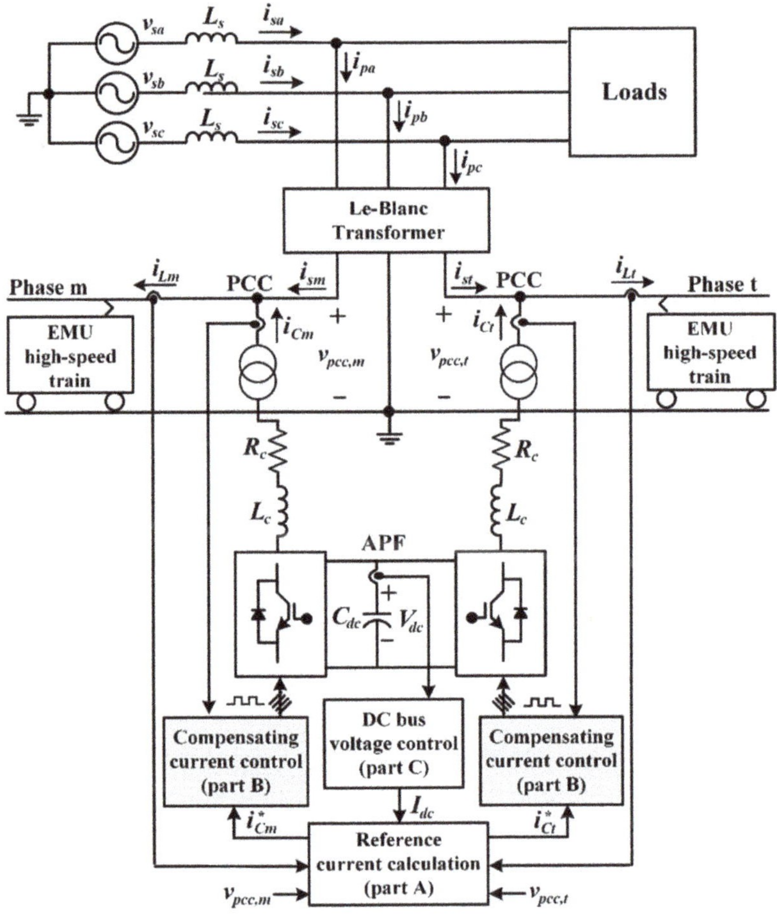

Figure 1. The structure of electric railway system and the control strategy of the harmonic mitigation system using an active power filter (APF). EMU: electric multiple unit (EMU); PCC: point of common coupling.

From the presentation in [9,10], the load current of electric railway systems fluctuates along with the traffic situation. From the load current fluctuation, a suitable compensating Current controller of the APF in the electric railway system must have a fast dynamic response according to the load current behavior. In recent years, the MPC is extensively used for power electronics converters because of the advantages of high control accuracy and fast dynamic response [21]. Therefore, the MPC is suitable for compensating the APF current control. Processing and sampling delays exist in a digital control. Here, the MPC can compensate for these delays. The mechanism of this controller can be designed to predict future switching states for the converter. The stability-constrained of the MPC controller is well-proven in [22]. In this work, the MPC was used to improve the performance of the APF compensating current control in an electric railway system. Moreover, the MPC compensating current control in the electric railway system was not found in any of the previous research.

This paper is structured as follows: Section 2 describes the structure of electric railway systems and the APF control strategy. The reference current calculation using the SD method is presented in Section 3. Then, the compensating current control based on the

MPC is discussed in detail in Section 4. Section 5 presents the configuration of the HIL simulation technique. The testing results of the harmonic mitigation and discussions are described in Section 6. Finally, Section 7 concludes this work.

2. Active Power Filter Topology in an Electric Railway System.

The structure of an electric railway system is depicted in Figure 1. The Le-Blanc transformer is utilized to step down the voltage level from 69 kV (three-phase voltage source) to 26 kV (two single-phase voltage source). The phase angle difference of the two single-phase power system between phase m and phase t is 90 degrees. These voltages are used to drive the EMU high-speed train. The advantage of using a Le-Blanc transformer is the voltage source balancing at phase m and phase t. The non-linear load characteristics of the electric railway system can be found in [23].

The APF circuit topology in the electric railway systems consists of voltage source inverters (VSIs). The two VSIs are connected back -to-back through a common DC capacitor (C_{dc}). The VSIs are connected to the step-down side of the transformer through the line impedance at phase m and phase t ($R_{C(m,t)}$, $L_{C(m,t)}$). These VSIs act as a source to inject the compensating currents into the point of common coupling (PCC). An APF control strategy would directly have an impact on the performance of the power quality enhancement. Therefore, this point has been continuously developed. An APF control strategy in an electric railway system, as shown in Figure 1, consists of the reference current calculation (part A), the DC bus voltage control (part B), and the compensating current control (part C). By considering "part A" from previous publications, there are many methods for the reference calculation [24–26]. In this work, the SD method was proposed. This method can calculate an accurate harmonic current. The calculation procedure of the SD method is simple. It results in a fast computational time. In the literature review, the DC bus voltage level for APFs [12,15,16,18] and power converter circuits [27,28] is commonly regulated by a PI controller because the steady-state accuracy for the DC bus voltage control was considered. From the previous work [29], the DC bus voltage control loop using the PI controller was suitably designed. Therefore, the PI controller parameters and the reference DC bus voltage in [29] are defined for DC bus voltage control in part B. Finally, part C is the compensating current control, which significantly affected the injection performance of the compensating current. The purpose of this part was to design a controller to track the compensating current with the least error possible. The harmonic mitigation, power factor correction, and load balancing can be achieved efficiently by the improvement of the control strategy.

3. Reference Current Calculation

The SD method is used to calculate the reference currents ($i^*_{C(m,t)}$) for the compensating current controls. The SD method applied in the three-phase power system was firstly proposed in 1992 by Lin et al. [30]. For this method, the computational capability of the reference currents is simple. Here, it can provide a fast calculation time. This approach is suitable for the real-time implementation on a digital signal processing (DSP) board. From the prominent point explained above, the SD method is applied to calculate the $i^*_{C(m,t)}$ in this work. The calculation procedure of the reference currents by the SD method can be summarized by the block diagram in Figure 2. There are four steps to calculate the $i^*_{C(m,t)}$:

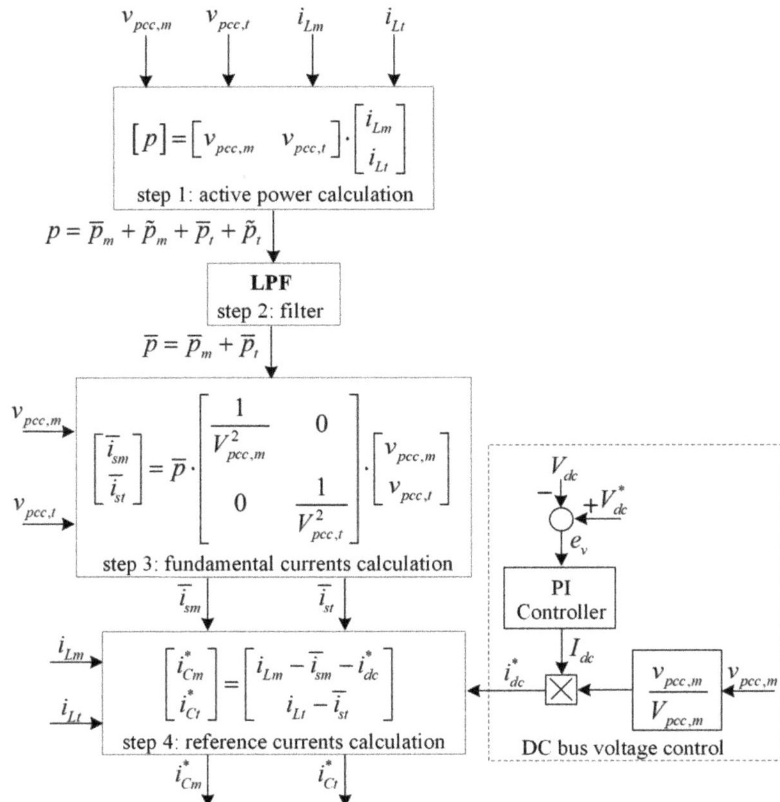

Figure 2. The block diagram of SD method for the APF in electric railway system.

Step 1: Calculate the instantaneous power of the EMU high-speed train (p) using Equation (1). The assumption of this method is that the equal average power have to be obtained after compensation. The p can be separated in terms of DC components ($\overline{p}_{m,t}$) and AC components ($\tilde{p}_{m,t}$). The DC and AC terms mean that there are the fundamental and harmonic components, respectively:

$$\begin{cases} p = (v_{pcc,m} \times i_{Lm}) + (v_{pcc,t} \times i_{Lt}) \\ p = \overline{p}_m + \tilde{p}_m + \overline{p}_t + \tilde{p}_t \end{cases} \tag{1}$$

Step 2: Calculate the fundamental component of the instantaneous power ($\overline{p}$). The low-pass filter (LPF) is chosen to draw the $\overline{p}$ from the p. In this work, the cutoff frequency of LPF is equal to 60 Hz.

Step 3: Calculate the fundamental currents ($\overline{i}_{s(m,t)}$) by Equation (2):

$$\begin{cases} \overline{i}_{sm} = \frac{\overline{p} \times v_{pcc,m}}{V_{pcc,m}^2} \\ \overline{i}_{st} = \frac{\overline{p} \times v_{pcc,t}}{V_{pcc,t}^2} \end{cases} \tag{2}$$

where $v_{pcc,(m,t)}$ are the amplitude of the single-phase source voltages for each phase.

Step 4: Calculate the $i^*_{C(m,t)}$ for the compensating current controls using Equation (3):

$$\begin{cases} i^*_{Cm} = i_{Lm} - \bar{i}_{sm} - i^*_{dc} \\ i^*_{Ct} = i_{Lt} - \bar{i}_{st} \end{cases} \quad (3)$$

4. Compensating Current Control System with Model Predictive Control

4.1. Principle of a MPC

The controller's delay time is an important issue with digital control because of the performance improvement in the transient condition. Therefore, the time delay compensation is concerned with the digital controller. The MPC has an ability to predict the next switching states of the converter over a fixed set time horizon, and also compensate the digital time delay. The operation of MPC is based on a system behavior prediction using a mathematical model. The measured values of the system in their present states are required for the predictive data. In addition, the minimization of cost function is a part of the MPC mechanism. This function determines suitably the converter's switching states. Therefore, this controller can provide a fast dynamic response and high control performance. The MPC approach, when applied to the APF compensating current control, is depicted in Figure 3, and can be divided into three principle components: the reference current prediction, the compensating current prediction, and the minimization of cost function. In this work, the MPC design is based on the APF switching state prediction using a mathematical model. The computational burden of the control platform and the DSP board operation leads to a processing delay of two sampling periods [31]. To compensate for the delay time in the computational of the control platform and the DSP board operation, the predictions are determined at the time n + 2. The prediction process uses the measured values of the system at the time n (present states). In addition, the minimization of cost function is a part of MPC mechanism.

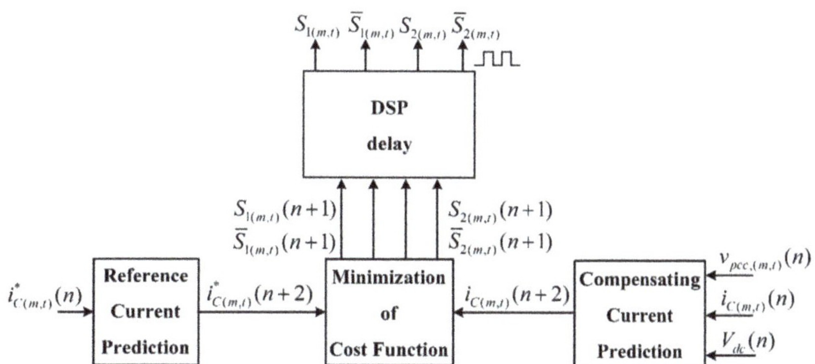

Figure 3. The model predictive control for compensating current control system. DSP: digital signal processing.

4.2. The Reference Current Prediction

The $i^*_{C(m,t)}$ calculated via the SD method is the reference currents at the time instant n ($i^*_{C(m,t)}(n)$). However, the computational burden of the control platform and the DSP board operation leads to a processing delay for the two sampling periods. In order to compensate for the processing delay, the reference currents prediction at the time instant n + 2 ($i^*_{C(m,t)}(n+2)$) must be taken into account in the MPC approach. Moreover, ref. [32] presents the use of the first-order Lagrange method, which can provide a good performance for the reference current prediction. Thus, the first-order Lagrange method is used to predict the $i^*_{C(m,t)}(n+2)$ in this paper. By considering the $i^*_{C(m,t)}(n)$ calculated from the SD method, the reference currents at the previous time instant $n-1$ ($i^*_{C(m,t)}(n-1)$) and $i^*_{C(m,t)}(n+2)$

are derived using the linear-type prediction. The $i^*_{C(m,t)}(n+2)$ can be expressed as Equation (4):

$$i^*_{C(m,t)}(n+2) = 3i^*_{C(m,t)}(n) - 2i^*_{C(m,t)}(n-1) \quad (4)$$

4.3. The Compensating Current Prediction

The APF circuit structure, as shown in Figure 4, is firstly considered to study the compensating current prediction. According to Figure 4, the APF operation is to inject the compensating currents ($i_{C(m,t)}$) from the APF bus to the PCC bus through a linear transformer. The differential equation of the compensating currents ($\frac{di_{CP,(m,t)}}{dt}$) are adopted for the process of the compensating the current prediction. Therefore, the differential equation of the APF in Equation (5) is considered:

$$\frac{di_{CP,(m,t)}}{dt} = \frac{v_{inv,(m,t)} - R_{C(m,t)}i_{CP,(m,t)} - v_{P(m,t)}}{L_{C(m,t)}} \quad (5)$$

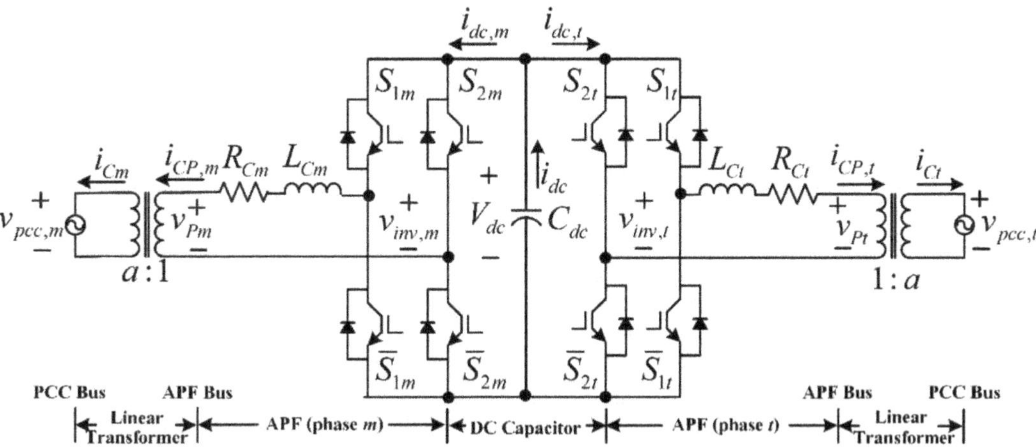

Figure 4. The structure of APF in electric railway systems.

Here, the forward Euler formula is used to approximate the $\frac{di_{CP,(m,t)}}{dt}$ term. The $\frac{di_{CP,(m,t)}}{dt}$ can be considered as the discrete time model in Equation (6). The compensating current of the next sampling period ($i_{CP,(m,t)}(n+1)$) are obtained, as in Equation (7), where the $v_{P(m,t)}(n)$ and $v_{inv,(m,t)}(n)$ are the APF output voltage and the inverter output voltage at the time instant n, respectively. $v_{P(m,t)}(n)$ are equal to $v_{inv,(m,t)}(n)/a$. Further, $1:a$ is a conversion ratio of the linear transformer:

$$\frac{di_{CP,(m,t)}}{dt} \approx \frac{i_{CP,(m,t)}(n+1) - i_{CP,(m,t)}(n)}{T_s} \quad (6)$$

$$i_{CP,(m,t)}(n+1) = \left(1 - \frac{R_{C(m,t)}T_s}{L_{C(m,t)}}\right)i_{CP,(m,t)}(n) + \frac{T_s}{L_{C(m,t)}}\left(v_{inv,(m,t)}(n) - v_{P(m,t)}(n)\right) \quad (7)$$

We note that the processing delay compensation is necessary to achieve the fast dynamic response and precise current tracking. The compensating current prediction at the time instant $n+2$ ($i_{CP,(m,t)}(n+2)$) are to compensate for the processing delay. Here, it has the ability to eliminate the current tracking error at the next sampling period. Therefore, $i_{CP,(m,t)}(n+1)$ is shifted forward to $i_{CP,(m,t)}(n+2)$, as in Equation (8):

$$i_{CP,(m,t)}(n+2) = \left(1 - \frac{R_{C(m,t)}T_s}{L_{C(m,t)}}\right)i_{CP,(m,t)}(n+1) + \frac{T_s}{L_{C(m,t)}}\left(v_{inv,(m,t)}(n+1) - v_{P(m,t)}(n+1)\right) \quad (8)$$

In addition, $v_{P(m,t)}(n+1)$ can follow a similar procedure as the one mentioned above. These values are calculated by Equation (9):

$$v_{P(m,t)}(n+1) = v_{inv,(m,t)}(n+1) + \frac{L_{C(m,t)}}{T_s} a i_{C(m,t)}(n) - \left(\frac{R_{C(m,t)} T_s + L_{C(m,t)}}{T_s}\right) i_{CP,(m,t)}(n+1) \quad (9)$$

From Equation (9), the inverter output voltage prediction at the time instant $n + 1$ ($v_{inv,(m,t)}(n+1)$) in Equation (10) relates to the switching states of the APF ($S_{1(m,t)}(n)$,$S_{2(m,t)}(n)$) and the DC bus voltage ($V_{dc}(n)$) at the time instant n:

$$v_{inv,(m,t)}(n+1) = \left(S_{1(m,t)}(n) - S_{2(m,t)}(n)\right) \times V_{dc}(n) \quad (10)$$

4.4. The Minimization of Cost Function

The main objective of this work was to reduce the tracking errors between $i_{C(m,t)}$ and $i^*_{C(m,t)}$. Therefore, the absolute errors between $i^2_{C(m,t)}(n+2)$ and $i^{*2}_{C(m,t)}(n+2)$ are the cost function ($g_{MPC,(m,t)}$). The smallest of $g_{MPC,(m,t)}$ value is used to select the switching state for the minimum errors between $i_{C(m,t)}$ and $i^*_{C(m,t)}$. The time horizon of the cost function is the possible switching states of APF. The $g_{MPC,(m,t)}$ can be defined by Equation (11), which the $i_{C(m,t)}(n+2)$ equal to $i_{CP,(m,t)}(n+2)/a$:

$$g_{MPC,(m,t)} = \left| i^{*2}_{C(m,t)}(n+2) - i^2_{C(m,t)}(n+2) \right| \quad (11)$$

The compensating currents are predicted by considering the APF switching states. The VSIs was studied. The possible switching states can be categorized into four cases, as listed in Table 1. This assumes that an inverted pair of signals is provided to the two switches connected to each converter leg. The switching state at the time instant $n + 2$ can provide the smallest value of $g_{MPC,(m,t)}$. This switching state is applied at the next sampling period. The whole process of MPC can be summarized using a flow diagram, as shown in Figure 5.

Table 1. The possible switching states of APF.

Switching States (k)	$S_{1(m,t)}$	$S_{2(m,t)}$	$V_{inv,(m,t)}$ (V)
1	0 (off)	0 (off)	0
2	1 (on)	0 (off)	V_{dc}
3	0 (off)	1 (on)	$-V_{dc}$
4	1 (on)	1 (on)	0

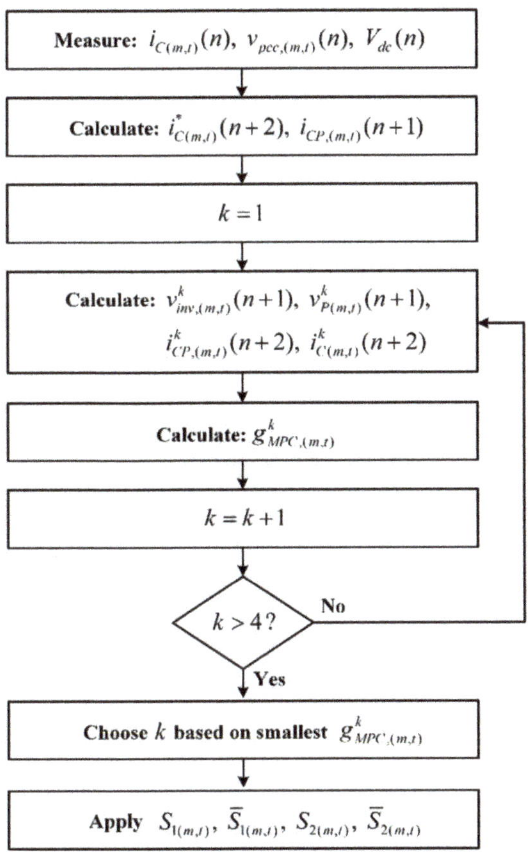

Figure 5. The MPC flow diagram for each sampling period.

5. Hardware-in-the-Loop (HIL) Simulation

In this work, we focused on the design and development of the MPC to improve current control APF performance. We introduced a harmonic mitigation system for electric railway systems. To first prove the enhanced approach and to clearly investigate its performance before the real experimental platform setup, the HIL technique was applied for validation. The overall control strategy improved in this work was implemented digitally in the eZdsp F28335 board. The considered harmonic mitigation system in the electric railway system was simulated using the MATLAB/Simulink program, while the reference current calculation was simulated using the SD method, the compensating current controls with the MPC controller, and the DC bus voltage control with the PI controller written in C programming languages using the code composer studio (CCStudio) software on the eZdsp F28335 board. The MATLAB/Simulink program in the host computer and the CCStudio software in the eZdsp F28335 board were interfaced by the JTAG emulator, as depicted in Figure 6. The blocks in Simulink called "From RTDX" and "To RTDX" were used to send and receive data between MATLAB/Simulink program and eZdsp F28335 board, respectively. Note that the HIL simulation setup was achieved by opening the operating mode of the TMS320 F28335 card and the J9 jumper on the docking station.

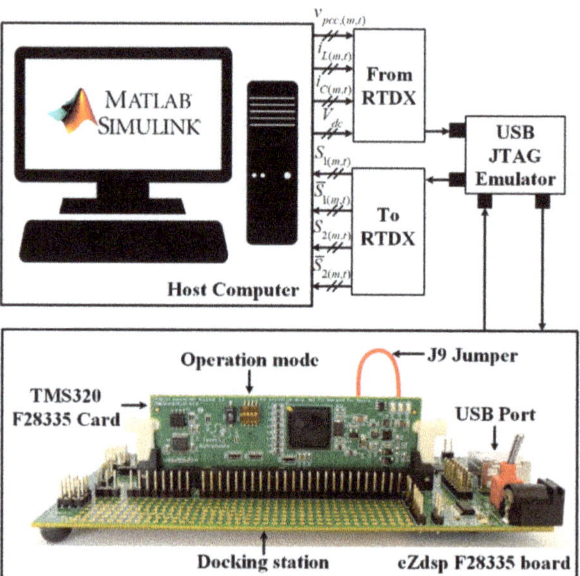

Figure 6. The configuration of the hardware-in-the-loop (HIL) simulation.

The HIL simulation model used to test the performance of the harmonic mitigation is shown in Figure 7. The PCC voltages ($v_{pcc,(m,t)}$), load currents ($i_{L(m,t)}$), compensating currents ($i_{C(m,t)}$), and DC bus voltage (V_{dc}) were measured from the considered power systems. These data were sent to the target of eZdsp F28335 with the "From RTDX" block, and were calculated in the overall control strategy to generated the pulses ($S_{m,t}$).$S_{m,t}$ were transferred into the host computer via the "To RTDX" block. $S_{m,t}$ were used to switch the IGBTs of the APF for the compensating current injection control.

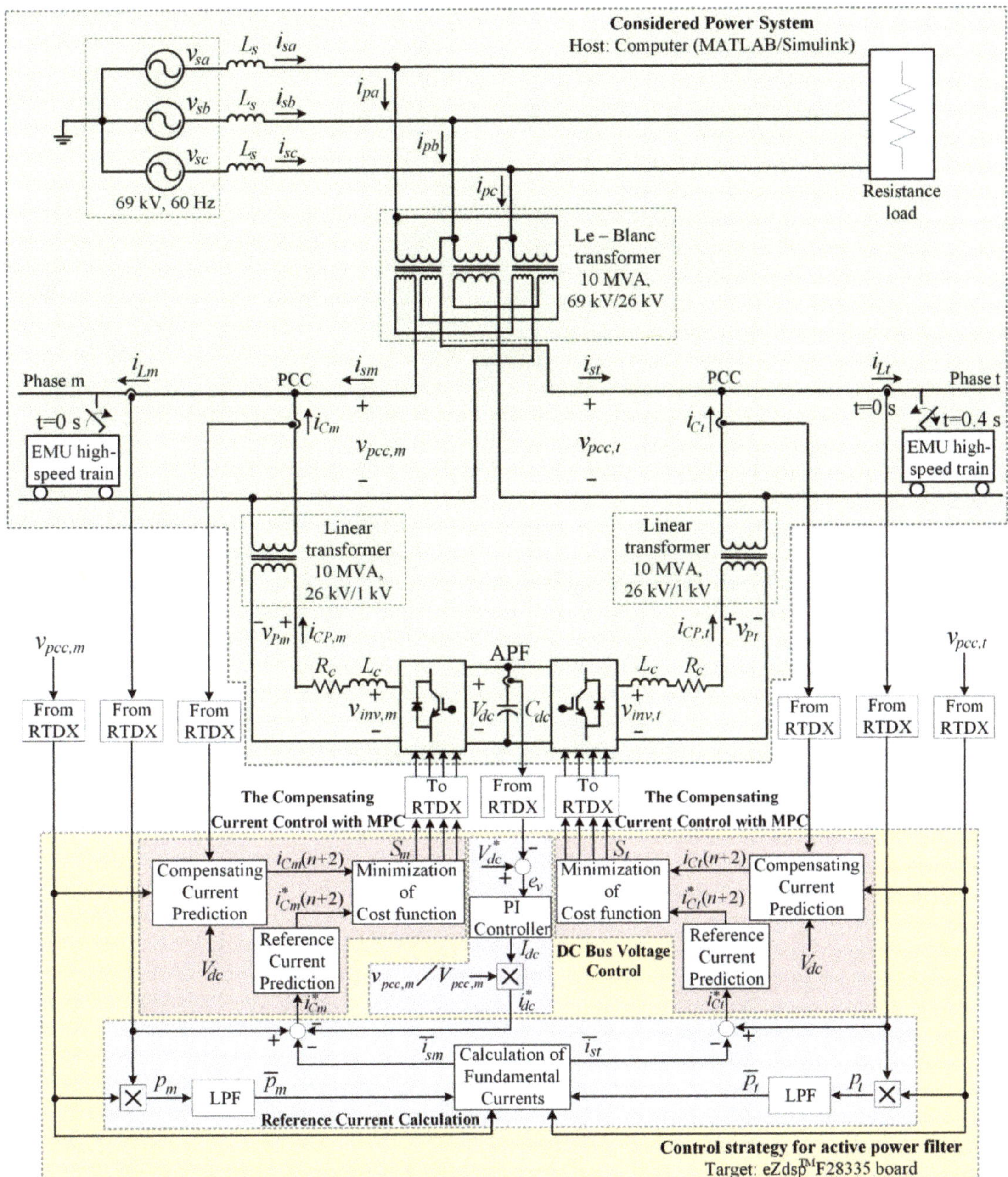

Figure 7. The HIL simulation model for the harmonic mitigation using the APF in the electric railway system.

6. Results and Discussion

From the HIL model presented above, the performance comparison study between the conventional PI controller [29] and the proposed MPC for the compensating current control is herein discussed. The harmonic mitigation in the electric railway system was tested in two situations. In situation 1, at t = 0–0.4 s, the balanced loads of EMU high-speed train was considered. In situation 2, at t = 0.4–0.7 s, the balanced loads were changed to the unbalanced loads. The harmonic producing load in this situation was extremely changed for the electric railway system. This was found in the case of a fault of the electrical system. The performance evaluation of the controllers tested in this system were classified into two categories: the tracking of the compensating current control and APF performance.

6.1. The Tracking Performance of the Compensating Current Control

The aim of the compensating current control was to reduce the errors between the $i_{C(m,t)}$ and the $i^*_{C(m,t)}$. The tracking performance of the currents between $i_{C(m,t)}$ and $i^*_{C(m,t)}$ controlled by the PI controller and the MPC can be seen in Figure 8. According to the waveforms in Figure 8, the MPC can control $i_{C(m,t)}$ to track $i^*_{C(m,t)}$, as calculated the SD method. The results confirmed that the MPC can provide a small tracking error compared with the PI controller, even though the load currents of the EMU high-speed train were changed. In addition, the MPC provided a faster dynamic response than the PI controller, which can be seen from Figure 8 at the t = 0.4 s. In the load changing situation, the high oscillating current from PI controller can cause a problem in the system.

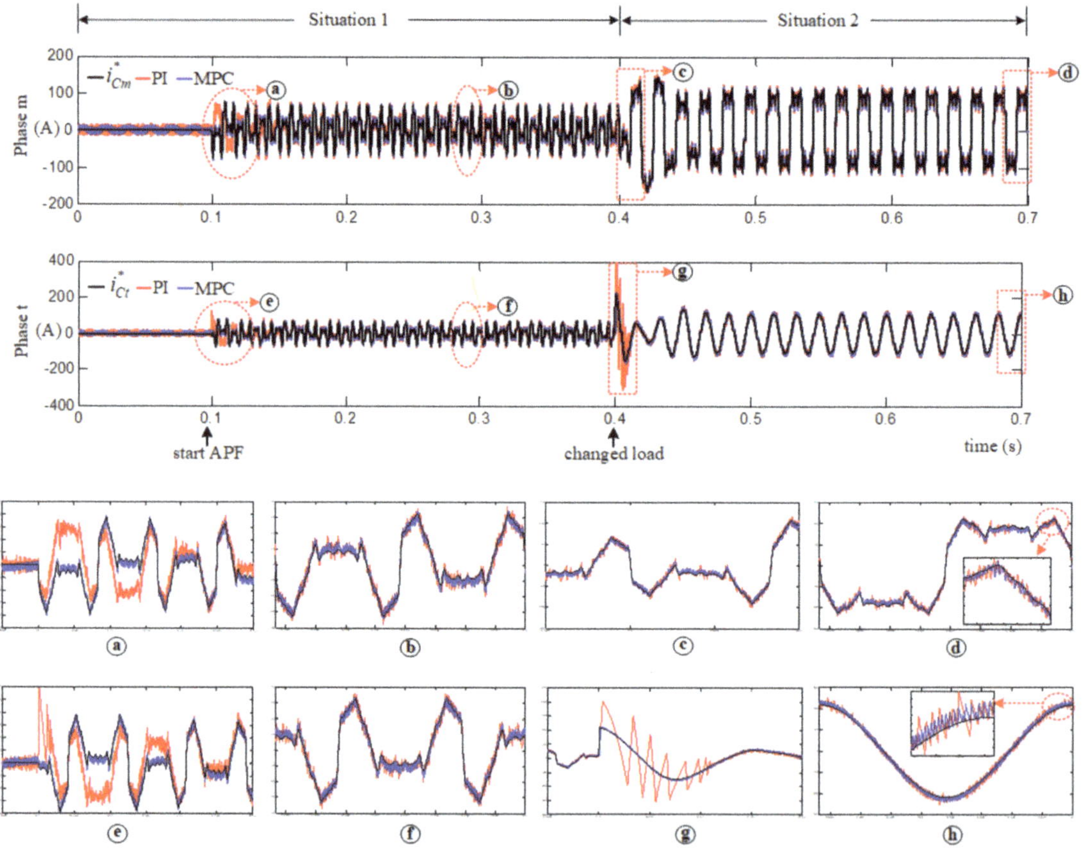

Figure 8. The tracking performance of the compensating currents.

6.2. The Performance of APF

EMU-high speed train loads can generate harmonic and unbalanced source currents. The performance indices for the harmonic mitigation in the electric railway system are the average total harmonic distortion of source currents (%THD_{av}) [33], the current unbalanced factor (%CUF) [34], and the power factor (PF). The testing results of the harmonic mitigation based on the MPC in Figure 9 show that the proposed controller can efficiently control the compensating currents. The performance indices of the harmonic mitigation are completely addressed in Table 2.

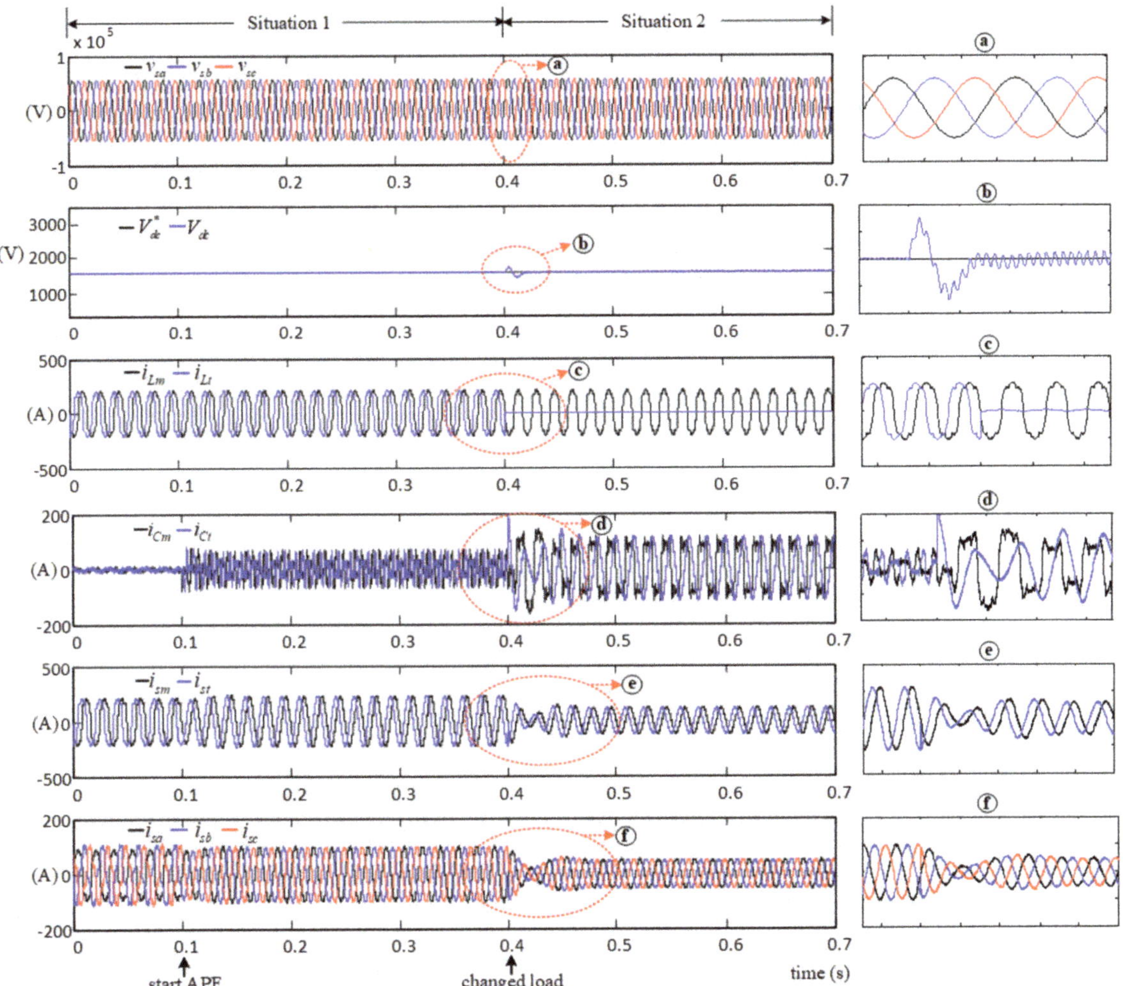

Figure 9. The testing results of the harmonic mitigation performance in the electric railway system using the MPC.

Table 2. The performance indices of the harmonic mitigation.

Performance Indices	Compensation	Situation 1		Situation 2	
		PI	MPC	PI	MPC
$\%THD_{av}$	Before	21.46		21.63	
	After	3.38	2.19	6.94	4.06
$\%CUF$	Before	0.00		95.45	
	After	0.00	0.00	2.25	1.56
PF	Before	0.97		0.70	
	After	0.99	0.99	0.99	0.99

Before compensation, the waveforms of $i_{L(m,t)}$ depended on nonlinear loads. These resulted in the source current distortion. The $\%THD_{av}$ values of these currents in situations 1–2 were equal to 21.46% and 21.63%, respectively. These values were much higher than the IEEE Std. 519–2014. Before compensation, the $\%CUF$ values for situations 1–2 were equal to 0.00% and 95.45%, respectively. The $\%CUF$ for the situation 2 was much greater than the IEEE Std. 1459–2010. In situations 1–2, the PF before compensation were equal to 0.97 and 0.70, respectively.

After compensation, the APF injected the $i_{C(m,t)}$ into the PCC at 0.1 s. The $i_{s(m,t)}$ became a more sinusoidal waveform, which caused the waveforms of $i_{s(a,b,c)}$ to nearly become a sinusoidal waveform, as shown in Figure 9. The $\%THD_{av}$ results of the controller testing are shown in Table 2. As a result, in situations 1–2, the MPC provides a small $\%THD_{av}$ compared with the PI controller. The unbalanced source current in situation 2 became balanced source current. According to situation 2, after compensation, the $\%CUF$ value from the MPC was less than the PI controller. Nevertheless, the PI controller and MPC provided good PF correctio. For the DC bus voltage control, the PI controller was sufficient to regulate the DC bus voltage (V_{dc}). The PI controller parameters ($K_p = 0.0554$ and $K_i = 0.7895$) [29] were defined for the DC bus voltage loop control. From Figure 9, it can be seen that the V_{dc} is kept constant at the V^*_{dc} using the PI controller.

7. Conclusions

The APF control based on the MPC in the electric railway system was completely presented in this paper. The SD method was applied for an APF reference current calculation. In order to compensate the processing delay for the MPC, the compensating current prediction at the time instant n+2 was designed. A fast transient response and steady-state tracking accuracy for the compensating current control was accomplished. The overall control strategy of APF was implemented on the eZdsp F28335 board. The HIL results were mentioned. The proposed MPC design clearly provides an excellent dynamic response in the load changing condition. The $\%THD_{av}$ and $\%CUF$ follows in the IEEE standard 519 and IEEE standard 1459, respectively. Moreover, this approach can provide the unity power factor. In future work, the control platform implemented on the eZdsp F28335 board is necessary for the experimental setup of APF in electric railway systems.

Author Contributions: C.P., K-L.A. (Kongpol Areerak) and S.S.Y. proposed the idea, completed the methodology analysis, and improved the current controller; C.P. analyzed data and performed the simulation; All authors participated in the conceptualization, methodology, and writing—review and editing. All authors have read and agreed to the published version of the manuscript.

Funding: This work was supported by the Thailand Research Fund (TRF), the Royal Golden Jubilee Ph.D program (RGJ: grant number PHD/0019/2560).

Conflicts of Interest: The authors declare no conflict of interest.

References

1. Weijun, L.; Qinghao, W.; Jingzhong, L.; Chenyang, L.; Zhitong, L.; Yi, W.; Xiangqun, Z. Research on transmission line power losses effected by harmonics. In Proceedings of the 2016 China International Conference on Electricity Distribution (CICED), Xi'an, China, 10–13 August 2016. [CrossRef]
2. Song, W.; Fang, J.; Jiang, Z.; Staines, M.; Badcock, R. AC Loss Effect of High-Order Harmonic Currents in a Single-Phase 6.5 MVA HTS Traction Transformer. *IEEE Trans. Appl. Supercond.* **2019**, *29*, 550145. [CrossRef]
3. Guo, L.; Gao, X.; Li, Q.; Huang, W.; Shu, Z. Online Antiicing Technique for the Catenary of the High-Speed Electric Railway. *IEEE Trans. Power Del.* **2015**, *30*, 1569–1576. [CrossRef]
4. Song, Y.; Liu, Z.; Rønnquist, A.; Nåvik, P.; Liu, Z. Contact Wire Irregularity Stochastics and Effect on High-Speed Railway Pantograph–Catenary Interactions. *IEEE Trans. Instrum. Meas.* **2020**, *69*, 8196–8206. [CrossRef]
5. Foley, F.J. The impact of electrification on railway signalling systems. In Proceedings of the 5th IET Professional Development Course on Railway Electrification Infrastructure and Systems, London, UK, 6–9 June 2011; pp. 146–153. [CrossRef]
6. Charalambous, C.A.; Demetriou, A.; Lazari, A.L.; Nikolaidis, A.I. Effects of Electromagnetic Interference on Underground Pipelines Caused by the Operation of High Voltage AC Traction Systems: The Impact of Harmonics. *IEEE Trans. Power Del.* **2018**, *33*, 2664–2672. [CrossRef]
7. Brumsickle, W.E.; Divan, D.M.; Luckjiff, G.A.; Freeborg, J.W.; Hayes, R.L. Power Quality and Reliability. *IEEE Ind. Appl. Mag.* **2005**, *11*, 48–53. [CrossRef]
8. IEEE. IEEE Recommended Practices and Requirements for Harmonic Control in Electrical Power Systems. *IEEE Standard 519–1992.* 1993. Available online: https://ieeexplore.ieee.org/servlet/opac?punumber=2227 (accessed on 1 April 2021).
9. Yu-quan, L.; Guo-pei, W.; Huang-sheng, H.; Li, W. Research for the effects of high-speed electrified railway traction load on power quality. In Proceedings of the 2011 4th International Conference on Electric Utility Deregulation and Restructuring and Power Technologies (DRPT), Weihai, China, 6–9 July 2011; pp. 569–573. [CrossRef]
10. Gazafrudi, S.; Langerudy, A.; Fuchs, E.; Al-Haddad, K. Power Quality Issues in Railway Electrification: A Comprehensive Perspective. *IEEE Trans. Ind. Electron.* **2015**, *62*, 3081–3090. [CrossRef]
11. Zhu, G.; Chen, J.; Liu, X. Compensation for the negative-sequence currents of electric railway based on SVC. In Proceedings of the 2008 3rd IEEE Conference on Industrial Electronics and Applications, Singapore, 3–5 June 2008; pp. 1958–1963. [CrossRef]
12. ElGebaly, A.E.; Hassan, A.E.-W.; El-Nemr, M.K. Reactive Power Compensation by Multilevel Inverter STATCOM for Railways Power Grid. In Proceedings of the 2019 IEEE Conference of Russian Young Researchers in Electrical and Electronic Engineering (EIConRus), Saint Petersburg and Moscow, Russia, 28–31 January 2019; pp. 2094–2099. [CrossRef]
13. Hu, H.; He, Z.; Gao, S. Passive Filter Design for China High-Speed Railway With Considering Harmonic Resonance and Characteristic Harmonics. *IEEE Trans. Power Del.* **2015**, *30*, 505–514. [CrossRef]
14. Sun, Z.; Jiang, X.; Zhu, D.; Zhang, G. A Novel Active Power Quality Compensator Topology for Electrified Railway. *IEEE Trans. Power Electron.* **2004**, *19*, 1036–1042. [CrossRef]
15. Lao, K.-W.; Wong, M.-C.; Dai, N.Y.; Wong, C.-K.; Lam, C.-S. A Systematic Approach to Hybrid Railway Power Conditioner Design with Harmonic Compensation for High-Speed Railway. *IEEE Trans. Power Electron.* **2015**, *62*, 930–942. [CrossRef]
16. He, Z.; Zheng, Z.; Hu, H. Power quality in high-speed railway systems. *Int. J. Rail Transp.* **2016**, *4*, 71–97. [CrossRef]
17. Luo, A.; Wu, C.; Shen, J.; Shuai, Z.; Ma, F. Railway Static Power Conditioners for High-speed Train Traction Power Supply Systems Using Three-phase V/V Transformers. *IEEE Trans. Power Electron.* **2011**, *26*, 2844–2856. [CrossRef]
18. Trinh, Q.-N.; Lee, H.-H. An Advanced Current Control Strategy for Three-Phase Shunt Active Power Filters. *IEEE Trans. Ind. Electron.* **2013**, *60*, 5400–5410. [CrossRef]
19. Herman, L.; Papic, I.; Blazic, B. A proportional-resonant current controller for selective harmonic compensation in a hybrid active power filter. *IEEE Trans. Power Del.* **2014**, *29*, 2055–2065. [CrossRef]
20. Santiprapan, P.; Booranawong, A.; Areerak, K.-L.; Saito, H. Adaptive repetitive controller for an active power filter in three-phase four-wire systems. *IET Power Electron.* **2020**, *13*, 2756–2766. [CrossRef]
21. Kouro, S.; Perez, M.A.; Rodriguez, J.; Llor, A.M.; Young, H.A. Model Predictive Control: MPC's Role in the Evolution of Power Electronics. *IEEE Ind. Electron. Mag.* **2015**, *9*, 8–21. [CrossRef]
22. Cheng, X.; Krogh, B.H. Stability-constrained model predictive control. *IEEE Trans. Auto. Control.* **2001**, *46*, 1816–1820. [CrossRef]
23. Huang, S.-R.; Chen, B.-N. Harmonic study of the Le Blanc transformer for Taiwan railway's electrification system. *IEEE Trans. Power Del.* **2002**, *17*, 495–499. [CrossRef]
24. Akagi, H.; Kanazawa, Y.; Nabae, A. Instantaneous Reactive Power Compensators Comprising Switching Devices without Energy Storage Components. *IEEE Trans. Ind. Appl.* **1984**, *IA-20*, 625–630. [CrossRef]
25. Sujitjorn, S.; Areerak, K.-L.; Kulworawanichpong, T. The DQ Axis with Fourier (DQF) Method for Harmonic Identification. *IEEE Trans. Power Del.* **2007**, *22*, 737–739. [CrossRef]
26. Santiprapan, P.; Areerak, K.-L.; Areerak, K.-N. A Novel Harmonic Identification Algorithm for the Active Power Filters in Non-Ideal Voltage Source Systems. *J. Power Electron.* **2017**, *17*, 1639–1649. [CrossRef]
27. Hosseinzadeh, M.; Salmasi, F.R. Robust Optimal Power Management System for a Hybrid AC/DC Micro-Grid. *IEEE Trans. Sust. Energy* **2015**, *6*, 675–687. [CrossRef]
28. Schonbergerschonberger, J.; Duke, R.; Round, S.D. DC-Bus Signaling: A Distributed Control Strategy for a Hybrid Renewable Nanogrid. *IEEE Trans. Power Electron.* **2006**, *53*, 1453–1460. [CrossRef]

29. Panpean, C.; Areerak, K.-L.; Areerak, K.-N.; Udomsuk, S.; Santiprapan, P. The Harmonic Detection for Co-Phase Railway System in Distorted Voltage Source Condition. In Proceedings of the 2019 16th International Conference on Electrical Engineering/Electronics, Computer, Telecommunications and Information Technology (ECTI-CON), Pattaya, Thailand, 10–13 July 2019; pp. 545–548. [CrossRef]
30. Lin, C.E.; Chen, C.L.; Huan, C.L. Calculating approach and implementation for active filters in unbalanced three phase system using synchronous detection method. In Proceedings of the 1992 International Conference on Industrial Electronics, Control, Instrumentation, and Automation, San Diego, CA, USA, 13 September 1992; pp. 374–380. [CrossRef]
31. Cortes, P.; Rodriguez, J.; Silva, C.; Flores, A. Delay Compensation in Model Predictive Current Control of a Three-Phase Inverter. *IEEE Trans. Ind. Electron.* **2012**, *59*, 1323–1325. [CrossRef]
32. Odavic, M.; Biagini, V.; Zanchetta, P.; Sumner, M.; Degano, M. One-sample-period-ahead predictive current control for high-performance active shunt power filters. *IET Power Electron.* **2011**, *4*, 414–423. [CrossRef]
33. IEEE. IEEE Recommended Practice and Requirements for Harmonic Control in Electric Power Systems. IEEE Standard 519. 2014. Available online: https://ieeexplore.ieee.org/servlet/opac?punumber=6826457 (accessed on 1 April 2021).
34. IEEE. *IEEE Standard Definitions for the Measurement of Electric Power Quantities under Sinusoidal, Nonsinusoidal, Balanced, or Unbalanced Conditions*; IEEE Standard 1459; IEEE: Piscataway, NJ, USA, 2010; p. 7586362.

Article

Application of a Non-carrier-Based Modulation for Current Harmonics Spectrum Control during Regenerative Braking of the Electric Vehicle

Marcin Steczek [1,*], Piotr Chudzik [2] and Adam Szeląg [1]

[1] Faculty of Electrical Engineering, Institute of Electrical Power Engineering, Warsaw University of Technology, 00-661 Warszawa, Poland; adam.szelag@ee.pw.edu.pl

[2] Faculty of Electrical Engineering, Institute of Automatic Control, Lodz University of Technology, 90-924 Łódź, Poland; piotr.chudzik@p.lodz.pl

* Correspondence: marcin.steczek@ien.pw.edu.pl

Received: 21 November 2020; Accepted: 14 December 2020; Published: 18 December 2020

Abstract: The regenerative braking of railway vehicles is widely used in DC railway systems all over the world. This mode of operation provides an opportunity to reuse part of the energy consumed by vehicles, and makes the railway system more energy efficient. During regenerative braking, not only energy management is an issue, but also Electromagnetic Compatibility EMC issues, such as interference of generated current harmonics with a railway signaling system. In this paper, the selective harmonic elimination modulation technique (SHE-PWM) was introduced to the traction drive with a three-level inverter to reduce specific catenary current harmonics generated during regenerative braking. The simulation model of a traction drive appropriate for harmonics analysis was proposed and verified by the measurements in the low-power laboratory drive system. The model was re-scaled to the 3 kV DC system for further study. The model of an induction motor with electromotive force and the method of its calculation was proposed. Furthermore, an analysis of the braking chopper operation was carried out. The asymmetric control of braking chopper was proposed to reduce the current harmonics below limits during chopper operation.

Keywords: selective harmonic elimination; regenerative braking; railway vehicle; EMC

1. Introduction

The regenerative braking of the rolling stock gives an opportunity to reduce energy consumption in electric transportation systems like railways, trams, and metro. The higher traffic density, the higher energy-saving can be achieved. However, the regenerative braking of a vehicle in the system changes its character from being a load to being a source. Therefore, this operating mode requires a separate approach to the issues solved for the tractive operation, such as the generation of current harmonics and compatibility with the railway signaling system.

One of the issues in the electric transportation system, with vehicles equipped with regenerative braking, is energy management. Studies can be found for metros [1], DC railway systems [2], and AC high-speed railway [3] where authors analyze and optimize the energy flow between braking vehicles, accelerating vehicles, and energy storage systems (ESS) [4,5]. The development of ESS is one of the options for increasing the reuse of regenerated energy. Thus, the development of topology and sizing of the ESS is crucial for application [6–8]. In DC systems, the flow of the regenerated energy is terminated by rectifiers at the substations, which is the cause of limited receptivity of the DC systems [9]. To unblock the flow of the energy from the DC railway or metro system reversible substations with inverters are being used [10,11].

Another system level approach to the problem of increasing efficiency of the energy regeneration is the optimization of an Automatic Train Operation (ATO) system [12]. The adjustment of vehicles' trajectories in multivehicle systems can result in energy-saving up to 19% [13].

Energy efficiency and performance of the rolling stock require the use of high power converters for traction and regenerative braking [14], leading to a wide range of EMC problems between the vehicles and the supply infrastructure (e.g., harmonic distortion [15], instability [16]) and towards command-control systems as well [17] (e.g., resonances [18], radio interference, including wireless systems [19]). It was observed that harmonics present in traction currents flowing in rails could cause disturbances in the operation of signaling systems, track circuits, and axle counters [20], in particular. Research works on new solutions applied in traction drives supplied by 3 kV DC could reduce energy losses and EMC problems [21]. It is a complex issue due to the variability of applied technical solutions and phenomena which are to be taken into account (resonances, and non-linearities of elements [22]), and different systems which could be disturbed (signaling, command and control, and especially track circuits). European regulations require confirmation of compatibility between rolling stock and track infrastructure [23]. Therefore railway infrastructure operators were imposing limits of electric vehicle current harmonics in order to assure the safety of traffic on railway lines. It was significantly important in counties when the modern rolling stock was put into service on railway lines with track circuits, as for example in Poland [24]. Some examples of such limits versus frequency are shown in Figure 1 [25] and Figure 2. [26]. Limits imposed by Polish railways in the 1990s of the XX century were too strict comparing with other railway operators; they were too difficult to be fulfilled by newly introduced rolling stock, leading to these limits being eased—as shown in Figure 3. [27]. Typically, research on compatibility issues was focused on traction currents taken by a vehicle operating in a driving mode. However, during recuperative braking, harmonics in current delivered from a vehicle to catenary must be within the defined limits [27].

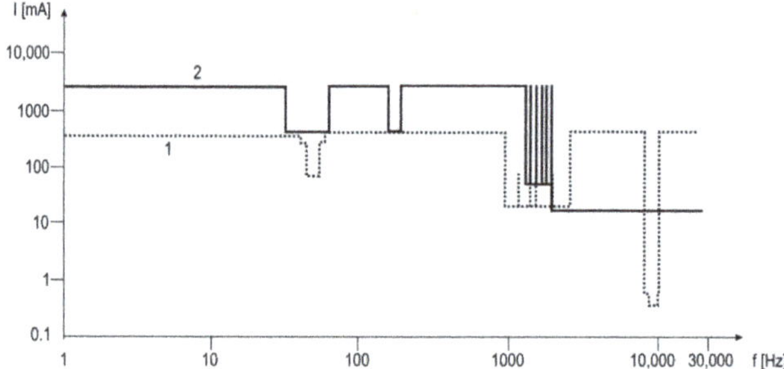

Figure 1. Comparison of limits imposed on harmonics in traction vehicles current supplied by 3 kV DC system on Polish railways in the 90s of the XX century (1) and on Italian railways (2) [25].

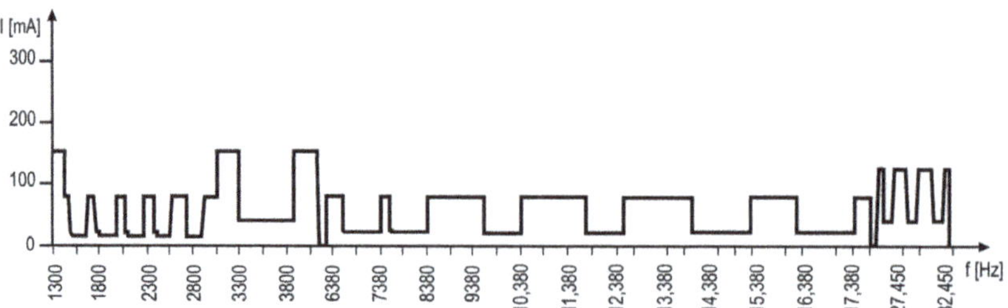

Figure 2. Modified limits of harmonics in vehicle/s current supplied by 3 kV DC voltage [26].

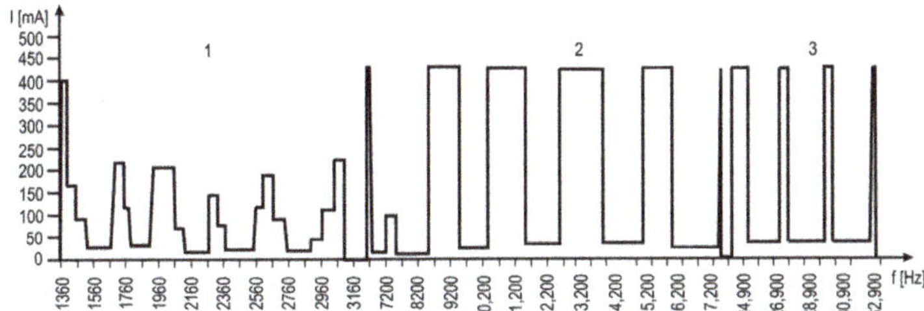

Figure 3. Currently applicable limits of harmonics in vehicle's current on Polish railway lines [27] for a range of frequencies according to types of track circuits applied (1,2,3).

This paper concerns theoretical, and simulation analyses of a spectrum of recuperation current harmonics. This research was also confirmed by measurements using a laboratory stand.

Figure 4 presents the power supply system of the DC railway line where the braking vehicle is transferring the regenerated energy via catenary to an accelerating vehicle. The area of possible interference between current harmonics generated by a drive of a braking vehicle (I_{cat}—catenary current) and the railway signaling (RS) systems (I_s—control signal).

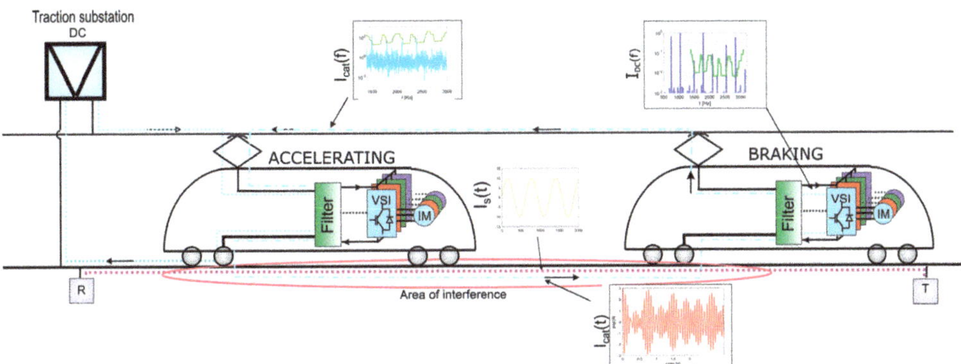

Figure 4. Simplified scheme of the railway power supply system and the area of possible interference between harmonics from braking vehicle and RS system.

2. Methodology

The methodology used for the study presented in this article was based on the use of the simulation model of a traction drive (3 kV DC–500 kW) verified in the laboratory scale (0.6 kV DC–2.5 kW). The simulation model, after being verified, was rescaled to the 3 kV DC and used to analyze the circulation of the current harmonics during regenerative braking in the 3 kV DC traction system. Only steady states were taken under consideration, and an Fast Fouries Transform (FFT) algorithm was used to determine harmonics amplitudes.

2.1. Simulation Model of the Drive

Figure 5 presents a simulation model of the drive used for the purpose of this work. It consists of a Voltage Source Inverter (VSI) with a braking circuit and a model of an induction machine. Figure 5b presents two types of the equivalent circuit of one phase of an induction motor. Circuit MOD1 from Figure 5b is suitable only for simulations for a fundamental component of a phase current. In MOD1 the load is represented by an additional series resistance R_{load} in the rotor's branch. The R_{load} should affect only the fundamental component of the rotor's current I'r. However, it will suppress all higher harmonics. For higher harmonics, R_{load} should be close to zero, and due to the slip is close to 1. In this work, the authors tested the MOD2 model from Figure 5b, where resistance R_{load} was replaced by Electromotive Force (EMF) modelled as a voltage source where V_{EMF} is a modulo and φ_{EMF} is an argument. Solving the Figure 5a (MOD2) equivalent circuit, the EMF is represented by the following formula:

$$\overline{EMF} = \left(\frac{\overline{V_{ph}} - \overline{Z_s} \cdot \overline{I_{ph}}}{\left(\overline{Z'_r} + R_{load}\right)} \right) \cdot R_{load} \tag{1}$$

where:

$\overline{Z'_r} = R'_r + j\omega L'_r$ is rotor's impedance,

$\overline{Z_s} = R_s + j\omega L_s$ is stator's impedance,

$R_{load} = R'_r \cdot \left(\frac{1-s}{s}\right)$ is additional resistance in rotor replaced by EMF,

$\overline{I_{ph}} = \frac{\overline{V_{ph}}}{\overline{Z}}$ is motor's phase current.

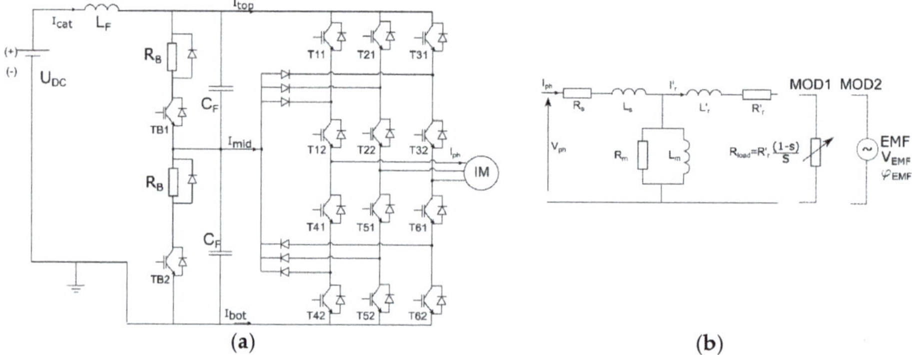

Figure 5. The simulation model (**a**) 3-lvl VSI a with braking circuit (**b**) equivalent circuit of the one phase of an induction model (MOD1: load modelled as resistance; MOD2: load modelled as a voltage source).

In this work, two drives have been considered (Table 1). The real-size drive (3KVM motor model) has been used for mathematical simulations of 3 kV drive system. The laboratory-size drive (two Siemens 1LA7106 motors-breaking and driving) has been used for real model verification in small scale.

Table 1. Parameters of the laboratory-size and full size-traction drives.

Type	Siemens 1LA7106 (Motor)	3KVM (Model)
Rated power—Pn	2.2 kW	500 kW
Rated current—In	4.85 A	170 A
Rated voltage—Un	400 V	1900 V
Stator leakage inductance per-phase—Ls	10.8 mH	1.56 mH
Stator winding resistance per phase—Rs	2.84 Ω	0.107 Ω
Rotor leakage inductance per-phase—L'r	10.6 mH	1.6 mH
Rotor resistance per phase—R'r	2.73 Ω	0.07 Ω
Core loss resistance—Rm	1200 Ω	∞
Magnetizing inductance—Lm	275 mH	53 mH

2.2. Selective Harmonic Elimination

The selective harmonic elimination technique is well studied and described in the literature [28]. This technique is an arbitrary type modulation which imposes switching times for each transistor (switching angles). It has no reference or carrier signal, like in other modulation methods. Switching angles determines the shape of the switching function for transistors which develops the inverter's output voltage waveform. There are numerous techniques for determining switching angles. One of the most popular is based on the Particle Swarm Optimization (PSO) algorithm [29]. However, most of the techniques [30,31] are based on the development of the Fourier Series (2) into the set of nonlinear Equation (6).

$$f(\omega t) = a_0 + \sum_{n=1}^{\infty} [a_n \sin(n\omega t) + b_n \cos(n\omega t)] \tag{2}$$

where: a_0, a_n and b_n: coefficients described by the Euler's formulas.

Assuming quarter-wave symmetry, the function has to fulfil the following conditions:

$$f(\omega t) = -f(\omega t + \pi) \tag{3}$$

$$f(\omega t) = f(\pi - \omega t) \tag{4}$$

For quarter-wave symmetry the coefficients a_0 and b_n are equal to zero and for the three-level (Figure 2), the coefficient a_n is given by:

$$a_n = \begin{cases} \frac{2 \cdot U_{DC}}{2n\pi}\left[1 + \sum\limits_{i=1}^{N} (-1)^i \cdot \cos(n \cdot k_i)\right] & for\ odd\ n \\ 0 & for\ even\ n \end{cases} \tag{5}$$

The set of nonlinear equations for $N = 5$ switching angles and 3 level inverter is as follows:

$$\begin{cases} \frac{2}{\pi}[1 - \cos(\alpha_1) + \cos(\alpha_2) - \cos(\alpha_3) + \ldots \\ \quad \ldots \cos(\alpha_4) - \cos(\alpha_5)] = M1 \\ \frac{2}{5\pi}[1 - \cos(5\alpha_1) + \cos(5\alpha_2) - \cos(5\alpha_3) + \ldots \\ \quad \ldots \cos(5\alpha_4) - \cos(5\alpha_5)] = 0 \\ \frac{2}{7\pi}[1 - \cos(7\alpha_1) + \cos(7\alpha_2) - \cos(7\alpha_3) + \ldots \\ \quad \ldots \cos(7\alpha_4) - \cos(7\alpha_5)] = 0 \\ \frac{2}{11\pi}[1 - \cos(11\alpha_1) + \cos(11\alpha_2) - \cos(11\alpha_3) + \ldots \\ \quad \ldots \cos(11\alpha_4) - \cos(11\alpha_5)] = 0 \\ \frac{2}{13\pi}[1 - \cos(13\alpha_1) + \cos(13\alpha_2) - \cos(13\alpha_3) + \ldots \\ \quad \ldots \cos(13\alpha_4) - \cos(13\alpha_5)] = 0 \end{cases} \tag{6}$$

where: $M1$ is for modulation index:

$$V_1 = M1\,\frac{U_{DC}}{2}; \quad for\ M1\langle 0,\ \frac{4}{\pi}\rangle \tag{7}$$

The set of Equation (6) can be solved using an optimization algorithm. In such a case the fitness function must be formulated and minimized. The example of a fitness function for $N = 5$ switching angles is the following:

$$\begin{gathered} Minimize,\ f_{fit}(\alpha_1, \alpha_2, \alpha_3, \alpha_4, \alpha_5) = \sigma_1 \cdot \left(V_1 - V_1^*\right)^2 + \sigma_5 \cdot (V_5)^2 + \\ \sigma_7 \cdot (V_7)^2 + \sigma_{11} \cdot (V_{11})^2 + \sigma_{13} \cdot (V_{13})^2 \\ \text{subject to: } 0 < \alpha_1 < \alpha_2 < \alpha_3 < \alpha_4 < \alpha_5 < \frac{\pi}{2} \end{gathered} \tag{8}$$

where: V_1 is a fundamental component, and V_5, V_7, V_{11}, V_{13} are 5th, 7th, 11th and 13th voltage harmonics (p.u.) respectively, σ_x are penalty weights for the optimization process.

The result of the optimization process is determination of the switching angles. In the three-level Neutral Point Clamped Voltage Source Inverter (NPC VSI) the switching function is realized by four transistors for each phase (T_{n1}, T_{n2}, T_{n3}, T_{n4}—where n is phase number). T_{n1} transistors operates in positive half of the period. T_{n4} transistors operates in negative half of the period and T_{n2}, T_{n3} transistors operate as negations of T_{n4}, T_{n1}, respectively. Figure 6 presents the division of the switching function developed of upper transistors in the inverter's leg (refer to Figure 5).

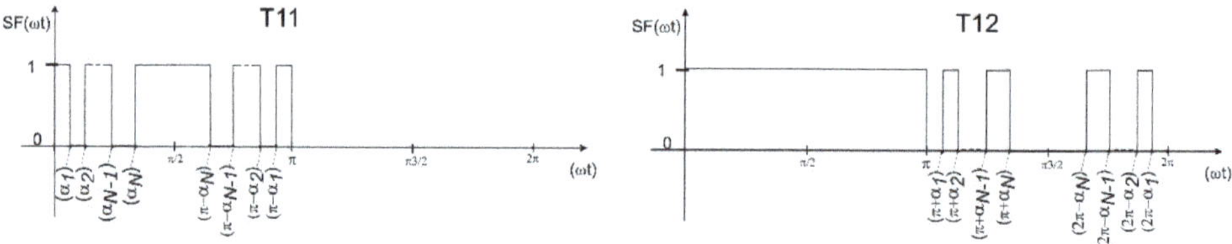

Figure 6. Implementation of switching angles in the switching function for three-level inverter.

In this work, the SHE-PWM was used to affect the spectrum of catenary current (I_{cat}) by control of harmonics of the inverter's output current (phase current of the motor).

2.3. *Laboratory Setup and Measurements*

The model verification was conducted with the laboratory setup (Figure 7). The parameters of the equivalent circuits are listed in Table 1.

The stand consists of two drive systems:

- Driving system built from the Invertec P2 inverter, which drives the first Siemens 1LA7106 motor.
- Breaking system, built of a specially designed laboratory 3-level inverter, which controls the voltage of the second Siemens 1LA7106 motor. The control of transistors, in the case of the Selective Harmonic Elimination (SHE) algorithm, is carried out in the DSPACE 1104 card, using the Look UP tables, created in the MATLAB. The control of transistors, in the case of the Sinus Pulse Width Modulation (SPWM) algorithm, is carried out in the DSPACE 1104 card, using a timer system, created in the MATLAB–Simulink.

The verification consisted of the DC-link currents (I_{bot}, I_{mid}, I_{top}—Figure 5a) comparison between results of measurements and simulations. The spectra of inverter's DC-link currents and motor's phase currents have been compared. The first comparison was made for SPWM modulation. The drive was performing regenerative braking with phase current I_{ph} = 7.4 A. During the experiment, the braking motor was driven with constant rotating speed by an induction motor (Driving motor and fed with a Driving inverter). Regenerated energy was dispersed in a resistor (120 Ω). Braking choppers were not active, and a voltage measured in DC-link was 600 V. Figures 8 and 9 present the comparison between the experimental results and simulations. Presented results have been obtained for modulation index M1 = 0.9, carrier frequency f_c = 850 Hz and fundamental frequency f_f = 50 Hz.

The same methodology of comparison between measurements and simulations was applied for SHE modulation. The SHE-PWM used in this part of research represents M1 = 0.9 and f_f = 50 Hz with N = 9 switching angles in quarter period. The following voltage harmonics: 5th, 7th, 11th, 13th, 17th, 19th, 29th, 31st have been eliminated from output voltage waveform. The fundamental of the phase current was I_{ph} = 7 A. Harmonics in three wires of a 3-level inverter were compared in terms of simulations and measurements (Figure 9). Thus, on the basis of results presented in this section, the authors claim that the proposed model and methodology has been validated and can be applied for further analysis.

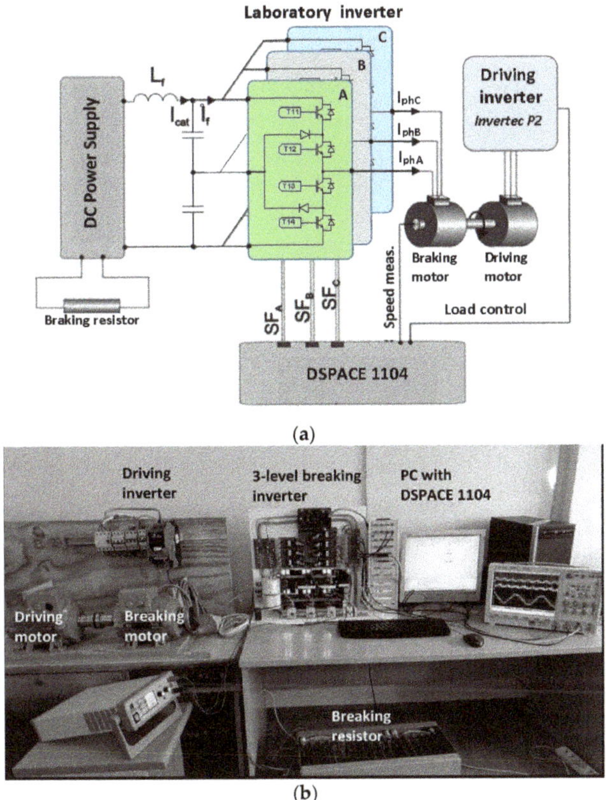

Figure 7. The laboratory setup: (**a**) simplified scheme, (**b**) laboratory stand.

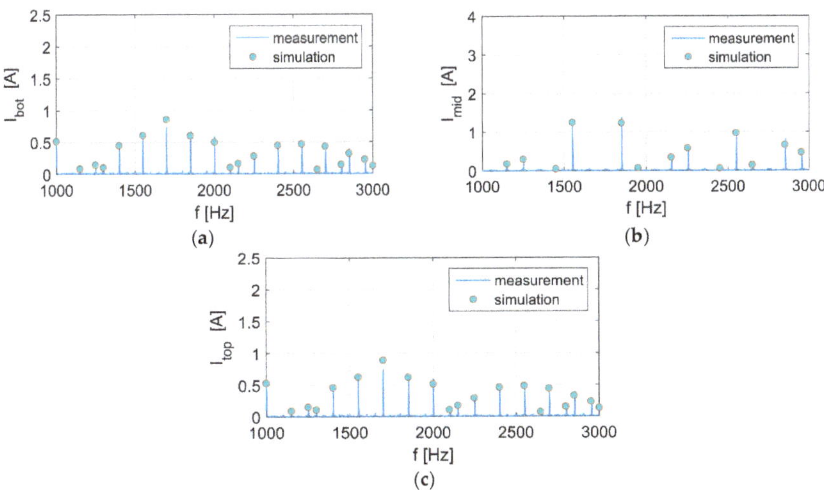

Figure 8. Spectrum of DC-link currents in VSI NPC inverter with SPWM (**a**) current I_{bot} (ref to Figure 5a), (**b**) current I_{mid} (ref to Figure 5a), (**c**) current I_{top} (ref to Figure 5a).

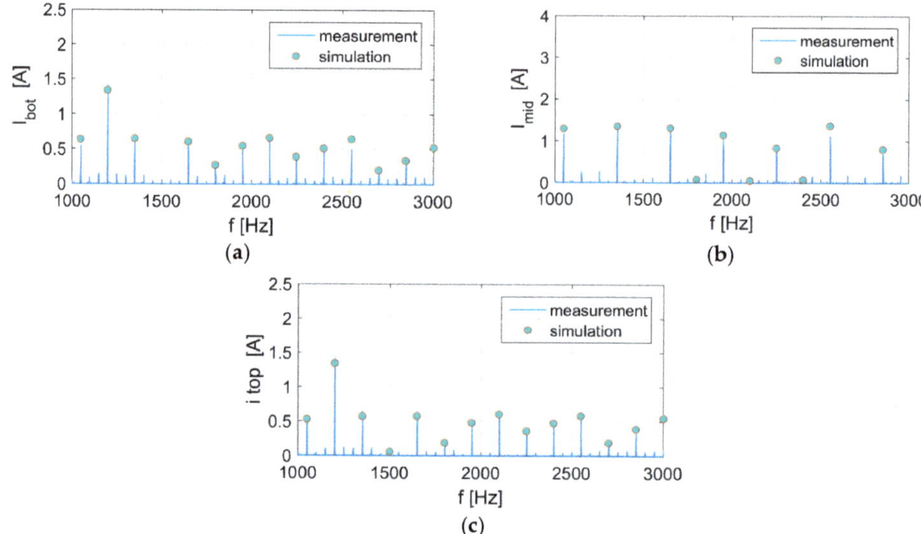

Figure 9. Spectrum of DC-link currents in VSI NPC inverter with SHE-PWM (**a**) current I_{bot} (ref to Figure 5a) (**b**) current I_{mid} (ref to Figure 5a) (**c**) current I_{top} (ref to Figure 5a).

3. Current Spectrum Generated by 3-lvl Laboratory Traction Drive

In this section, the results of simulations for a three-level, laboratory-size inverter during regenerative braking have been presented. The spectra of DC-link harmonics has been analyzed to reveal the differences between SPWM and SHE-PWM during vehicles braking. The aim of this section is to present the influence of the SHE application on current harmonics in the band of frequency 1500–3000 Hz, generated by a traction vehicle during braking.

3.1. Regenerative Braking with SPWM

Figures 10 and 11 present the simulations of DC-link currents in an NPC 3 level inverter during regenerative operation with SPWM modulation. Simulations were conducted for modulation index M1 = 0.9, and fundamental frequency f_f = 50 Hz, and carrier frequency f_c = 850 Hz. The third harmonic of a fundamental component ($3f_f$) was observed in all three wires. The first harmonic of the carrier has significant value as well. Figure 12 presents the result of the spectral analysis of the catenary current. The characteristic feature of presented modulation is that the carrier harmonic (f_c) is present in DC-link and it is canceled in the catenary current. The pattern of generated harmonics is typical for SPWM with a natural sampling technique.

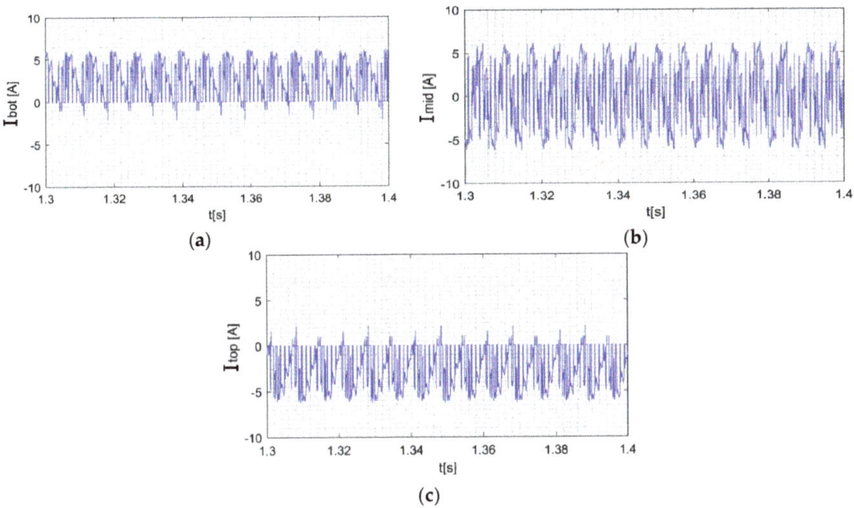

Figure 10. Waveform of DC-link currents during regenerative operation of 3-level NPC inverter with SPWM (current description refers to Figure 5) (**a**) current I_{bot} (**b**) current I_{mid} (**c**) current I_{top}.

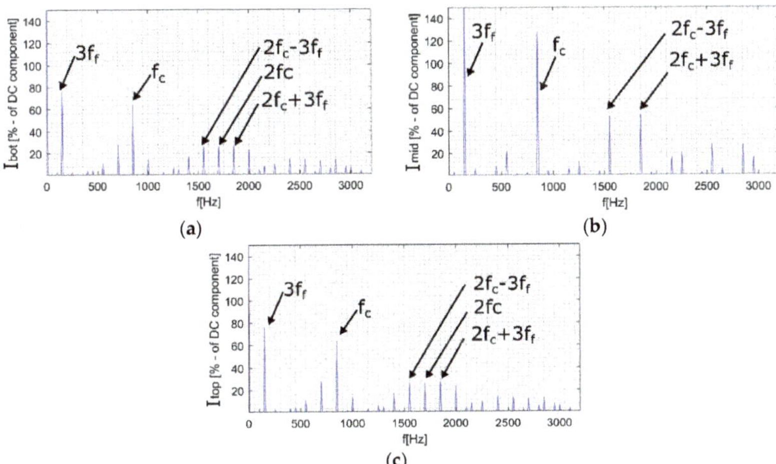

Figure 11. Spectra of DC-link currents during regenerative operation of 3-level NPC inverter with SPWM (current description refer to Figure 5) (**a**) current I_{bot}, (**b**) current I_{mid}, (**c**) current I_{top}.

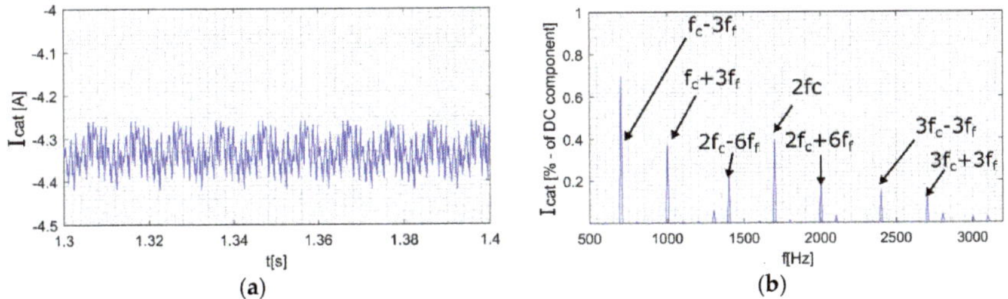

Figure 12. Simulated catenary current of 3-level NPC inverter with SPWM (**a**) waveform, (**b**) spectrum.

3.2. Regenerative Braking with SHE

The similar study, compared to the ones presented in Section 3.1, has been carried out for the SHE modulation. The goal was to identify which harmonic circulates in the DC-link and which transfers to catenary during regenerative braking with SHE modulation. The operating point studied in this section was defined by M1 = 0.9, f_f = 50 Hz, and N = 9, and the following harmonics were eliminated from output voltage: 5th, 7th, 11th, 13th, 17th, 19th, 29th, 31st. Figures 13 and 14 presents simulated DC-link current waveform and spectrum respectively. Elimination of 29th, 31st harmonics from the output voltage results with elimination of 30th harmonics from catenary current (Figure 15).

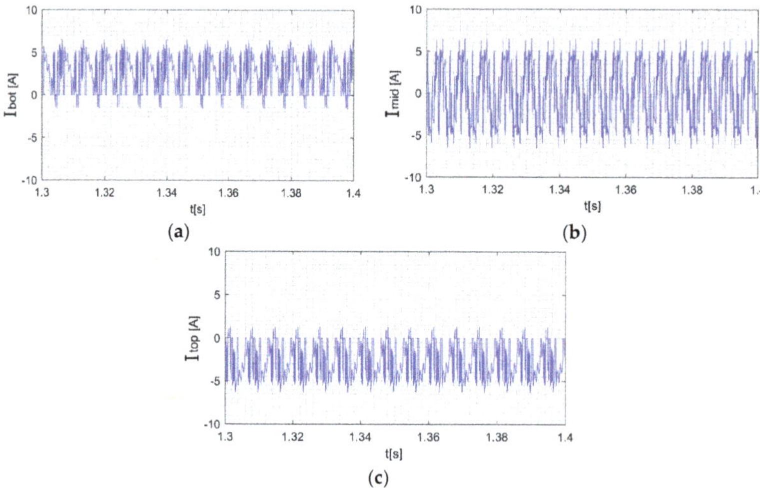

Figure 13. Waveform of DC-link currents during regenerative operation of a 3 lvl NPC inverter with SHE-PWM (current description refers to Figure 5) (**a**) current I_{bot}, (**b**) current I_{mid}, (**c**) current I_{top}.

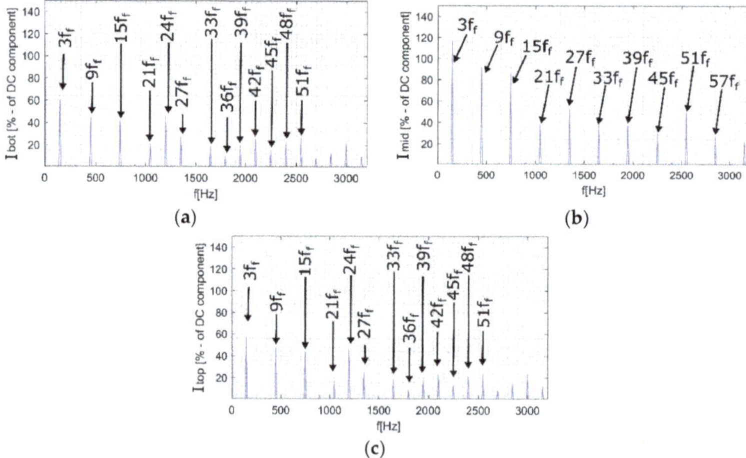

Figure 14. Spectra of DC-link currents during regenerative operation of a 3-level NPC inverter with SHE-PWM (current description refer to Figure 5) (**a**) current I_{bot}, (**b**) current I_{mid}, (**c**) current I_{top}.

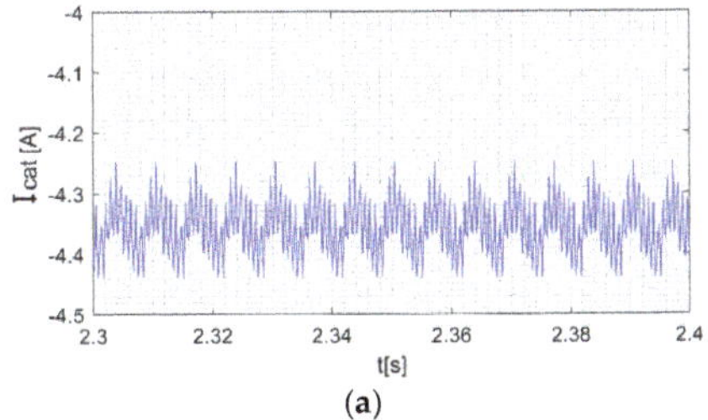

(a)

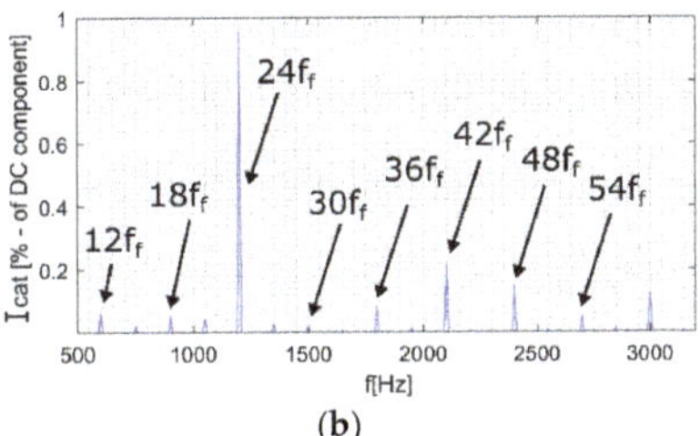

(b)

Figure 15. Simulated catenary current of a 3-level NPC inverter with SHE-PWM (**a**) waveform, (**b**) spectrum.

4. Results for a 3 kV DC Railway System

4.1. Model Scalling for 3 kV DC

The attention in this section is focused on the study of current harmonics circulation in the 3 kV DC railway system with a vehicle operating in a regenerative braking mode (Figure 16). The results were achieved by using the model of the braking drive, presented and verified in Section 2, and scaled to the 3 kV by application of parameters from Table 1. The load for the braking energy was modelled as the DC current source (I_L) with an input gamma filter (L_L, C_L). This model is able to represent both the accelerating vehicle and the energy source or storage device. The I_{LOAD} current represents the energy drawn by the LOAD from the system, without harmonics generated by the device. Thus, the harmonics generated by the braking vehicle will not be interfered by other sources. It makes a clear picture of harmonics generated by the vehicle and circulating in the system without any additional disturbances. The model of the traction substation (TS) was simplified to the voltage source with internal resistance (R_{IN}) and diode blocking the reverse circulation of energy through a rectifier. To reduce the computational effort a single drive was modelled (500 kW). Results have been referred to the limits recalculated by the following formula:

$$i_{\text{ln}} = \frac{i_{\text{lo}}}{\left(\frac{n_m}{n_{inv}}\right) \cdot \sqrt{n_v}} \tag{9}$$

where:

i_{ln}—limits recalculated,
i_{lo}—original limits,
n_m—number of motors on-board single vehicle,
n_{inv}—number of inverters on-board single vehicle,
n_v—number of vehicles in the traction set.

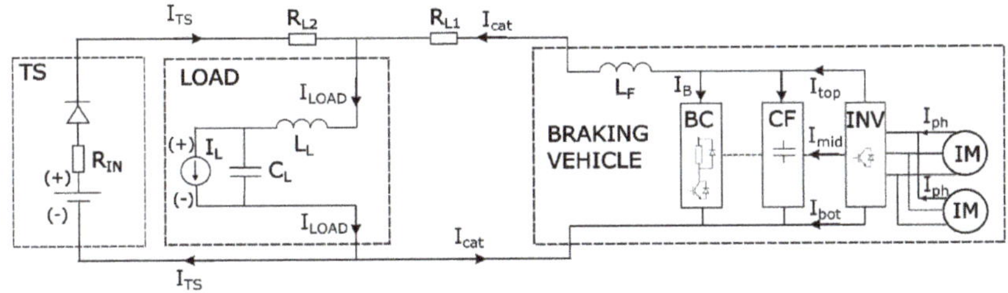

Figure 16. The simulation model of 3 kV DC system with braking vehicle and the load for regenerated energy.

Formula (9) was developed with the following assumptions: algebraic summation of current harmonics generated on-board single vehicle and geometric summation of current harmonics generated on-board different vehicles in the same traction set (based on general law for summation of harmonic current from different harmonic sources presented in IEC 61000-3-6).

4.2. Results of Simulation

The traction drive operates in variable conditions, such as variable load. In the proposed methodology, the load is modelled by the electromotive force (EMF) as the perfect AC voltage source. Figure 17 presents the EMF calculated for f_f = 52 Hz and M1 = 0.9 for different load conditions of a 500 kW induction motor. Both, modulo and argument of EMF were determined for motoring (positive values of a fundamental phase current, I_{ph1}) and braking (negative values of a fundamental phase current, I_{ph1}). The fundamental phase current was calculated to represent a load from 50% to 150% of nominal value. The parameters of EMF are not symmetrical for motor and braking operation. The reason is that during motor operation, the reactive power consumed by the magnetization branch is taken from the inverter, and in braking operation it is taken from the EMF. Thus, this phenomena gives asymmetry between motor and braking operation and can be observed in Figure 17.

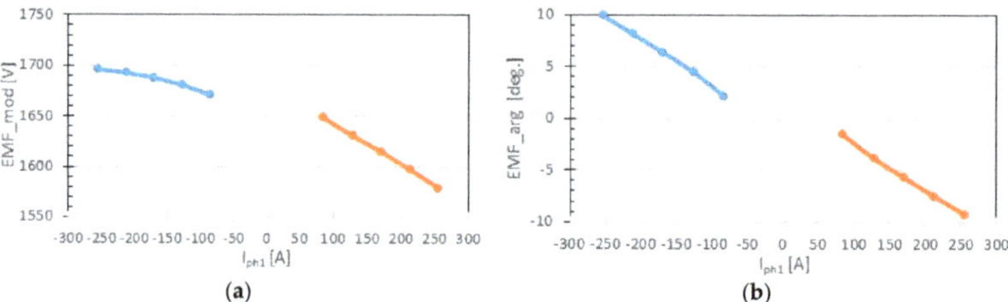

Figure 17. The electromotive force determined for motor and braking operation (**a**) modulo, (**b**) argument.

Figures 18 and 19 present the simulated spectra of phase current for different load conditions. It can be observed that phase current harmonics do not depend on load conditions (fundamental component) which refers to the assumption that slip for higher harmonics is close to zero. The proposed simulation model correctly models this phenomenon for both SPWM and SHE-PWM. The next step was to compare the catenary current (I_{cat}) harmonics for the motor and braking operating mode. The results of this comparison, for unlimited system receptivity for regenerated power, are presented in Figure 20 for SPWM and Figure 21 for SHE-PWM. It can be noticed that for perfect conditions, the catenary current harmonics are similar for motor and braking operation.

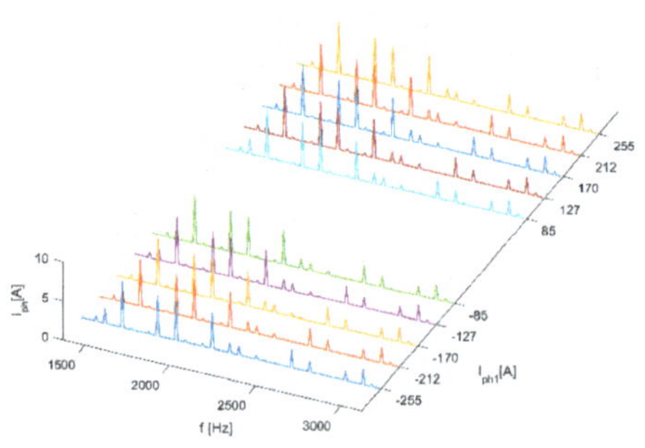

Figure 18. Spectrum of the phase current (I_{ph}) as the function of fundamental component (I_{ph1}) for SPWM.

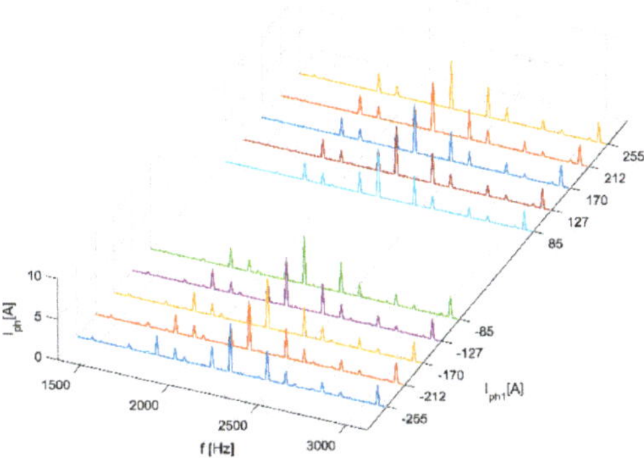

Figure 19. Spectrum of the phase current (I_{ph}) as the function of fundamental component (I_{ph1}) for SHE.

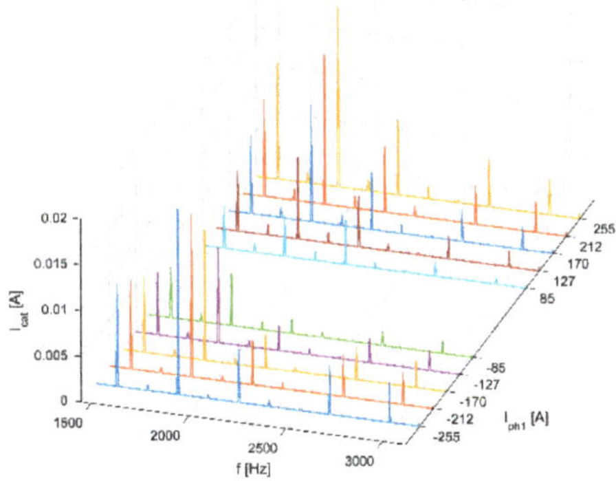

Figure 20. Spectrum of the catenary current (I_{cat}) as the function of fundamental component (I_{ph1}) for SPWM.

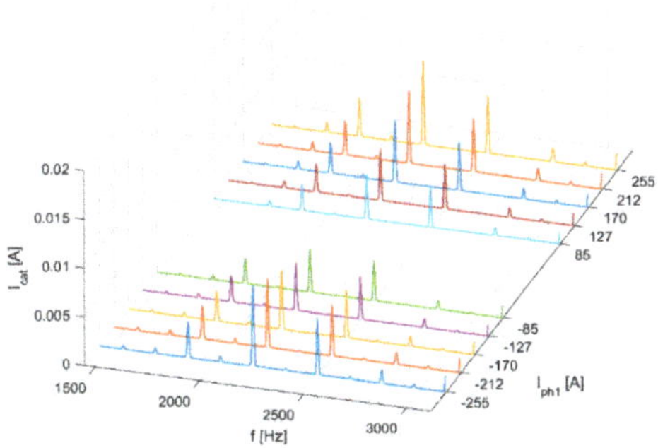

Figure 21. Spectrum of the catenary current (I_{cat}) as the function of fundamental component (I_{ph1}) for SHE.

The next experiment was developed to prove that SHE-PWM is beneficial in the application for regenerative braking mode regarding control of a catenary current harmonic. Figure 22 presents simulation results for the model from Figure 16 where M1 = 0.9; f_f = 52 Hz; f_c = 18·ff; I_{LOAD} = 400 A; I_{ph1} = −255 A (per motor). The parameters of the vehicle on-board filters were as follows L_F = 4.7 mH; C_F = 6.5 mF. The receiver for the recuperated energy was assumed to be an energy storage device with the input filter L_L= 4 mH; C_L= 0.4 mF. Results presented in Figure 22 refer to the limits recalculated using Formula (9) for two vehicles (n_v = 2) with three 3-level inverters on-board each vehicle (n_{inv} =3) and six induction motors on-board each vehicle (n_m = 6). It can be observed that the violation of the limits is present for this operating point. For the same conditions, the SHE-PWM was applied, and the results are presented in Figure 23. The following voltage harmonics were eliminated: 5th, 7th, 11th, 13th, 17th, 19th, 29th, 31st. Elimination of these voltage harmonics provides the modulation which generates I_{CAT} harmonics below assumed limits (Figure 23) The SHE gives the possibility to control current harmonics during regenerative braking in such a way as to avoid assumed limits. However, the 42nd harmonic (2184 Hz) of I_{CAT} is very close to the limit. In practice, the margin is required. Assuming 10% of the required margin (between harmonic amplitude and the limit), the SHE was not able to find feasible solution. To solve this issue with the same number of switching angles, the Selective Harmonic Mitigation was applied (SHM–PWM). The following pattern of elimination and mitigation of voltage harmonics was developed: 5th = 0, 7th = 0, 11th = 0, 13th = 0, 17th = 0, 19th = 5%, 25th = 20%, 29th = 5%. Figure 24 presents results for SHM-PWM and the margin between current harmonic and limit is visible.

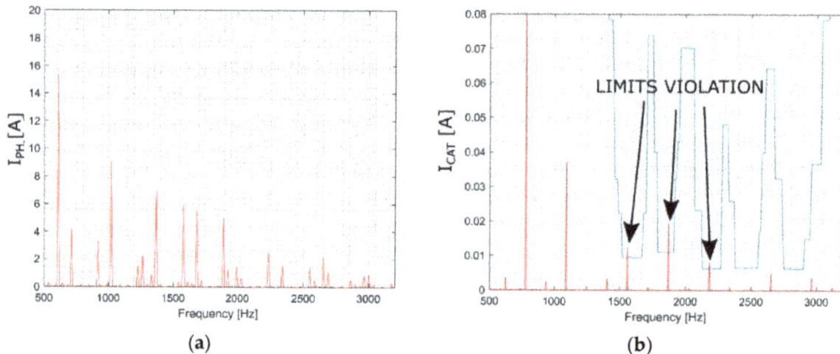

Figure 22. Simulation results for regenerative braking with SPWM (**a**) I_{ph}—phase current, (**b**) I_{cat}—catenary current.

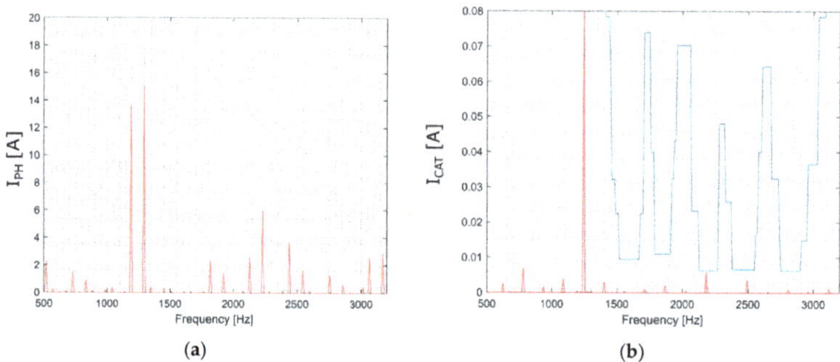

Figure 23. Simulation results for regenerative braking with SHE-PWM (**a**) I_{ph}—phase current, (**b**) I_{cat}—catenary current.

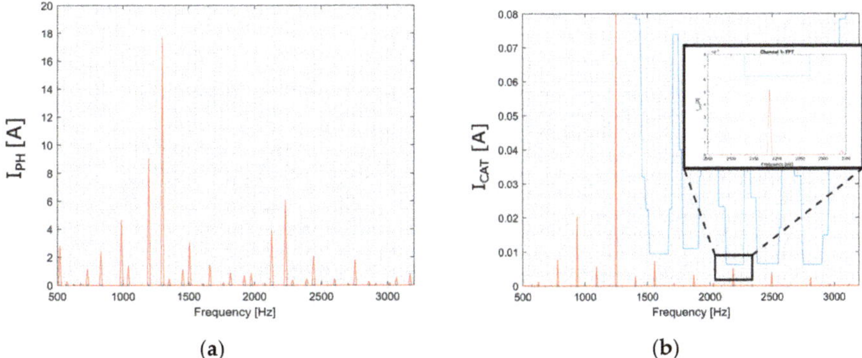

Figure 24. Simulation results for regenerative braking with SHM-PWM (**a**) I_{ph}—phase current, (**b**) I_{cat}—catenary current.

4.3. The Influence of a Braking Chopper

The previous study was carried out with an assumption that the DC-power supply is fully receptive. The case presented in this section assumes that DC system cannot absorb all regenerated energy. Thus, the current of the energy storage device is limited to 300 A (Figure 25) In such a case, the braking chopper onboard the vehicle must be activated. In this section, the influence of a braking chopper on the catenary current harmonics is presented. In a three-level NPC inverter a braking chopper is divided into two sections. The regular operation of the chopper generates harmonics on the fundamental chopper frequency and its multiplications.

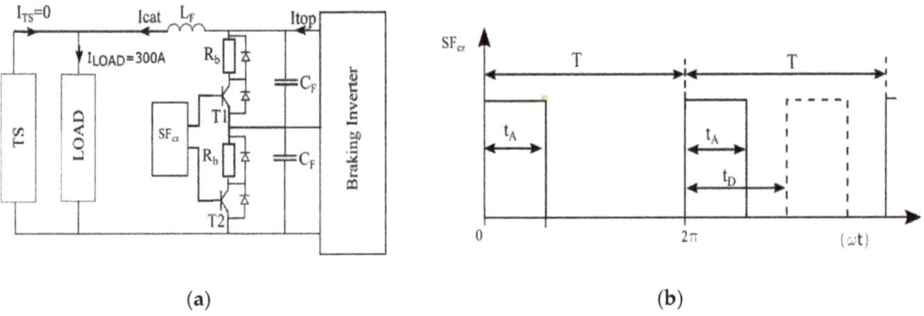

Figure 25. Schema of braking circuit (**a**) topology of braking chopper in 3 lvl NPC inverter, (**b**) switching function for chopper transistors.

In following part of this section an influence of a braking chopper on I_{cat} spectrum is presented. The chopper frequency is fb = 900 Hz. and braking resistor was R_b = 5 Ω each. The I_{LOAD} was limited to 300 A to model the limited receptivity of the DC-system. The assumption was to keep the DC-link voltage close to 3600 V. The choppers were operating during this experiment with duty cycle 0.3. Different values of the shift (shifted braking pulsation (SBP)) were implemented between chopper pulses (Figure 26) to investigate the possibilities of reducing current harmonics generated by the chopper below assumed limits. The shift was defined by the following formula:

$$SHIFT = \frac{t_D}{T} \tag{10}$$

where:

t_D—is a time delay of every second pulse,

T—period of the chopper pulsation.

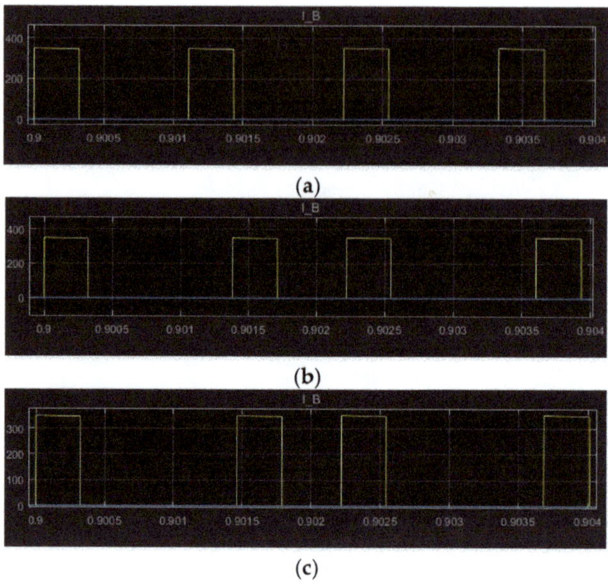

Figure 26. Current of braking choppers (I_B [A]) vs time [s] for (**a**) shift =0, (**b**) shift = 0.25, (**c**) shift = 0.32.

Figure 27 presents the influence of a braking chopper on I_{cat} spectrum. The implementation of shift = 0.25 eliminates harmonic 2xfb, however it generates inter-harmonics 1.5xfb and 2.5xfb. The optimum value of shift, in this case, was 0.32 (Figure 27c). The inter-harmonics were generated, but their values were reduced below limits. The 2fb harmonic was present but reduced as well. The proposed SBP was efficient for the presented operating point and it is worth further study and development. Moreover, to satisfy the margin between amplitudes of generated current, harmonics and imposed limits the SHM-PWM was implemented (Figure 27d) with the same pattern as proposed in Section 4.2. However, the problem of the lack of margin for harmonics generated by chopper remains. To solve this problem, the variable mixed cycle techniques (MDC). The idea of MDC was explained in Figure 28. In MDC the assumed duty cycle of braking chopper is achieved as the mean value of two duty cycles k_A and k_B e.g., k = 0.3 can be generated by k_A = 0.1 and k_B = 0.5. In Figure 29 it is presented that application of MDC provides margins between amplitudes of current harmonics generated by chopper operation and assumed limits.

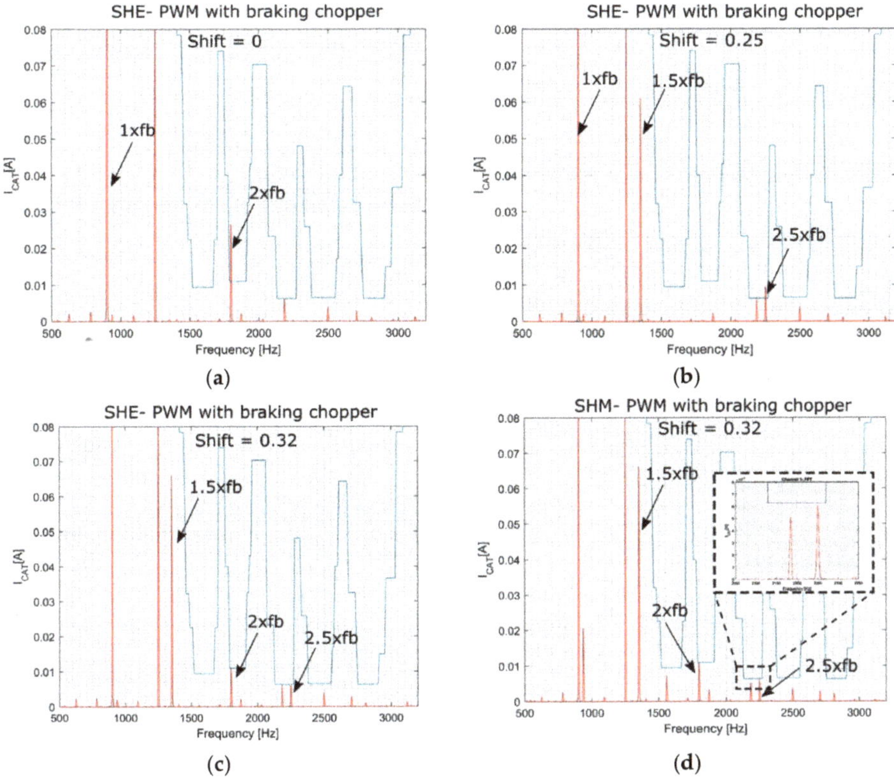

Figure 27. Catenary current harmonics for chopper braking operation (frequency of braking chopper fb = 900 Hz, duty cycle k = 0.3) (**a**) SHE-PWM with regular chopper braking, (**b**) SHE-PWM with chopper pulses with shift = 0.25, (**c**) SHE-PWM with chopper pulses with shift = 0.32, (**d**) SHM-PWM with chopper pulses with shift = 0.32.

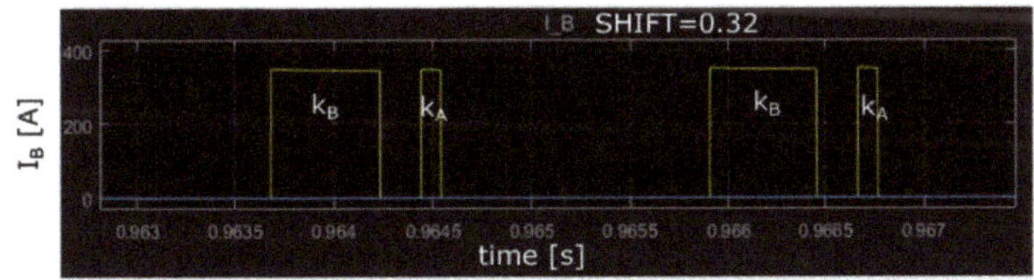

Figure 28. Current of braking choppers (I_B [A]) vs time [s] for mixed duty cycle (MCD) k_A = 0.1, k_B = 0.5 (effective duty cycle k = 0.3).

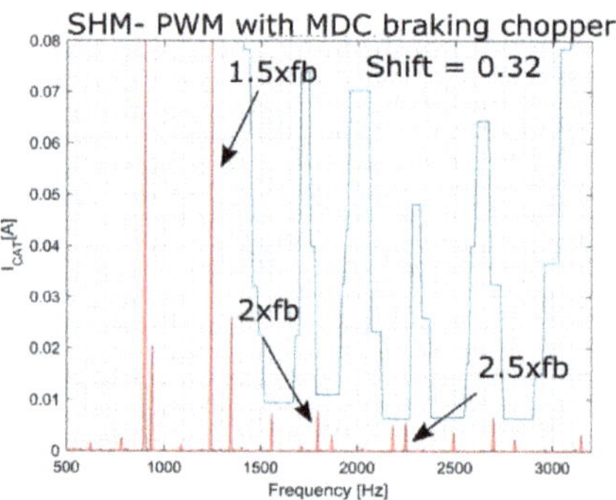

Figure 29. Catenary current harmonics for chopper braking operation (frequency of braking chopper fb = 900 Hz) for mixed duty cycle (MCD) $k_A = 0.1$, $k_B = 0.5$ (effective duty cycle k = 0.3).

More results for different duty cycles are presented in Figure 30. For every duty cycle of chopper there is a combination of Shift and MDC which allows to reduce amplitudes of harmonics generated by braking chopper below assumed limits. The proposed control over braking chopper, combined with SHM-PWM, gives control over current harmonics generated by the vehicle and allows them to not violate assumed limits for their amplitudes.

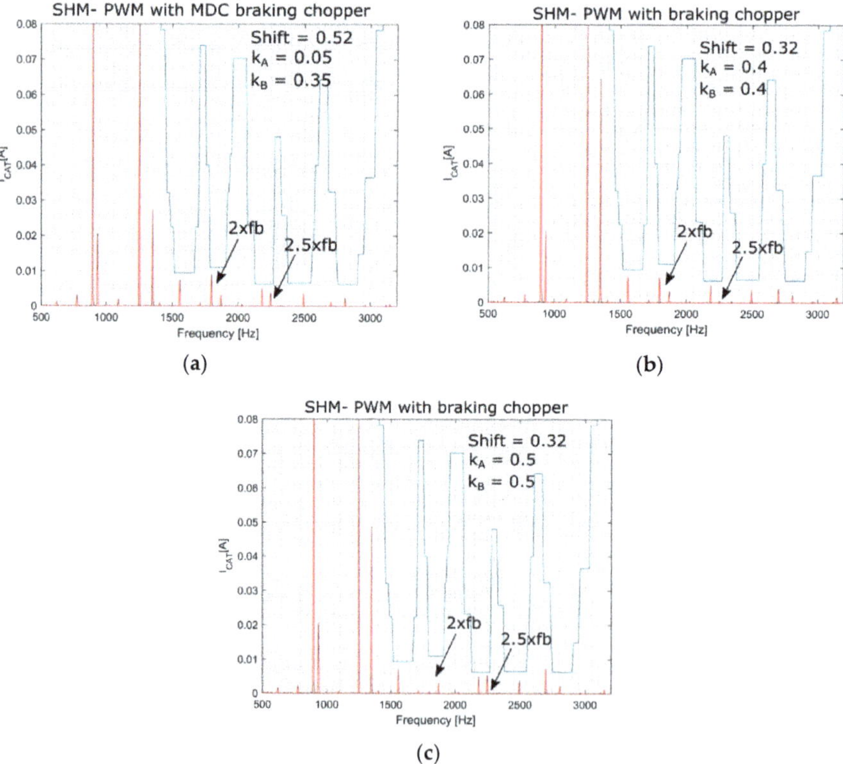

Figure 30. Catenary current harmonics for chopper braking operation (frequency of braking chopper fb = 900 Hz) for different duties (**a**) k = 0.2 with MDC; (**b**) k = 0.4; (**c**) k = 0.5.

5. Conclusions

The aim of this paper was to present the benefits of the application of SHE-PWM in a traction drive with a three-level inverter, to reduce specific current harmonics generated during regenerative braking. Moreover, the operation of braking choppers was taken into consideration. One operating point was chosen as a reference for the case study. The authors claim that SHE-PWM can be used in inverter's operating areas only where the typical modulations used in traction drives generate too high amplitudes of current harmonics and violate limits imposed by signaling system. Thus, there is no reason to study it in a whole range of operation during regenerative braking. It should be used as the interventional technique, where the probability of interference is high. However, the concern might be the applicability of SHE-PWM in real traction dives. Nowadays the implementation of SHE in traction drives by combining it with SVM is not an issue, due to rapid development of inverter controllers. Thus, its implementation was not a scope of this work. However, it is suggested to use SHE-PWM as the off-line technique, which means the switching angles should be pre-processed and stored as the lookup tables. It gives the opportunity to utilize a highly developed optimization algorithm without time pressure, typical for on-line techniques. The innovation of this paper is based on the presentation that SHE-PWM and SHM-PWM can be applied to control current harmonics as well during regenerative braking of a railway vehicle. Moreover the asymmetric technique of controlling a braking chopper in a three-level inverter was proposed to avoid the generation of current harmonics by a chopper in bands of frequency where limits are imposed on them. For low values duty cycles, the mixed duty cycle technique (MDC) was proposed to avoid generation of current harmonics above assumed limits. The proposed combination of SHM-PWM and chopper operation on braking operation of the vehicle allows the tuning of the amplitudes of catenary current harmonic below assumed limits to meet compatibility requirements. In this paper only examples have been presented. The further analysis are required for generalization of proposed technique. The dynamic performance of the proposed SHM-PWM will be investigated in the next research.

Author Contributions: M.S.: Methodology, theoretical analysis, software, simulations; P.C.: Development of laboratory setup, measurements; A.S.: Supervision, literature research, text editing. All authors have read and agreed to the published version of the manuscript.

Funding: This research received no external funding.

Conflicts of Interest: The authors declare no conflict of interest.

References

1. Kleftakis, V.A.; Hatziargyriou, N.D. Optimal control of reversible substations and wayside storage devices for voltage stabilization and energy savings in metro railway networks. *IEEE Trans. Transp. Electrif.* **2019**, *5*, 515–523. [CrossRef]
2. Alfieri, L.; Battistelli, L.; Pagano, M. Impact on railway infrastructure of wayside energy storage systems for regenerative braking management: A case study on a real Italian railway infrastructure. *IET Electr. Syst. Transp.* **2019**, *9*, 140–149. [CrossRef]
3. Aguado, J.A.; Racero, A.J.S.; De La Torre, S. Optimal operation of electric railways with renewable energy and electric storage systems. *IEEE Trans. Smart Grid* **2018**, *9*, 993–1001. [CrossRef]
4. Cui, G.; Luo, L.; Liang, C.; Hu, S.; Li, Y.; Cao, Y.; Xie, B.; Xu, J.; Zhang, Z.; Liu, Y.; et al. Supercapacitor Integrated Railway Static Power Conditioner for Regenerative Braking Energy Recycling and Power Quality Improvement of High-Speed Railway System. *IEEE Trans. Transp. Electrif.* **2019**, *5*, 702–714. [CrossRef]
5. Jiang, Y.; Liu, J.; Tian, W.; Shahidehpour, M.; Krishnamurthy, M. Energy Harvesting for the Electrification of Railway Stations: Getting a charge from the regenerative braking of trains. *IEEE Electrif. Mag.* **2014**, *2*, 39–48. [CrossRef]
6. De La Torre, S.; Sánchez-Racero, A.J.; Aguado, J.A.; Reyes, M.; Martínez, O. Optimal Sizing of Energy Storage for Regenerative Braking in Electric Railway Systems. *IEEE Trans. Power Syst.* **2015**, *30*, 1492–1500. [CrossRef]
7. Radu, P.V.; Szelag, A.; Steczek, M. On-Board Energy Storage Devices with Supercapacitors for Metro Trains—Case Study Analysis of Application Effectiveness. *Energies* **2019**, *12*, 1291. [CrossRef]

8. Jefimowski, W.; Nikitenko, A.; Drążek, Z.; Wieczorek, M. Stationary supercapacitor energy storage operation algorithm based on neural network learning system. *Bull. Pol. Acad. Sci. Tech. Sci.* **2020**, 68. [CrossRef]
9. Jabr, R.A.; Džafić, I. Solution of DC railway traction power flow systems including limited network receptivity. *IEEE Trans. Power Syst.* **2018**, *33*, 962–969. [CrossRef]
10. Zhang, G.; Tian, Z.; Tricoli, P.; Hillmansen, S.; Wang, Y.; Liu, Z. Inverter Operating Characteristics Optimization for DC Traction Power Supply Systems. *IEEE Trans. Veh. Technol.* **2019**, *68*, 3400–3410. [CrossRef]
11. Jefimowski, W.; Szeląg, A. The multi-criteria optimization method for implementation of a regenerative inverter in a 3 kV DC traction system. *Electr. Power Syst. Res.* **2018**, *161*, 61–73. [CrossRef]
12. Domínguez, M.; Fernández-Cardador, A.; Cucala, A.P. Pecharromán RR. Energy savings in metropolitan railway substations through regenerative energy recovery and optimal design of ATO speed profiles. *IEEE Trans. Autom. Sci. Eng.* **2012**, *9*, 496–504. [CrossRef]
13. Liu, J.; Guo, H.; Yu, Y. Research on the Cooperative Train Control Strategy to Reduce Energy Consumption. *IEEE Trans. Intell. Transp. Syst.* **2017**, *18*, 1134–1142. [CrossRef]
14. Ronanki, D.; Singh, S.A.; Williamson, S.S. Comprehensive Topological Overview of Rolling Stock Architectures and Recent Trends in Electric Railway Traction Systems. *IEEE Trans. Transp. Electrif.* **2017**, *3*, 724–738. [CrossRef]
15. Szelag, A.; Patoka, M. Issues of low frequency electromagnetic disturbances measurements in traction vehicles equipped with power electronics drive systems. *Przegląd Elektrotech.* **2013**, *89*, 290–296.
16. Tao, H.; Hu, H.; Wang, X.; Blaabjerg, F.; He, Z. Impedance-Based Harmonic Instability Assessment in a Multiple Electric Trains and Traction Network Interaction System. *IEEE Trans. Ind. Appl.* **2018**, *54*, 5083–5096. [CrossRef]
17. Mariscottii, A.; Ogunsola, A. *Lecture Notes in Electrical Engineering Series: Electromagnetic Compatibility in Railways: Analysis and Management*; Springer: Berlin, Germany, 2012.
18. Hu, H.; Shao, Y.; Tang, L.; Ma, J.; He, Z.; Gao, S. Overview of Harmonic and Resonance in Railway Electrification Systems. *IEEE Trans. Ind. Appl.* **2018**, *54*, 5227–5245. [CrossRef]
19. Baboszin, V.A.; Gavrilovich, B.; Jakovlev, A.A. To the study of analysis on how harmonic radio interference affects the power supply systems of railway transport, Mathematical modeling. *Autom. Control Process* **2020**, *2*, 36–45.
20. Adamski, D.; Ortel, K.; Zawadka, Ł. Unified verification method of electromagnetic compatibility between rolling stock and train detection systems. In Proceedings of the Global Debate on Mobility Challenges for the Future Society, Warsaw, Poland, 15–16 November 2018.
21. Adamowicz, M.; Szewczyk, J. Research works on new solutions applied in traction drives supplied by 3 kV DC coulud reduce energy losses and EMC problems [SiC-Based Power Electronic Traction Transformer—(PETT) for 3 kV DC Rail Traction]. *Energies* **2020**, *13*, 5573. [CrossRef]
22. Mariscotti, A. Impact of Rail Impedance Intrinsic Variability on Railway System Operation, EMC and Safety. *Int. J. Electr. Comput. Eng.* **2020**, *11*, 17–26.
23. BSI Standards Publication. *EN 50238–1:2019, Railway Applications—Compatibility between Rolling Stock and Train Detection Systems*; BSI Standards Publication: London, UK, 2019.
24. Furman, J.; Białoń, A. *Influence of Commutation Disturbances on Analysis of Harmonics in Traction Current (In Polish: Wpływ zakłóceń Komutacyjnych na Analizę Harmonicznych w Prądzie Trakcyjnym)*; Railway Report, No 187; Instytut Kolejnictwa: Warsaw, Poland, 2020. [CrossRef]
25. Durzyński, Z.; Łatowski, M. PKP requirements on electromagnetic disturbances versus other railway infrastructure operators (In Polish: Wymagania PKP w zakresie zakłóceń elektromagnetycznych na tle innych zarządów kolei). In Proceedings of the Research-Technical Conference—Influence of Electric Traction on Environment, Zakopane, Poland, 14–16 October 1999; pp. 14–16.
26. Białoń, A. *Deriving Limits of Disturbances Parameters for Command, Control and Signaling Systems and Traction Vehicles, Research Work CNTK 6915/23*; CNTK: Warsaw, Poland, 1999.
27. Requirements towards Allowed Limits and Parameters of Disturbances for Devices of Track Occupancy Control on Railway Lines Operated by PKP PLK S.A., Ie-115, Warsaw, Poland. 2015. Available online: https://www.plk-sa.pl/files/public/user_upload/pdf/Akty_prawne_i_przepisy/Instrukcje/Wydruk/Wymagania_Ie_-115_internet.pdf (accessed on 1 November 2020).

28. Dahidah, M.S.A.; Konstantinou, G.; Agelidis, V.G. A Review of Multilevel Selective Harmonic Elimination PWM: Formulations, Solving Algorithms, Implementation and Applications. *IEEE Trans. Power Electron.* **2015**, *30*, 4091–4106. [CrossRef]
29. Steczek, M.; Chudzik, P.; Lewandowski, M.; Szeląg, A. PSO-Based Optimization of DC-Link Current Harmonics in Traction VSI for Electric Vehicle. *IEEE Trans. Ind. Electron.* **2020**, *67*, 8197–8208. [CrossRef]
30. Napoles, J.; Leon, J.I.; Portillo, R.; Franquelo, L.G.; Aguirr, M.A. Selective Harmonic Mitigation Technique for High-Power Converters. *IEEE Trans. Ind. Electron.* **2009**, *57*, 2315–2323. [CrossRef]
31. Leon, J.L.; Kouro, S.; Franquelo, L.G.; Rodrigez, J.; Wu, B. The Essential Role and the Continuous Evolution of Modulation Techniques for Voltage-Source Inverters in the Past, Present, and Future Power Electronics. *IEEE Trans. Ind. Electron.* **2016**, *63*, 2688–2701. [CrossRef]

Publisher's Note: MDPI stays neutral with regard to jurisdictional claims in published maps and institutional affiliations.

Article

Experimental Validation of a Reduced-Scale Rail Power Conditioner Based on Modular Multilevel Converter for AC Railway Power Grids

Mohamed Tanta [1,*], Jose Cunha [1], Luis A. M. Barros [1], Vitor Monteiro [1], José Gabriel Oliveira Pinto [1], Antonio P. Martins [2] and Joao L. Afonso [1]

[1] School of Engineering, Campus of Azurém, University of Minho, 4800-058 Guimarães, Portugal; jcunha@dei.uminho.pt (J.C.); lbarros@dei.uminho.pt (L.A.M.B.); vmonteiro@dei.uminho.pt (V.M.); gpinto@dei.uminho.pt (J.G.O.P.); jla@dei.uminho.pt (J.L.A.)

[2] Faculty of Engineering, University of Porto, Rua Roberto Frias, 4200-465 Porto, Portugal; ajm@fe.up.pt

* Correspondence: mtanta@dei.uminho.pt; Tel.: +351-253-510-190

Abstract: Rail power conditioner (RPC) has the ability to improve the power quality in AC railway power grids. This power conditioner can increase the loading capacity of traction substations, balance the active power between the feeder load sections, and compensate for reactive power and current harmonics. At present, there is increasing use of multilevel converter topologies, which provide scalability and robust performance under different conditions. In this framework, modular multilevel converter (MMC) is emerging as a prominent solution for medium-voltage applications. Serving that purpose, this paper focuses on the implementation, testing, and validation of a reduced-scale laboratory prototype of a proposed RPC based on an MMC. The developed laboratory prototype, designed to be compact, reliable, and adaptable to multipurpose applications, is presented, highlighting the main control and power circuit boards of the MMC. In addition, MMC parameter design of the filter inductor and submodule capacitor is also explained. Experimental analysis and validation of a reduced-scale prototype RPC based on MMC topology, are provided to verify the power quality improvement in electrified railway power grids. Thus, two experimental case studies are presented: (1) when both of the load sections are unequally loaded; (2) when only one load section is loaded. Experimental results confirm the RPC based on MMC is effective in reducing the harmonic contents, solving the problem of three-phase current imbalance and compensating reactive power.

Keywords: electrified railway systems; modular multilevel converter; power quality; rail power conditioner

Citation: Tanta, M.; Cunha, J.; Barros, L.A.M.; Monteiro, V.; Pinto, J.G.O.; Martins, A.P.; Afonso, J.L. Experimental Validation of a Reduced-Scale Rail Power Conditioner Based on Modular Multilevel Converter for AC Railway Power Grids. *Energies* **2021**, *14*, 484. https://doi.org/10.3390/en14020484

Received: 4 December 2020
Accepted: 14 January 2021
Published: 18 January 2021

Publisher's Note: MDPI stays neutral with regard to jurisdictional claims in published maps and institutional affiliations.

1. Introduction

Electric traction power systems are the main suppliers of electrical energy to electric locomotives, which normally operate with a higher power-to-weight ratio than conventional diesel locomotives [1]. Moreover, electric locomotives are less noisy and require less frequent maintenance than diesel locomotives [2]. However, when considering AC traction power systems, electric locomotives represent non-linear single-phase loads and adversely influence the power quality of the three-phase power grid [3]. Some of the problems related to power quality deterioration are the harmonic distortion produced by the electric locomotives and the negative sequence components (NSC) of currents created by the three-phase current imbalance of the power grid [4]. Numerous approaches to improve power quality have been developed and investigated in recent decades to overcome the deterioration of power quality in electrified railway systems [5,6]. In this regard, active power compensator-based power electronics converters are still in development for consistency with the latest evolution of high-speed electrified railway systems that broadly use 25 kV, 50 Hz power supplies [7]. For this reason, a power conditioner installed on a traction feeder, called a rail power conditioner (RPC), was introduced in [8]. The RPC system consists of two power converters (one power converter is connected for each traction feeder or load section) that provide the reactive power required by the traction loads, and maintain a unitary power factor on the three-phase power grid. In addition, the RPC shifts the active power from the highly loaded section to the lightly loaded section to maintain balanced active power between phases [9]. The RPC also compensates for the harmonics and the NSC of currents, consequently providing a robust power grid without perturbation and with a higher overloading capacity [10].

The typical RPC, presented in Figure 1, is composed of two back-to-back single-phase converters that are connected to the traction feeders (load sections) via two coupling transformers [10,11]. These transformers are important to step-down the load voltages and to achieve the necessary isolation between the RPC and the single-phase traction power grid [12]. However, these transformers significantly increase the total cost of the typical RPC solution. Moreover, the two-level back-to-back power converters are unscalable and should hold the total power of the RPC [13]. In this case, due to the low switching frequency of the power switching devices, harmonic content in the synthetized waveforms is presented in the low-frequency domain. These reasons may weaken the RPC performance, leading to a lower quality of synthesized waveforms and bulky grid-connected passive filters. Consequently, it is recognized that multilevel power converters are important to synthesize high quality waveforms and reduce the size of grid-connected passive filters [14].

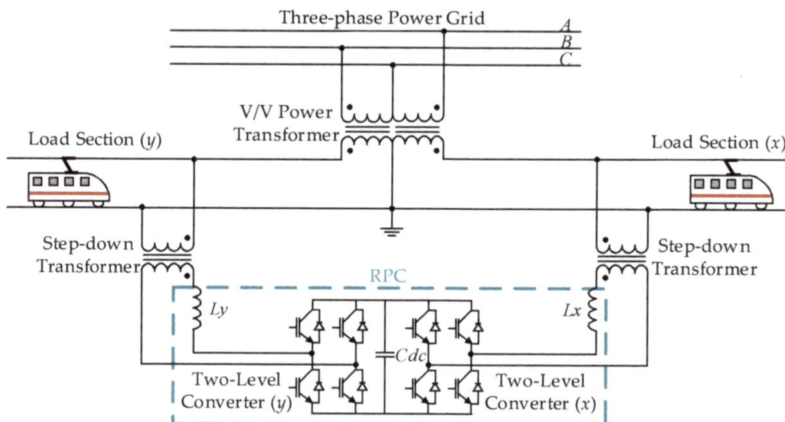

Figure 1. Typical rail power conditioner (RPC) connection with two-level back-to-back converters and open-delta (V/V) power transformer.

In this context, the modular multilevel converter (MMC) is an attractive and promising solution for high power and custom power applications [15]. This is due to its salient features in terms of modularity, superior reliability, and high efficiency, and the reduced size of passive filters resulting from the low harmonic content in the multilevel voltage and current waveforms [16]. The modular structure means that MMC can scale to different power and voltage levels, and, in some applications, it can be a transformer-less configuration for grid-connected applications [17]. The MMC, in this paper, is composed of several low-voltage two-level converters, called submodules (SMs), connected in a cascade manner. Thus, the total MMC switching frequency is divided between the SMs, which results in low switching frequency for each device and reduced total switching losses. In addition, these cascade-connected SMs are capable of generating high-quality waveforms due to the reduced voltage stress and the utilization of low-voltage switching devices [18,19]. The MMC topology was first introduced in 2001 by the Marquardt Group for medium-voltage transmission applications. After 19 years of improvement, the MMC became the preferable multilevel converter topology for medium- and high-power applications [20].

Several MMC applications based on power electronics transformers in railway traction systems are presented in [21].

Many RPC topologies have been proposed in the literature using the MMC with high compensation capacity. For instance, an RPC based on a full-bridge back-to-back AC/DC/AC MMC and an open-delta (V/V) power transformer is presented in [22], an RPC based on a two-phase three-wire AC/DC/AC MMC and V/V power transformer is presented in [14,23], and an RPC based on an AC/AC MMC [24] and V/V power transformer is presented in [25]. In addition, authors in [26] present a classification of the RPC topologies. This paper mainly discusses the RPC based on a half-bridge AC/DC/AC MMC and V/V power transformer. The MMC is equipped with half-bridge SMs that only introduce positive voltage with a bidirectional current. At present, the majority of the marketed MMC projects are based on the AC/DC/AC MMC with half-bridge SMs. This is due to its lower costs, higher efficiency, and simple structure of the half-bridge SM compared to other types of MMC SMs [27].

The main contributions of the current paper are: (a) design and confirm the functionality of one half-bridge SM, including driver, protection, and power boards; (b) description of the implementation of a reduced-scale laboratory workbench of an RPC based on a half-bridge MMC and V/V power transformer; (c) experimental analysis of the proposed RPC based on MMC system considering two case studies—when two load sections are loaded and when only one load section is loaded; and (d) Presenting the main advantages of the implemented reduced-scale prototype in terms of power quality improvement in electrified railway systems. Validation of this prototype helps to predict the performance of the RPC based on MMC under high-power applications. The novelty of this paper is in the experimental analysis and validation of the proposed reduced-scale prototype RPC based on MMC.

This paper is organized as follows: Section 2 explains the rail power conditioner based on a modular multilevel converter (RPC based on MMC) and the related control algorithm. Section 3 presents the implementation of the proposed reduced-scale RPC based on MMC laboratory prototype. Section 4 presents the experimental results of the proposed reduced-scale RPC based on MMC prototype, considering two case studies: when two load sections are loaded and when only one load section is loaded. Finally, Section 5 summarizes the main conclusions of the developed work.

2. Rail Power Conditioner Based on an AC/DC/AC Modular Multilevel Converter and V/V Power Transformer

2.1. Rail Power Conditioner Topologies

In the literature, three types of RPC based on AC/DC/AC MMC exist: (a) an RPC based on a full-bridge back-to-back AC/DC/AC MMC and V/V power transformer as shown in Figure 2a; (b) an RPC based on a two-phase three-wire AC/DC/AC MMC and V/V power transformer as shown in Figure 2b; and (c) an RPC based on a half-bridge AC/DC/AC MMC and V/V power transformer as shown in Figure 2c. Many studies recommend using a half-bridge SM in the AC/DC/AC MMC and a full-bridge SM in the AC/AC MMC [28,29]. Therefore, the design and implementation process of the half-bridge SM is discussed in this paper. Reference [30] presents various MMC SM topologies, such as half-bridge SM, full-bridge SM, multilevel neutral point clamped SM, and multilevel flying capacitor SM. In the half-bridge SM, which is the simplest approach and the core of this study, each half-bridge SM, as shown in Figure 3, involves two power switching devices; in this example, we use the isolated gate bipolar transistor (IGBT), and one capacitor that could be inserted or bypassed. By comparison, filter inductors are important to limit the circulating currents between MMC arms and other harmonic content [31].

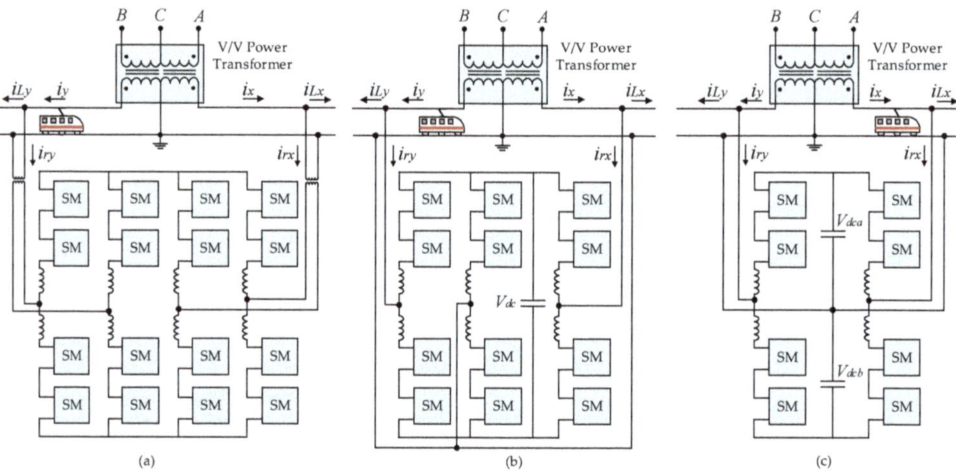

Figure 2. Rail power conditioner (RPC) topologies based on an AC/DC/AC modular multilevel converter (MMC) and V/V power transformer: (**a**) RPC based on full-bridge back-to-back MMC; (**b**) RPC based on two-phase three-wires MMC; (**c**) RPC based on half-bridge MMC.

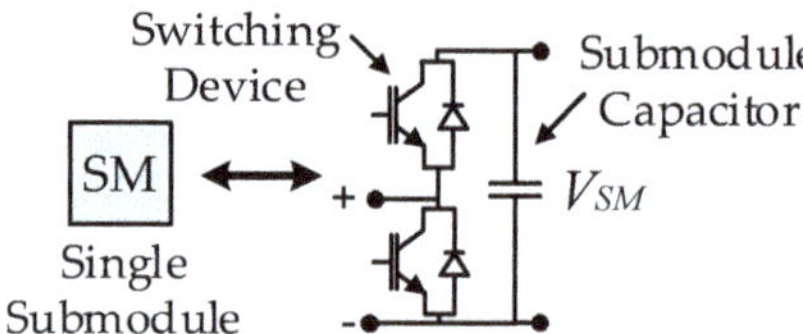

Figure 3. Schematic of a half-bridge submodule (SM).

The RPC based on full-bridge back-to-back MMC, presented in Figure 2a, has the highest number of hardware components and the most complex control among the RPC based on MMC topologies. In addition, it requires isolation transformers to interface with the single-phase traction power grid because two of the RPC phases are connected to the same wire [23]. As a result, this solution is expensive when implemented as a reduced-scale laboratory prototype. In contrast, according to [26], the RPC based on a two-phase three-wire MMC and the RPC based on a half-bridge MMC have fewer components and require a smaller area for installation. They also can be classified as a special-purpose RPC because they are recommended to be used with only the V/V power transformer. For instance, using these topologies with a Scott power transformer requires a higher DC-link voltage and, therefore, using power switching devices with higher power ratings. Moreover, isolation transformers are not mandatory in these solutions because each phase of the RPC is connected to a single wire [26].

Due to the lower control complexity and the simple structure of the RPC based on half-bridge MMC, and because this topology saves 50% of the RPC based on full-bridge back-to-back MMC power devices, an implementation of a reduced-scale laboratory prototype was selected to validate the RPC based on half-bridge MMC configuration. In this context, [31] presents the control algorithm of the RPC based on half-bridge MMC using deadbeat predictive control and the associated simulation results. Reference [32] presents another possible control algorithm for the RPC based on half-bridge MMC using proportional-integral (PI) controllers and the related simulation results. In this paper, the control algorithms of the proposed RPC based on MMC, published in [31], are experimentally validated. The operation principle of the RPC and other operation modes are presented in [26]. In addition, a comprehensive comparison between the RPC based on MMC topologies is presented in [23]. Consequently, authors in [26,32,33] presented the control theory, operation principle, simulation results, and other comparative analyses of the RPC based on half-bridge MMC.

2.2. Rail Power Conditioner Based on Half-Bridge Modular Multilevel Converter Control Algorithm

One of the main objectives of the RPC based on MMC control algorithm is to calculate the compensation current references, i_{rx}^*, i_{ry}^*. Figure 4 shows the RPC based on half-bridge MMC and V/V power transformer control strategy. The compensation currents can be determined using the instantaneous load section currents i_{Lx}, i_{Ly}, where these currents are the most important variables because they are considered as the input signals for the control algorithm. There are two current sensors used on both load sections (x and y) to acquire the currents, i_{Lx} and i_{Ly}. Furthermore, a phase-locked loop (PLL) is essential to obtain the phase angles for both catenary voltages of u_x and u_y. More information about the PLL used can be found in [34]. The low-pass filter (LPF) is necessary to extract the DC current component resulting from multiplying the load section currents i_{Lx}, i_{Ly} with the correspondent sine waves in the V/V traction system, $\sin(\omega t - \pi/6)$ and $\sin(\omega t - \pi/2)$, respectively. In this case, the instantaneous currents of phase x and phase y after compensation i_{x2}, i_{y2} are generated and, then, using Equation (1) the compensation current references, i_{rx}^*, i_{ry}^* are obtained. Detailed mathematical analysis and explanation of the establishment of the compensation current references can be found in [31,32].

$$\begin{aligned} i_{rx}{}^* &= i_{x2} - i_{Lx} \\ i_{ry}{}^* &= i_{y2} - i_{Ly} \end{aligned} \tag{1}$$

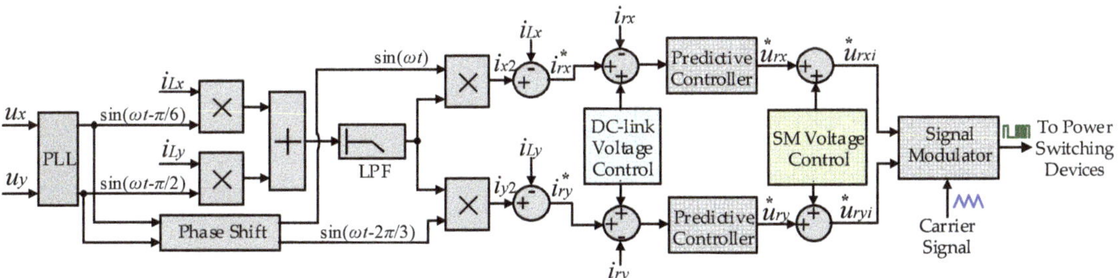

Figure 4. RPC based on half-bridge MMC and V/V power transformer control strategy.

The MMC DC-link voltage control and SM voltage control are crucial to guarantee a good compensation performance. Because the DC-link voltage in the RPC based on half-bridge MMC is applied in two capacitors placed in series, voltage balance between these capacitors is necessary to obtain the neutral point between them. If the two DC-link capacitor voltages are unbalanced, it may cause voltage fluctuations and distortion in current waveforms [35]. A PI controller is thus used to correct the mismatch between the actual value of the DC-link voltage with its reference value. Another PI controller is applied to balance the voltages between the two DC-link capacitors, as shown in Figure 5a. The output of the DC-link voltage control is added to the compensation current references i_{rx}^*, i_{ry}^*. Then, a predictive controller is used to produce the compensation currents. Mathematical analysis of the RPC based on half-bridge MMC, including circulating current and predictive control analysis can be found in [31].

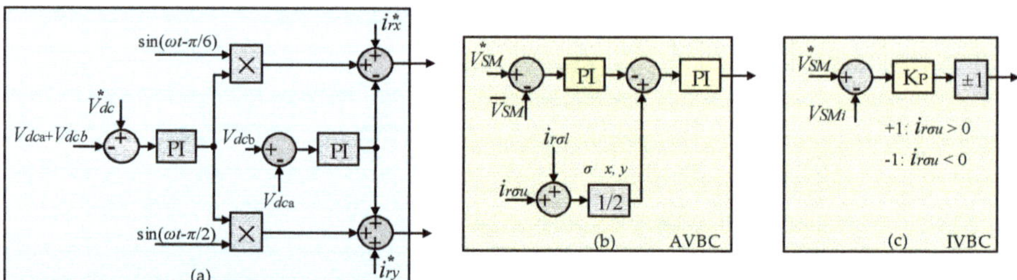

Figure 5. RPC based on half-bridge MMC control blocks: (**a**) DC-link voltage control; (**b**) averaging voltage balancing control (AVBC); (**c**) individual voltage balancing control (IVBC).

The SM voltage control in MMC consists of two control parts: averaging voltage balancing control (AVBC) and individual voltage balancing control (IVBC) [32]. The AVBC presented in Figure 5b ensures that the voltage for each SM in the MMC phase is close to the average voltage that is provided as a reference. It is implemented by adding the measured SM voltages for each MMC phase and dividing the result by the number of SMs per phase. The actual average voltage value, in this case, is calculated and compared to a reference average voltage, V_{SM}^*. Then, a PI controller is used to correct the difference between the actual and the reference values of the SM average voltage per MMC phase. The output of that PI controller is considered as a reference for a circulating current controller, which is implemented by summing the upper and the lower arm currents for each MMC phase $i_{r\sigma u}$ and $i_{r\sigma l}$, where σ belongs to phase x or phase y. The second PI controller allows a low circulating current between the MMC phases [32]. By comparison, the IVBC forces the capacitor voltage of each SM to follow its reference, which is performed by a proportional controller for each SM. The output of the IVBC is multiplied by +1 if the arm's current direction is to charge the capacitors, or by -1 if its direction is to discharge the capacitors [33]. The final signals of SM voltage control block are added to the final reference waveforms u_{rx}^*, u_{ry}^* to give the voltage command generation for each SM in phase x and phase y, u_{rxi}^*, u_{ryi}^*, where i refers to the number of SMs. Finally, the voltage command for each SM is applied to a signal modulator to drive the MMC power switching devices. Further details of the RPC control strategy can be found in [31] and [32].

3. Implementation of a Reduced-Scale Laboratory Prototype RPC Based on Half-Bridge MMC

This section presents the main steps for the development of the reduced-scale RPC based on half-bridge MMC. The implemented prototype is developed to validate the RPC based on half-bridge MMC concept and the associated control algorithms. The main specifications of this prototype are selected and some of the hardware components are described in this section. Furthermore, design of the parameters of the reduced-scale RPC based on half-bridge MMC, such as the size of the MMC SM capacitor, the size of the filter inductors, and the size of the MMC main DC-link capacitor, are also presented in this section. Additionally, this section explains the SM structure, the control system hardware, and the supplementary power equipment. As is demonstrated, each SM consists of three boards: the driver circuit board, the protection circuit board, and the power circuit board. It should be noted that some of the components and systems have been oversized in order to ensure that, in the event of abnormal operation, safety is not compromised.

3.1. Parameters Design

The RPC based on the half-bridge MMC system, presented in Figure 2c, designed as a laboratory reduced-scale prototype, has eight SMs in total (four SMs in one MMC leg). The main objective of the implementation is to validate the RPC based on the half-bridge MMC topology, the proposed control algorithm, and the proposed MMC protection system.

Consequently, a reduced-scale prototype with a few voltage levels is sufficient for a proof of concept. The prototype is implemented considering its operation in a two-phase V/V connection. In addition, the implementation is carried out considering that the converter will be used for the purpose of power quality improvement. In the V/V traction system configuration, the total DC-link voltage of the RPC based on half-bridge MMC should have a value at least twice that of the load section peak voltages due to the half-bridge configuration [32]. In this section, the main guidelines and principles for design of the MMC filter inductance, the main DC-link capacitor, and the SM capacitor are presented.

The design of the filter inductance directly influences the control system capability to track the compensation current references. In addition, the current ripples depend on the voltage applied on the inductance, the value of the inductance, and the time that the voltage is applied. The inductance can be designed by considering the factors of current ripple suppression and the speed of signal tracking [36]. The filter inductance operates as an inner MMC filter to attenuate high-frequency harmonics in the arm current and to limit the DC-link short circuit current [16]. Moreover, sizing the filtering inductance requires taking into account the suppression of undesired low-frequency harmonics, such as the MMC circulating current that is mainly composed of the 2nd order harmonics [37,38]. According to [32,39], the equation that determines the value of the filter inductance, $L_{\sigma(u,l)}$; $\sigma \in \{x, y\}$ is presented in Equation (2), where, V_{dc} is the total DC-link voltage of the MMC, V_{SM} is the SM DC-link voltage, f is the traction power grid fundamental component frequency, $I_{r\sigma(u,l)}$; $\sigma \in \{x, y\}$ is the MMC arm current, f_{sw} is the switching frequency, and $\Delta\, I_{r\sigma(u,l)}$ is the MMC arm current ripple.

$$\frac{V_{SM}}{8\, f_{sw}\, \Delta\, I_{r\sigma(u,l)}} \leq L_{\sigma(u,l)} \leq \frac{\sqrt{\frac{V_{dc}{}^2}{4} - V_{SM}{}^2}}{2\pi\, f\, I_{r\sigma(u,l)}} \tag{2}$$

A low value of the SM capacitor implies high-voltage ripples. Alternatively, a high value of capacitance results in an expensive and bulky MMC prototype [40]. The bulky converter requires a bigger area for installation. Thus, the previous principle is taken into account in the selection process of the SM capacitor value, where this value is estimated by considering the trade-off between the size, costs, and voltage ripple [16]. Furthermore, according to a study presented in [39], there is a resonance point related to the filter inductance and the DC-link capacitor of each SM, in which the maximum value of the resonance angular frequency should always be smaller than the fundamental frequency to avoid the resonance phenomenon. As a result, the resonance frequency that is related to the filter inductance and the SM capacitor should be considered in the MMC design process. According to [32,39], the equation that calculates the minimum capacitance of each SM capacitor is presented in Equation (3), where $L_{\sigma(u,l)}$ is the filter inductance, N is the MMC voltage level, $(N-1)$ is the number of SMs in one MMC arm, n is the modulation index, $U_{\sigma\ (peak)}$ is the peak voltage value of the catenary voltage, V_{dc} is the MMC DC-link voltage, and f is the power grid fundamental component frequency.

$$C_{SM} > \frac{3(N-1) + 2(N-1)\, n^2}{48\, (2\pi\, f)^2\, L_{\sigma(u,l)}} \tag{3}$$

$$n = \frac{2U_{\sigma\ (peak)}}{V_{dc}};\ \sigma \in \{x,\ y\} \tag{4}$$

Because the main DC-link voltage is always equal to $(N-1)$ times of a single SM voltage as presented in Equation (5), in the half-bridge MMC, the size of the main DC-link capacitor should be at least $(N-1)$ times the SM capacitor value [39]. Choosing the right value of the main DC-link capacitor is important because this value determines the converter power capacity, thus guaranteeing the ability of the RPC based on half-bridge MMC to compensate reactive power and harmonics, and to shift half of the active power difference between the load sections. In this regard, the main DC-link power should cover the MMC power losses, p_{losses}, and the AC power of the converter, p_{ac}, as presented in Equation (6) [16].

$$V_{dc}^{*} = V_{SM}^{*}\,(N-1);\ N\text{: MMC voltage levels} \tag{5}$$

$$p_{dc} = p_{ac} + p_{losses} \Leftrightarrow v_{dc}\, i_{dc} = \sum_{\sigma = x,\, y} u_{\sigma}\, i_{r\sigma} + p_{losses} \tag{6}$$

The instantaneous power provided by each capacitor of the main DC-link, p_{dca}, can be expressed by Equation (7), where W_{dca} is the energy stored in the DC-link capacitor, C_{dca} is the capacitance of the DC-link capacitor, and v_{dca} is the DC-link voltage.

$$p_{dca} = \frac{dW_{dca}}{dt} = \frac{d\left(\frac{1}{2}\, C_{dca}\, v_{dca}^{2}\right)}{dt} = v_{dca}\, C_{dca}\, \frac{dv_{dca}}{dt} \tag{7}$$

Then, it is possible to define Equation (8), which gives the minimum capacitance of the main DC-link capacitor, where f_d is the DC-link voltage ripple frequency (twice the fundamental frequency in the full-bridge converter and equal to the fundamental frequency in the half-bridge converter), $C_{dca(min)}$ is the minimum capacitance of the DC-link capacitor, and ΔV_{dca} is the DC-link voltage ripples (peak-to-peak).

$$C_{dca(\min)} \geq \frac{P_{dca}}{V_{dca}\, \Delta V_{dca}\, f_d} \tag{8}$$

The active power difference between the load sections in the reduced-scale prototype varies between 0 and 1500 W. The converter shifts half of the active power difference between sections. Therefore, the worst-case scenario is when only one load section is loaded when the converter shifts a higher value of active power between the load sections. In this context, it is possible to consider the parameters (P_{dca} = 750 W, f_d = 50 Hz; V_{dca} = 200 V; ΔV_{dca} = 30 V), then, the minimum capacitance for one DC-link capacitor, as presented in Equation (8), should be equal to or greater than 2.5 mF.

The final MMC parameters of the reduced-scale prototype are presented in item 4. These are calculated with regard to Equations (2), (3) and (8). By considering the root-mean square (RMS) voltage value of the load section voltages 100 V and fundamental grid frequency of 50 Hz, and by supposing the maximum RMS value of the MMC arm current $I_{r\sigma(u,l)}$ = 20 A, then the filter inductance value should be within the range of 0.1 mH $\leq L_{\sigma(u,l)}$ $\leq$ 10 mH. Consequently, by applying this range in Equation (3), the SM capacitance should be higher than 100 μF.

In the RPC based on half-bridge MMC reduced-scale prototype, only eight SMs with high switching frequency will be used to reduce the MMC complexity. The used switching frequency is 40 kHz, which allows a sinusoidal waveform to be synthesized with very good quality and fast dynamic response when compensating load current harmonics. However, in high-power applications, a few tens of SMs with low switching frequency can be used to achieve the same objectives.

3.2. Supplementary Power Equipment

The implemented reduced-scale prototype mainly consists of the power system and control system hardware. This section describes the supplementary power hardware used to obtain a reduced-scale traction power grid. Figure 6 shows a diagram of the supplementary power equipment used to obtain the three-phase public power grid and the two-phase traction power grid after using a V/V transformer. The objective is to have two-phase voltages, U_x and U_y with 60° out-of-phase due to the V/V connection [41].

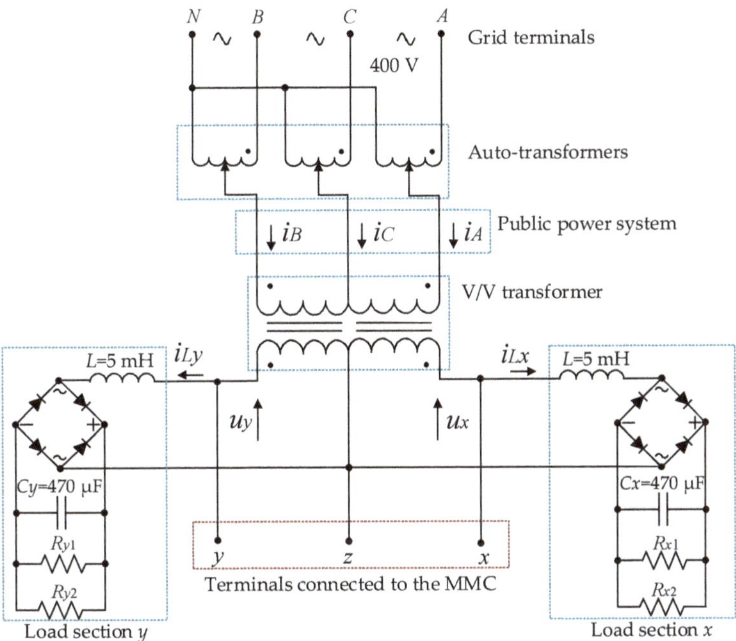

Figure 6. Supplementary power equipment diagram.

Figure 7 shows the supplementary power equipment setup. Three single-phase auto-transformers are used to change the voltage magnitudes of U_x and U_y. The connection must be carried out after respecting the phase order as shown in Figure 6. The output voltages of the auto-transformers (input voltages of the V/V transformer) must be in phase with the line-to-line voltages, U_{AC} and U_{BC}, to have a phase shift of 60°, as presented in Equation (9).

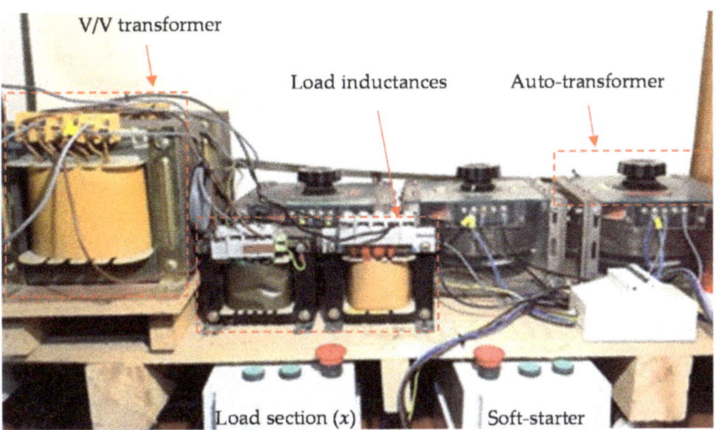

Figure 7. Supplementary power equipment setup.

One of the V/V connection advantages is its easy implementation in the laboratory. Hence, single-phase step-down transformers are used to implement the V/V connection and to obtain two-phase voltages, Ux and U_y, as presented in Equation (9). Attention to the polarity of the single-phase transformers must be considered to obtain the correct phase shift. These single-phase transformers have a turns ratio of 0.5. In this case, the point z, as shown in the Figure 6, is the neutral of the reduced-scale traction power grid. Each load section is connected to an inductance and a full-bridge rectifier. The inductances are important to represent the non-linear inductive traction load. The indicated values in Figure 6 are used in the real implementation.

Because the single-phase auto-transformers have a large magnetization inductance value, a soft-starter is required to avoid high inrush currents at the moment of connection to the power grid. This soft-starter inserts series-connected resistors into the auto-transformers input, and a timer contactor is used to bypass the series-connected resistors after a predefined period.

The main purpose of the supplementary power equipment is to have different load power values to show the capability of the RPC based on half-bridge MMC in improving the three-phase grid power quality. The resistors, R_{x1} and R_{y1}, have a small value and mainly determine most of the consumed power in each load section. On the contrary, the resistors, R_{x2} and R_{y2} have a high value because the main purpose of these resistors is to dissipate the stored energy in the capacitors for safety reasons after turning off the power supply or disconnecting the load resistors of R_{x1} and R_{y1} using contactors.

$$\begin{array}{c} \dot{U}_A = 230 \angle 0°; \dot{U}_B = 230 \angle{-120°}; \dot{U}_C = 230 \angle{-240°} \text{ (V)} \\ \dot{U}_{AC} = 400 \angle{-30°}; \dot{U}_{BC} = 400 \angle{-90°} \text{ (V)} \end{array} \tag{9}$$

3.3. *Implementation of a Reduced-Scale Modular Multilevel Converter*

This section presents the implementation of the reduced-scale MMC, including the SM components. Initially, the main aim is to implement a well-designed SM, then replicate the work to have a full MMC. For this purpose, various tests should be employed on the designed SM before implementing the final MMC. This is essential to guarantee a robust performance under abnormal conditions when several cascade-connected SMs are under operation. Consequently, this item presents the MMC design process, including the SM implementation and validation.

3.3.1. IGBT Driver Circuit Board

The main application of the driver board is to drive the SM switching devices. In this context, an isolated driver with two complementary channels in a single package is used (SI824x from Silicon Labs [42]). This driver is specifically targeted to drive complementary switching devices (as in the case of half-bridge power converter developed). In addition, its main feature is the integrated deadtime generator between the high-side/low-side drivers that allow highly precise control for achieving optimal total harmonics distortion (THD). Another driver, HCPL-3120, is employed to drive an additional IGBT switch for the purpose of overvoltage protection, as explained in Section 3.3.2. Figure 8 shows the driver circuit board hardware. The pulse width modulation (PWM) input signals are supplied from the main central control unit and the outputs of this board are connected to the gate-emitter of each IGBT that composes the SM.

The driver circuit board is designed using an optimized layout to reduce the length of the board traces. That is, the design should respect the minimum compulsory dimensions to maintain the isolation between the channels. In addition, the optimized layout is important to minimize the board parasitics, such as parasitic inductance and capacitance, thus reducing noise and improving the board performance. This is possible by decreasing some signal loops and keep connections as short as possible. Moreover, decoupling capacitors located close to the main power supply signals are used to avoid erroneous operation. Furthermore, extra jumpers are added, allowing the voltage value (either 0 V or −15 V) of the IGBT switching device to be turned off. The turn-off negative voltage applied to the gate-emitter junction helps to decrease the turn-off time of the IGBT, thus improving the overall performance and obtaining a faster switching device.

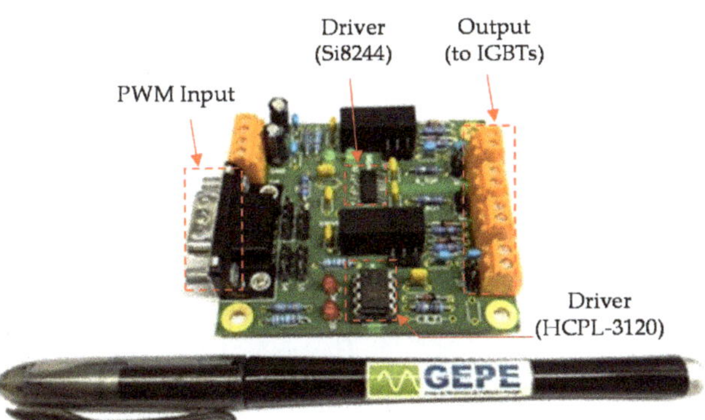

Figure 8. Isolated gate bipolar transistor (IGBT) driver circuit board.

3.3.2. Protection Circuit Board

MMC reliability and stability are some of the most important demands. This includes protection against consequences that may appear due to the MMC malfunctioning, bearing in mind that the MMC topology contains many power switching devices, such as an IGBT and diodes, and each device can be a possible failure point. Because the MMC is composed of multiple cascade-connected SMs, each SM should have its own protection system; thus, the developed protection circuit board is essential for each MMC SM. Figure 9 shows the developed protection board, which is applied to each MMC SM. This protection consists of two main parts: overcurrent protection and overvoltage protection. A detailed description of this overvoltage and overcurrent protection with experimental results is presented in [43].

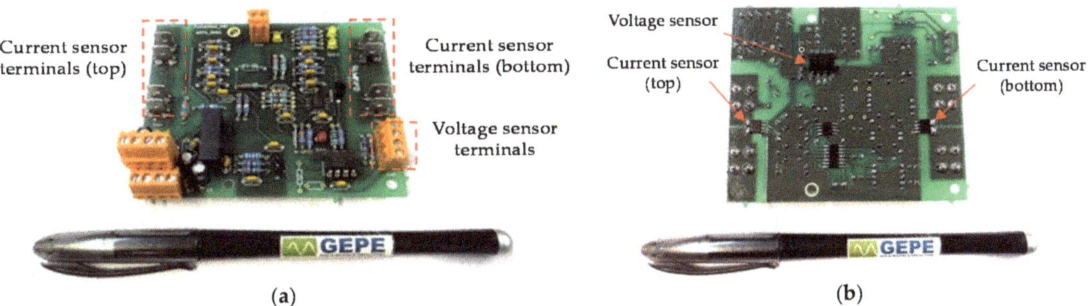

Figure 9. Protection circuit board: (**a**) top view; (**b**) bottom view.

It is worth noting that several protection systems are implemented, either by means of implemented boards, such as the protection circuit board, or through verification functions in the developed supervision software. In addition, four semiconductor fuses are used as ultimate protection, where each MMC arm has one semiconductor protection fuse.

3.3.3. Power Circuit Board

The power circuit board includes the half-bridge SM components, such as DC-link capacitors, coupling capacitors, power switching devices, discharge resistors, and power terminals. Three power switching devices, IGBTs, are mounted on the top surface of the power circuit board, where two of these switching devices commutate at 40 kHz switching frequency to synthesize output waveform signal of 50 Hz. The third switching device (the one that uses the HCPL-3120) is used for overvoltage protection purposes. In this context, IGBT switching device with fast free-wheeling diode is used in the SM power circuit board.

A bank of electrolytic capacitors, composed of 21 parallel-connected electrolytic capacitors of 47 μF, is used to increase the SM capacitor lifetime because it is possible to reduce both the equivalent series resistance and the equivalent series inductance of the SM capacitor. All aluminum electrolytic capacitors have the same rating, and they are connected in parallel to increase the equivalent SM capacitance, thus storing a greater amount of electrical energy. Furthermore, reducing the total equivalent series resistance of the SM capacitor is important to reduce the power dissipation in the electrolytic capacitors. It also helps to obtain better stability of the control loop, in addition to boosting the overall performance and reliability [44]. The power circuit board is presented in Figure 10. This board contains power terminals to connect with the driver circuit and the protection circuit boards. In addition, these power terminals are beneficial to accomplish the cascade connection of the MMC SMs. Table 1 offers a list of the power circuit board component values.

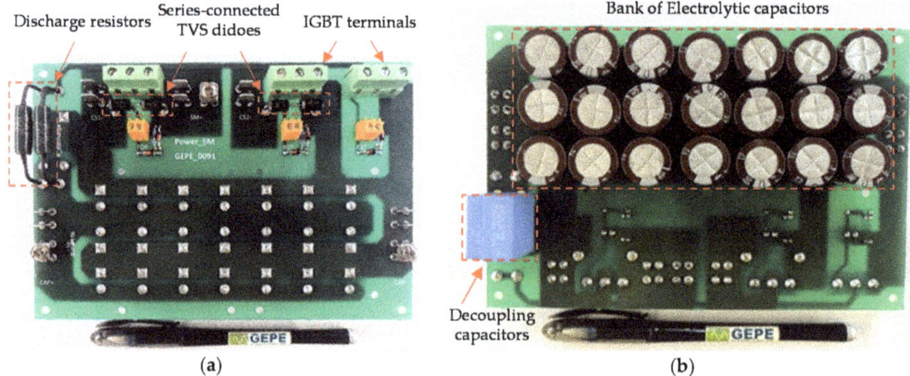

Figure 10. Power circuit board: (**a**) top view; (**b**) bottom view.

Table 1. Power circuit board components.

Components	Symbols	Values
Electrolytic capacitor	C_{sm}	21 × 47 μF = 987 μF
Decoupling capacitor	C_{de}	1.1 μF
Discharge resistor	R_{dis}	5 kΩ

After the implementation of one SM, it is important to validate the SM operation conditions before implementing a complete reduced-scale MMC prototype. The full implementation of an MMC prototype is not a good strategy to test the power circuit board because it is difficult to predict the SMs performance when they are connected in a cascade configuration. In this case, the power switching devices, and the electromagnetic and thermal characteristics for each SM, are different. Consequently, an experimental setup to test the half-bridge SM power circuit board is required. Authors in [45] introduced a simplified scheme to test the half-bridge MMC SM. In this study, the same testing scheme presented in [45] is used to validate one MMC SM. However, presenting the experimental results of this test is outside of the scope of this paper because the main contribution of this study is to validate the RPC based on half-bridge MMC in terms of control theory and operation.

3.3.4. Final Modular Multilevel Converter Submodule

Figure 11 presents the final power submodule (SM) structure, including the driver circuit board, the protection circuit board, the power circuit board and the IGBT heatsink.

It is worth noting that the final SM hardware has a compact, extensible, and cubic design to build the reduced-scale RPC based on half-bridge MMC prototype. Each four cascade-connected SMs are composed of a single MMC leg/phase, as shown in Figure 12. The total number of SMs for the RPC based on half-bridge MMC reduced-scale prototype is eight. Section 4 presents the experimental analysis of the reduced-scale prototype RPC based on half-bridge MMC under two case studies: (a) when two load sections are loaded; and (b) when only one load section is loaded. Then, a discussion of experimental results of a reduced-scale prototype is provided to link with the performance of the RPC based on MMC under high-power applications.

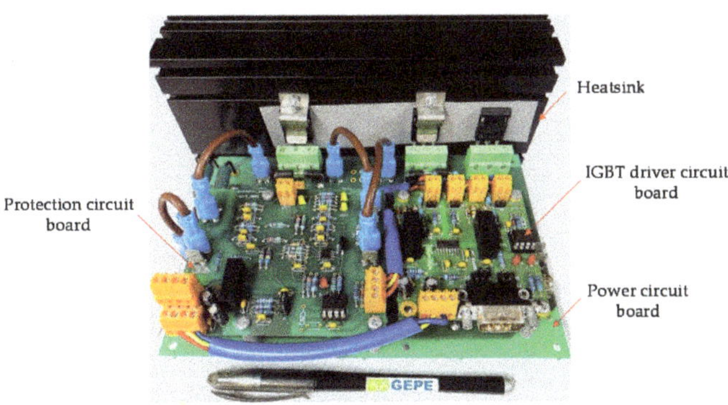

Figure 11. Final submodule (SM) structure.

Figure 12. Single MMC leg/phase (four submodules).

3.4. Control System Hardware

The control system hardware has several output flat cable connectors to link between the digital signal processor (DSP) and the auxiliary control units, such as sensors and actuators. The digital system loop always exchanges information between the DSP and the auxiliary control units (such as signal conditioning unit, sensors unit, and the PWM control unit). The DSP board is presented in Figure 13a, and integrates several connectors to facilitate the connection between different control boards and the DSP control unit. Figure 13b presents a digital-to-analogue converter (DAC) board, which is important to verify the accuracy of the digital signals generated in the DSP. Two signal conditioning boards are used; the one presented in Figure 13c is dedicated to the external analogue-to-digital converter (ADC) channels, and the other presented in Figure 13d,e is dedicated to the internal ADC channels in the DSP. This ADC board contains two main parts: the voltage sensors placed at the bottom side, and the signal conditioning part placed at the top side. PWM adapters are used to convert from ribbon cable terminals to DB-9 connector terminals, as presented in Figure 13f.

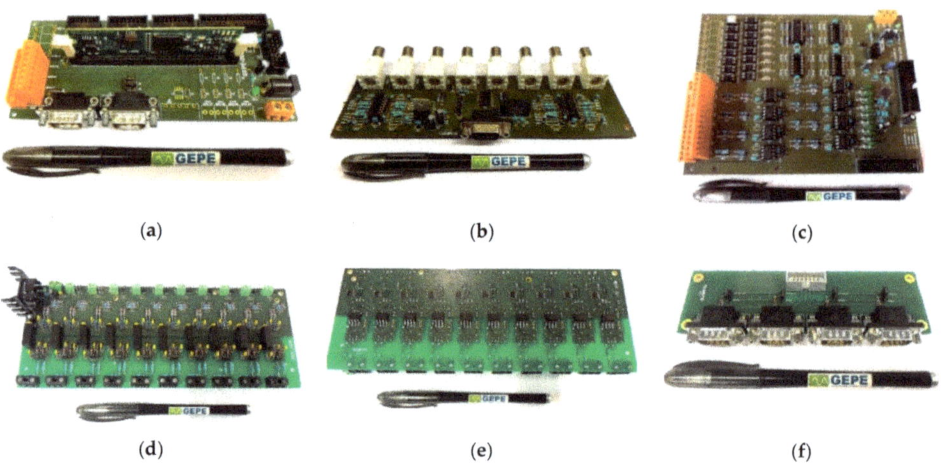

Figure 13. Some of the control system boards: (**a**) digital signal processor (DSP) board; (**b**) digital-to-analogue converter (DAC) board; (**c**) Signal conditioning board for the external analogue-to-digital converter (ADC); (**d**) Signal conditioning board for the internal ADC—top view; (**e**) Signal conditioning board for the internal ADC—bottom view; (**f**) pulse width modulation (PWM) adapter.

The control system hardware is enclosed in a metallic box, as shown in Figure 14. The metallic box also includes several input/output connectors to interface between the control system and the power system hardware, such as the driver circuit and the protection circuit boards. A command circuit board is used to interface and isolate between the driver circuit boards and the protection circuit boards. This command board also saves the error when an overcurrent or overvoltage condition is detected, stopping the IGBT switching devices, until a reset command is released. In addition, several power connectors in the metallic box are used to connect the SM DC-link voltages with the associated voltage sensors. A DC power supply is also used to convert the 230 V AC to a regulated low-voltage (+15 V, +5 V, −15 V) to supply the electronic components of the control system hardware. It is worth noting that the majority of the control system hardware is installed in the metallic box. This is very important to maintain the distance between the control system hardware and the power system hardware, thus reducing the electromagnetic interference effects. The electromagnetic fields generated due to the power system hardware can affect the correct functionality of the DSP, where these fields may induce currents loops in the electronic components.

Figure 15 shows the global communication structure of the reduced-scale RPC based on half-bridge MMC prototype. A depicted hardware architecture of the control system is presented, where a Delfino DSP controller from Texas Instruments TMS320f28335 is used as a central control unit [46]. The converter requires 18 ADC channels (12 voltage sensors and six current sensors), a DAC unit to check the correctness of digital signals in the DSP, and eight PWM channels (one PWM channel is used for each SM, then the driver circuit board generates two complementary PWM signals to derive the IGBT switching devices). A user interface unit based on the RS-232 data interface is used to send and receive commands, such as commands for pre-charging and discharging of SM capacitors. Two ADC units are used to convert the analogue signals to digital signals. Subsequently, the digital signals can be analyzed and processed by the DSP. However, the ADC units have different bit resolutions; the internal ADC channels of the DSP are unipolar and have 12 bits of resolution. Therefore, they are dedicated to the purpose of DC signal processing, such as the MMC DC-link voltages (SM voltages and main DC-link voltage). The external ADC integrated circuit (IC) has 8 bipolar ADC channels with 14 bits of resolution. Therefore, they are reserved for the purpose of AC signal processing, such as the load section voltages waveforms, as shown in Figure 15.

Figure 14. Control system hardware fitted in a metallic box.

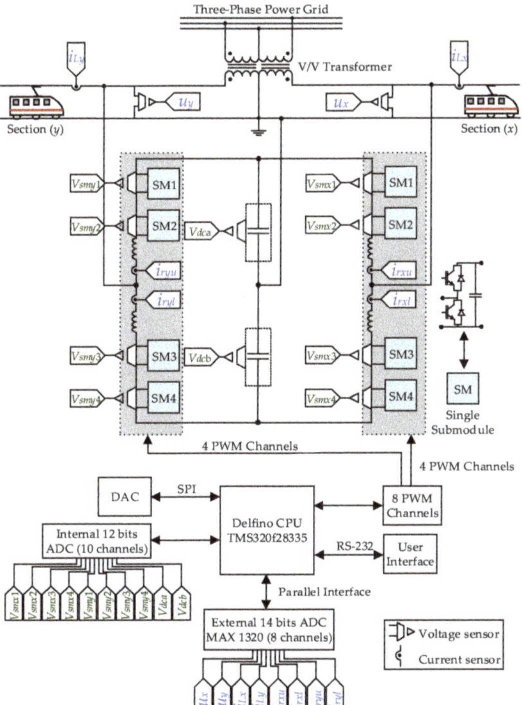

Figure 15. Global communication structure of the reduced-scale prototype RPC based on half-bridge MMC.

4. Experimental Results of the Laboratory Prototype RPC Based on Half-Bridge MMC

This section presents the RPC based on half-bridge MMC experimental results. Figure 16 shows the implemented MMC workbench, which includes the control system hardware, isolated-channel oscilloscopes, and a spare SM. A computer device is used that allows user interface with the DSP. The experimental results were obtained using the TPS2000B oscilloscope from Tektronix. In addition, a Fluke 434 Power Quality Analyzer was used to measure the THD, the ratio of NSC, and the frequency spectrum of the three-phase currents. Figure 17 shows the experimental setup schematic, which consists of the supplementary power equipment (auto-transformers, V/V transformer, and full-bridge rectifiers) and the RPC based on half-bridge MMC. For this purpose, full-bridge rectifiers with filtering capacitors are used to create harmonic contents and current imbalance in the three-phase public power grid. The parameters used in this experimental validation are presented in Table 2, where load section y active power is selected to be 150% of the active power of the load section x. Therefore, as presented in Table 2, the value of the resistor R_{x1} is 150% the value of the resistor R_{y1}.

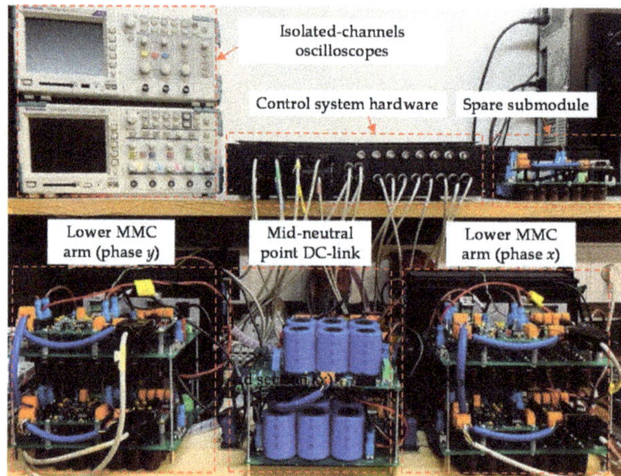

Figure 16. Workbench with the developed modular multilevel converter (MMC).

The resistors of R_{x2} and R_{y2} have a high value because their main function is to discharge the filtering capacitors after disconnecting the load power resistors, R_{x1} and R_{y1}, using contactors. The reduced-scale RPC based on half-bridge MMC consists of eight half-bridge SMs, in which each MMC leg or phase has four SMs. The MMC parameters, such as capacitance of the SM capacitor, the capacitance of the main DC-link capacitor, the MMC leg filter inductors, and the PWM switching frequency, are also presented in Table 2. Two case studies are presented in this experimental test: The first case study is when the load section y active power is 150% of the load section x active power, and the second case study is when only load section y is loaded.

4.1. Experimental Results When Two Load Sections are Loaded

This item presents the RPC based on half-bridge MMC prototype experimental results when the load section y active power is 150% of the active power of the load section x. Figure 18a shows the three-phase currents at the secondary windings of the V/V transformer before compensation, and when both load sections are unequally loaded. The currents are imbalanced and contain harmonic content. Figure 19a shows the phase difference angle between the phase x and phase y currents or the load section currents before compensation and the phase-to-neutral voltage u_A, which is used as a reference waveform with a phase angle equal to zero. There are almost 30° out-of-phase between u_A and i_x, and nearly 90° out-of-phase between u_A and i_y before compensation.

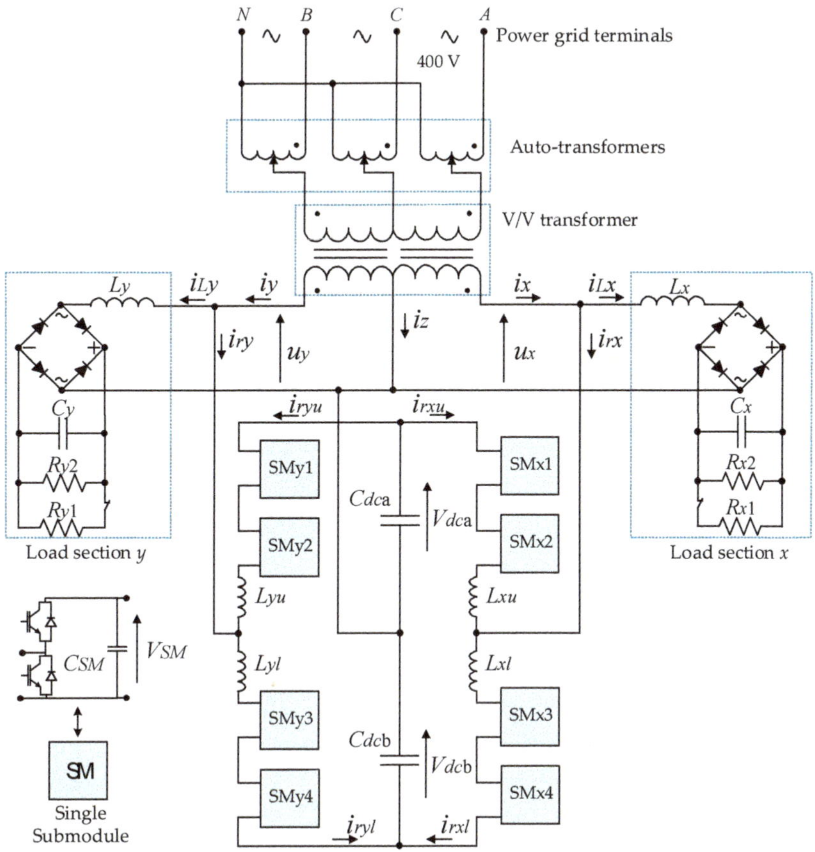

Figure 17. Schematic of the RPC based on half-bridge MMC experimental setup.

Table 2. Experimental parameters of the RPC based on half-bridge MMC.

Parameters	Symbols	Values
Power grid RMS phase voltage	u_A, u_B, u_C	230 V
Phase x and phase y RMS voltage	u_x, u_y	40 V
SM voltage	V_{SM}	80 V
MMC main DC-link voltage	$V_{dca} + V_{dcb}$	160 V
Filter inductance of the arm	L_{xu}, L_{xl}, L_{yu}, L_{yl}	1.6 mH
Load section inductance	L_x, L_y	5 mH
Load resistor (section x)	R_{x1}	6.5 Ω
Load resistor (section y)	R_{y1}	4.34 Ω
Resistors to discharge the capacitors	R_{x2}, R_{y2}	33 kΩ
Capacitance of the filtering capacitors	C_x, C_y	470 µF
Capacitance of the SM capacitor	C_{SM}	987 µF
Capacitance of the main DC-link capacitors	$C_{dca} = C_{dcb}$	2820 µF
PWM SM switching frequency	f_{isw}	40 kHz
Modulating signal fundamental frequency	f	50 Hz
Highest/Lowest duty-cycle	–	0.92/0.08

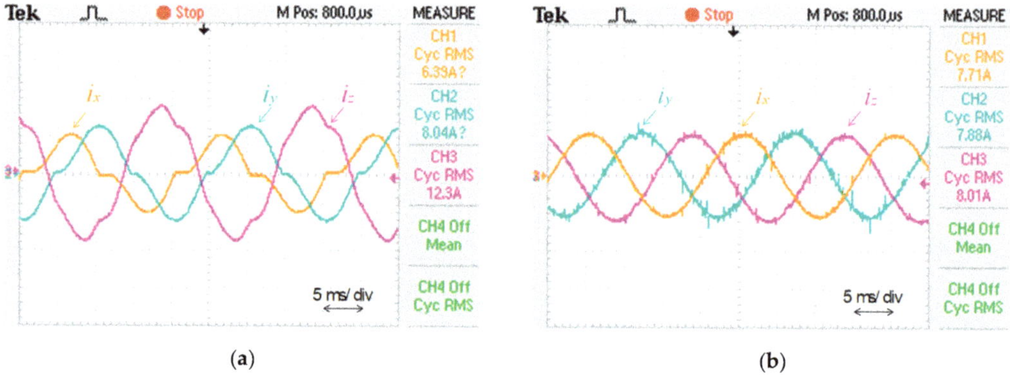

Figure 18. Experimental results when two load sections are loaded: (**a**) before compensation; (**b**) after compensation. Phase x current (i_x: 10 A/div); phase y current (i_y: 10 A/div); phase z current (i_z: 10 A/div).

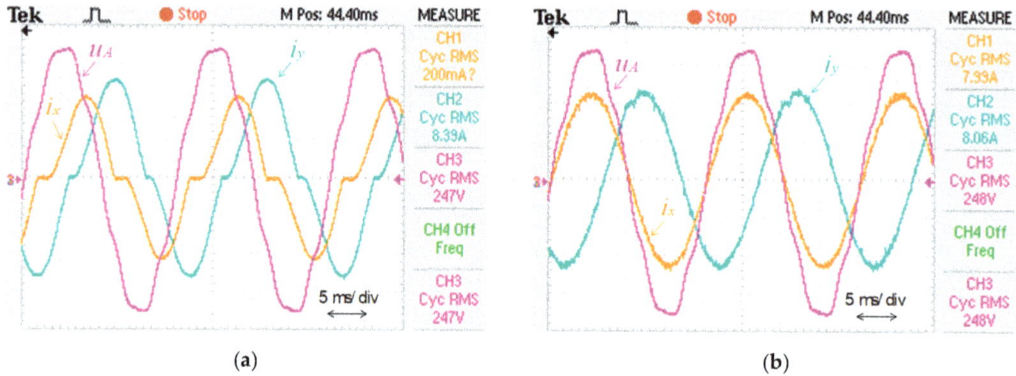

Figure 19. Experimental results when two load sections are loaded: (**a**) before compensation; (**b**) after compensation. Phase x current (i_x: 5 A/div); phase y current (i_y: 5 A/div); phase A voltage (u_A: 100 V/div).

Figure 18b presents the three-phase currents at the secondary windings of the V/V transformer after compensation. The currents are balanced with lower harmonic content and without NSCs. In this case, the RPC compensates reactive power and balances the active power between the load sections. Consequently, a unitary power factor is obtained at the three-phase power grid. In addition, there are almost 0° out-of-phase between u_A and i_x, and almost 120°out-of-phase between u_A and i_y after compensation, as presented in Figure 19b. This means the reactive power is totally exchanged between the load sections and the power compensator. There is no reactive power exchanged with the three-phase power grid and the power factor is unitary.

Figure 20 presents the frequency spectrum of the three-phase currents at the secondary windings of the V/V transformer. Figure 20a shows the harmonic contents before compensation with a THD ratio close to 23.1%. Figure 20b shows the harmonic contents after compensation with a THD ratio close to 3.1%. These results are obtained at a fundamental frequency component of 50 Hz.

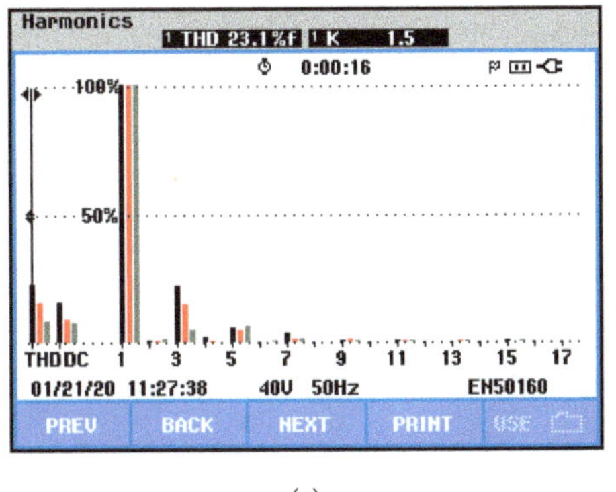

(**a**)

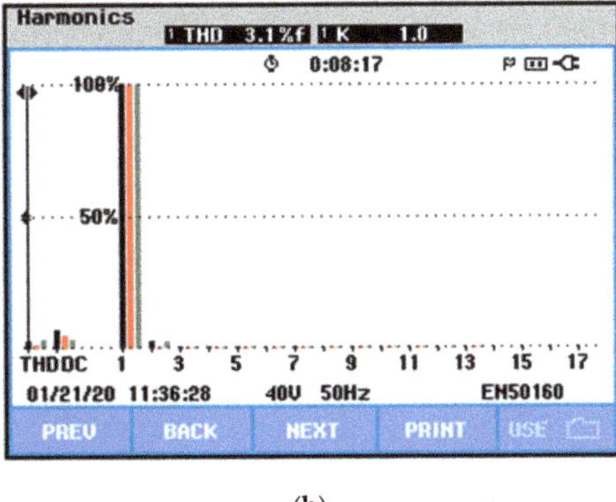

(**b**)

Figure 20. Frequency spectrum of the three-phase currents at the secondary windings of the V/V transformer (when two load sections are loaded): (**a**) before compensation; (**b**) after compensation.

Figure 21a presents the values of the harmonic contents before compensation. Phase x current i_x has the highest ratio of THD, with a value close to 22.9%. The third-order harmonic is the highest harmonic content before compensation. Figure 21b shows the values of the harmonic contents after compensation. In this case, the THD ratio is significantly reduced, particularly the third- and fifth-order harmonics. It is worth noting that the even-order harmonics circulate between the MMC phases and do not contribute to the RPC compensation currents.

HARMONICS TABLE (0:00:16)

Amp	L1	L2	L3
THD%f	22.9	15.7	8.2
H3%f	22.0	14.8	4.8
H5%f	5.5	4.7	6.2
H7%f	3.6	1.5	1.7
H9%f	0.8	1.5	1.0
H11%f	0.9	0.4	0.4
H13%f	0.2	0.7	0.5
H15%f	0.5	0.3	0.4

01/21/20 11:27:38 40V 50Hz EN50160
PREV BACK NEXT PRINT USE

(**a**)

HARMONICS TABLE (0:08:17)

Amp	L1	L2	L3
THD%f	2.8	1.6	3.3
H3%f	1.0	0.8	0.8
H5%f	0.4	0.5	0.7
H7%f	0.4	0.8	1.0
H9%f	0.3	0.4	0.4
H11%f	0.2	0.4	0.5
H13%f	0.3	0.4	0.6
H15%f	0.2	0.3	0.3

01/21/20 11:36:28 40V 50Hz EN50160
PREV BACK NEXT PRINT USE

(**b**)

Figure 21. Harmonic contents value of the three-phase currents at the secondary windings of the V/V transformer (when two load sections are loaded): (**a**) before compensation; (**b**) after compensation.

Figure 22 shows the phasors diagram and the unbalance ratio (NSC ratio) of the three-phase currents at the secondary of the V/V transformer. The unbalance ratio before compensation is close to 46.8%, as shown in Figure 22a. The phasors diagram, in this case, shows three-phase current vectors that have different magnitude values. By considering the phase voltage u_A as a reference and due to the nature of the V/V connection, the phase x current lags the reference voltage by nearly 30°, whereas the phase y current lags the reference voltage by nearly 90°. Figure 22b shows the unbalance ratio and the phasors diagram of the three-phase currents after compensation. The unbalance ratio, in this case, is significantly reduced and has a value close to 2.9%. As observed, the three-phase current phasors are balanced, with similar magnitude values and 120° out-of-phase.

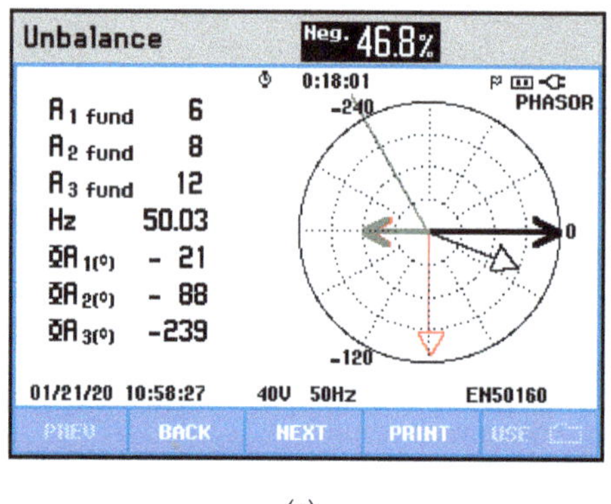

(a)

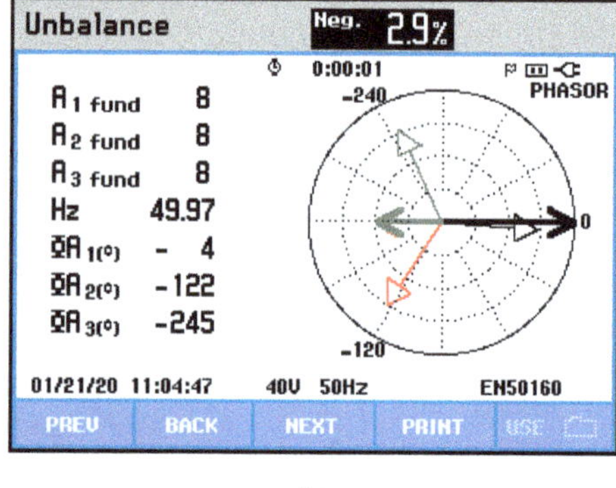

(b)

Figure 22. Unbalance ratio and phasors diagram of three-phase currents at the secondary of the V/V transformer (when two load sections are loaded): (**a**) before compensation; (**b**) after compensation.

The compensation currents synthesized by the RPC are presented in Figure 23. The compensation current of the load section x, i_{rx}, is higher than the compensation current of the load section y, i_{ry}. This is due to the unequal loading power and the fact that the section x converter injects a higher amount of energy than the section y converter. In a V/V connection, the section x converter compensates a quantity of a capacitive reactive power, whereas the section y converter compensates a quantity of an inductive reactive power.

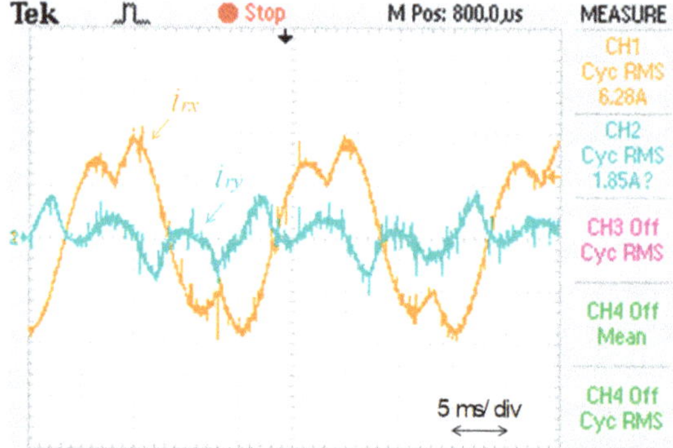

Figure 23. Experimental results (when two load sections are loaded): phase x compensation current (i_{rx}: 5 A/div); phase y compensation current (i_{ry}: 5 A/div).

Figure 24a,b shows the SM voltage waveforms of the section x converter and the section y converter, respectively. The SM voltages are close to the reference value of 80 V. However, due to the half-bridge topology of the SM, the voltage ripple frequency is equal to the fundamental frequency of 50 Hz. Furthermore, because the section x converter injects a higher amount of power, the SM voltages of section x converter V_{SMx1}, V_{SMx2}, V_{SMx3}, and V_{SMx4}, have higher voltage ripples than the SM voltages of section y converter V_{SMy1}, V_{SMy2}, V_{SMy3}, and V_{SMy4}. These results confirm the effectiveness of the MMC leg averaging voltage balancing control and the MMC SM individual voltage balancing control to maintain a balanced voltage of the MMC SMs.

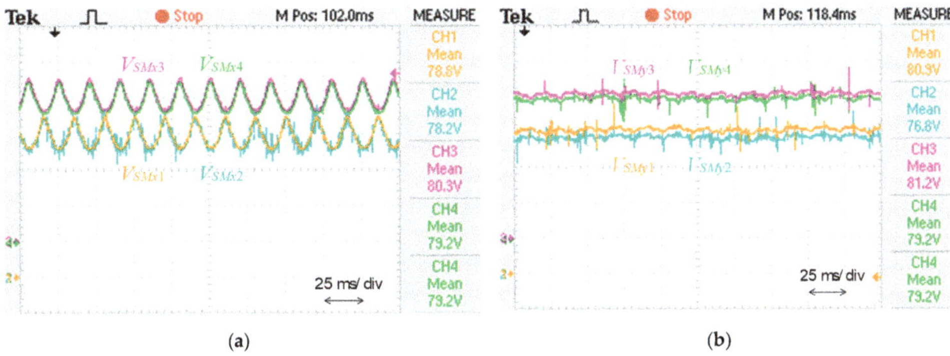

Figure 24. Experimental results (when two load sections are loaded): (**a**) SM voltages of section x converter; (**b**) SM voltages of section y converter. SM DC voltage (V_{SM}: 20 V/div).

4.2. Experimental Results When One Load Section is Loaded

This item presents the RPC based on half-bridge MMC prototype experimental results when only the load section y is loaded (load section x has no loads). Figure 25a shows the secondary windings three-phase currents of the V/V transformer before compensation, and when only load section y is loaded. The currents are imbalanced and contain harmonic contents. Phase x current i_x, has a zero RMS value because the load section x is not loaded. Consequently, phase y current i_y and phase z current i_z, have 180° out-of-phase. The imbalance ratio, in this case, is higher than the case when both of the load sections were loaded. Figure 25b presents the secondary windings three-phase currents of the V/V transformer after compensation. The currents are balanced with lower harmonic contents and lower NSCs of currents. In this case, the RPC compensates reactive power and balances the active power between the load sections, thus a unitary power factor is obtained at the three-phase power grid side.

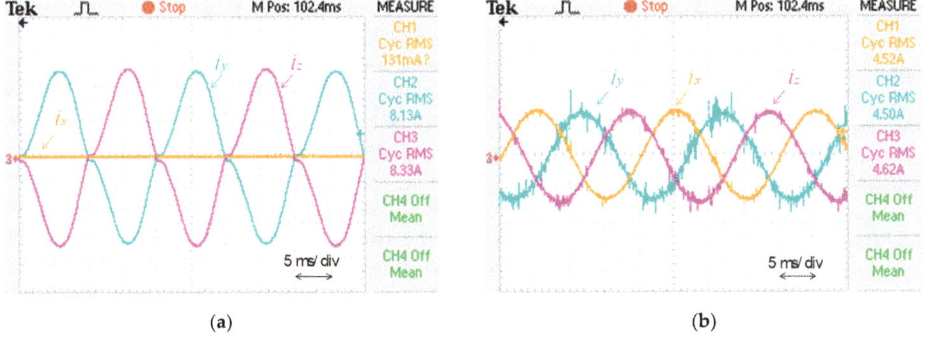

Figure 25. Experimental results (one load section is loaded): (**a**) before compensation; (**b**) after compensation. Phase x current (i_x: 5 A/div); phase y current (i_y: 5 A/div); phase z current (i_z: 5 A/div).

Figure 26 presents the frequency spectrum of the three-phase currents at the secondary windings of the V/V transformer. In this context, Figure 26a shows the harmonic contents before compensation with a THD ratio close to 15.1% at a fundamental frequency of 50 Hz. Figure 26b shows the harmonic contents after the compensation with a THD ratio close to 2.5%. This confirms the effectiveness of the RPC based on half-bridge MMC system in improving the power quality of the three-phase power grid

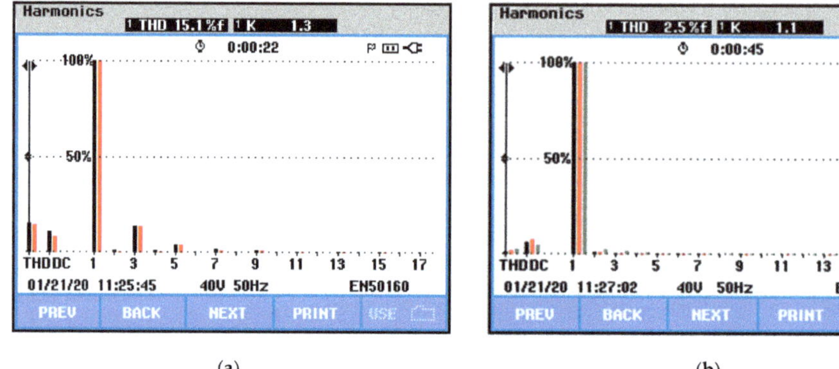

(**a**) (**b**)

Figure 26. Frequency spectrum of the three-phase currents at the secondary windings of the V/V transformer (when one load section is loaded): (**a**) before compensation; (**b**) after compensation.

Figure 27a presents the values of harmonic contents before compensation. Phase y and phase z currents have 180° out-of-phase, and have almost the same ratio of the THD. The third-order harmonic is the highest harmonic content before compensation, with a ratio close to 14.2%, considering a fundamental frequency of 50 Hz. Figure 27b shows the values of the harmonic contents after compensation. In this case, the THD is significantly reduced, and the third-order harmonics does not exceed the ratio of 2%. It is worth noting that the even-order harmonics circulate between the MMC phases and do not contribute to the RPC compensation currents.

HARMONICS TABLE (0:00:22)

Amp	L1	L2
THD%f	15.2	15.0
H3%f	14.2	14.3
H5%f	4.0	4.5
H7%f	1.8	1.4
H9%f	1.4	1.2
H11%f	0.4	0.4
H13%f	0.7	0.7
H15%f	0.4	0.4

01/21/20 11:25:45 40V 50Hz EN50160

(**a**)

HARMONICS TABLE (0:00:45)

Amp	L1	L2	L3
THD%f	1.7	2.0	2.8
H3%f	1.0	1.0	1.8
H5%f	0.6	0.6	1.0
H7%f	0.5	0.7	0.6
H9%f	0.4	0.7	0.7
H11%f	0.2	0.4	0.4
H13%f	0.2	0.5	0.5
H15%f	0.2	0.4	0.5

01/21/20 11:27:02 40V 50Hz EN50160

(**b**)

Figure 27. Harmonic contents value of the three-phase currents at the secondary windings of the V/V transformer (when one load section is loaded): (**a**) before compensation; (**b**) after compensation.

Figure 28 shows the phasors diagram and the unbalance ratio of the three-phase currents at the secondary windings of the V/V transformer. The unbalance ratio (NSC ratio) before compensation is close to 97.7%, as shown in Figure 28a. The phasors diagram, in this case, shows only two-phase currents that have equal magnitudes and 180° out-of-phase. By considering the phasor of the phase voltage u_A as a reference, due to the nature of the V/V connection, phase y current lags the reference by nearly 90°, whereas phase z current lags the reference by nearly 270°. Figure 28b shows the unbalance ratio and the phasors diagram of the three-phase currents after compensation. The unbalance ratio, in this case, is significantly reduced and has a value close to 1.6%. As observed, the three-phase current phasors are balanced, with similar magnitude values and 120° out-of-phase.

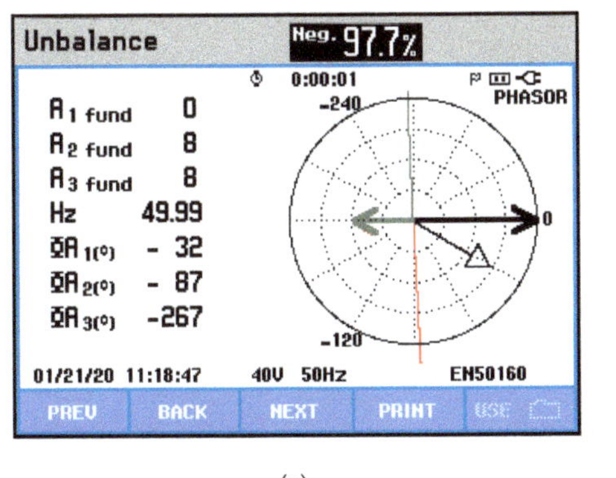

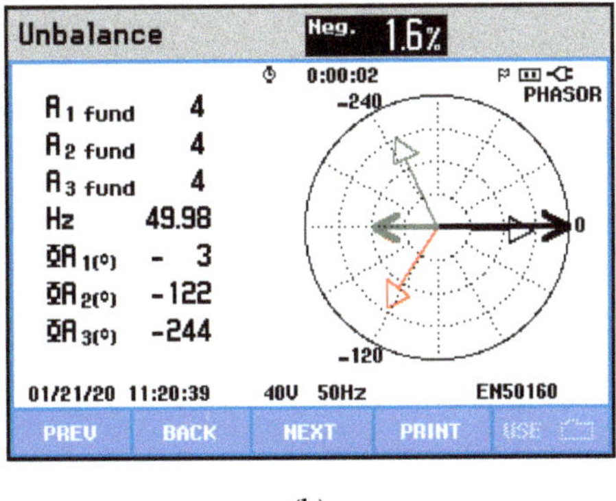

(a) (b)

Figure 28. Unbalance ratio and phasors diagram of the three-phase currents at the secondary of the V/V transformer (when one load section is loaded): (**a**) before compensation; (**b**) after compensation.

The compensation currents synthesized by the RPC are presented in Figure 29. The compensation current of the load section x, i_{rx}, is completely sinusoidal because there are no loads connected to the load section x. In this case, the section x converter compensates the reactive power and balances the active power between the load sections. The section y converter compensates harmonic contents and reactive power, and also balances the active power between the load sections.

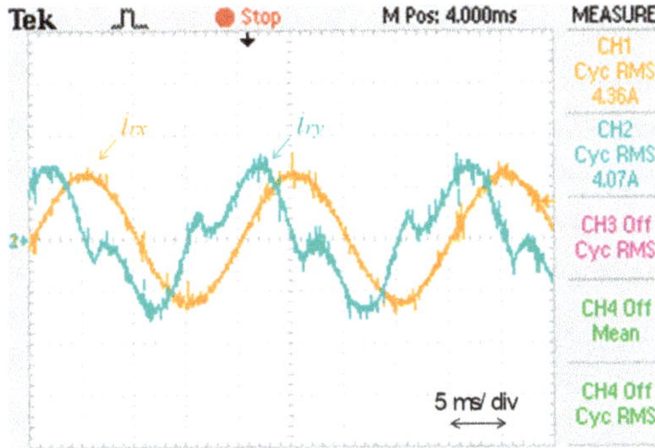

Figure 29. Experimental results (when one load section is loaded): phase x compensation current (i_{rx}: 5 A/div); phase y compensation current (i_{ry}: 5 A/div).

Figure 30a,b shows the SM voltage waveforms of the section x converter and the section y converter, respectively. The SM voltages are close to the reference value of 80 V. However, due to the half-bridge topology of the SM, the voltage ripple frequency is equal to the fundamental frequency of 50 Hz. Furthermore, because the section x converter and the section y converter compensate an almost equal amount of power, the SM voltages of the MMC have almost equal voltage ripples for the section x and section y converters. These results confirm the effectiveness of the MMC leg averaging voltage balancing control and the MMC SM individual voltage balancing control to maintain a balanced voltage of the MMC SMs.

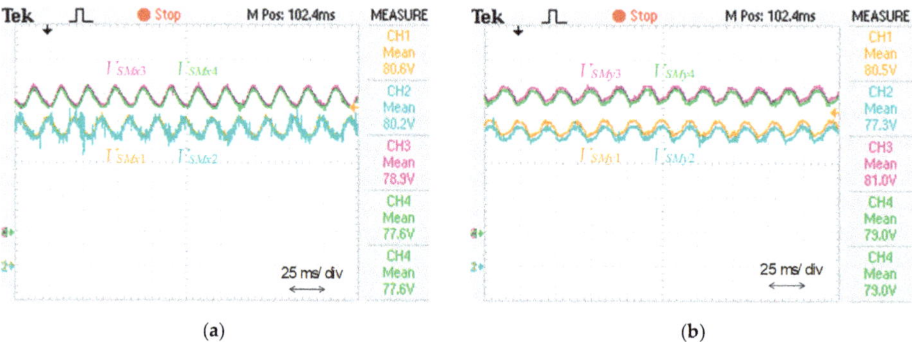

Figure 30. Experimental results (when one load section is loaded): (**a**) SM voltages of section x converter; (**b**) SM voltages of section y converter. SM DC voltage (V_{SM}: 20 V/div).

4.3. Discussion and Experimental Analysis

The presented experimental results confirm the validity of the compensation strategy and the associated capacitor voltage balancing control of the MMC SMs presented in this paper. In addition, the experimental results show that the RPC based on half-bridge MMC is a viable solution for the purpose of power quality improvement in electrified railway systems. In the experimental validation, only eight SMs with a 40 kHz switching frequency were used. However, for high-power applications, a few tens of SMs with a lower SM switching frequency should be employed, which and would allow the MMC to synthesize sinusoidal waveforms with better quality and lower harmonic contents. Subsequently, the experimental case reported in this paper involves a low number of SMs with a high SM switching frequency (this choice was selected to reduce the MMC complexity and costs). It was verified that the compensation currents synthesized by the reduced-scale RPC based on half-bridge MMC satisfy the requirements of NSC compensation and harmonics cancelation. As a result, it is important to highlight that, in high-power applications, the number of MMC SMs should increase, but the switching frequency could be reduced to achieve the same objectives.

5. Conclusions

This paper presented the implementation, testing, and experimental validation of a reduced-scale laboratory prototype of a proposed rail power conditioner (RPC) based on a modular multilevel converter (MMC). The design of the parameters of the MMC, such as filter inductors and submodule (SM) capacitors, were also described in the paper. Moreover, SM power components and the associated power hardware were explained in detail. This power hardware involved three main circuit boards: the driver circuit board, protection circuit board, and power circuit board. Furthermore, the control system hardware of the reduced-scale RPC based on MMC was introduced in the paper, in addition to the related communication structure between the MMC power and control circuit boards.

Experimental results were presented considering two case studies of electrified railway systems: when both load sections (x and y) are loaded and when only the load section y is loaded. In the first case study, the load section y active power is 150% of the active power of the load section x, whereas, in the second case study, the load section x has no load and the load section y maintains its power value. Experimental results show that the RPC based on half-bridge MMC system injects a higher amount of reactive power and shifts a higher amount of active power in the second case study, which is the worst-case scenario. In this case, the compensation currents synthesized by the RPC are higher than those presented in the first case study.

In this developed prototype, only eight SMs with high switching frequency are used, to reduce the MMC complexity. The switching frequency is 40 kHz, which allows sinusoidal current waveforms to be synthesized with very good quality, and also permits a fast dynamic response to be obtained. However, in high-power applications, a few tens of SMs with low switching frequency should be used to achieve the same objectives. This will lead to higher scalability and reliability of the MMC in high power applications. Thus, if one SM breaks, it will not fully affect the MMC functionality. The higher the number of SMs, the better the quality and robustness of the generated waveforms. Consequently, increasing the total number of SMs is a relevant suggestion for future work. The presented experimental analysis proves the effectiveness of the proposed RPC based on MMC and its principle of operation, thus improving the power quality in electrified railway power grids.

Author Contributions: M.T., J.C. and L.A.M.B.: performed the analysis, design, testing and validating of the reduced-scale laboratory prototype explained in this paper. M.T.: involves in the writing—original draft preparation. All authors participated in the conceptualization, methodology, and writing—review and editing. All authors have read and agreed to the published version of the manuscript.

Funding: This work has been supported by the Portuguese Foundation of Science and Technology (FCT), in Portuguese, Fundação para a Ciência e Tecnologia within the R&D Units Project Scope: UIDB/00319/2020. Mohamed Tanta was supported by FCT grant with a reference PD/BD/127815/2016. Jose Cunha is supported by FCT grant with a reference PB/BD/143005/2018. Luis A. M. Barros is supported by FCT grant with a reference PD/BD/143006/2018.

Institutional Review Board Statement: Not applicable.

Informed Consent Statement: Not applicable.

Data Availability Statement: Not applicable.

Conflicts of Interest: The authors declare no conflict of interest.

References

1. Farnesi, S.; Marchesoni, M.; Vaccaro, L. Advances in Locomotive Power Electronic Systems Directly Fed through AC Lines. In Proceedings of the 2016 International Symposium on Power Electronics, Electrical Drives, Automation and Motion (SPEEDAM), Anacapri, Italy, 22–24 June 2016; pp. 657–664.
2. Nesterov, N.; Bykov, A. Locomotive Noise Reduction. In *MATEC Web of Conferences*; EDP Sciences: Moscow, Russia, 2020; Volume 320, p. 00013.
3. Perin, I.; Nussey, P.F.; Cella, U.M.; Tran, T.V.; Walker, G.R. Application of Power Electronics in Improving Power Quality and Supply Efficiency of AC Traction Networks. In Proceedings of the 2015 IEEE 11th International Conference on Power Electronics and Drive Systems, Sydney, Australia, 9–12 June 2015; pp. 1086–1094.
4. He, Z.; Zheng, Z.; Hu, H. Power Quality in High-Speed Railway Systems. *Int. J. Rail Transp.* **2016**, *4*, 71–97. [CrossRef]
5. Langerudy, A.T.; Mariscotti, A.; Abolhassani, M.A. Power Quality Conditioning in Railway Electrification: A Comparative Study. *IEEE Trans. Veh. Technol.* **2017**, *66*, 6653–6662. [CrossRef]
6. Morais, V.A.; Afonso, J.L.; Carvalho, A.S.; Martins, A.P. New Reactive Power Compensation Strategies for Railway Infrastructure Capacity Increasing. *Energies* **2020**, *13*, 4379. [CrossRef]
7. Gazafrudi, S.M.M.; Langerudy, A.T.; Fuchs, E.F.; Al-Haddad, K. Power Quality Issues in Railway Electrification: A Comprehensive Perspective. *IEEE Trans. Ind. Electron.* **2015**, *62*, 3081–3090. [CrossRef]
8. Luo, A.; Wu, C.; Shen, J.; Shuai, Z.; Ma, F. Railway Static Power Conditioners for High-Speed Train Traction Power Supply Systems Using Three-Phase V/V Transformers. *IEEE Trans. Power Electron.* **2011**, *26*, 2844–2856. [CrossRef]
9. Oso, H.; Kaneko, T.; Suzuki, A. Railway Static Power Conditioner for Shin-Kurobe Substation of Hokuriku Shinkansen. *Fuji Electr. Rev.* **2015**, *61*, 52–57.
10. Tanta, M.; Pinto, G.; Monteiro, V.; Martins, A.P.; Carvalho, A.S.; Afonso, J.L. A Comprehensive Comparison of Rail Power Conditioners Based on Two-Level Converters and a V/V Power Transformer in Railway Traction Power Systems. In Proceedings of the 7th Transport Research Arena (TRA 2018), Vienna, Austria, 16–19 April 2018; pp. 1–10.
11. Langerudy, A.T.; Mariscotti, A. Tuning of a Railway Power Quality Conditioner. In Proceedings of the 2016 18th Mediterranean Electrotechnical Conference (MELECON), Lemesos, Cyprus, 18–20 April 2016; pp. 1–7.
12. Serrano-Jiménez, D.; Abrahamsson, L.; Castaño-Solís, S.; Sanz-Feito, J. Electrical Railway Power Supply Systems: Current Situation and Future Trends. *Int. J. Electr. Power Energy Syst.* **2017**, *92*, 181–192. [CrossRef]

13. Afonso, J.L.; Cardoso, L.A.L.; Pedrosa, D.; Sousa, T.J.C.; Machado, L.; Tanta, M.; Monteiro, V. A Review on Power Electronics Technologies for Electric Mobility. *Energies* **2020**, *13*, 6343. [CrossRef]
14. Tanta, M.; Barros, L.A.M.; Pinto, G.; Martins, A.P.; Afonso, J.L. Modular Multilevel Converter in Electrified Railway Systems: Applications of Rail Static Frequency Converters and Rail Power Conditioners. In Proceedings of the 2020 International Young Engineers Forum (YEF ECE), Costa da Caparica, Portugal, 3 July 2020; pp. 1–6.
15. Kouro, S.; Malinowski, M.; Gopakumar, K.; Pou, J.; Franquelo, L.G.; Wu, B.; Rodriguez, J.; Pérez, M.A.; Leon, J.I. Recent Advances and Industrial Applications of Multilevel Converters. *IEEE Trans. Ind. Electron.* **2010**, *57*, 2553–2580. [CrossRef]
16. Debnath, S.; Qin, J.; Bahrani, B.; Saeedifard, M.; Barbosa, P. Operation, Control, and Applications of the Modular Multilevel Converter: A Review. *IEEE Trans. Power Electron.* **2015**, *30*, 37–53. [CrossRef]
17. Knaak, H.J. Modular Multilevel Converters and HVDC/FACTS: A Success Story. In Proceedings of the 2011 14th European Conference on Power Electronics and Applications, Birmingham, UK, 30 August–1 September 2011; pp. 1–6.
18. Ronanki, D.; Williamson, S.S. Modular Multilevel Converters for Transportation Electrification: Challenges and Opportunities. *IEEE Trans. Transp. Electrif.* **2018**, *4*, 399–407. [CrossRef]
19. Barros, L.A.M.; Tanta, M.; Martins, A.P.; Afonso, J.L.; Pinto, J.G. Opportunities and Challenges of Power Electronics Systems in Future Railway Electrification. In Proceedings of the 2020 IEEE 14th International Conference on Compatibility, Power Electronics and Power Engineering (CPE-POWERENG), Setubal, Portugal, 8–10 July 2020; Volume 1, pp. 530–537.
20. Mohammadi, P.H.; Bina, M.T. A Transformerless Medium-Voltage STATCOM Topology Based on Extended Modular Multilevel Converters. *IEEE Trans. Power Electron.* **2011**, *26*, 1534–1545. [CrossRef]
21. Feng, J.; Chu, W.Q.; Zhang, Z.; Zhu, Z.Q. Power Electronic Transformer-Based Railway Traction Systems: Challenges and Opportunities. *IEEE J. Emerg. Sel. Top. Power Electron.* **2017**, *5*, 1237–1253. [CrossRef]
22. Xu, Q.; Ma, F.; He, Z.; Chen, Y.; Guerrero, J.M.; Luo, A.; Li, Y.; Yue, Y. Analysis and Comparison of Modular Railway Power Conditioner for High-Speed Railway Traction System. *IEEE Trans. Power Electron.* **2016**, *32*, 6031–6048. [CrossRef]
23. Tanta, M.; Pinto, G.; Monteiro, V.; Martins, A.P.; Carvalho, A.S.; Afonso, J.L. Cost Estimation of Rail Power Conditioner Topologies Based on Indirect Modular Multilevel Converter in V/V and Scott Power Transformers. In Proceedings of the 4th International Conference on Energy and Environment: Bringing together Engineering and Economics, Guimarães, Portugal, 16 May 2019; pp. 365–370.
24. Qin, F.; Hao, T.; Gao, F. A Railway Power Conditioner Using Direct AC-AC Modular Multilevel Converter. In Proceedings of the 2019 2nd International Conference on Smart Grid and Renewable Energy (SGRE), Doha, Qatar, 19–21 November 2019; pp. 1–5.
25. Ma, F.; He, Z.; Xu, Q.; Luo, A.; Zhou, L.; Li, M. Multilevel Power Conditioner and Its Model Predictive Control for Railway Traction System. *IEEE Trans. Ind. Electron.* **2016**, *63*, 7275–7285. [CrossRef]
26. Tanta, M.; Pinto, J.G.; Monteiro, V.; Martins, A.P.; Carvalho, A.S.; Afonso, J.L. Topologies and Operation Modes of Rail Power Conditioners in AC Traction Grids: Review and Comprehensive Comparison. *Energies* **2020**, *13*, 2151. [CrossRef]
27. Lu, M.; Hu, J.; Zeng, R.; He, Z. Fundamental-Frequency Reactive Circulating Current Injection for Capacitor Voltage Balancing in Hybrid-MMC HVDC Systems During Riding Through PTG Faults. *IEEE Trans. Power Deliv.* **2018**, *33*, 1348–1357. [CrossRef]
28. Krastev, I.; Tricoli, P.; Hillmansen, S.; Chen, M. Future of Electric Railways: Advanced Electrification Systems with Static Converters for AC Railways. *IEEE Electrif. Mag.* **2016**, *4*, 6–14. [CrossRef]
29. He, X.; Peng, J.; Han, P.; Liu, Z.; Gao, S.; Wang, P. A Novel Advanced Traction Power Supply System Based on Modular Multilevel Converter. *IEEE Access* **2019**, *7*, 165018–165028. [CrossRef]
30. Perez, M.A.; Bernet, S.; Rodriguez, J.; Kouro, S.; Lizana, R. Circuit Topologies, Modeling, Control Schemes, and Applications of Modular Multilevel Converters. *IEEE Trans. Power Electron.* **2015**, *30*, 4–17. [CrossRef]
31. Tanta, M.; Pinto, J.G.; Monteiro, V.; Martins, A.P.; Carvalho, A.S.; Afonso, J.L. Deadbeat Predictive Current Control for Circulating Currents Reduction in a Modular Multilevel Converter Based Rail Power Conditioner. *Appl. Sci.* **2020**, *10*, 1849. [CrossRef]
32. Tanta, M.; Monteiro, V.; Exposto, B.; Pinto, J.G.; Martins, A.P.; Carvalho, A.S.; Meléndez, A.A.N.; Afonso, J.L. Simplified Rail Power Conditioner Based on a Half-Bridge Indirect AC/DC/AC Modular Multilevel Converter and a V/V Power Transformer. In Proceedings of the IECON 2017—43rd Annual Conference of the IEEE Industrial Electronics Society, Beijing, China, 29 October–1 November 2017; pp. 6431–6436.
33. Hagiwara, M.; Akagi, H. Control and Experiment of Pulsewidth-Modulated Modular Multilevel Converters. *IEEE Trans. Power Electron.* **2009**, *24*, 1737–1746. [CrossRef]
34. Karimi-Ghartemani, M. Linear and Pseudolinear Enhanced Phased-Locked Loop (EPLL) Structures. *IEEE Trans. Ind. Electron.* **2014**, *61*, 1464–1474. [CrossRef]
35. Ma, F.; Luo, A.; Xu, X.; Xiao, H.; Wu, C.; Wang, W. A Simplified Power Conditioner Based on Half-Bridge Converter for High-Speed Railway System. *IEEE Trans. Ind. Electron.* **2013**, *60*, 728–738. [CrossRef]
36. Dai, N.; Wong, M. Design Considerations of Coupling Inductance for Active Power Filters. In Proceedings of the 2011 6th IEEE Conference on Industrial Electronics and Applications, Beijing, China, 21–23 June 2011; pp. 1370–1375.
37. Tu, Q.; Xu, Z.; Huang, H.; Zhang, J. Parameter Design Principle of the Arm Inductor in Modular Multilevel Converter Based HVDC. In Proceedings of the 2010 International Conference on Power System Technology, Hangzhou, China, 24–28 October 2010; pp. 1–6.
38. Harnefors, L.; Antonopoulos, A.; Norrga, S.; Angquist, L.; Nee, H. Dynamic Analysis of Modular Multilevel Converters. *IEEE Trans. Ind. Electron.* **2013**, *60*, 2526–2537. [CrossRef]

39. Liu, L.; Dai, N. Hybrid Railway Power Conditioner Based on Half-Bridge Modular Multilevel Converter. In Proceedings of the 2016 IEEE Energy Conversion Congress and Exposition (ECCE), Milwaukee, WI, USA, 18–22 September 2016; pp. 1–7.
40. Sharifabadi, K.; Harnefors, L.; Nee, H.-P.; Norrga, S.; Teodorescu, R. *Design, Control and Application of Modular Multilevel Converters for HVDC Transmission Systems*; Wiley-IEEE Press: Hoboken, NJ, USA, 2016; ISBN 978-1-118-85156-2.
41. Barros, L.A.M.; Tanta, M.; Martins, A.P.; Afonso, J.L.; Pinto, J.G. STATCOM Evaluation in Electrified Railway Using V/V and Scott Power Transformers. In Proceedings of the Sustainable Energy for Smart Cities, SESC 2019, Braga, Portugal, 4–6 December 2019; Afonso, J.L., Monteiro, V., Pinto, J.G., Eds.; Springer International Publishing: Cham, Switzerland, 2020; pp. 18–32.
42. Silicon Labs Si824x—Class D Audio Driver with Precision Dead-Time Generator. Available online: https://www.silabs.com/documents/public/data-sheets/Si824x.pdf (accessed on 9 October 2018).
43. Tanta, M.; Cunha, J.; Monteiro, V.; Martins, A.P.; Carvalho, A.S.; Afonso, J.L. A Novel Hardware Protection Scheme for a Modular Multilevel Converter Half-Bridge Submodule. In Proceedings of the IECON 2019—45th Annual Conference of the IEEE Industrial Electronics Society, Lisbon, Portugal, 14–17 October 2019; Volume 1, pp. 6043–6048.
44. Wang, H.; Blaabjerg, F. Reliability of Capacitors for DC-Link Applications in Power Electronic Converters—An Overview. *IEEE Trans. Ind. Appl.* **2014**, *50*, 3569–3578. [CrossRef]
45. Gruson, F.; Kadri, R.; Colas, F.; Guillaud, X.; Delarue, P.; Bergé, M.; Dennetiere, S.; Bachir, T.O. Design, Implementation and Testing of a Modular Multilevel Converter. *EPE J.* **2017**, *27*, 153–166. [CrossRef]
46. Texas Instruments TMS320F2833x, TMS320F2823x Digital Signal Controllers (DSCs). Available online: http://www.ti.com/lit/ds/symlink/tms320f28335.pdf (accessed on 18 October 2018).

Article

Complex Design Method of Filtration Station Considering Harmonic Components

Vaclav Kus, Bohumil Skala and Pavel Drabek *

Department of Power Electronics, FEE, University of West Bohemia, 301 00 Plzeň 3, Czech Republic; kus@fel.zcu.cz (V.K.); skalab@fel.zcu.cz (B.S.)
* Correspondence: drabek@fel.zcu.cz; Tel.: +420-377634437

Abstract: The paper deals with a new methodology for calculating the filter parameters. The basis is respect for the fact that the real filter current consists of other harmonic components, which filter is tuned. The proposed methodology was used to design filters for traction substation 25 kV/50 Hz. The operation of the locomotives in the AC supply systems of 25 kV/50 Hz leads to a rising of higher order harmonic currents. Due to the 1-phase supply system, these are mainly the 3rd and 5th harmonics. By simulation and subsequent measurement of the proposed traction power station filters the proposed methodology was verified. Thus, the filter design can also be used for filter compensating stations of the standard 3-phase distribution network. The described method presents an optimal filter design without unnecessary oversizing. This fact reduces the size and cost of the filter. It is shown that it is possible to design a filter that meets the requirements for power quality under extreme load and to minimise distortion of line voltage.

Keywords: AC traction systems; traction substation; harmonics; power harmonic filters; total harmonic distortion

Citation: Kus, V.; Skala, B.; Drabek, P. Complex Design Method of Filtration Station Considering Harmonic Components. *Energies* **2021**, *14*, 5872. https://doi.org/10.3390/en14185872

Academic Editors: Andrea Mariscotti and Leonardo Sandrolini

Received: 28 July 2021
Accepted: 8 September 2021
Published: 16 September 2021

1. Introduction

The amount and power of non-linear loads are growing steadily. Although the connection and converter control are also completed with respect to the reduction of harmonic currents, it is often necessary to equip the substations with equipment for harmonics minimization and power factor compensation. The classic solution is to install filters tuned to the desired harmonics. The integral (physical) property of the filters is also the power factor compensation.

The advantage of tuned filters is their low cost. Filters are usually tuned to the 5th and 7th harmonic, in exceptional cases, on the 11th and 13th harmonic. For single-phase systems, it is usually on the 3rd and 5th harmonic. The design of filters also for high powers is well described in the literature, for example [1,2]. They can also be used in a combination with a broadband filter, or other connection for high order harmonics [3,4]. In articles, filters are tuned to a characteristic harmonic. Their designs are based on fundamental relations of Theoretical Electrical Engineering. As the power filters also has compensation effect, overcompensation of the system can occur at high loads. Therefore, it is necessary to equip the substation with a quick decompensation unit [5,6]. Using filters and decompensation unit in the traction system is described in [7]. Another way is to use active filters or other types of electronic converters [8–10]. The advantage of these systems is a better filtering and compensating effect. The disadvantage is a considerably higher price. The effects of filters designed for resonant frequency, in addition to extreme loads, are described in [11]. Verification of the effect of the filter is achieved from the impedance ratio of the filter and the system or through simulations. At the same time, one can check the current load of the filter. Due to the load, it is necessary to change the parameters of the filter capacitor and repeat the calculations. Consequently, its compensation power is also checked. The second option is an under-tuning of the filter. This will reduce the current stress on the filter, but also reduces the filtering effect.

This article describes the derivation of a new filter design method for a traction power station. Electric traction is specific in its power level and single-phase character. In extreme cases, traction overhead lines can reach up to 50 km in length. This also produced harmonic voltages whose values significantly distorted the AC waveform [12].

For new types of locomotives, harmonic currents are significantly reduced. The reference [13] describes measurement of harmonic components of modern locomotives in the UK. In operation, however, a number of older types of locomotives are equipped with diode and thyristor 1-phase rectifiers, which take a considerably distorted current from an overhead line. It is estimated that 32% of diode and thyristor locomotives are still in operation. In the section where the proposed filters were installed (see Figure 1), the amount of these locomotives is an even 80%. Since it is necessary to accept new power quality requirements and reduce losses, many traction stations need to be supplemented by equipment for harmonics reduction and power factor compensation.

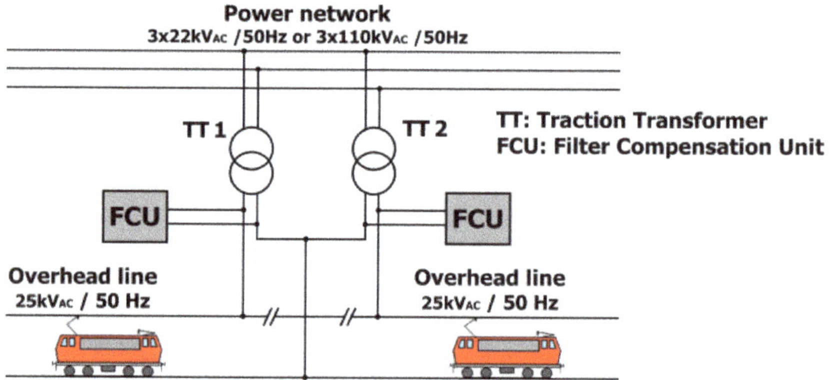

Figure 1. Block scheme of traction substation with filter compensation unit.

The basis of the new methodology is to determine the power of the filter capacitor so that the required filtering and compensating effect is already included directly in the filter design. Using this method does not overload the filter. The filter then has the greatest filtering effect at minimal financial costs.

Since the harmonic currents of the other orders are presented in the overhead line currents and the filter design does not respect this fact, due to the impedance conditions, these currents also pass through some filters. To avoid overloading the filters, it is necessary to design the filters even with respect to these harmonic currents. In the described calculations, these harmonics are also accepted in the design.

The proposed methodology describes the optimum filter design without unnecessary oversizing; thereby it reduces the price, weight, size, and filter losses. The filter tuned to a specific harmonic is also loaded by other harmonics. The standard design oversizes the filter to protect its damage. The described methodology enables optimal filter design considering the calculation of a specific harmonic and other main harmonics.

The method is applied in the design of the substation for traction overhead lines of 25 kV/50 Hz. The proposed methodology is verified simulations and measurements in real traffic.

2. The Method of Filter Compensation Unit Design

The issue of filter design for high voltage and high power (in order MVAr) is relatively difficult. It is solved e.g., in [14]. At present, it may be constructing the new traction power station also with the active filters, e.g., [15–17].

2.1. The Initial Data for the Filter Design

The design requires the following data (see Figure 2):

1. The individual harmonic currents to be filtered, especially those of h-orders;
2. The total compensating power required, Q_{kreq}, to be supplied to the traction power network;
3. The voltage U_1 at the point where filters are going to be installed.

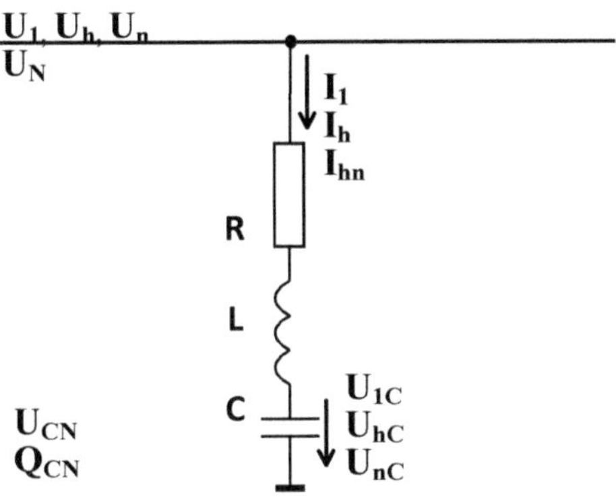

Figure 2. Schematic diagram of proposed symbols.

It is necessary to determine the following basic parameters:

1. The nominal voltage U_{CN} and the nominal installed power Q_{CN} of the capacitor;
2. The current and inductance of the filter reactor, and the effective resistance R of the filter reactor.

2.2. Derivation of Basic Relations

The resonance condition for the h-th harmonic is:

$$L = \frac{1}{h^2 \omega_1^2 C} \tag{1}$$

Assuming that $R = 0$, the voltage of the first harmonic across the capacitor C for the filter of the h-th harmonic (important for the U_{CN} determination → capacitor design) is:

$$\begin{aligned} U_{1C} &= I_1 X_C = I_1 \tfrac{1}{\omega_1 C_1} = \tfrac{U_1}{X_c - X_L} \tfrac{1}{\omega_1 C_1} \\ &= \tfrac{U_1}{\left(\frac{1}{\omega_1 C_1} - \omega_1 \frac{1}{h^2 \omega_1^2 C_1}\right)} \cdot \tfrac{1}{\omega_1 C_1} = \tfrac{U_1}{1 - \frac{1}{h^2}} \end{aligned} \tag{2}$$

$$U_{1C} = U_1 \frac{h^2}{h^2 - 1} \tag{3}$$

By introducing the voltage increase coefficient of the first harmonic across the capacitor:

$$a_h = \frac{h^2}{h^2 - 1} \quad (4)$$

The following is valid:

$$U_{1C} = a_h U_1 \quad (5)$$

The parameters of power capacitors are usually determined by the value of the nominal power Q_{CN} at the nominal voltage U_{CN}. Then the nominal power of the capacitor is:

$$Q_{CN} = \omega_1 C U_{CN}^2 \quad \Rightarrow \quad \frac{1}{X_{1C}} = \omega_1 C = \frac{Q_{CN}}{U_{CN}^2} \quad (6)$$

For the first harmonic current passing through the filter of the h-th harmonic, the following is valid:

$$I_1 = \frac{U_{1C}}{X_C} = \frac{a_h U_1}{X_C} = a_h \frac{U_1 Q_{CN}}{U_{CN}^2} \quad (7)$$

This current has a capacitive character and the filter of the h-th harmonic supplies the traction overhead line with the compensating power:

$$Q_k = U_1 I_1 = a_h \left(\frac{U_1}{U_{CN}} \right)^2 Q_{CN} \quad (8)$$

2.3. Simplified Design of the Capacitor Battery

The equations mentioned above do not lead to the device dimensioning. The current passing through the individual filter elements determines the current dimensioning of the filter parts. Voltage drops across the filter impedances, caused by this current together with the supply voltage, determine the voltage load and thus the capacitor battery dimensioning.

In order to formulate the mathematical relations more easily, the following are first assumed:

1. Only the compensating current and the h-order current, for which the filter is designed, pass through each filter;
2. The traction overhead line voltage U_1 can be permanently higher than the nominal voltage U_N;
3. The voltage across the capacitor U_C can be permanently higher than its nominal voltage U_{CN};
4. The capacitor current I_C can be permanently higher than its nominal current I_{CN}.

Based on these assumptions, it is necessary to prevent the capacitor from either current or voltage overload. It must comply with both current and voltage conditions of the solution. The following is valid:

$$U_1 = k_{us} U_N \;\; U_C = \kappa_u U_{CN} \;\; I_C \leq \kappa_i I_{CN} \quad (9)$$

where

k_{us}—the nominal voltage increase factor of the traction overhead line;
κ_i—the nominal current increase factor of the capacitor;
κ_u—the nominal voltage increase factor of the capacitor.

2.4. Basic Design of Capacitor Banks

In Equation (9) the k_{us} constant is set by a standard (EN 50, 160 or national standards), the other two constants are set by the manufacturers of capacitors.

2.4.1. Calculation Based on the Current Conditions

The calculation is based on the assumptions mentioned above, i.e., only the compensating current and the h-order harmonic current which is to be eliminated pass through the filter. Then the following is valid:

$$\sqrt{I_1^2 + I_h^2} \leq \kappa_i I_{CN} \tag{10}$$

By inserting the above-mentioned relations and after formal rearrangements, the relation for the value of the capacitor minimal power Q_{CNmin} is obtained:

$$a_h^2 k_{US}^2 U_N^2 Q_{CN}^2 + I_h^2 U_{CN}^4 \leq \kappa_i^2 Q_{CN}^2 U_{CN}^2, \text{ and} [Q_{CN}]_i \geq [Q_{CN\text{min}}]_i = \frac{U_{CN} I_h}{\sqrt{\kappa_i^2 - a_h^2 k_{us}^2 \left(\frac{U_N}{U_{CN}}\right)^2}} \tag{11}$$

The calculation of Q_{CNmin} ensures that the capacitor will not be overloaded with current. The resulting power of the capacitor then will be higher.

2.4.2. Calculation Based on the Voltage Condition

As in the previous section, it is possible to define the voltage condition for the determination of the minimal power of the capacitor battery:

$$\sqrt{U_{1C}^2 + U_{hC}^2} \leq \kappa_u U_{CN} \tag{12}$$

Yielding the equation defining the value of the minimal power of the capacitor:

$$[Q_{CN}]_U \geq [Q_{CN\text{min}}]_U = \frac{U_{CN} I_h}{h\sqrt{\kappa_u^2 - a_h^2 k_{us}^2 \left(\frac{U_N}{U_{CN}}\right)^2}} \tag{13}$$

NB: The index "i" in (11) denotes the power calculated on the basis of the current condition. Similarly, the index "u" in (13) denotes the power calculated on the basis of the voltage condition.

2.4.3. Determination of the Minimal Installed Power

In order to achieve further simplification, it is possible to introduce auxiliary quantities, i.e., the apparent power S_h, caused by the harmonic current across an ideal voltage source, and coefficients k_u and k_i, taking into account the capacitors' load:

$$S_h = U_N I_h \tag{14}$$

$$k_{iC} = \frac{1}{\frac{U_N}{U_{CN}} \sqrt{\kappa_i^2 - a_h^2 k_{us}^2 \left(\frac{U_N}{U_{CN}}\right)^2}} \tag{15}$$

$$k_{uC} = \frac{1}{h\frac{U_N}{U_{CN}} \sqrt{\kappa_u^2 - a_h^2 k_{us}^2 \left(\frac{U_N}{U_{CN}}\right)^2}} \tag{16}$$

When applying the introduced coefficients, the following are valid for the minimum values of power of capacitor:

$$[Q_{CN}]_{.i} \geq k_{iC} S_h \text{ and } [Q_{CN}]_{.U} \geq k_{uC} S_h \tag{17}$$

The minimal value of the installed power is then acquired as the higher value obtained from Equation (17).

2.5. Calculation Taking into Account Non-Filtered Harmonics

The equations derived in the previous sections are derived for an ideal situation, when only the compensating current of the first harmonic I_1 and the current of the h-order to which the filter is tuned pass through the filter. The actual voltage and current are distorted by a whole spectrum of characteristic and non-characteristic harmonics. This results in an additional load each capacitor is subjected to, as part of the current of other harmonics I_n to which the filter is not tuned, passes through the filter. The following section demonstrates the derivation of the influence of the current non-filtered harmonics on the power load of the filter.

The filter is tuned to the frequency $f = h.f1$. Here, the frequency f can be arbitrary. Part of the harmonic currents of other (non-filtered) orders pass through the filter at the same time. Now it is possible to rearrange Equation (10), and subsequently Equation (12), that they meet the new conditions for the determination of the minimal power of the capacitor:

$$\sqrt{I_1^2 + I_h^2 + \sum_{\substack{n=2\\ n\neq h}}^{40} I_n^2} \leq \kappa_i I_{CN} \tag{18}$$

$$\sqrt{U_1^2 + U_h^2 + \sum_{\substack{n=2\\ n\neq h}}^{40} U_{nC}^2} \leq \kappa_u U_{CN} \tag{19}$$

The third part of the radicand in Equation (18) and Equation (19) corresponds to the proportion of the higher order harmonics (i.e., the non-filtered harmonics) which influence the filter load.

Taking into account the load of the capacitors, the coefficients k_{uC} and k_{iC} f from Equation (15) and Equation (16) can be rearranged as follows:

$$k_{iC} = \frac{1}{\frac{U_N}{U_{CN}}\sqrt{\kappa_i^2 - a_h^2 k_{us}^2\left(\frac{U_N}{U_{CN}}\right)^2 - \left(\frac{U_N}{U_{CN}}\right)^2 \sum_{\substack{n=2\\ n\neq h}}^{40} \frac{\Delta u_{Nn}^2}{\left(\frac{n}{h^2}-\frac{1}{h}\right)^2}}} \tag{20}$$

$$k_{uC} = \frac{1}{h\frac{U_N}{U_{CN}}\sqrt{\kappa_u^2 - a_h^2 k_{us}^2\left(\frac{U_N}{U_{CN}}\right)^2 - \left(\frac{U_N}{U_{CN}}\right)^2 \sum_{\substack{n=2\\ n\neq h}}^{40} \frac{\Delta u_{Nn}^2}{\left(\frac{n}{h^2}-1\right)^2}}} \tag{21}$$

The subsequent procedure is the same; Equation (17) is used.

2.6. Selection of the Capacitor according to the Required Compensating Power

The relations mentioned above are used for the determination of the minimal power of a capacitor. The minimal compensating power expressed by means of Equation (8) is as follows:

$$Q_{k\min} = a_h\left(\frac{U_1}{U_{CN}}\right)^2 Q_{CN} = a_h\left(k_{uS}\frac{U_N}{U_{CN}}\right)^2 k_h S_h \tag{22}$$

where k_h is the coefficient determined on the basis of the current or voltage condition by means of the relations of Equations (16) and (17), and Equations (20) and (21).

2.7. Calculation of Filter Inductance

From the resonance condition and the relation for the capacitor power, it is possible to derive:

$$L = \frac{1}{h^2\omega_1}\frac{U_{CN}^2}{Q_{CN}} \tag{23}$$

3. Design of Filters for Traction Supply Stations

3.1. Determination of Key Parameters

3.1.1. The Required Compensating Power

The determination of the minimal compensating power of filters is based on the length of the traction overhead line and the track structure (thus the number of locomotives is determined together with their maximum power on the track). Based on the requirements of the track operator, as much as 110 A of inductive current needs to be smoothed out. The corresponding value of the compensating power required is Q_{kreq} = 2970 kVAr.

3.1.2. Requirements for Filtering of Harmonic Currents

Harmonic currents can be determined by a calculation based on the amplitude law and an assumed load, or by measurements. If measurements cannot be performed, a method based on experience with similar existing devices is used. For a single-phase load, the selected filters will be tuned close to the third and fifth harmonics. The filter selected for the third harmonic will filter the third harmonic current and at the same time, it will exhibit the highest percentage of compensating power. In contrast, the fifth harmonic filter will exhibit low compensation, but it will also reduce currents of higher harmonic orders to some extent.

It is also necessary to install a decompensation reactor in the traction supply station. With respect to the required compensating power of filters, the selected nominal current is I_{DC} = 120 A. This switching reactor is also a source of other harmonic currents, which need to be added to the harmonics in the traction overhead line. Table 1 shows the values of the currents measured under a maximum load of the traction overhead line, which need to be added to the values of harmonics produced by the decompensation element. Other harmonics exhibit negligible values from the viewpoint of the filter load.

Table 1. Values of harmonic currents measured in the traction overhead line and across the decompensation element.

h [-]	3	5	7	9	11	13
I_h [A]	42.7	17.5	14.2	12.1	11.4	1.8

3.1.3. Proportional Values of Higher Harmonic Voltages

In addition to the effects of the compensating current and the fifth harmonic current, the dimensioning of the fifth harmonic filter will also be affected by values of harmonic currents of higher orders. Table 2 shows values obtained by measurements performed at a maximum load. The values for the 7th to 13th harmonics are crucial for the design of a filter, taking into account the effect of the non-filtered harmonics.

Table 2. Proportional values of harmonic voltages (without filters).

h [-]	3	5	7	9	11	13
u_h [%]	4.24	2.99	3.02	3.45	4.08	1.03

3.2. Simplified Calculation of a Capacitor

3.2.1. Calculation of the Nominal Voltage U_{CN} of a Capacitor

The nominal output voltage of the traction supply station is U_N = 27,000 V. According to Equations (3) and (4), the first harmonic voltage across the capacitor is U_{1C3} = 30,375 V for the third harmonic filter and U_{1C5} = 28,125 V for the fifth harmonic filter. Taking into account the voltage and the requirement for the lowest possible number of capacitors connected in series, the selected value of the nominal voltage of capacitors is $U_{CN} = 2 \times 20{,}000 = 40{,}000$ V.

3.2.2. Calculation of the Minimum Power Q_{CNmin} of a Capacitor

First, coefficients are determined according to Equations (15) and (16):

k_{ic3} = 1.594; k_{ic5} = 1.508; k_{uc3} = 1.635; k_{uc5} = 0.508;

These values indicate that the third harmonic filter will be designed on the basis of the voltage condition in Equation (12), and the fifth harmonic filter on the basis of the current condition in Equation (10).

According to Equation (14), the apparent power of the filter for the selected harmonics is S_{h3} = 1611.9 kVA; S_{h5} = 567 kVA. By applying coefficients k_{uc3} and k_{ic5}, the value of the minimum installed power of a capacitor is Q_{CN3} = 2635.4 kVAr; Q_{CN5} = 855 kVAr.

3.3. Calculation of a Capacitor with Respect to Non-Filtered Harmonics

In this section, the calculation described in Section 2.4 is modified, e.g., in respect to the effects of harmonics to which the filter is not tuned.

Most harmonic currents of the fifth order and higher orders will pass through the fifth harmonic filter. According to Equations (20) and (21), the design takes into account a certain share of these harmonics in the load of the filter (actually, the third harmonic filter will not be loaded with these currents). To make the calculation clear, the value of the denominator of the fraction in the third part of Equation (20) is calculated first. Then the values of the whole fraction for the individual harmonics are calculated by applying Table 2. Table 3 shows the results. With respect to the designed power of the third harmonic filter, the design now focuses on the filter of the fifth harmonic.

Table 3. Auxiliary table used for the calculation of coefficients k.

n [-]	7	9	11	13
$1/\left(\frac{n}{5^2}-\frac{1}{n}\right)^2$	53.16	16.14	8.2	5.09
$\Delta u_{Nn}^2/\left(\frac{n}{5^2}-\frac{1}{n}\right)^2$	0.0501	0.0204	0.0157	0.0005

The values in the second line of the table in fact show the proportional value of the filtration of the listed harmonics. The values that are taking into account the actual voltage distortion are listed in the bottom line of the table. By applying results from Table 3, the values of coefficients k_i and k_u are: k_{uc5} = 0.505; k_{ic5} = 1.557;

Now the capacitor design is based on the current condition (18). By applying Equations (17) and (20), the calculated value of the minimum installed power of a capacitor is Q_{CN5} = 883 kVAr.

3.4. Compensating Power of Filters

By means of Equation (22) and the necessary minimal power of the filter capacitor, the compensating power is determined: Q_{k3} = 1485.8 kVAr; Q_{k5} = 459 kVAr. The total compensating power of the traction supply station is Q_k = 1485.8 + 459 = 1918 kVAr. This value is lower than the value required Q_{kreq} = 2970 kVAr. As the main portion of the compensation will be achieved by the third harmonic filter (as required by the investor), its compensating power has to be the following: Q_{k3} = 2970 − 459 = 2511 kVAr. Then, by means of Equation (20), the following is true:

$$Q_{CN3} = \frac{1}{a_3}\left(\frac{U_{CN}}{k_{U_s}U_N}\right)^2 Q_{k3} = \frac{1}{1,125}\left(\frac{40}{1,1.27}\right)^2 .2511 = 4048 \qquad kVAr \tag{24}$$

3.5. Compensating Power of Filters

Taking into account U_{CN} = 40 kV and the minimal Q_{CN} required, the following are selected according to datasheets:

Filter F3: 8 units connected, U_{CN} = 20 kV and Q_{CN} = 535.6 kVAr each. Then Q_{CN3} = 4284.8 kVAr.

Filter F5: 4 units connected, U_{CN} = 20 kV and Q_{CN} = 223.7 kVAr each. Then Q_{CN5} = 894.8 kVAr.

For further calculations based on the voltages and powers mentioned above, the following is true: C_{F3} = 8.52 μF and C_{F5} = 1.78 μF.

The actual compensating power of installed filters will be Q_k = 3171 kVAr, which is a higher value than the required Q_{kreq}.

As the principle of the proposed method is to ensure the maximum exploitation of the filter power without concerns over overloading the filter with other harmonics, it is in this case possible to tune the filters to frequencies relatively close to the resonance frequency of harmonic currents. Filters will therefore be tuned to the frequencies of: f_{r3} = 147 Hz; f_{r5} = 247 Hz. Based on the resonance condition, values of filter inductance are then calculated: L_{F3} = 137.6 mH; L_{F5} = 233.25 mH.

Figure 3 gives an overall view of the traction substation, where the filters were designed according to this new method and installed. The filters themselves are housed with components in separate cells in a substation.

Figure 3. View of traction substation 25 kV/50 Hz.

4. Simulation of Effects of Filters in the Traction System

4.1. Description of the Traction System and Its Parameters

Before the installation itself, the effects of filters were verified by simulations. The methods of modelling and the traction specifics are described in [18]. Article [19] is concerned with resonances occurring in the traction system, which can significantly influence the results. The following sections will deal with steady-state simulations and measurements. Transient phenomena are the concern of article [20].

Basic parameters of the system:

- Short-circuit power on high voltage (110 kV) line: S_{KS}'' = 19,702 MVA;
- The value calculated for the selected voltage level is then 27 kV: L_{ES} = 2.15 mH;
- The transformer in the substation: S_{NT} = 12.5 MVA, 110/27 kV, 50 Hz, u_k = 12.5%, P_k = 80 kW;
- By calculation, the following values are obtained: R_{NT} = 0.373 [Ω], L_{NT} = 22.88 mH.

The values of harmonic currents used in simulations are obtained from the measurements performed and are listed in Table 1. The length of the traction overhead line is extreme and reaches a length of up to 50 km from the traction substation. The line parameters per one kilometre are: R_{vk} = 0.26 Ω/km; L_{vk} = 1.43 mH/km; C_{vk} = 20.5 nF/km. The modelling principles and details are described in [21].

Overview scheme of the overall system presents the Figure 4. For research of a filtration traction station design, it is necessary to specify the individual parts of the system:

- Power supply network;
- Overhead line;
- Locomotive (train);
- Filter compensation unit (FCU).

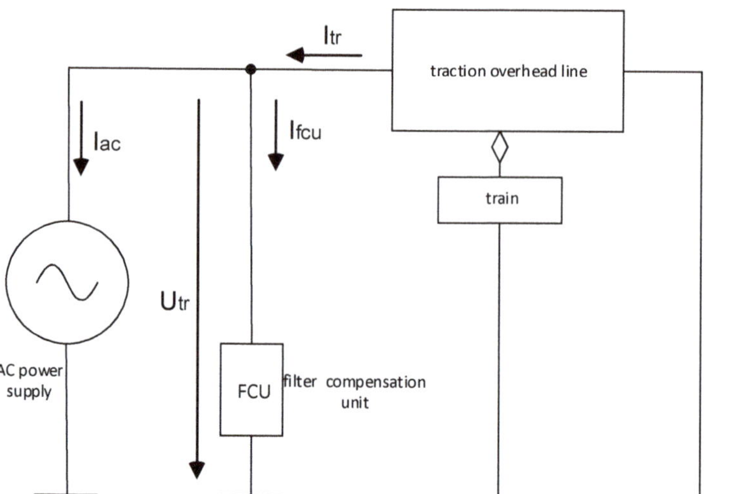

Figure 4. Overview scheme for simulations.

4.1.1. Power Supply Network (AC Power Supply)

Power supply network is characterised by inductance, which can be calculated by knowledge of the short-circuit power (at HV line 110 kV). Supply system 110 kV has a short-circuit power of S_{KS}'' = 10,000 MVA. The nominal overhead line voltage is considered 27.5 kV. The appropriate inductance of the overhead line is L_{ES} = 2.15 mH. The AC power supply 25 kV is assumed as ideal sinewave source by its first harmonic and constant amplitude.

The proportion of any harmonic voltage would be low in comparison to the voltage resulting from harmonic currents by locomotives. The traction substation transformer rated parameters are: S_{NT} = 12.2 MVA, 110/27.5 kV, 50 Hz, u_k = 12.5%, P_k = 80 kW. By the calculation, there are: Z_{NT} = 7.23 Ω, R_{NT} = 0.373 Ω, L_{NT} = 22.88 mH.

The model of the power supply system is shown in the Figure 5.

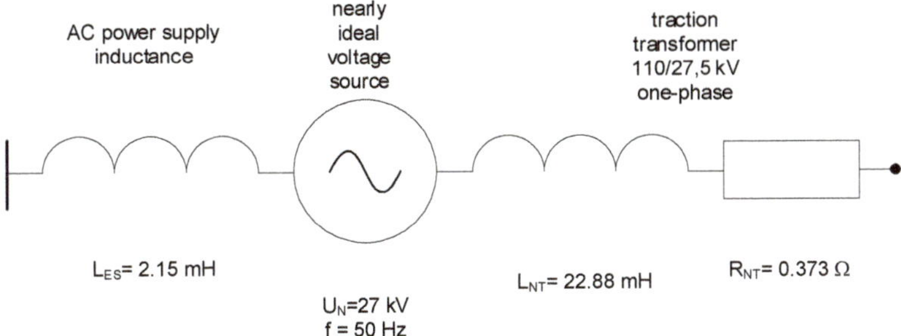

Figure 5. Power supply system model.

4.1.2. Overhead Line

The overhead line modelling is very specific. It is standardly one-way power supply; compare to distribution network the length is approx. 40 km. There are two ways to simulate the overhead line: use the distributed parameter theory or π-cells model.

The π-cells model enables to simulate the locomotive at any point of the overhead line; therefore, the π-cells model was selected. It is recommended to use one π-cell per each 4 km of the overhead line. We considered the total length of the overhead line up to 40 km; therefore, the model of the overhead line consists of 10 π-cells. Figure 6 shows the recalculated values of each parts of the model.

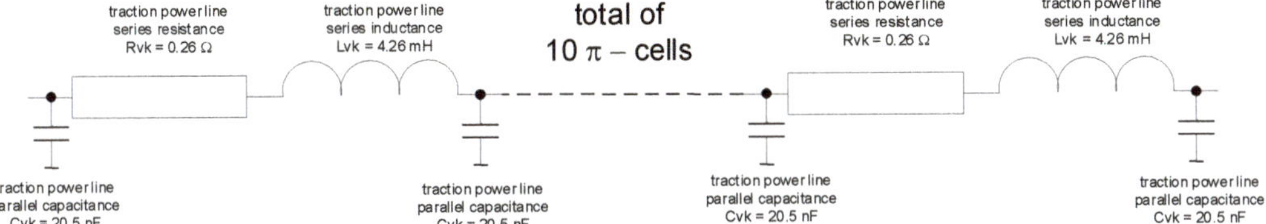

Figure 6. The overhead line simulation model.

4.1.3. Locomotive (Train)

The locomotive (train) is modelled for harmonic detection purposes only as a current source. This current source contains a first harmonic of 157 A and other characteristic harmonics measured on the locomotives. The values of these harmonics are given in Table 1. The overhead line model described in Section 4.1.2 allows the connection of the locomotive at any point of the traction overhead line during the simulation.

4.1.4. Filter Compensation Unit (FCU)

The basic scheme of the filter is shown in Figure 2. As the filter always filters and compensates by its nature, it is referred to as FCU (filter compensation unit) in Figures 1 and 4. The simulations were involved filters tuned to h = 3 and 5. The filter unit operates as a decompensation unit as well.

Calculated values of filter components are described in Section 3.5, including compensation filter power.

4.2. Results of the Simulation

The voltage is monitored in the place of the connection to the traction overhead line and for all cases, it is denoted as U_{tr}. The monitored currents are depicted in Figure 1, i.e., the current passing through the traction overhead line I_{tr}, the current drawn from the AC power supply system I_{ac} and the current passing through the filters I_{fcu}.

Figure 7 shows the waveforms of currents and voltages without the filters installed. Figure 8 depicts the waveforms with the filters installed. The waveforms indicate an active, reactive and distortion power, thus subjecting the traction system to an extreme load. What is apparent in the case of the filter application is both the system overcompensation and the significant reduction in the distortion of voltages and currents caused by higher-order harmonics. Figure 9 shows the values of the individual voltage harmonics and the total harmonic distortion (THD) factor for the traction overhead line. The values for the current drawn from the AC power supply network are depicted in Figure 10. Exact values and their comparison with the values obtained by measurements can be found in tables in Section 6.

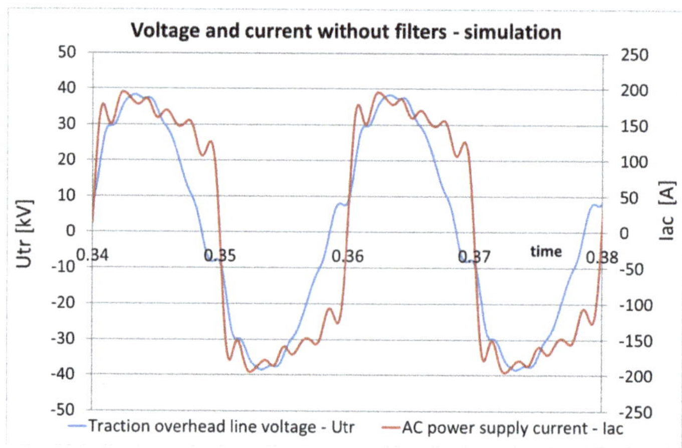

Figure 7. Voltage and current waveforms of the system without filter—simulation.

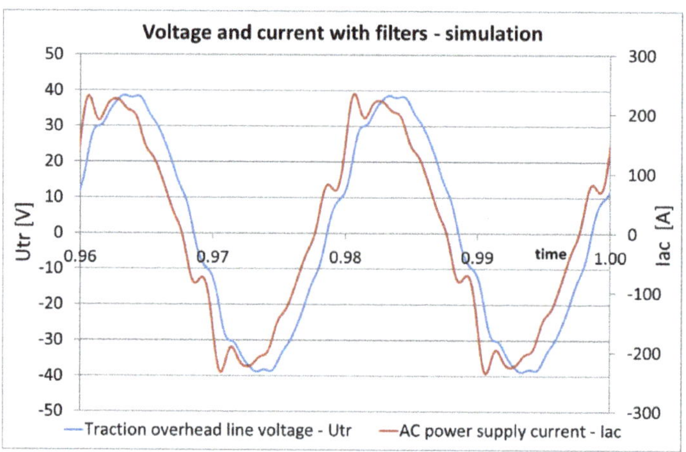

Figure 8. The voltage and current waveforms of the system in operation with connected filters—simulation.

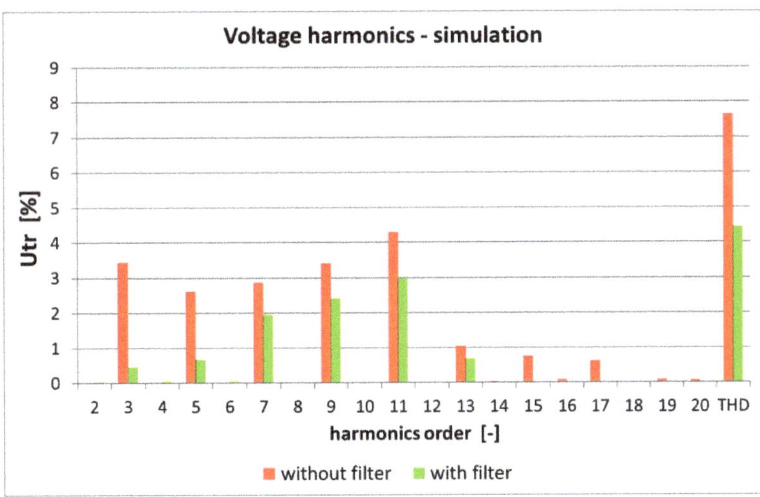

Figure 9. Filters influence on harmonic voltages of the substation system—simulation.

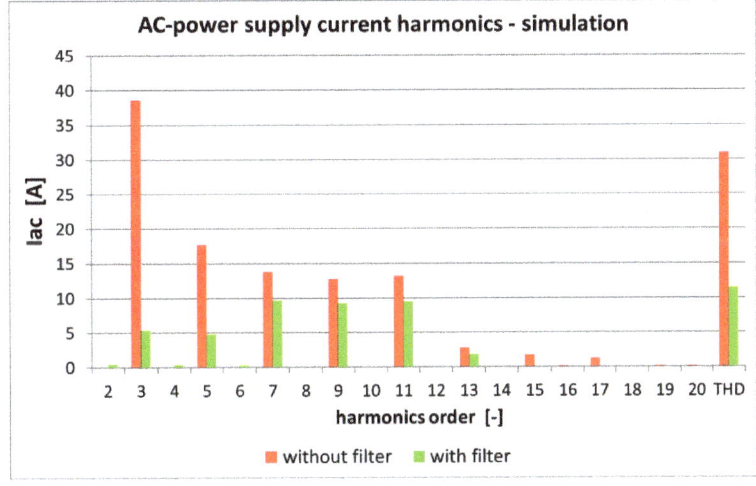

Figure 10. Filters Influence on harmonic currents of the substation system—simulation.

5. The Harmonic Measurement of the Railway Substation

In order to verify the efficiency of filters at extreme load and at the same time their current stresses, a series of measurements were selected for load current harmonics of about 150 A. At a phase shift of 26 degrees, this corresponds to the active power $P = 3.6$ MW and reactive power $Q = 1.77$ MVAr. By using the filters, the total harmonic distortion (THD) of voltage is reduced from 7.63% to 4.45% and the I_{ac} current from 33.6% to 12.65%. The measured voltages without the filters are shown in Figure 11 and with the attached filter in Figure 12. Figure 13 then presents the measurement equipment.

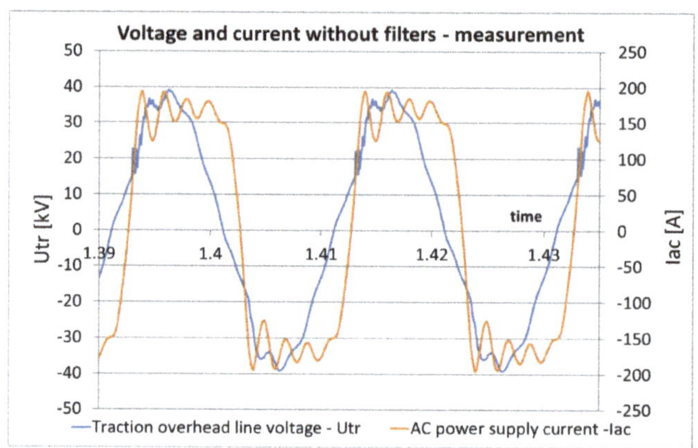

Figure 11. Voltage and current waveforms of the system without filter—measurement.

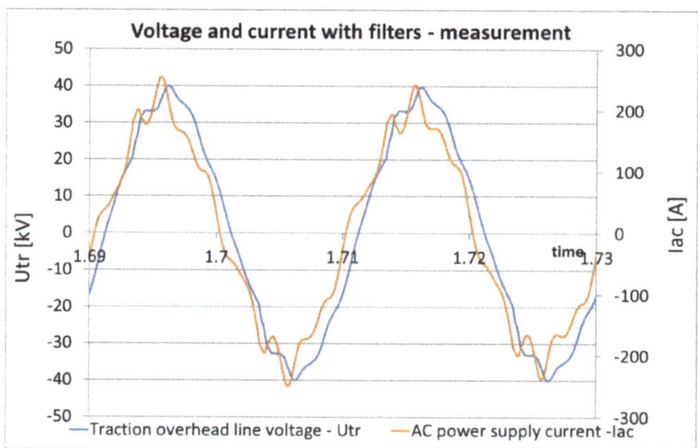

Figure 12. Voltage and current waveforms of the system in operation with connected filters—measurement.

Figure 13. View of the measuring stand at the traction substation.

6. Evaluation of Filtration Efficiency

The effect of the filter on the current waveform is shown in the waveform in Figure 14. There is a significant reduction in the third and the fifth harmonic current. The fifth harmonic filter has also an effect on the reduction currents of other harmonics. In Table 4 the percent harmonic voltages in the system are listed without filter (index NF) and with filter (index WF). At the same time, the values of permitted distortion according to the standard [22] are shown here. Monitoring of harmonics of order h = 21 or higher has no practical significance. The filtration effect of the filters is evident from the tables. Measurements and simulations show excellent filtration efficiency at the third and fifth harmonic, and the influence of the fifth harmonic filter to reduce the values for the seventh, ninth and the eleventh harmonic components.

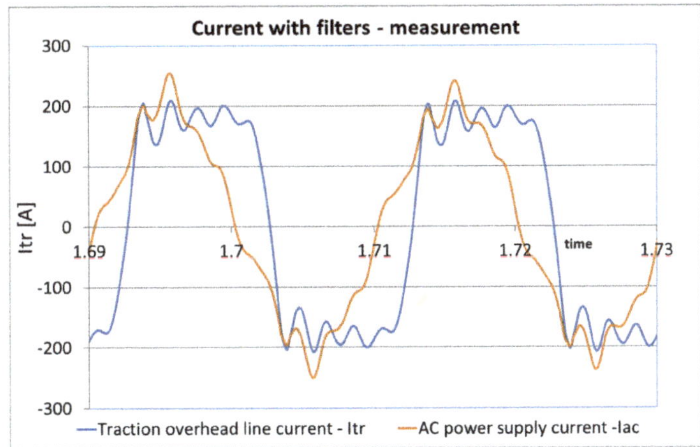

Figure 14. Comparison of taken current under load without filters and with filters.

Table 4. Percent values of harmonic voltages by measurement and simulation and limits by standard IEC.

Percentages.	Measurement		Simulation		IEC Standard
harmonics order	Utrnf	Utrwf	Utrnf	Utrwf	Uh
2	0.03	0.03	0.00	0.03	2
3	4.24	0.42	3.42	0.45	5
4	0.05	0.08	0.00	0.05	1
5	2.99	0.66	2.62	0.67	6
6	0.11	0.08	0.01	0.04	0.5
7	3.02	2.00	2.85	1.95	5
8	0.11	0.11	0.01	0.02	0.5
9	3.45	2.34	3.40	2.40	1.5
10	0.14	0.16	0.01	0.01	0.5
11	4.08	2.79	4.28	2.98	3.5
12	0.11	0.16	0.01	0.01	0.46
13	1.03	0.68	1.05	0.68	3
14	0.14	0.13	0.03	0.00	0.43
15	0.27	0.29	0.75	0.00	0.4
16	0.14	0.12	0.08	0.02	0.41
17	0.24	0.24	0.60	0.02	2
18	0.11	0.09	0.02	0.02	0.39
19	0.22	0.20	0.07	0.01	1.76
20	0.11	0.11	0.06	0.01	0.37
THD	7.63%	4.45%	7.64%	4.42%	8%

note: nf = filter off, wf = filter on.

7. Conclusions

The paper described a new methodology for optimum filter design for harmonic low-order filtering of traction substations and reactive power compensation. A specific feature of the traction system is the single-phase system, which is the cause of the 3rd harmonic current. Another typical feature is the operation of older types of locomotives (diode or thyristor), which are a source of high harmonics and reactive power consumption.

The main goal of the paper was to propose a novel method for the design of a traction filtration unit. The mentioned method performed the design of the filter of the whole spectrum of harmonic loading filter and prevented unnecessary oversizing of the design of the filter, which, of course, had a major impact on minimizing the filter capacitor power and thus on the size and price of the capacitor. The methodology is designed in such a way that we accept the proportional part of harmonic currents of higher orders, which will flow through the filter and thus can overload the filter. Using filters greatly reduced the harmonic currents that were taken from the traction substation and reduced the voltage distortion.

The described method was applied for the design of filters for a 27 kV traction substation. The methodology was verified by simulations and measurements.

The results show that the harmonic currents taken from the supply system (see Table 1) and the voltage (see Table 2) are high without the use of filters. With the use of filters, these harmonics decrease. As can be seen from Figure 11, the current of the 3rd and 5th harmonics is significantly reduced. The total THDI current distortion coefficient is reduced from 33% to 12% using a filter.

It is shown that the filter of the 5th harmonic has a filter effect even at higher frequencies. The decrease in voltage distortion is evident up to the 11th harmonic. For this frequency the value allowed by the standard was exceeded when measuring without a filter (allowed 3.5%, measured 4.08%, see Table 4). Using a 5th harmonic filter, the 11th harmonic voltage is reduced to 2.79%.

The Table 4 shows the evident filtering effect of the 5th harmonic filter.

By comparing with other design methods, it is evident that the filters according to the proposed methodology show the highest filtration effect.

The proposed methodology was tested on the traction substation 25 kV/50 Hz, where dynamic states with a high proportion of harmonics occur. In the same way it can be used for design of common (three-phase) filter compensation stations.

Author Contributions: Investigation, V.K.; Methodology, B.S.; Validation, P.D. All authors have read and agreed to the published version of the manuscript.

Funding: This research has been supported by the Ministry of Education, Youth, and Sports of the Czech Republic under the project OP VVV Electrical Engineering Technologies with High-Level of Embedded Intelligence CZ.02.1.01/0.0/0.0/18_069/0009855 and project No. SGS-2021-021.

Institutional Review Board Statement: Not Applicable.

Informed Consent Statement: Not Applicable.

Acknowledgments: This research has been supported by the Ministry of Education, Youth, and Sports of the Czech Republic under the project OP VVV Electrical Engineering Technologies with High-Level of Embedded Intelligence CZ.02.1.01/0.0/0.0/18_069/0009855 and project No. SGS-2021-021.

Conflicts of Interest: The authors declare no conflict of interest.

References

1. Chang, W.; Lin, H.W.; Chen, S.K. Modelling characteristics of harmonic currents generated by high-speed railway traction drive converters. *IEEE Trans. Power Deliv.* **2004**, *19*, 766–773. [CrossRef]
2. Foiadelli, F.; Lazaroiu, G.C.; Zaninelli, D. Probabilistic method for harmonic analysis in railway system. In Proceedings of the IEEE Power Engineering Society General Meeting, San Francisco, CA, USA, 12–17 June 2005.
3. Arrillaga, J. *Power System Harmonic Analysis*; John Wiley & Sons: New York, NY, USA, 1997.
4. Sollazzo, A.; Testa, A.; Carbone, R.; Rosa, F. Modeling Waveform Distortion Produced by High Speed AC Locomotive Converters. In Proceedings of the IEEE Bologna Power Tech Conference, Bologna, Italy, 23–26 June 2003.

5. Shu, Z.; Xie, S.; Li, Q. Single-Phase Back-To-Back Converter for Active Power Balancing, Reactive Power Compensation, and Harmonic Filtering in Traction Power System. *IEEE Trans. Power Electron.* **2010**, *26*, 334–343. [CrossRef]
6. Cecati, C.; Dell Aquila, A.; Monopoli, V.G. Design of h-bridge multilevel active rectifier for traction systems. *IEEE Trans. Ind. Appl.* **2003**, *39*, 1541–1550. [CrossRef]
7. Boora, A.A.; Zare, F.; Ghosh, A.; Ledwich, G. Applications of power electronics in railway systems. In Proceedings of the Australasian Universities Power Engineering Conference, Perth, WA, Australia, 9–12 December 2007; pp. 1–9.
8. Luo, A.; Ma, F.; Wu, C.; Ding, S.Q.; Zhong, Q.-C.; Shuai, Z.K. A Dual-Loop Control Strategy of Railway Static Power Regulator Under V/V Electric Traction System. *IEEE Trans. Power Electron.* **2011**, *26*, 2079–2091. [CrossRef]
9. Luo, A.; Wu, C.; Shen, J.; Shuai, Z.; Ma, F. Railway Static Power Conditioners for High-speed Train Traction Power Supply Systems Using Three-phase V/V Transformers. *IEEE Trans. Power Electron.* **2011**, *26*, 2844–2856. [CrossRef]
10. Zanotto, L.; Piovan, R.; Toigo, V.; Gaio, E.; Bordignon, P.; Consani, T.; Fracchia, M. Filter Design for Harmonic Reduction in High-Voltage Booster for Railway Applications. *IEEE Trans. Power Deliv.* **2005**, *20*, 258–263. [CrossRef]
11. Hu, H.; He, Z.; Gao, S. Passive Filter Design for China High-Speed Railway With Considering Harmonic Resonance and Characteristic Harmonics. *IEEE Trans. Power Deliv.* **2015**, *30*, 505–514. [CrossRef]
12. Hammond, P.W. A harmonic filter installation to reduce voltage distortion from static power converters. *IEEE Trans. Ind. Appl.* **1988**, *24*, 53–58. [CrossRef]
13. Seferi, Y.; Blair, S.M.; Mester, C.; Stewart, B.G. Power Quality Measurement and Active Harmonic Power in 25 kV 50 Hz AC Railway Systems. *Energies* **2020**, *13*, 5698. [CrossRef]
14. Yao, X. The method for designing the third order filter. In Proceedings of the 8th International Conference on Harmonics and Quality of Power (ICHQP), Athens, Greece, 14–16 October 1998; Volume 1, pp. 139–142.
15. Tyll, H.K.; Schettle, F. Historical overview on dynamic reactive power compensation solutions from the begin of AC power transmission towards present applications. In Proceedings of the IEEE/PES Power Systems Conference and Exposition, Seattle, WA, USA, 15–18 March 2009.
16. Hafner, J.; Aredes, M.; Heumann, K.A. A shunt active power filter applied to high voltage distribution lines. *IEEE Trans. Power Deliv.* **1997**, *12*, 266–272. [CrossRef]
17. Tan, P.C.; Morrison, R.E.; Holmes, D.G. Voltage Form Factor Control and Reactive Power Compensation in a 25-kV Electrified Railway System Using a Shunt Active Filter Based on Voltage Detection. *IEEE Trans. Ind. Appl.* **2003**, *39*, 575–581.
18. Han, Z.; Zhang, Y.; Liu, S.; Gao, S. Modeling and Simulation for Traction Power Supply System of High-Speed Railway. In Proceedings of the Asia-Pacific Power and Energy Engineering Conference (APPEEC), Wuhan, China, 25–28 March 2011.
19. Kolář, V.; Paleček, J.; Kocman, S. Interference between Electric Traction Supply Network and Distribution Power Network–Resonance Phenomenon. In Proceedings of the 14th International Conference on Harmonics and Quality of Power—ICHQP, Politecnico di Milano, Bergamo, Italy, 26–29 September 2010; ISBN 978-1-4244-7245-1.
20. Varetsky, Y. Transient Overvoltages during Filter Circuit Switching-off. In Proceedings of the International Symposium Modern Electric Power Systems (MEPS), Wroclaw, Poland, 20–22 September 2010; ISBN 978-83-921315-7-1.
21. Kus, V.; Skala, B. Effect of Filter and Compensation Unit in the Traction Power Supply at Extreme Distortion. In Proceedings of the CPE—International Conference-Workshop Compatibility And Power Electronics, Ljubljana, Slovenia, 5–7 June 2013; pp. 1–6.
22. Standard: IEC 61000-2-12: Electromagnetic compatibility (EMC)—Part 2-12. Environment—Compatibility Levels for Low-Frequency Conducted Disturbances and Signaling in Public Medium-Voltage Power Supply Systems. Available online: https://webstore.iec.ch/publication/4130 (accessed on 14 September 2021).

Article

Harmonic Overvoltage Analysis of Electric Railways in a Wide Frequency Range Based on Relative Frequency Relationships of the Vehicle–Grid Coupling System

Qiujiang Liu, Binghan Sun, Qinyao Yang, Mingli Wu * and Tingting He

School of Electrical Engineering, Beijing Jiaotong University, Beijing 100044, China; qjliu@bjtu.edu.cn (Q.L.); 19117020@bjtu.edu.cn (B.S.); 19126189@bjtu.edu.cn (Q.Y.); hetingting@bjtu.edu.cn (T.H.)

* Correspondence: mlwu@bjtu.edu.cn; Tel.: +86-136-2135-9126

Received: 20 November 2020; Accepted: 15 December 2020; Published: 17 December 2020

Abstract: Harmonic overvoltage in electric railway traction networks can pose a serious threat to the safe and stable operation of the traction power supply system (TPSS). Existing studies aim at improving the control damping of grid-connected converters, neglecting the impedance frequency characteristics (IFCs) of the actual TPSS. The applicable frequency range of these studies is relatively low, usually no more than half of the switching frequency, and there is a large gap with the actual traction network harmonic overvoltage frequency range of 750 Hz–3750 Hz. In this paper, first, the IFCs of the actual TPSS in the wide frequency range of 150 Hz–5000 Hz are obtained through field tests, and the resonant frequency distribution characteristics of TPSS are analyzed. After that, the aliasing effect of the sampling process and the sideband effect of the modulation process of the digital control of the grid-connected converter are considered. Based on the relative relationships among the inherent resonant frequency of the TPSS, sampling frequency and switching frequency, an impedance matching analysis method is proposed for the wide frequency range of the vehicle–grid coupling system. By this method, the sampling frequency and switching frequency can be decoupled, and the harmonic overvoltage of traction network in the frequency range of two times switching frequency and above can be directly estimated. Finally, the method proposed in this paper is validated by the comparative simulation analysis of seven different cases.

Keywords: electric railway; harmonic resonance; harmonic overvoltage; stability analysis; resonant frequency; traction power supply system

1. Introduction

Since alternating current-direct current-alternating current (AC-DC-AC) Electric Multiple Units (EMUs) and electric locomotives have been adopted widely in China, the traction power supply system (TPSS) and electric locomotives' electrical matching instability caused by high-order harmonic overvoltage traction network problems occur from time to time. Harmonic overvoltage of traction networks can cause the substation feeder to trip. In severe cases, it will cause damage to the high-voltage equipment of the catenary, the on-board high-voltage equipment, and the electrical equipment in the substation. In addition, it can even result in the disruption of the locomotive operation [1]. A widely applied method is that the harmonic overvoltage problem of traction networks is mostly regarded as a harmonic instability problem of the vehicle–grid coupling control system. The method utilizes Re $\{Zin(j\omega)\} > 0$ as stability criterion, where Re $\{Zin(j\omega)\} > 0$ denotes the real part of input impedance of grid-connected converters port. When $Zin(j\omega)$ does not satisfy the condition that the real part is greater than 0 at a certain frequency, the risk of harmonic instability is considered to exist [2–6].

The European Norm EN 50388, which is enforced by railway administrations in several countries, requires all new elements added to the system checked for compatibility [7]. The standard also specifies that harmonic overvoltage is classified into three main categories: overvoltage caused by system instability, harmonics and other phenomena. However, the standard only requires that in a 25,000 V, 50 Hz network, the peak voltage is always below 50,000 V, which is a lax standard with regard to harmonics. There are two deficiencies in the method that the harmonic overvoltage problem is completely regarded as an equivalent control stability analysis of the grid-connected converter, which makes it difficult to solve practical cases.

The impedance frequency characteristics (IFCs) of the TPSS are deliberately simplified. In the existing literatures, input impedance of traction network is usually represented by a parallel LC circuit. This simple circuit can introduce a parallel resonance. However, the IFCs of the actual TPSS may exhibit multiple parallel resonances, and the resonant frequency is affected by power supply mode, the length of railway, location of the locomotive and other factors. This implies a wide range of resonance frequencies. The discrepancy between the simplified circuit model in a laboratory setting and actual system requires more detailed consideration when analyzing real problems with existing methods. Researchers reckoned that their proposed method may become invalid if there are two resonances on a TPSS with an approximately symmetrical distribution of Nyquist frequency [8]. The current research on this issue is still insufficient, mainly because of the lack of field test data and effective techniques that can test the IFCs of a 25 kV electric railway over a wide frequency range.

The applicable frequency range of the existing research is lower than the switching frequency. Although the influence of complex factors, such as time delay and LCL filters, is taken into consideration, most of the current research has focused on the relatively low frequency range, that is, below the Nyquist sampling frequency or half of the switching frequency. These studies do not apply to real on-board grid-connected converters switching frequency of only a few hundred hertz, while harmonic overvoltage frequency can reach thousands of hertz (several times the switching frequency). Though the published papers [8–10] attempt to broaden the frequency range applicable to the study by considering the aliasing effect of the converter-controlled sampling processes, these studies are mostly limited by the premise assumptions $f_{sam} = f_{sw}$ or $f_{sam} = 2f_{sw}$. The pulse-width modulation (PWM) element can be simplified into a proportional element. However, in practice, due to the multiplexed or modular cascade structure of the grid-connected converter, the multiplicative relationship between f_{sam} and f_{sw} can be 0.5-fold, 1-fold, 2-fold, 4-fold or even any frequency multiplicity [11]. Therefore, the current study is limited in terms of its applicable frequency range.

In the simulation or experiment of previous studies, the switching frequency of the converter or the sampling frequency of the controller is much higher than the resonant frequency of the impedance network. In most studies, the resonant frequency of the impedance network is less than half of the sampling frequency and switching frequency (see Table 1). That is, the sampling frequency and switching frequency are much larger than the Nyquist frequency. In addition, in almost all previous studies, the ratio of the switching frequency of the converter to the sampling frequency of the controller is fixed at 2:1, 1:1 or 1:2 and there is a lack of research on other ratios. Furthermore, the switching frequency of the converter often ranges from 2 kHz to higher than 10 kHz in previous studies, while in actual railway systems, the switching frequency of the train converter is usually in the range of 250–1250 Hz. There is a large difference between the two frequency ranges. In our earlier research, the resonance frequency of TPSS under different operation conditions is distributed in a wide frequency range of hundreds to more than 3000 Hz.

All in all, it can be deduced that, in reality, the resonant frequency of the traction network may be: 1. above the sampling frequency of the train converter; 2. below the sampling frequency and within the range of Nyquist frequency (0—half of the sampling frequency); 3. below the sampling frequency but between the Nyquist frequency and the sampling frequency.

Table 1. Comparison among different frequencies.

References	f_{sw}/kHz	f_{sam}/kHz	$f_{resonant}$/kHz	Comments
[1]	10	10	1 or 2	$f_{resonant} < 0.5f_{sw} = 0.5f_{sam}$
[2]	15	15	3.08, 2.03	$f_{resonant} < 0.5f_{sw} = 0.5f_{sam}$
[3]	20	10	1.4	$f_{resonant} < 0.5f_{sam} < 0.5f_{sw}$
[4,8]	0.8	1.6	0.95 or 1.45	$f_{sw} < f_{resonant} < f_{sam}$ Only the aliasing effect is considered.
[9]	4	4	0.476	$f_{resonant} < 0.5f_{sw} = 0.5f_{sam}$
[11]	10	10	1.6	$f_{resonant} < 0.5f_{sw} = 0.5f_{sam}$
[12]	20	20	5	$f_{resonant} < 0.5f_{sw} = 0.5f_{sam}$
[13,14]	6	12	2.7	$f_{resonant} < 0.5f_{sw} = 0.25f_{sam}$
[15]	15	10	4.4	$f_{resonant} < 0.5f_{sw} < f_{sam}$
[16]	10	10	0.4	$f_{resonant} < 0.5f_{sw} = 0.5f_{sam}$
[17]	10	10	1	$f_{resonant} < 0.5f_{sw} = 0.5f_{sam}$

Our work is more effective than previous research. This paper investigates the mechanism of resonant overvoltage in a wider frequency range, that is, all of the above three cases. In addition, this paper also considers more combinations of the ratio of the sampling frequency (f_{sam}) to the switching frequency (f_{sw}), which can be categorized into four situations: (a.) $2f_{sam} \le f_{sw}$; (b.) $f_{sw} \le f_{sam} < 2f_{sw}$; (c.) $2f_{sw} \le f_{sam} < 4f_{sw}$; (d.) $4f_{sw} \le f_{sam}$. Taking the above mentioned three cases and four situations into account, this paper provides a wider vision to the TPSS resonant overvoltage problem.

To address the above research deficiencies, this paper mainly focuses on harmonic overvoltage of the traction network with a wide frequency range. First of all, this paper presents the results of the impedance frequency characteristics test and analyzes the inherent resonant frequency distribution characteristics of the traction network. Moreover, the aliasing effect of the converter control processes and the PWM sideband effect are taken into consideration in this paper. The relative magnitude relationship of the traction network impedance resonant frequency, sampling frequency and PWM switching frequency is considered. The obtained model is suitable for any multiple relationship between f_{sw} and f_{sam}, and can suppress harmonic overvoltage of the traction network by changing f_{sw} and f_{sam}. This paper presents the link between the steady-state waveform of traction network harmonic overvoltage and the control stability of the vehicle–grid coupling system, and proposes a formula for the harmonic overvoltage of the traction network, which makes a comprehensive description of the harmonic overvoltage of the traction network of the network coupling system.

$$\begin{aligned} \sup_{\omega}|V_{hv}(j\omega)| &= \|\boldsymbol{Z}_{hv}\boldsymbol{H}_{hv}\boldsymbol{I}_{hv}\|_{\infty} \\ &\le \|\boldsymbol{Z}_{hv}\|_{\infty}\ \|\boldsymbol{H}_{hv}\|_{\infty}\ \|\boldsymbol{I}_{hv}\|_{\infty} \end{aligned} \tag{1}$$

where $\boldsymbol{V}_{hv}$ represents the harmonic overvoltage, $\boldsymbol{Z}_{hv}$ is the impedance matrix of vehicle–grid coupling system, $\boldsymbol{H}_{hv}$ is the control coupling coefficient matrix, $\boldsymbol{I}_{hv}$ is the harmonic source matrix. Equation (1) has the form and dimension of Ohm's law, which can take all the variables and parameters of the vehicle–grid coupling system into account. Equation (1) will be elaborated on below.

2. Vehicle–Grid Coupling System

Vehicle–Grid Coupling System Model

In Figure 1, it is shown that the electrified railway vehicle–grid coupling system is mainly composed of TPSS and on-board converter.

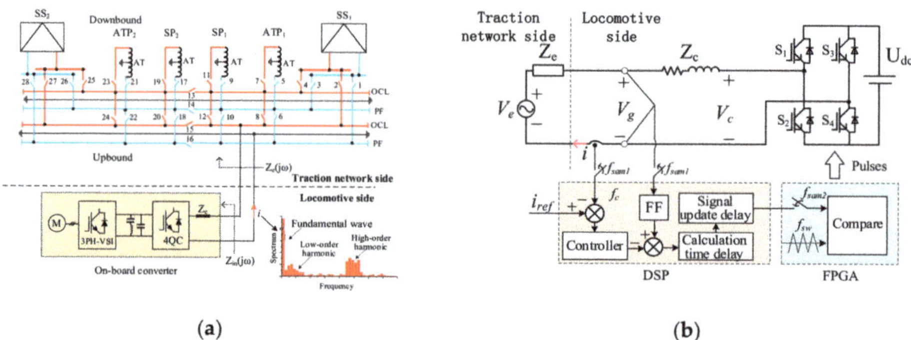

Figure 1. Model of the vehicle–grid coupling system. (**a**) Actual railways; (**b**) the converter model.

The TPSS is made up of multiple subsystems: substation (SS), auto-transformer post (ATP), section post (SP), rail, over contact line (OLC), positive feeder (PF), and multi-conductor transmission line traction network with complex structure and parameters. The model of the on-board converter is shown in Figure 1b. The design of the converter digital control system on control logic includes digital signal processor (DSP) and field-programmable gate array (FPGA). The control algorithm is implemented on DSP. FPGA is used to execute modulation algorithms. The actual power supply operation mode is AT/double traction/parallel lines/overhead contact line–classic–hanging. The railroad section researched was not covered by a tunnel.

The Root Mean Square (RMS) and instantaneous value waveforms of voltage, before and after the tripping of the T-bus of a traction substation, are depicted in Figure 2. The measuring data of the traction substation are from one case of [1]. The general procedure for evaluating harmonic overvoltage based on measured waveforms is: Fourier analysis is performed on the measured instantaneous waveforms to obtain the magnitude and frequency of each order harmonics of traction network voltage, and then the frequency of the harmonic component with the largest amplitude is taken as the traction network voltage resonant frequency [1,12,13]. According to this procedure, the resonant frequency of the traction network voltage shown in Figure 2 is 950 Hz. In order to distinguish the terms' resonant frequency in the control system from the inherent resonant frequency of the TPSS, this paper refers to the frequency of the harmonic component with the largest amplitude in the traction network voltage as the traction network resonant overvoltage frequency f_{hv}.

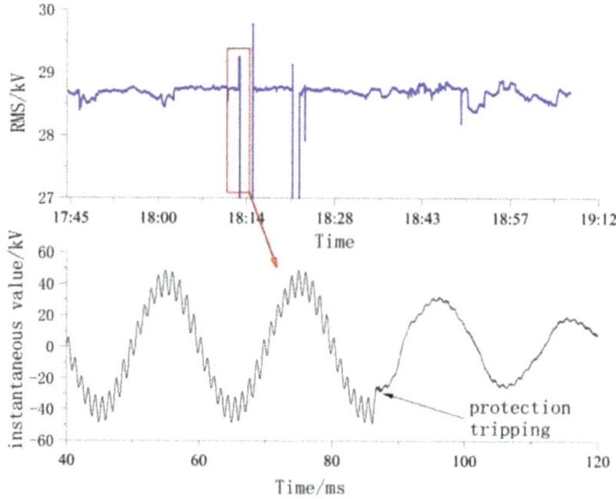

Figure 2. Harmonic overvoltage waveforms of the traction network in actual railways.

3. IFCs of TPSSs

The authors' previous research has proposed a harmonic impedance testing method based on a harmonic generator (HG) constructed with cascaded H-bridge converters, which can directly measure the input harmonic impedance of the 25 kV electric railway traction network port in the range of 5000 Hz. The test method and implementation process have been presented in detail in the literature [14,15]. Operation of the telecontrol switches 1–28 in Figure 1a to switch the different power supply modes, during the test, is shown in Table 2. Here, we inject the harmonic current with a harmonic source and obtain harmonic impedance by the method of harmonic voltage and current test [1]. Eventually, the IFCs of the TPSSs are obtained at SP1 under different operating modes. The test results are depicted in Figure 3.

Table 2. Switch operation table during the impedance frequency characteristic (IFC) tests.

No.	Switch Connected	Switch Disconnected	Power Supply Operation Mode
1	1~12	13~28	Autotransformer (AT)/Double track/Non-over-zone
2	1~24	25~28	AT/Double track/Over-zone
3	1~5,7,9,11	6,8,10,12~28	AT/Single track/Non-over-zone
4	1~5,7,9,11,13, 14,17,19,21,23	6,8,10,12,15, 16,18,20,22,24~28	AT/Single track/Over-zone
5	1~4	5~28	Direct supply/Double track/Non-over-zone
6	2,4	1,3,5~28	Direct supply/Double track/Non-over-zone/Without PF
7	4	1~3,5~28	Direct supply/Single track/Non-over-zone/Without PF

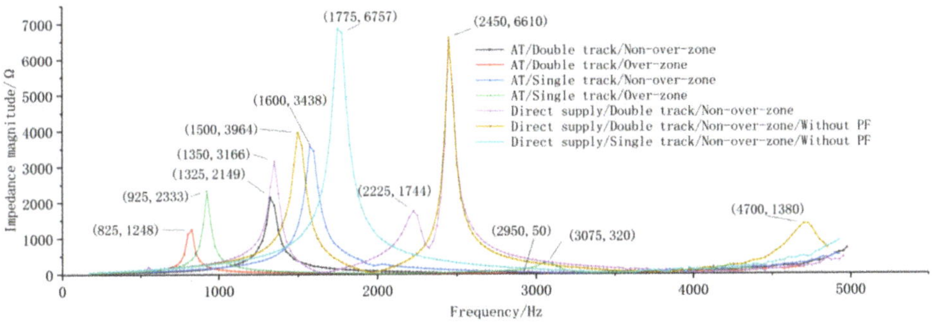

Figure 3. Impedance frequency characteristics of the actual railway by field tests.

The test results shown in Figure 3 indicate that the inherent resonant frequency of the TPSS itself is affected by factors such as length of the traction network (over-zone vs. non-over-zone test), auto-transformer (AT vs. direct supply test), line type (single track vs. double track railway test), and the structure of the traction network (with or without a PF comparison test), which are subject to complex variations. In addition, the existence of more than two resonances for the over-zone feeding with a long power supply section provides measured data and supporting evidence for the potential problems identified in [8]. The above results indicate that the complex IFCs of the actual traction network and its own inherent resonance bring great challenges to the analysis of the vehicle–grid coupling system and the control design and parameter setting of the grid-connected converter.

4. Control Modeling of Converters

4.1. Sampling Process and Aliasing Effect

Consider the sampling process of voltage $\mathbf{V}_g$, and current $\mathbf{i}$, as shown in the Figure 4. When the original signal frequency $X(j\omega)$ is less than the Nyquist sampling frequency, the sampling output $R(j\omega)$ will be attached to generate a high-frequency interference signal. If the frequency distance between the signals is large, the low-pass filter can be used to recover the original signal from the sampled signal, that is $V(j\omega)$. If the frequency of $X(j\omega)$ is greater than or equal to the Nyquist sampling frequency, the original signal will additionally generate a low-frequency sampling signal. Low-pass filter will not be used to recover the original signal from the sampled signal. In Figure 4, it leads to the effect of spectrum aliasing.

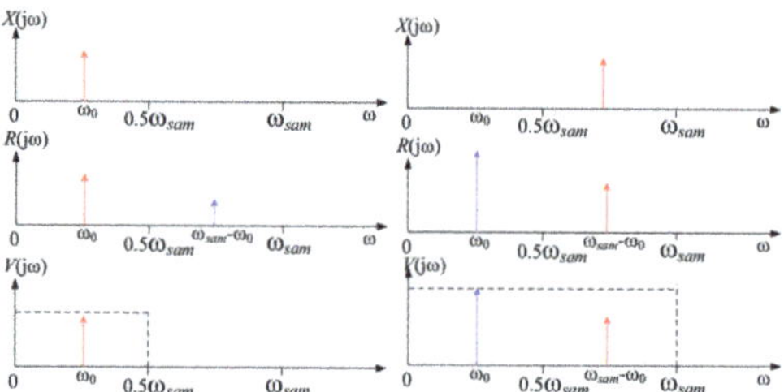

Figure 4. Diagram of the aliasing effect.

Considering the effect of the aliasing effect, there are multiple frequency signals in the sampling element at the same time, and signals of various frequencies couple each other. The sampling element can be equivalent to a multi-input multi-output system. The relationship of the input and output system can be described as Equation (2) by matrix. Assuming that the input signal is composed of n frequency components, the output signal will also be composed of n components.

$$\begin{pmatrix} r(\omega_x) \\ r(\omega_{\text{sam}}-\omega_x) \\ r(\omega_{\text{sam}}+\omega_x) \\ r(2\omega_{\text{sam}}-\omega_x) \\ \cdots \end{pmatrix} = \begin{pmatrix} a_{11} & a_{12} & a_{13} & a_{14} & \cdots \\ a_{21} & a_{22} & a_{23} & a_{24} & \cdots \\ a_{31} & a_{32} & a_{33} & a_{34} & \cdots \\ a_{41} & a_{42} & a_{43} & a_{44} & \cdots \\ \cdots & \cdots & \cdots & \cdots & \cdots \end{pmatrix} \begin{pmatrix} x(\omega_x) \\ x(\omega_{\text{sam}}-\omega_x) \\ x(\omega_{\text{sam}}+\omega_x) \\ x(2\omega_{\text{sam}}-\omega_x) \\ \cdots \end{pmatrix} \tag{2}$$

The elements on the main diagonal in (2) are the self-coupling conductivity coefficients, which characterize the input–output relationship of the same frequency signal. The elements on the non-diagonal are the mutual coupling conductivity coefficients, which characterize the input–output relationship of signals of different frequencies.

As shown in Figure 5, in order to obtain the coefficients of $\boldsymbol{A} = (a_{ij})_{n \times n}$ by the superposition theorem, let the input signal have only one non-zero element at a time, implying:

$$a_{kl} = \frac{x_l}{r_k}|x_i = 0, \ \forall i \neq l \tag{3}$$

where $x_l(\omega_l) = \delta(\omega - \omega_l)$. Take the solution of the conduction coefficients of the first component as an example. As shown in Figure 5, at first, set the amplitude of frequency ω_ξ of the input signal as 1, and the other components are 0, that is, $\delta(\omega - \omega_\xi)$. According to the signal equation $x_p(t)$ affected after the impulse train and the transfer function of the zero-order holder (ZOH), the amplitude and phase angle curves of each component of the output signal r(t) can be depicted in Figure 5 | r(ω) | and ∠ r(ω), respectively. According to the component represented by the red curves in Figure 5, the self-coupling conductivity coefficient a11, as well as the mutual coupling conductivity coefficients a12, a13, etc., can be obtained. Therefore, the first column element of the conduction matrix in (2) can be obtained.

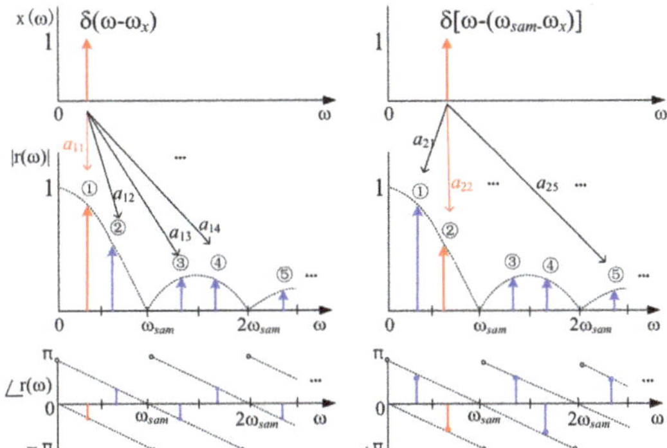

Figure 5. Derivation of the sampling transfer function at multiple frequencies.

Repeat the above steps until all the elements of the entire matrix are found. It can be written as (4):

$$A(\omega_x) = \begin{pmatrix} G(\omega_x) & G(\omega_x)e^{j\pi} & G(\omega_x) & \cdots \\ G(\omega_{sam} - \omega_x)e^{j\pi} & G(\omega_{sam} - \omega_x) & G(\omega_{sam} - \omega_x)e^{j\pi} & \cdots \\ G(\omega_{sam} + \omega_x) & G(\omega_{sam} + \omega_x)e^{j\pi} & G(\omega_{sam} + \omega_x) & \cdots \\ \cdots & \cdots & \cdots & \cdots \end{pmatrix} \quad (4)$$

where $G(\omega)$ is defined as the transfer function of ZOH.

4.2. Modulation Process and Aliasing Effect

4.2.1. Natural Sampling PWM (NSPWM)

According to the analytical expression [16,18] for the output voltage of a single H-bridge converter NSPWM, the frequency spectrum of the bridge port output voltage varies as the frequency of the modulation waveform gradually increases, and its trajectory is depicted in Figure 6, where the y-axis represents the frequency of the modulation waveform signal and the x–axis represents the frequency spectrum of $u_c(t)$. Taking Figure 6 as an example, make tangent 1 parallel to the x–axis at y = 2250 Hz. The tangent plane and trajectory intersect at two points. The value of the x–axis corresponding to the intersection point represents the frequency of the main component of the output voltage. As shown in Figure 6, the pair of natural numbers (m,k) indicates the kth harmonic component of the mth group of sideband harmonics. The baseband trace in Figure 6 represents the component change of the $u_c(t)$ at the same frequency as the modulation waveform and is the most dominant component of the PWM modulation process.

Through the analysis in Figure 6, it can be found that the NSPWM input and output segments are also similar to the sampling process with the aliasing effect. If the frequency of the input modulation wave signal is less than the switching frequency, the output signal will additionally generate a high-frequency interference signal, then the input modulation wave signal can be recovered with a low-pass filter. On the contrary, if the frequency of the input modulation wave signal is greater than the switching frequency, the output signal will additionally produce low-frequency interference. The frequency of the uppermost low-frequency interference component is $\omega_\sigma - \omega_o$, and the input modulation wave signal cannot be fully recovered by the low-pass filter, as shown in Figure 7. The above NSPWM process can be intuitively understood from the perspective of signal transmission, compared with the switching frequency, the low-frequency original signal will produce additional high-frequency interference, and the high-frequency original signal will produce additional low-frequency interference.

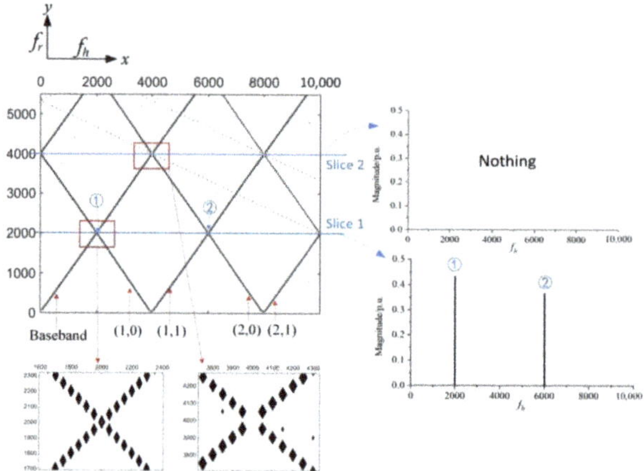

Figure 6. Locus of the spectral distribution (f_s = 2000 Hz, d_r = 0.25).

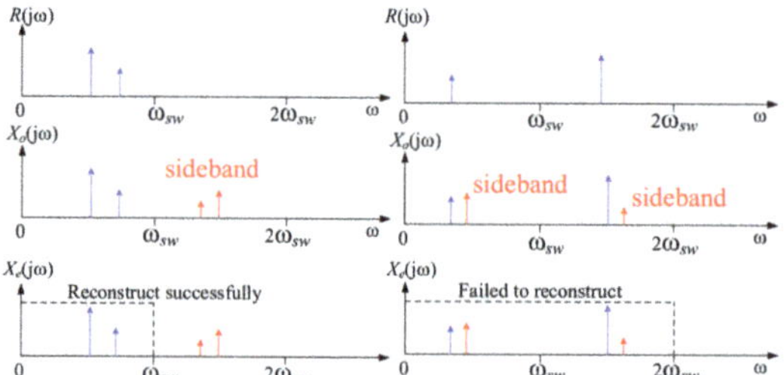

Figure 7. Diagram of the natural sampling pulse-width modulation (NSPWM) sideband effect.

According to the above analysis, the relationship between the input and output signal of the NSPWM segment can be approximated by (5). ∂_m^k represents the relationship between the kth harmonic component of the mth group of sideband harmonics in the output voltage spectrum and the modulation waveform signal. In order to simplify the analysis, only the main 0th and 1st sideband harmonics are considered in Equation (5).

$$\begin{pmatrix} u(\omega_z) \\ u(2\omega_s-\omega_z) \\ u(2\omega_s+\omega_z) \\ u(4\omega_s-\omega_z) \\ \cdots \end{pmatrix} = \begin{pmatrix} 1 & \partial_1^0 & \partial_1^0 & \partial_2^0 & \cdots \\ \partial_1^0 & 1 & \partial_2^0 & \partial_1^0 & \cdots \\ \partial_1^1 & \partial_2^0 & 1 & \partial_3^0 & \cdots \\ \partial_2^0 & \partial_1^1 & \partial_3^0 & 1 & \cdots \\ \cdots & \cdots & \cdots & \cdots & \cdots \end{pmatrix} \begin{pmatrix} r(\omega_z) \\ r(2\omega_s-\omega_z) \\ r(2\omega_s+\omega_z) \\ r(4\omega_s-\omega_z) \\ \cdots \end{pmatrix} \quad (5)$$

4.2.2. Regular Sampling PWM (RSPWM)

In the method of NSPWM, sinusoidal modulation wave and triangular carrier are compared in real time. Their intersection points are defined by the transcendental equation, which is complicated to solve. In order to facilitate the application in digital control system, the modulation wave is usually sampled and kept constant for a certain period of time; this modulation method is called RSPWM, which can be equated with a sampling segment in series with NSPWM segment. As shown in Figure 8, a simulation is built in the software PSCAD to test the RSPWM input/output signal description function, set $f_{sam} = f_{sw}$ = 2000 Hz, and the cutoff frequency is 5500 Hz.

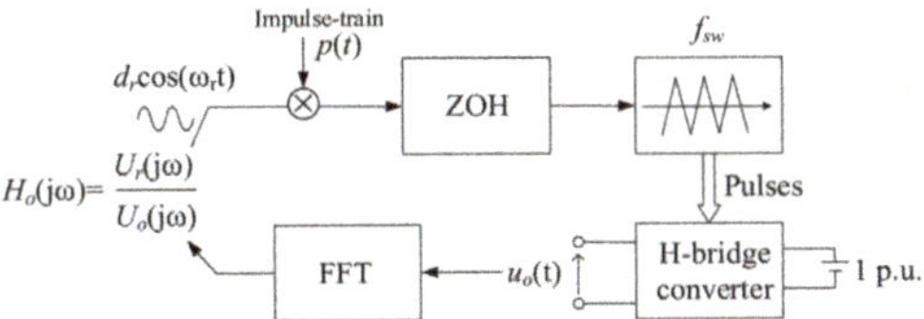

Figure 8. Test about the signal input–output relation of regular sampling (RSPWM).

Figure 9 shows the results of the test. Spikes within the switching frequency range are due to the aliasing effect of the sampling process. Above the switching frequency range, the aliasing effect and the sideband effect are combined. Figure 9 also shows that the existing research mostly treats the modulation segment as a ZOH segment [9,17], which only performs well in the low-frequency range. However, there are large errors in the high-frequency range.

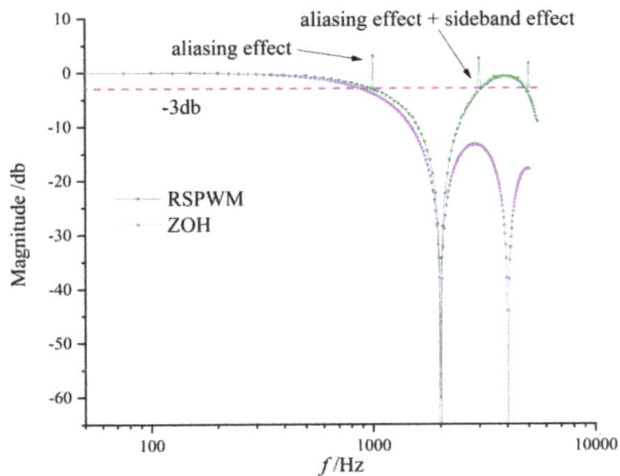

Figure 9. Amplitude cure of RSPWM in a wide frequency.

5. Harmonic Overvoltage Analysis of Traction Network in a Wide Frequency Range

5.1. Relative Relationship among TPSS Subsection Inherent Resonant Frequency, Sampling Frequency and Switching Frequency

The relative relationship among the inherent resonant frequency (IRF), sampling frequency and switching frequency of the TPSS is shown in Figure 10. As presented in the existing research [11], the switching frequency and sampling frequency can take different ratios. Moreover, inherent resonant frequency of the actual TPSSs is distributed over a wide frequency range, there are different ways of combining the relative value of the three frequency types. According to the analysis in Sections 4.1 and 4.2, the low-pass filter can be used to recover the original signal at half of f_{sam} and f_{sw}. Therefore, when the IRF of the TPSSs exists in the range of 0.5 $f_{sam} \cap f_{sw}$, the low-frequency single-component model can be used for analysis. When IRF2 ε (0.5 f_{sam}, f_{sw}), the aliasing effect of sampling should be considered. When IRF3 ε (f_{sw},∞), the aliasing effect of sampling and sideband effect should be considered simultaneously. By decomposing the relative relationship of IRF, f_{sam} and f_{sw}, the frequency range of harmonic overvoltage in the traction network can be greatly expanded for research.

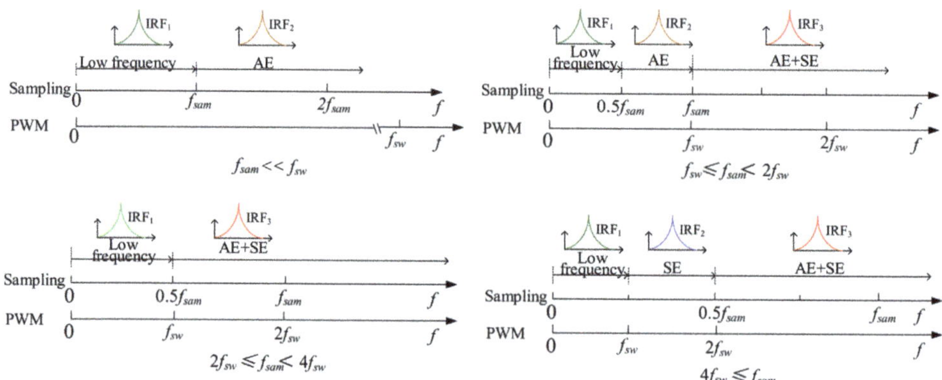

Figure 10. Relative relations among the IRF, sampling frequency, and switching frequency.

Figure 11 shows a block diagram of a wide frequency range control system considering sampling aliasing effects and modulation sideband effects, where $H_{cv}(s)$ represents the capacitor voltage controller, $H_{ci}(s)$ represents the current loop controller, $H_{sam}(s)$ represents the transfer function of the sampling segment, $H_{ffv}(s)$ represents the transfer function of the network voltage feed-forward shown in (4), $G_d(s)$ represents the total delay of the system, $H_{pwm}(s)$ represents the transfer function of the NSPWM segment shown in (5), $H_{fbi}(s)$ represents correcting the transfer function of the current sampling sensor.

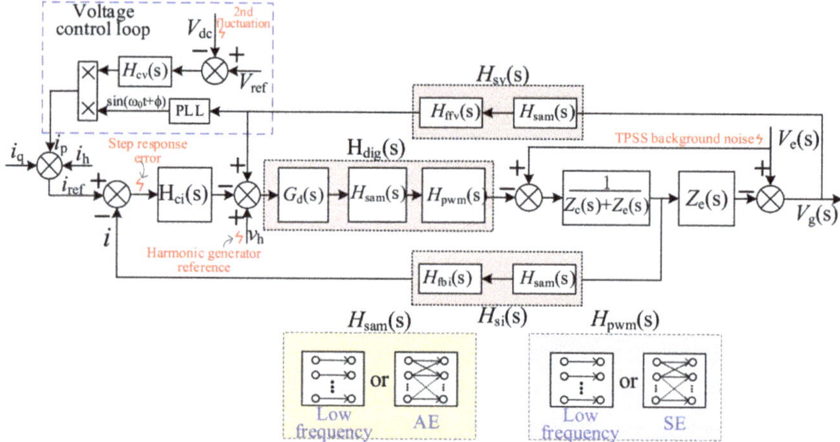

Figure 11. Control block diagram of the vehicle–grid coupling system at wide frequencies.

5.2. Low-Frequency Range Analysis

In the range of low frequency, the system closed-loop transfer function matrix $H_c(s)$ shown in Figure 11 can be described as:

$$\begin{pmatrix} I \\ V_g \\ V_c \end{pmatrix} = \begin{pmatrix} \frac{G_{ref}}{1+Z_eY_g} & \frac{Y_g}{1+Z_eY_g} & \frac{Y_h}{1+Z_eY_g} \\ \frac{-Z_eG_{ref}}{1+Z_eY_g} & \frac{1}{1+Z_eY_g} & \frac{-Z_eY_h}{1+Z_eY_g} \\ \frac{-(Z_c+Z_e)G_{ref}}{1+Z_eY_g} & \frac{1-Y_gZ_c}{1+Z_eY_g} & \frac{-(Z_c+Z_e)Y_h}{1+Z_eY_g} \end{pmatrix} \begin{pmatrix} I_{ref} \\ V_e \\ V_h \end{pmatrix} \tag{6}$$

where Y_g is the input admittance of the converter port, G_{ref} is the current control equivalent gain, Y_h is the equivalent admittance of voltage fluctuation. According to (6), these three parameters can be measured by small signal frequency scanning for amplitude frequency and phase frequency curves.

$$I = Y_g V_g + G_{ref} I_{\text{ref}} + Y_h V_h \tag{7}$$

Y_g, G_{ref}, Y_h can be described as:

$$\begin{cases} Y_g = \frac{1-H_{\text{dig}}H_{\text{sv}}}{Z_c+H_{\text{dig}}H_{ci}H_{si}} \\ G_{ref} = \frac{H_{\text{dig}}H_{ci}}{Z_c+H_{\text{dig}}H_{ci}H_{si}} \\ Y_h = \frac{-H_{\text{dig}}}{Z_c+H_{\text{dig}}H_{ci}H_{si}} \end{cases} \tag{8}$$

The traction network voltage $\boldsymbol{V}_{\text{hv}}$ of the vehicle–grid coupling system can be obtained by (6), which can be described as (9):

$$\begin{aligned} \boldsymbol{V}_{\text{hv}} &= \begin{pmatrix} -Z_{\text{e}} \\ Z_{\text{g}} \\ -Z_{\text{e}} \end{pmatrix}^T \begin{pmatrix} H_{\text{c11}} & 0 & 0 \\ 0 & H_{\text{c12}} & 0 \\ 0 & 0 & H_{\text{c13}} \end{pmatrix} \begin{pmatrix} I_{\text{ref}} \\ V_{\text{e}} \\ V_{\text{h}} \end{pmatrix} \\ &= \boldsymbol{Z}_{\text{hv}} \boldsymbol{H}_{\text{hv}} \boldsymbol{I}_{\text{hv}} \end{aligned} \tag{9}$$

where H_{c11}, H_{c12}, H_{c13} are the elements in the corresponding position in $Hc(s)$. The harmonic overvoltage of the traction network can be described as (1).

In (1), it is illustrated that the value of harmonic voltage is closely related to the harmonic impedance matrix $\boldsymbol{Z}_{\text{hv}}$, harmonic gain matrix $\boldsymbol{H}_{\text{hv}}$, and harmonic noise $\boldsymbol{I}_{\text{hv}}$: (1) When $||\boldsymbol{H}_{\text{hv}}||$ gets the maximum value at a certain frequency, it means (represents) that this frequency is the resonant frequency of $\boldsymbol{H}_{\text{hv}}$. In this case, the control of vehicle–grid coupling system will be unstable. With the excitation of limited harmonics, a harmonic overvoltage of the traction network can be generated. (2) When $||\boldsymbol{Z}_{\text{hv}}||$ gets the maximum value at a certain frequency, it means that this frequency is the inherent resonant frequency of the TPSSs. Due to the extremely large input impedance amplitude of the TPSS port, even if the grid-connected converter satisfies Re{Y(jω)} > 0, the problem of traction network harmonic overvoltage may have occurred under the excitation of limited harmonics. In this case, the grid-connected converter can be treated as a harmonic current source. (3) $||\boldsymbol{I}_{\text{hv}}||$ stands for harmonic excitation. The larger its value, the more serious the harmonic overvoltage accident of the traction network will occur. Theoretically, after reducing $\boldsymbol{I}_{\text{hv}}$ to 0 with effective measures, harmonic overvoltage of the traction network will not occur, even if $||\boldsymbol{Z}_{\text{hv}}\boldsymbol{H}_{\text{hv}}||$ is large.

5.3. High-Frequency Range Analysis

In the range of high-frequency, each signal in Figure 11 no longer has single frequency, but becomes matrix or vector multi-frequency form. In order to simplify the analysis, the sampling process in $\text{H}_{\text{sv}}(\text{s})$ and $\text{H}_{\text{si}}(\text{s})$ is not considered for the time being. The signals of each node can be described as (10):

$$\begin{cases} \vec{i}(\omega) = [i(\omega)\, i(\omega_{\text{s}}-\omega)\, i(\omega_{\text{s}}+\omega)\, \ldots]^T \\ \vec{i}_{\text{ref}}(\omega) = [i_{\text{ref}}(\omega)\, i_{\text{ref}}(\omega_{\text{s}}-\omega)\, i_{\text{ref}}(\omega_{\text{s}}+\omega)\, \ldots]^T \\ \vec{v}_{\text{e}}(\omega) = [v_{\text{e}}(\omega)\, v_{\text{e}}(\omega_{\text{s}}-\omega)\, v_{\text{e}}(\omega_{\text{s}}+\omega)\, \ldots]^T \\ \vec{v}_{\text{g}}(\omega) = [v_{\text{g}}(\omega)\, v_{\text{g}}(\omega_{\text{s}}-\omega)\, v_{\text{g}}(\omega_{\text{s}}+\omega)\, \ldots]^T \\ \vec{v}_{\text{h}}(\omega) = [v_{\text{h}}(\omega)\, v_{\text{h}}(\omega_{\text{s}}-\omega)\, v_{\text{h}}(\omega_{\text{s}}+\omega)\, \ldots]^T \\ \vec{v}_{\text{c}}(\omega) = [v_{\text{c}}(\omega)\, v_{\text{c}}(\omega_{\text{s}}-\omega)\, v_{\text{c}}(\omega_{\text{s}}+\omega)\, \ldots]^T \end{cases} \tag{10}$$

The high-frequency form of (8) can be described as:

$$\begin{cases} \boldsymbol{Y}_{\mathrm{g}} = (\boldsymbol{Z}_{\mathrm{c}} + \boldsymbol{H}_{\mathrm{dig}}\boldsymbol{H}_{\mathrm{ci}}\boldsymbol{H}_{\mathrm{si}})^{-1}(\boldsymbol{I} - \boldsymbol{H}_{\mathrm{dig}}\boldsymbol{H}_{\mathrm{sv}}) \\ \boldsymbol{G}_{\mathrm{ref}} = (\boldsymbol{Z}_{\mathrm{c}} + \boldsymbol{H}_{\mathrm{dig}}\boldsymbol{H}_{\mathrm{ci}}\boldsymbol{H}_{\mathrm{si}})^{-1}\boldsymbol{H}_{\mathrm{dig}}\boldsymbol{H}_{\mathrm{ci}} \\ \boldsymbol{Y}_{\mathrm{h}} = -(\boldsymbol{Z}_{\mathrm{c}} + \boldsymbol{H}_{\mathrm{dig}}\boldsymbol{H}_{\mathrm{ci}}\boldsymbol{H}_{\mathrm{si}})^{-1}\boldsymbol{H}_{\mathrm{dig}} \end{cases} \tag{11}$$

where:

$$\begin{cases} \boldsymbol{Z}_{\mathrm{c}} = \mathrm{diag}(Z_{\mathrm{c}}(\omega)\ Z_{\mathrm{c}}(\omega_{\mathrm{s}} - \omega)\ Z_{\mathrm{c}}(\omega_{\mathrm{s}} - \omega)\ Z_{\mathrm{c}}(2\omega_{\mathrm{s}} - \omega) \ldots) \\ \boldsymbol{Z}_{\mathrm{e}} = \mathrm{diag}(Z_{\mathrm{e}}(\omega)\ Z_{\mathrm{e}}(\omega_{\mathrm{s}} - \omega)\ Z_{\mathrm{e}}(\omega_{\mathrm{s}} + \omega)\ Z_{\mathrm{e}}(2\omega_{\mathrm{s}} - \omega) \ldots) \\ \boldsymbol{Z}_{\mathrm{ce}} = \boldsymbol{Z}_{\mathrm{c}} + \boldsymbol{Z}_{\mathrm{e}} \\ \boldsymbol{Y}_{\mathrm{c}} = \boldsymbol{Z}_{\mathrm{c}}^{-1} \\ \boldsymbol{Y}_{\mathrm{e}} = \boldsymbol{Z}_{\mathrm{e}}^{-1} \\ \boldsymbol{Y}_{\mathrm{ce}} = \boldsymbol{Z}_{\mathrm{ce}}^{-1} \end{cases} \tag{12}$$

The high-frequency form of (6) can be described as:

$$\boldsymbol{H}_{\mathrm{c}} = \begin{pmatrix} (\boldsymbol{I} + \boldsymbol{Y}_{\mathrm{g}}\boldsymbol{Z}_{\mathrm{e}})^{-1}\boldsymbol{G}_{\mathrm{ref}} & (\boldsymbol{I} + \boldsymbol{Y}_{\mathrm{g}}\boldsymbol{Z}_{\mathrm{e}})^{-1}\boldsymbol{Y}_{\mathrm{g}} & (\boldsymbol{I} + \boldsymbol{Y}_{\mathrm{g}}\boldsymbol{Z}_{\mathrm{e}})^{-1}\boldsymbol{Y}_{\mathrm{h}} \\ -\boldsymbol{Z}_{\mathrm{e}}(\boldsymbol{I} + \boldsymbol{Y}_{\mathrm{g}}\boldsymbol{Z}_{\mathrm{e}})^{-1}\boldsymbol{G}_{\mathrm{ref}} & \boldsymbol{Y}_{\mathrm{g}}^{-1}(\boldsymbol{I} + \boldsymbol{Y}_{\mathrm{g}}\boldsymbol{Z}_{\mathrm{e}})^{-1}\boldsymbol{Y}_{\mathrm{g}} & -\boldsymbol{Z}_{\mathrm{e}}(\boldsymbol{I} + \boldsymbol{Y}_{\mathrm{g}}\boldsymbol{Z}_{\mathrm{e}})^{-1}\boldsymbol{Y}_{\mathrm{h}} \\ -\boldsymbol{Z}_{\mathrm{ec}}(\boldsymbol{I} + \boldsymbol{Y}_{\mathrm{g}}\boldsymbol{Z}_{\mathrm{e}})^{-1}\boldsymbol{G}_{\mathrm{ref}} & (\boldsymbol{I} - \boldsymbol{Z}_{\mathrm{c}})\boldsymbol{Y}_{\mathrm{g}}^{-1}(\boldsymbol{I} + \boldsymbol{Y}_{\mathrm{g}}\boldsymbol{Z}_{\mathrm{e}})^{-1}\boldsymbol{Y}_{\mathrm{g}} & -\boldsymbol{Z}_{\mathrm{ec}}(\boldsymbol{I} + \boldsymbol{Y}_{\mathrm{g}}\boldsymbol{Z}_{\mathrm{e}})^{-1}\boldsymbol{Y} \end{pmatrix} \tag{13}$$

The high-frequency solution of (1) can be obtained by bringing (10)–(13) into (9). So far, the harmonic overvoltage analysis of the wide frequency range traction network is integrated by (1).

6. Discussions

Build the simulation model in PSCAD as shown in Figure 1b. The key parameters used in the model are given in Table 3. There are two inherent parallel resonance points (21st- and 36th-order harmonic resonance) in TPSS. The current controller only uses the proportional controller $\mathbf{K}_{\mathrm{ci}}$. Analogous to Figure 10, it is necessary to consider the case where the switching frequency and sampling frequency take different ratios. In case 7, only the ninth-order harmonic current is applied as a reference to simulate the low-order harmonic interference of the voltage outer loop output in Figure 11.

Table 3. List of simulating cases.

Case	K_{ci}	f_{sam}/Hz	f_s/Hz	Low-Order Harmonic	f_{sam}:f_s	Stability	Overvoltage
1	40	1500	6000	N	1:4	N	Y
2	5	1500	3000	N	1:2	Y	N
3	40	1500	1500	N	1:1	N	Y
4	5	1500	1500	N	1:1	Y	N
5	40	1500	750	N	2:1	N	Y
6	40	3000	1500	N	2:1	Y	N
7	40	3000	750	9th	4:1	Y	Y

Simulation results are shown in the last two columns of Table 3. It can be observed that when there is no passive filter on the traction network side, the control of cases 1, 3 and 5 is unstable and the harmonic overvoltage of the traction network occurs at the same time. However, the control of cases 2, 4 and 6 is stable, and no harmonic overvoltage of the traction network occurs. In particular, the system of case 7 is stable, but still produces harmonic overvoltage. When the passive filter on the traction network side is put into operation, all cases have no stability and no harmonic overvoltage problems.

The impedance–frequency curves of Cases 1–4 are depicted in Figure 12. Assuming that $f_{sw} \geq f_{sam}$ in these four cases, the first inherent resonant frequency fr1 of the TPSSs satisfies $0.5f_{sam} < f_{r1} < f_{sam}$, the second one f_{r2} satisfies $f_{sam} < f_{r1}$. These frequencies are outside the frequency range of most existing studies. The simulation results of Figure 12 and Table 3 illustrate that (Case 1 vs. 3, Case 2 vs. 4) increasing the switching frequency has no effect on the input impedance of the converter in the low frequency range, however, this can improve the input impedance characteristics in the high frequency range. The vehicle–grid coupling system will be unstable when the input impedance of the TPSSs port and the input impedance of the grid-connected converter meet the conditions shown in (14).

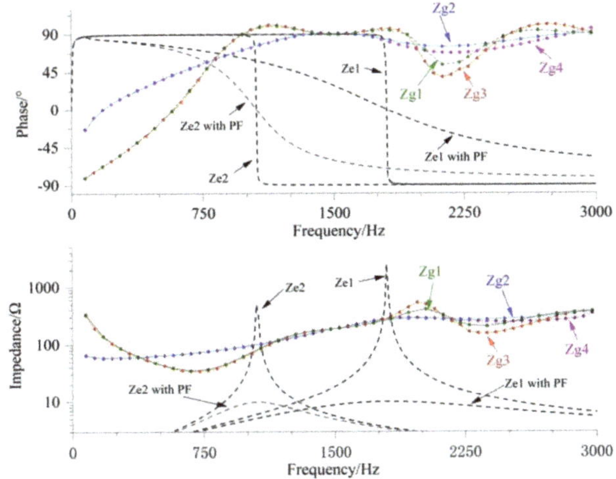

Figure 12. Impedance–frequency curves of Cases 1–4.

After the passive filter is put into use in the traction network, the amplitude of Z_e is reduced sharply, the phase change becomes flat at the same time, and the amplitude and phase conditions required for instability in (14) are destroyed, thus suppressing the harmonic overvoltage of the traction network.

$$\begin{cases} |Z_e| = |Z_g| \\ \mathrm{agr}(Z_e) - \mathrm{agr}(Z_g) + 180^\circ < 0 \end{cases} \quad (14)$$

The simulation result of Case 4 is shown in Figure 13. Before 0.22 s, $\mathbf{K}_{ci}$ = 5. In this case, the condition of (14) is not satisfied, and there is no overvoltage in the traction network. $\mathbf{K}_{ci}$ is increased to 40 at 0.22 s. According to the vehicle–grid impedance frequency characteristics shown in Figure 12, the condition of (14) is satisfied. Control instability occurs in the system, and the harmonic voltage of the traction network is continuously amplified as a divergent waveform. After 0.44 s, when the harmonic voltage of the traction network is amplified to a certain degree, it will not continue to be amplified, showing a steady-state waveform. The reason is that the amplitude limiting nonlinear segment of the control loop reduces the equivalent open-loop gain and forms a nonlinear self-excited oscillation process. In actual cases, nonlinear factors are more complicated. When the passive filter on the traction network side is put into operation after 0.6 s, the harmonic overvoltage disappears instantaneously.

The impedance–frequency curves of Cases 5–7 are depicted in Figure 14 and the simulation result of Case 7 is shown in Figure 15. After applying the 9th-order harmonic disturbance of the current loop at 0.4 s, where analog is the output interference of the outer voltage loop. The high-order harmonic overvoltage appears in the voltage of the traction network, and the frequency of it is the same as the traction network inherent resonant frequency 1050 Hz. According to the impedance curve depicted in Figure 14, this system does not satisfy (14), hence, it is a stable system. However, harmonic overvoltage still occurs because at this time, f_{sam} = 3000 Hz, whereas f_{sw} = 750 Hz. Based on the sideband effect of modulation process mentioned in Section 4.2, the 9th-order low-harmonic will generate a high harmonic interference of 2 × 750 − 450 = 1050 Hz, which exactly coincides with the inherent resonant frequency of the TPSS (see curve Z_{e1} in Figure 12). Therefore, a traction network overvoltage occurs when the current is multiplied by extremely large impedance. The situation that controlled the stable vehicle–grid coupling system generates harmonic overvoltage in the traction network belonging to the second case discussed in (1). In this case, the grid-connected converter can be substituted by a harmonic current source model in the vehicle–grid coupling system.

The limiting link of the control loop, as a nonlinear link, is equivalent to a gain reduction when the modulating wave is overmodulated. In future studies, we should consider the nonlinearity of the transformer hysteresis loop in the actual traction power system, which also has the effect of reducing the gain. In addition, we should consider other nonlinear conditions that occur in the actual line, which is our approach to be carried out in the future in this area.

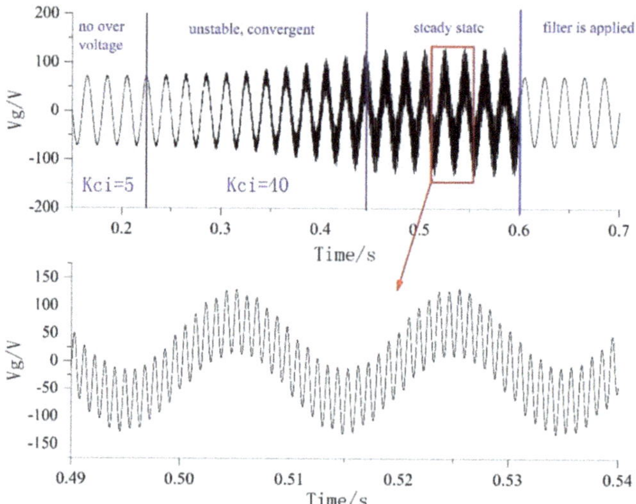

Figure 13. Simulating waveforms of Case 4.

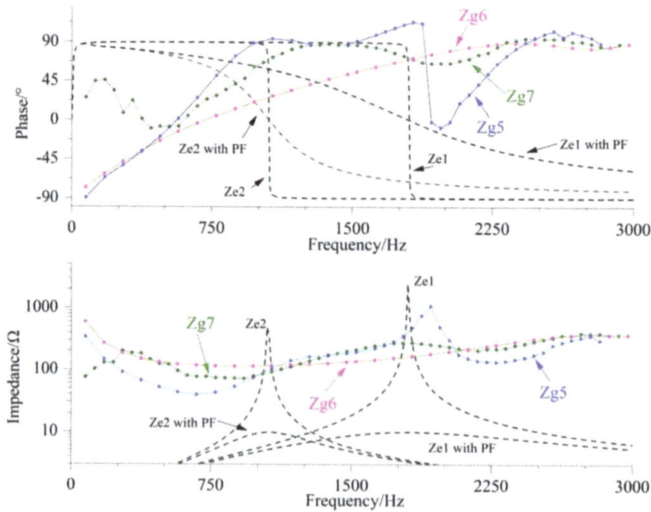

Figure 14. Impedance–frequency curves of Cases 5–7.

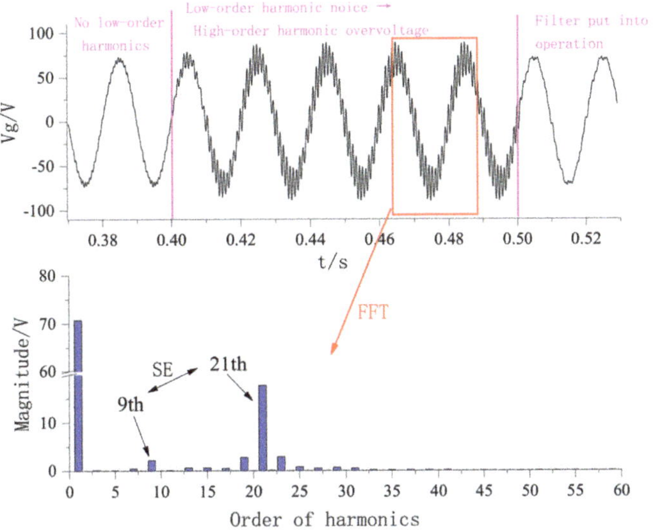

Figure 15. Simulating waveforms of Case 7.

7. Conclusions

In this paper, by studying the IFCs, sampling aliasing effect and modulation of the sideband effect model of the actual TPSSs, as well as considering the relative relationship of the inherent resonant frequency, sampling frequency and switching frequency of the TPSSs, harmonic overvoltage of traction network can be analyzed in a wide frequency range. Equation (1) is proposed to comprehensively elucidate the harmonic overvoltage mechanism of the traction network. The vehicle–grid coupling system instability model and the converter harmonic source model are unified and described in (1). Meanwhile, the application scope, distinction, and connection of these two models are discussed in detail.

Author Contributions: All authors contributed to the research in the paper. Conceptualization, M.W. and Q.L.; formal analysis, Q.L.; methodology, B.S. and Q.L.; project administration, M.W.; supervision, M.W.; validation, Q.Y.; writing—original draft, B.S.; writing—review and editing, T.H. and Q.L.; all the authors have read and approved the final manuscript. All authors have read and agreed to the published version of the manuscript.

Funding: This work was supported in part by the Fundamental Research Funds for the Central Universities (2020JBZD012) and the Project funded by China Postdoctoral Science Foundation (2020M670124).

Conflicts of Interest: The authors declare no conflict of interest.

References

1. Song, K.; Mingli, W.; Yang, S.; Liu, Q.; Agelidis, V.G.; Konstantinou, G. High-Order Harmonic Resonances in Traction Power Supplies: A Review Based on Railway Operational Data, Measurements, and Experience. *IEEE Trans. Power Electron.* **2020**, *35*, 2501–2518. [CrossRef]
2. Rodriguez-Diaz, E.; Freijedo, F.D.; Guerrero, J.G.M.; Marrero-Sosa, J.-A.; Dujic, D. Input-Admittance Passivity Compliance for Grid-Connected Converters With an LCL Filter. *IEEE Trans. Ind. Electron.* **2019**, *66*, 1089–1097. [CrossRef]
3. Wang, Y.; Liu, S.; Liu, J. The stability criterion and damping control strategy for grid-connected inverters and its impact on the global system. *Proc. CSEE* **2020**, *40*, 3008–3020.
4. Du, C.; Du, X.; Zou, X. Impedance modeling and stability analysis of grid-connected modular multilevel converter considering frequency coupling effect. *Proc. CSEE* **2020**, *40*, 2866–2876.
5. Yang, R.; Zhou, F.; Zhong, K. A Harmonic Impedance Identification Method of Traction Network Based on Data Evolution Mechanism. *Energies* **2020**, *13*, 1904. [CrossRef]

6. Zhou, F.; Liu, F.; Yang, R.; Liu, H. Method for Estimating Harmonic Parameters Based on Measurement Data without Phase Angle. *Energies* **2020**, *13*, 879. [CrossRef]
7. European Standard. *Railway Applications—Power Supply and Rolling Stock—Technical Criteria for the Coordination between Power Supply (Substation) and Rolling Stock to Achieve Interoperability*; EN 50388:2012; European Electro Technical Standardization Committee: Brussels, Belgium, 2012.
8. Harnefors, L.; Finger, R.; Wang, X.; Bai, H.; Blaabjerg, F. VSC Input-Admittance Modeling and Analysis above the Nyquist Frequency for Passivity-Based Stability Assessment. *IEEE Trans. Ind. Electron.* **2017**, *64*, 6362–6370. [CrossRef]
9. Freijedo, F.D.; Ferrer, M.; Dujic, D.; Ferrer-Duran, M. Multivariable High-Frequency Input-Admittance of Grid-Connected Converters: Modeling, Validation, and Implications on Stability. *IEEE Trans. Ind. Electron.* **2019**, *66*, 6505–6515. [CrossRef]
10. Liu, M.; Wei, Q.; Xie, S.; Qian, Q.; Zhang, Z.; Xu, J. Multifrequency Impedance Model of Single-Phase Grid-Connected Parallel Inverters for Analysis on Circulating Resonant Current. In Proceedings of the 2019 IEEE Energy Conversion Congress and Exposition (ECCE), Baltimore, MD, USA, 29 September–3 October 2019; pp. 3655–3658.
11. Yang, J.; Liu, J.; Shi, Y.; Zhao, N.; Zhang, J.; Fu, L.; Zheng, T.Q. Carrier-Based Digital PWM and Multirate Technique of a Cascaded H-Bridge Converter for Power Electronic Traction Transformers. *IEEE J. Emerg. Sel. Top. Power Electron.* **2019**, *7*, 1207–1223. [CrossRef]
12. Zhou, F.; Xiong, J.; Zhong, K. Research on the phenomenon of the locomotive converter output current spectrum move based on the coupling of the train net system. *Proc. CSEE* **2018**, *38*, 1818–1825.
13. Cui, H.; Feng, Y.; Lin, X. Simulation study of the harmonic resonance characteristics of the coupling system with a traction network and AC-DC-AC trains. *Proc. CSEE* **2014**, *34*, 2736–2745.
14. Liu, Q.; Wu, M.; Li, J.; Yang, S. Frequency-Scanning Harmonic Generator for (Inter)harmonic Impedance Tests and Its Implementation in Actual 2 × 25 kV Railway Systems. *IEEE Trans. Ind. Electron.* **2020**. [CrossRef]
15. Liu, Q. Research on Impedance-Frequency Characteristic Test Technology of Traction Power Supply Systems. Ph.D. Thesis, Beijing Jiaotong University, Beijing, China, 2018.
16. Mouton, H.D.T.; McGrath, B.; Holmes, D.G.; Wilkinson, R.H. One-Dimensional Spectral Analysis of Complex PWM Waveforms Using Superposition. *IEEE Trans. Power Electron.* **2014**, *29*, 6762–6778. [CrossRef]
17. Hans, F.; Oeltze, M.; Schumacher, W. A Modified ZOH Model for Representing the Small-Signal PWM Behavior in Digital DC-AC Converter Systems. In Proceedings of the IECON 2019 45th Annual Conference of the IEEE Industrial Electronics Society, Lisbon, Portugal, 14–17 October 2019; Volume 1, pp. 1514–1520.
18. Liu, Q.; Wu , M.; Li, J.; Zhang, J. Controllable harmonic generating method for harmonic impedance measurement of traction power supply systems based on phase shifted PWM. *J. Power Electron.* **2018**, *18*, 1140–1153.

Article

Detection of Harmonic Overvoltage and Resonance in AC Railways Using Measured Pantograph Electrical Quantities

Andrea Mariscotti [1,*] and Leonardo Sandrolini [2]

1 Department of Electrical, Electronics and Telecommunication Engineering and Naval Architecture (DITEN), University of Genova, 16145 Genova, Italy
2 Department of Electrical, Electronic and Information Engineering "Guglielmo Marconi" (DEI), Alma Mater Studiorum, University of Bologna, 40136 Bologna, Italy; leonardo.sandrolini@unibo.it
* Correspondence: andrea.mariscotti@unige.it

Abstract: Harmonic resonances are part of the power quality (PQ) problems of electrified railways and have serious consequences for the continuity of service and integrity of components in terms of overvoltage stress. The interaction between traction power stations (TPSs) and trains that causes line resonances is briefly reviewed, showing the dependence on infrastructure conditions. The objective is monitoring of resonance conditions at the onboard pantograph interface, which is new with respect to the approaches proposed in the literature and is equally applicable to TPS terminals. Voltage and current spectra, and derived impedance and power spectra, are analyzed, proposing a compact and efficient method based on short-time Fourier transform that is suitable for real-time implementation, possibly with the hardware available onboard for energy metering and harmonic interference monitoring. The methods are tested by sweeping long recordings taken at some European railways, covering cases of longer and shorter supply sections, with a range of resonance frequencies of about one decade. They give insight into the spectral behavior of resonances, their dependency on position and change over time, and the criteria needed to recognize genuine infrastructure resonances from rolling stock emissions.

Keywords: electrified railway; Fourier transform; harmonic resonance; overvoltage; power quality; traction power supply

Citation: Mariscotti, A.; Sandrolini, L. Detection of Harmonic Overvoltage and Resonance in AC Railways Using Measured Pantograph Electrical Quantities. *Energies* **2021**, *14*, 5645. https://doi.org/10.3390/en14185645

Academic Editor: Tek Tjing Lie

Received: 14 July 2021
Accepted: 2 September 2021
Published: 8 September 2021

1. Introduction

Railways are being used worldwide as an efficient and effective transportation infrastructure for people and goods, both long distance and within an urban context. Peculiarly, the traction supply arrangement differs from three-phase industrial networks in several aspects: AC railways are single phase operated at the medium voltage (MV) level, they have a physical extension typical of a transmission network but separated into smaller sections for most AC railways, the traction line feeds power to distributed moving loads (the trains), and the interaction between traction power stations (TPSs) and trains causes a variety of power quality (PQ) phenomena that are relevant to internal operation and disturbance to third parties [1]. In particular, as mainly focused on by the EN 50388 standards [2,3] regarding the interaction between rolling stock and the supply traction line, harmonic distortion and, in general, conducted emissions may cause local disturbances to signaling and control devices as well as supply line distortion, instability, and overvoltages. The first have been subject to extensive research for PQ indexes and criteria [1], as well as compensation strategies and implementations [4–8]. The latter are the objectives of recent research aimed at defining the conditions for line instability, harmonic resonances, consequences in terms of excessive distortion, and insulation breakdown [9–13].

Supply traction lines of some tens of km or longer, with significant capacitive loading due to MV cables, transformers, and trains (with their own onboard devices), exhibit a wide range of resonance phenomena, occurring in the frequency interval of a hundred of Hz up to some kHz, depending on the supply arrangement and length [14]. Traction systems in use today are mostly operated at 25 kV 50/60 Hz and 15 kV 16.7 Hz with different supply schemes. The supply sections of the latter are much longer, with extensive earthing bringing the resonant frequency of the line in the lower part of the considered frequency range, namely one or a few hundred Hz. The 2 $\times$ 25 kV systems have shorter supply sections with resonances in the kHz range [15–21], according to recent investigations for a series of resonance incidents [1,9]. As shown in [9,22], in a real scenario the response of the line may be quite complex.

The behavior of the traction line impedance with fixed feeding points (the TPSs) and moving loads (the trains) was initially analyzed, among others, in [14], proposing simple one-dimensional models based on transmission-line theory and indicating the high-level conditions for line resonance, as recalled by [9]. Basically, a line resonance occurs when the inductive and capacitive reactance terms with opposite signs are nearly equal in amplitude and compensate, leading to extreme conditions of very small or very large terminal impedance, depending on whether such elements are series or parallel connected. As for elementary resonant circuits, series and parallel resonances lead to situations of current or voltage amplification at the resonant frequency.

A real system is a more complex and articulated combination of series and parallel connected elements. The overall traction line can be subdivided into different sub-circuits that are relevant to define and analyze propagation of harmonics and coupling onto a range of affected systems. The harmonics propagate along the pantograph and catenary system and along the return circuit, pertaining to the so-called hot and cold paths. The cold path is relevant for induced and conducted disturbance onto signaling and communication systems that share the track with the traction supply circuit; the studied circuit must include an accurate model of the return circuit, of common to differential signal transformation, and of local resonances. The hot path, conversely, concerns mainly the traction supply and overhead distribution system with an overall return along the return circuit and earth; the studied circuit is relevant for distortion shared by the TPSs and trains, and propagated back onto the high voltage feeding line and then into the public grid.

Resonances can occur between the elements interconnected by the long traction line over a broad frequency range, concretizing as low-frequency and high-frequency oscillations (identified as LFOs and HFOs, respectively). Recalling the EN 50388 standards [2,3] and the interpretation given in [9], it may be said that LFOs are related to system instability [23], mostly caused by delay and phase rotation along the line [24] and interaction of active controls onboard rolling stock, applying power factor compensation [25,26]. This led to the catastrophic blackout of the Swiss network in April 1995 (as reported in [25,27]) and significant research activity among the infrastructure owners, manufacturers, and scholars. A comprehensive list of incidents caused by network resonances is reported in [1] and a detailed list of those which occurred in China is reported in [9].

HFOs, along with causing overvoltages, are sometimes interpreted as a PQ problem, where network resonances are excited by the rolling stock harmonic emissions and by transient phenomena, such as the electric arc at pantograph [28]. Network resonances are of course subject to variability depending on loading and the relative position and the type of trains in the supply section, although the theory of the one-dimensional transmission line suggests that main resonances do not depend on train position [9,14]. Such slight variability, in particular for secondary resonances, was observed during tests carried out in a well-documented system, the test ring at Velim in Czechia, where the influence of train position every 500 m, including traction supply cables, was investigated [29]. The longer the traction supply section between two TPSs, the lower the main resonant frequency, as theoretically demonstrated in [14] and taken up and commented by [9].

The consequence of parallel resonances occurring in the hot path is an increase in the distortion injected back into the public grid and, most of all, line voltage amplification, with potentially catastrophic effects:

- excessive instantaneous voltage may trigger TPS overvoltage protections, possibly causing cascaded tripping and the collapse of a large portion of the railway's electric network,
- overvoltage stress in terms of single or repeated overvoltages of varying intensity that may cause the failure of surge arresters, likely resulting in a short circuit through the failed unit and a sudden TPS out-of-service due to protection tripping.

Series resonances may also occur [15], where the propagation of distortion involves the TPS and the upstream network, resulting in an increased distortion injected into the public grid. Sainz et al. traced correspondences between these resonances and the zeros of the pantograph impedance Z_p. In general, major series resonance should occur in the lower part of the HFO range, where the series resistance is lower and the risk of excitation by low-order harmonics is higher, and consequences are thus worse.

As HFOs are caused by the resonance of TPS and the line while accounting for distributed and lumped capacitance terms, and are excited by rolling stock emissions, two approaches may be identified for harmonic and resonance suppression: ground-based suppression and on-board suppression [9]. Among ground-based suppression techniques, use of passive and active filters installed at TPSs is the most common solution [30], although some network's reconfigurations may also be considered in order to shift the resonance frequency and reduce the factor of merit. On-board suppression may be achieved by installing passive filters, which may be exposed to excessive stress when an entire line section with resonance excited by nearby trains occurs [9]. On-board suppression most often relies on the modification of the converter modulation, rather than a revolutionary change of topology [31].

Converter control can be initiated and adapted if continuous monitoring of the pantograph quantities and of incipient line resonances is provided. Similarly, critical situations and the necessity of suppression implementation for a given line may be assessed if pantograph quantities are available and evaluated. In fact, this work discusses practical conditions for detection of resonance conditions from a railway vehicle perspective, using information available at the pantograph electric interface. Other measurement techniques have been proposed in the past [9,30,32], using purposely developed equipment located at TPS. The focus here instead is on the observability of resonance phenomena from the pantograph interface, to provide a distributed monitoring system that can be, in principle, installed onboard all the trains of the network.

The work is structured as follows: Section 2 describes the quantities and conditions for resonance to occur; Section 3 goes into the details of the detection and interpretation of harmonic resonances; Section 4 reports the results obtained from measured data collected during test runs in Switzerland and France, thus covering both 16.7 Hz and 50 Hz traction supply systems.

2. Network Resonance

As shown in [9], the system may be analyzed by means of multiconductor transmission line (MTL) equations, representing the overall network as a meshed set of series and parallel connected branches, to which lumped circuits (such as the TPS and the train as well as auxiliary transformers, including the pole mounted step-down transformers considered in [16]) may be attached at nodes. Although often neglected, MV connecting cables between TPS and the traction line may give a significant additional capacitance, which can bring the "natural" resonance of the supply section to a lower frequency (see considerations on the relevance of such parameters for exact fitting of measured frequency response in [29]). Similarly, adding capacitance to the train, as for the roof's HV cable joining the two pantographs or a separate capacitor for arcing emission reduction [28], reduces the resonance frequency.

For the analysis of the hot path there is in general no need of a detailed model of the conductors forming the return path (the "cold path"); conversely, studying track-connected signaling circuits requires attention to track balance and rail-to-rail and rail-to-earth parameters. The hot path is often studied using simplification of conductors at the same potential into equivalent conductors [33], as considered also in [9] (see refs no. 92 and 93 there). However, the accuracy of the simplification worsens with increasing frequency [33,34], so that while this approach may be used for load flow studies and electromechanical simulation, it could lead to errors when frequencies of one to some kHz are considered, as they are for HFOs.

Network resonances depend on the studied transfer function and may be different if considering different quantities:

- the TPS voltage to evaluate the occurrence of overvoltages and the impact of TPS equipment,
- the line current flowing through the TPS transformer, as a PQ phenomenon impacting on the feeding network upstream,
- the pantograph voltage as a measure of parallel resonance along the line and the chance of interfering with the operation of the onboard converters,
- the pantograph current providing direct information on the exchanged active and reactive power and disturbance to track signaling, if interpreted as a return current.

Considering, in particular, the effect of the additional capacitance provided by the high-voltage cables wayside and onboard, the local train-based measurement of line resonances becomes more relevant.

Focusing on the source of the distortion exciting the traction line resonances and on the available measurements from the locomotive (or electro-train), the impedance is modeled at the pantograph-line interface, as carried out in [9,14]. The simplified schematic is shown in Figure 1.

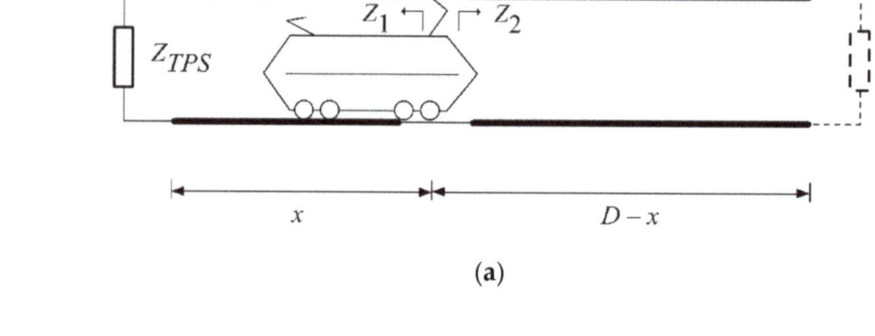

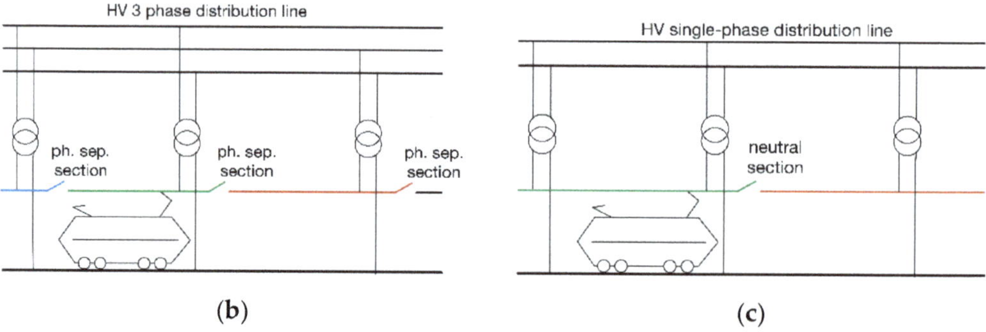

Figure 1. (**a**) Simplified equivalent circuit for the AC traction line seen at the pantograph port; (**b**) single-side supply scheme typical of 25 kV 50/60 Hz systems; (**c**) portion of an interconnected network typical of 15 kV 16.7 Hz systems.

The frequency response of a railway line depends first of all on its length, and in particular on the length of the supply section, when this is electrically separated from the adjacent sections. The 50/60 Hz systems have the necessity of strict separation of each supply section, being derived from the national grid by loading alternatively two of the three phases (each supply section has thus a phase rotation of 120° electrical degrees); this is achieved by the phase separation sections, more commonly altogether named "neutral sections". Conversely, 16.7 Hz systems have in several countries (such as Switzerland) a dedicated single-phase distribution and transmission network upstream, so that a phase separation is not, in principle, necessary. Then, mainly exigencies of network stability and continuity of service require the use of some neutral sections, more separated than for 50/60 Hz systems. To fix the ideas, the length for a 50/60 Hz high-speed line is in the order of 40 km, whereas for 16.7 Hz systems it may be in the order of a hundred km.

From the point of view of HFO modeling, this has two consequences:

- A longer supply section implies that HFOs begin in principle at a lower frequency, although a direct proportionality between length for the two systems is not correct, as there are differences in the traction conductors and per-unit-length parameters.
- The ideal equivalent circuit arrangement for 50/60 Hz and 16.7 Hz systems differs in that the former always has one TPS at one end of the line and the other end is left floating against the neutral section, whereas for the latter there may be configurations with more than one TPS with a piece of line terminated on a TPS at each end, without electrical separation. These two cases are shown below for completeness.

The impedance resulting from the parallel combination of the left and right sections at the pantograph connection point gives the following known expressions, having assumed a typical single-side supply scheme, as in 25 kV systems (see Figure 1a):

$$Z_1 = Z_c \frac{Z_{TPS}\cosh(\gamma x) + Z_c \sinh(\gamma x)}{Z_{TPS}\sinh(\gamma x) + Z_c \cosh(\gamma x)} \quad Z_2 = Z_c \frac{1}{\tanh(\gamma(D-x))} \tag{1}$$

$$Z = Z_1 // Z_2 = Z_c \cosh(\gamma(D-x)) \frac{Z_{TPS}\cosh(\gamma x) + Z_c \sinh(\gamma x)}{Z_{TPS}\sinh(\gamma D) + Z_c \cosh(\gamma D)} \tag{2}$$

The resonance condition results from:

$$Z_{TPS}\sinh(\gamma D) + Z_c \cosh(\gamma D) = 0 \quad \gamma D = \tan\mathrm{h}^{-1}\left(-\frac{Z_c}{Z_{TPS}}\right) \tag{3}$$

Now, in [9] a purely inductive Z_{TPS} is assumed and the tanh( ) function is simplified with its own argument, so that (3) reduces to:

$$j\omega L_{TPS} = -\frac{Z_c}{\gamma D} = -\frac{1}{j\omega C} \tag{4}$$

having indicated with C the total capacitance of the line. The first objection may be that the inductance of the line has disappeared, but neglecting the shunt conductance as sensible for the overhead conductors of the traction supply, the $r + j\omega l$ term in $Z_c = \sqrt{(r + j\omega l)/j\omega c}$ and in $\gamma = \sqrt{(r + j\omega l)(j\omega c)}$ simplify in the fraction $Z_c/\gamma D$ (small letters indicate per-unit-length parameters).

For a system without phase separation points and longer supply sections separated by some neutral sections, as in the case of 16.7 Hz railways (shown in Figure 1c), the two-terminal impedance condition at many locations are those of Z_{TPS} at each side. This is a simplification that neglects the two adjacent line sections beyond the considered TPSs that are assumed to represent a low-impedance termination.

$$Z_1 = Z_c \frac{Z_{TPS}\cosh(\gamma x) + Z_c \sinh(\gamma x)}{Z_{TPS}\sinh(\gamma x) + Z_c \cosh(\gamma x)} \tag{5}$$

$$Z_2 = Z_c \frac{Z_{TPS}\cosh[\gamma(D-x)] + Z_c \sinh[\gamma(D-x)]}{Z_{TPS}\sinh[\gamma(D-x)] + Z_c \cosh[\gamma(D-x)]} \tag{6}$$

$$Z = Z_1//Z_2 = \frac{Z_c(Z_{TPS}\cosh[\gamma(D-x)] + Z_c\sinh[\gamma(D-x)])(Z_{TPS}\cosh(\gamma x) + Z_c\sinh(\gamma x))}{2Z_cZ_{TPS}\cosh(\gamma D) + (Z^2{}_c + Z^2{}_{TPS})\sinh(\gamma D)} \quad (7)$$

$$\gamma D = \tan\mathrm{h}^{-1}\left(-\frac{2Z_cZ_{TPS}}{Z^2{}_c + Z^2{}_{TPS}}\right) \quad (8)$$

At resonance, inductive and capacitive terms compensate, leaving a line impedance with resistive behavior [35,36]. The resistive term should then be determined quite accurately, and the skin effect should be taken into account for a correct estimate of damping and height of the impedance peaks. Similarly, the depth of anti-resonance (or series resonance) points depends on the series resistance, chiefly influenced by skin effect in the traction supply conductors. For traction line conductors, skin effect is prominent in the running rails [37,38], whereas overhead conductors are not appreciably affected, being made of metals with negligible magnetic permeability in addition to their smaller cross section. Transformers at TPS and on-board can increase the overall line loss and may introduce some proximity effect within their windings; this explains the fact that stray inductance (that by definition represents the uncoupled magnetic field in air and is a linear term) is slightly frequency dependent.

3. Resonance Detection

Network impedance and resonance measurement methods may be broadly classified in active and passive methods, the former actively applying test signals to the network, whereas the latter listens passively to the accessible electrical quantities [39–41]. Active methods are invasive, necessitate precautions to interface the test signal generator to the high-voltage railway traction supply [9] and are more suited for ground installation, rather than onboard. Measuring line impedance and resonance effects without injecting a test signal, but exploiting the accessible electrical quantities at the pantograph interface, has the drawback instead of increased noise, incoherence between various spectral components of the same quantity, and a more jagged and noisy impedance curve.

The selected approach for onboard implementation is a passive method measuring electrical quantities at pantograph and mimicking an expert's behavior when observing an ideal display that contains a set of signal characteristics extracted from the original input quantities. Such characteristics correspond to the spectrum components of the pantograph voltage and current, together with other derived quantities (impedance and power terms). Investigated criteria regard the abnormal increase in some spectrum components, together with their specific behavior vs. time. The approach considers the analogous behavior of adjacent spectrum components in order to discard isolated components, possibly caused by onboard power converters, that do not match the assumption of a limited factor of merit and broader resonance peaks.

3.1. Selected Pantograph Quantities and Basic Fourier Analysis

With the objective of detecting incipient resonances in real time using on-board instrumentation, as discussed in Section 2, the quantities that are accessible at the pantograph interface are V_p and I_p, the frequency domain spectra of the pantograph voltage and current, obtained from the corresponding v_p and i_p time waveforms. The same approach can then be transferred to the TPS, where the available quantities are the line voltage V_l and each feeder current $I_{f,k}$, or the total line current I_l.

Spectra are calculated with a short-time Fourier transform (STFT) approach, with care to use $P \geq 2$ periods of the fundamental f_1 for demonstration purposes to set a frequency resolution $df = f_1/P$ that has the following advantages:

- it attenuates rapid transients lasting less than one cycle,
- it displays well even and odd harmonics with at least one intermediate bin between each of them; we should remember, however, that odd harmonics prevail and have a

behavior coherent with the train's operating conditions, so that they are even more separated and exempt from significant spectral leakage from the adjacent components,
- it allows the use of tapering windows with a broad main lobe (such as flat-top), as the resulting reduction in frequency resolution is less than a factor of 2 anyway with respect to the implicit rectangular window.

In any case, the signals can be analyzed with a percentage of overlap $p \geq 50\%$, ensuring a time resolution down to the fundamental period without hindering the desirable real-time response and frequency resolution.

3.2. Resonance Conditions

Resonance of an electrical circuit is defined as the situation at a given frequency where inductive and capacitive reactance in a loop are equal in magnitude and opposite in sign so that they cancel each other, resulting in an exchange of energy between the two, stored alternatively in the magnetic field of the inductive parts and the electric field of the capacitive parts. The resulting oscillation is internal to the loop and the externally visible effect is that of extremely small or large resistive impedance for series and parallel resonances, respectively.

A HFO condition occurs at the maxima of pantograph impedance Z_p as a parallel resonance; the line voltage resonance then occurs if there corresponds a significant current excitation close to the identified resonance frequency. A series resonance occurs at the zeros of the same Z_p, causing a maximization of the current flowing back down to the TPS, and possibly upstream. In both cases, the resistance of the return circuit due to the skin effect in the running rails [37,38] plays a major role and reduces the factor of merit, especially when observing the increase in minima and reduction in maxima of the Z_p curve.

Practically speaking, with values of the factor of merit Q in the order of 10, the bandwidth δf around the ideal resonance frequency f_r, which is proportional to $1/Q$, so about 10%, will represent an interval with non-negligible width, bracketing more than one harmonic component, when f_r is of the order of magnitude of one or more kHz. This is confirmed by the voltage spectra shown in [18,20,21] and discussed later in Section 4.

$$\delta f = f_r \ / \ Q \tag{9}$$

The lower the Q factor, the lower the peak at resonance and the less relevant the effect of such resonance and the necessity to detect it.

At resonance (parallel or series one) the reactive components compensate and the net resulting impedance has a real value. This condition translates into the voltage V_h and current I_h phasors at the resonance frequency being in phase.

From a harmonic power flow point of view this is equivalent to a cancellation of the harmonic reactive power term Q_h at the said resonance frequency, maximizing the harmonic active power fraction. Remember that the term "harmonic reactive power" results from the direct multiplication of voltage and current phasors at the same frequency, neglecting distortion power terms resulting from mixed multiplication of voltage and current phasors at different frequencies [35,36]. It is not meant by this that the active power term P_h is maximum overall, but its fraction taken with respect to the total harmonic apparent power $A_h = \sqrt{(P_h{}^2+Q_h{}^2)}$ (namely the harmonic displacement factor $d_h = P_h/A_h$) is. This is clear as all low-order harmonics are characterized by large values of apparent power and consequently, in proportion, active power as well, as demonstrated in [42]. The use of d_h allows a normalized weighting of the active power flow at all frequencies without problems of scale, with which resonance conditions may be characterized and identified.

What is also observed is that the harmonic power flow at the resonant frequency is prevalently active (theoretically only active power would flow in relation to circuit losses). As the analysis is carried out at the characteristic harmonics of the traction supply fundamental that might not coincide with the observed resonant frequency, the cancellation of the harmonic reactive power terms may be only approximate.

However, cancellation of reactive power at a given harmonic is a necessary, but not sufficient, condition for identification of network resonances, series or parallel. In fact, there are transient situations, as identified in the polar plots of the harmonic active and reactive power components in [42], for which the reactive power term may be temporarily very small with active power prevailing. This behavior was investigated in [43] to identify suitable PQ source indicators, focusing in particular on the sign and intensity of the active power indicator.

The harmonic active power P_h has thus proven itself as a valid indicator of power flow and behavior of harmonic power sources, as well as of resonance conditions, with a better discernibility than for voltage and current quantities alone. As the product of the two quantities, it has better scale properties than their quotient, i.e., the pantograph impedance Z_p.

The transient conditions mentioned, in any case, last namely for a limited amount of time and are often related to the train's operating conditions, whereas network resonances are more persistent and may depend only slightly on train position. Thus, an additional criterion to reject the vast majority of such transients is that the identified resonance condition last longer than the typical transient duration, that may be assumed to be some seconds, based on practice [42].

In addition, recalling the considerations on the factor of merit Q, a network resonance will affect several frequency bins at which there will be a significant increase in the harmonic power factor, whereas a loco emission is often limited to one or few harmonics.

From the definition of HFO it is understood also that a network parallel resonance as such should be accompanied by some amount of voltage amplification. Series resonances instead should be characterized by low voltage components and correspondingly current amplification.

3.3. Detection Criteria

Regarding criteria for detection, all harmonic components and the basic frequency resolution may be used, as well as grouped harmonics and a band-pass approach.

The most direct approach is based on detecting an excessive distortion of V_p that would cause the increase in the peak and rms values of the waveform. Alternatively, the attention may be focused on the peak and rms values of the waveform in an attempt to avoid the Fourier analysis. The peak value is nevertheless exposed to transients that would be detected as false positives; the rms value instead requires a calculation whose complexity is approximately that of a FFT implementation. In addition, the use of harmonic spectra allows the implementation of additional criteria, as the presence of adjacent harmonics and harmonic grouping.

Therefore, a time interval where some distortion threshold is exceeded is marked as a first candidate for resonance detection, although the excessive voltage distortion may be caused by an excess of current distortion due, e.g., to a particular operating point of the rolling stock power system. Confirmation comes from the corresponding assessment of current distortion at the same harmonics, verifying that it does not exceed a suitable threshold. This is equivalent to the verification of a sufficiently large Z_p to justify the increased distortion observed in V_p, in turn not caused by an increase in I_p.

A set of rules is given identifying the criteria for the identification of resonance conditions, HFO, and their time behavior.

Rule 1: a resonance condition configures around the peaks of Z_p, the local maxima of the network impedance curve at the pantograph interface; frequency bins satisfying such a condition form the set $H_{r,1}$ (set of indexes h satisfying rule 1).

Rule 2: HFO is triggered if the rolling stock has emissions exciting the resonance and if this is visible in the V_p spectrum, peaking around the resonance frequency.

Rule 3: an incipient HFO situation with an increase in a voltage component V_{p,h^*} can be prevented, once distinguished from a momentary increase in a current component I_{p,h^*}

(e.g., caused by a transient operation of onboard converters), for which the ratio $Z_p = V_p/I_p$ is taken into account.

In general, the identification of the resonance is made on a semi-quantitative basis, where the detection of a fractional increase in Z_p and V_p is sufficient (as it will be demonstrated in Section 5 with experimental data); the uncertainty requirement for the measurement of the pantograph quantities is thus not demanding and fits existing transducers and instrumentation already installed for monitoring purposes.

As observed in practice, converters' emissions in the kHz range may be accompanied by a slight increase in Z_p, due, for example, to the inductive behavior of the onboard transformer through which such emissions flow. However, such an increase is limited to a few spectral lines, so that it might be distinguished from a HFO due to the apparent larger Q factor. To make the interpretation of the Z_p spectrum easier and less prone to errors, a robust check is carried out by combining the Z_p spectrum with the distribution of d_h values, resulting in the filtered impedance Z'_p. A convenient threshold $d_{thr} = 0.9$ was determined with some trial and error, with respect to the discernibility and interpretability of the resulting graphs, but also inspired from the often adopted value of 0.9 as a limit for the fundamental displacement factor (or power factor in general) in AC railways and distribution networks. Such threshold value, as commented below, may be slightly increased [36,42], but its physical meaning remains; it cannot be too close to unity as the frequency resolution limits the capability of capturing the ideal reactive power cancellation at resonance.

Rule 4: a HFO is confirmed if $d_h \geq d_{thr}$ ($d_{thr} = 0.9$ is a convenient threshold, but other values, maybe slightly larger, are possible); frequency bins satisfying such condition form the set $H_{r,4}$ (set of indexes h satisfying rule 4).

$$Z'_p = Z_p \times (d_h \geq d_{thr}) \tag{10}$$

The intersection of sets $H_{r,1}$ and $H_{r,4}$ indicates the frequency interval that satisfies both rules 1 and 4: $H_r = H_{r,1} \cap H_{r,4}$. This effectively removes many extraneous points and makes easier the interpretation of 2-D STFT spectrum of Z'_p.

The confirmation of a HFO condition with parallel resonance comes from a local increase in voltage distortion components. To this aim, the V_p spectrum is scaled by normalizing it with respect to the fundamental value at the same time instant ($\hat{V}_{p,h}$), and for an exigency of scale, only harmonic values are displayed, discarding the fundamental and the larger low-frequency components.

$$\hat{V}_{p,h} = V_{p,h}/V_{p,1} \tag{11}$$

A further confirmation, as anticipated at the end of Section 3.2, is the time duration for which the same hypothetical resonance condition (holding all rules 1 to 4 so far) persists for a preset time interval.

Rule 5: a HFO is confirmed if the bins in the set H_r defined above remain in the set for a sufficient time duration set to $T_{H,\min}$.

From $\hat{V}_{p,h}$, it is possible to evaluate the overall distortion as total harmonic distortion (THD), just by taking the rooted sum of squares:

$$THD = \sqrt{\sum_{h>1} \left(\hat{V}_{p,h}\right)^2} \tag{12}$$

This approach is in line with [18], but it is ineffective in practice if the real behavior of the AC railway and rolling stock is considered:

- Low-order harmonics are ubiquitous in AC railways [1,4,42], and may or may not be produced by the rolling stock used for tests, depending on the type of the on-board converters; modern 4QCs (four-quadrant converters) are not, in general, a source of low-order harmonics. Low-order harmonics have the largest amplitude of all voltage spectrum components in AC railways, and they would mask the effect in the overall THD of higher-order components at resonance.

- High-order components at HFO frequencies do not always correspond to the main emission components of rolling stock, but are excited by lesser distortion components, such as some of the lateral bands of 4QC emissions.

For a scale problem (avoiding the influence of low-order harmonics) and for selectivity with respect to the emission patterns of various types of rolling stock, it is thus advisable to limit the calculation of THD to frequency bands, starting from a conveniently large minimum frequency and with an extension that preserves some accuracy and sensitivity for detection of incipient resonance conditions. This concept corresponds to the proposal of evaluating harmonics and supraharmonics in power systems implementing a wavelet bank [44] or, equivalently, the ripple of DC grids in bands, using intervals for the STFT indexes or, equivalently, a bank of pass-band filters [45].

$$THD_i = \sqrt{\sum_{h \in H_i} \left(\hat{V}_{p,h}\right)^2} \tag{13}$$

Such bands H_i may have an extension of some or several hundred Hz, that should be selected taking into account that the two AC railways with different fundamental frequency will populate differently each interval, with a denser harmonic sequence for 16.7 Hz systems. It is in fact unavoidable that 50/60 Hz components are more spaced apart and contribute less terms within the same bandwidth; a minimum number of harmonics per frequency band should be decided to bracket the whole harmonic group of a 4QC emission (whose spread as per pulse-width modulation theory depends on the output frequency).

Conversely, resonances are related to the geometry and electrical characteristics of the infrastructure and not to the fundamental frequency, for which the same band representation would fit both systems. From this, it is evident that a meaningful and effective representation of voltage distortion must trade off between these two exigencies.

In the following, the index $i = 0$ will indicate the THD for the first band between the fundamental and 500 Hz, selected as a convenient frequency value to separate low-order distortion from the first emission components of modern onboard converters. All other bands H_i are numbered consecutively starting from $i = 1$.

Frequency-limited harmonic distortion profiles calculated using real measured data are rarely smooth, as they collect adjacent spectrum components of mixed origin. Some deal of numeric smoothing is thus necessary to ease readability, as will be shown at the end of Section 4.

4. Results and Discussion

Long data records taken along some AC railway systems are considered for the verification of the rules and conditions discussed so far. The considered systems represent modern railways with quite different topologies, as described in [46], where the origin and the characteristics of data are also clarified:

- 15 kV 16.7 Hz system (Switzerland) with a passenger train in normal service hauled by a Re460 locomotive (nominal power about 6 MW, single pantograph), traveling at commercial speed of about 130 km/h and frequent stops,
- 2 × 25 kV 50 Hz system (France), featuring a large-power high-speed train (TGV Dasye) with almost double nominal power (about 12 MW), double pantograph and higher speed (about 250 km/h). Power demand in the French system is quite large and in some cases this leads to the installation of additional substations or booster solutions to compensate voltage drops, reducing the length of supply sections.

The collection through one or two pantographs does not influence the quantities subject to the present study. Pantograph and voltage current waveforms may occasionally be affected by arcing, whose effects disappear as long as spectra are the result of a calculation over many periods, or they are subject to averaging during post-processing; neutral sections instead have a clear impact as the two pantograph quantities drop to zero. Other mechanical oscillations have no direct effect on the analysis, except for the mentioned occasional arcing. For the double-pantograph TGV train, the overall current I_p is the result of the sum of the two individual pantograph currents I_{p1} and I_{p2}.

The measurement system was described in [47], together with a quantification of the uncertainty that is summarized in Table 1.

Table 1. Metrological characteristics of V_p and I_p sensors (Adapted with permission from ref. [47]. 2021 IEEE).

Metrological Performance	**V_p and I_p Channels**		
	I_p (Rog1)	**I_p (Rog2)**	**V_p (Cap. div.)**
Sensitivity	1.2/9 mArms	<10 mArms	<4 Vrms
Full scale	300/3000 Arms	>1000 Arms	>100 kVrms
Uncertainty (k = 1)	<3% rel. for 10 × sens. values	<1.4% full scale	<1.5% relative

Note: k indicates the coverage factor of uncertainty, that is, the multiplying factor of the estimated sample dispersion.

The contactless capacitive voltage divider was made of a helical winding making the secondary plate of the HV capacitor, with the first plate represented by a bare pantograph part at catenary voltage; such a device is intrinsically linear and has a large bandwidth, and its limitations are in the signal conditioning circuit that is discussed in [47]. The two Rogowski coils are derivative circuits corrected by the respective electronic integrator and signal conditioning circuitry. In each case, the uncertainty reported in Table 1 was estimated considering the characteristics of the signal conditioning circuitry and the available full-scale values.

The use of the condition of *Rule 4* allows for exclusion during post-processing points in the time-frequency space that are not relevant from a resonance-tracking viewpoint. Such points in the following graphs are excluded and their position is set as white as the background (this was implemented using the feature NaN, "not a number", in Matlab).

Three cases are considered in the following figures: a 1200 s run on the 16.7 Hz system in Figure 2, an 850 s run on the 50 Hz system in Figure 3, and a zoom of a 16.7 Hz system situation that shows a time-varying anti-resonance. In each figure the information from top to bottom, left to right is displayed with the following scheme:

- a diagram with voltage (black), current (green), and active power (red) profiles, scaled to accommodate them compactly in the same graph,
- two 2-D plots versus time and frequency of the harmonic impedance Z_p and harmonic displacement factor d_h using color-coded intensity,
- two post-processed 2-D plots, where the filtered impedance Z'_p and normalized harmonic voltage $\hat{V}_{p,h}$ are shown, using two different color maps for a matter of easy discernibility with the NaN arrangement mentioned.

In Figure 2a, a neutral section is clearly visible just before 600 s with V_p falling rapidly to 0, with zero current and power absorption as well. There are frequent phases of traction and braking that alternate during the trip, as this train was in normal passenger service.

It is possible to recognize some red areas for Z_p (Figure 2b) corresponding to large values of impedance and, in principle, to line resonance situations, if confirmed by the P_h map (Figure 2c) with application of *Rule* 4. These areas are located, for example, at about 9 and 12 kHz and extend intermittently before and after the neutral section; the resonance at about 14.5 kHz, instead, disappears after the neutral section.

The persistent horizontal line at 4 kHz is well evident as a voltage harmonic in Figure 2e, but is not backed up by a corresponding large impedance value Z'_p in Figure 2d, as this is, as known, an emission of another type of train running on the same line.

In Figure 3a two neutral sections are visible, at about 400 s and 600 s, again with a drop of pantograph current and absorbed power to zero. The phases of traction and braking appear to be milder (especially braking) than in the previous case of Figure 2a. Looking more closely, the absorbed power is almost double, as the train is a high-speed TGV train without intermediate train stops.

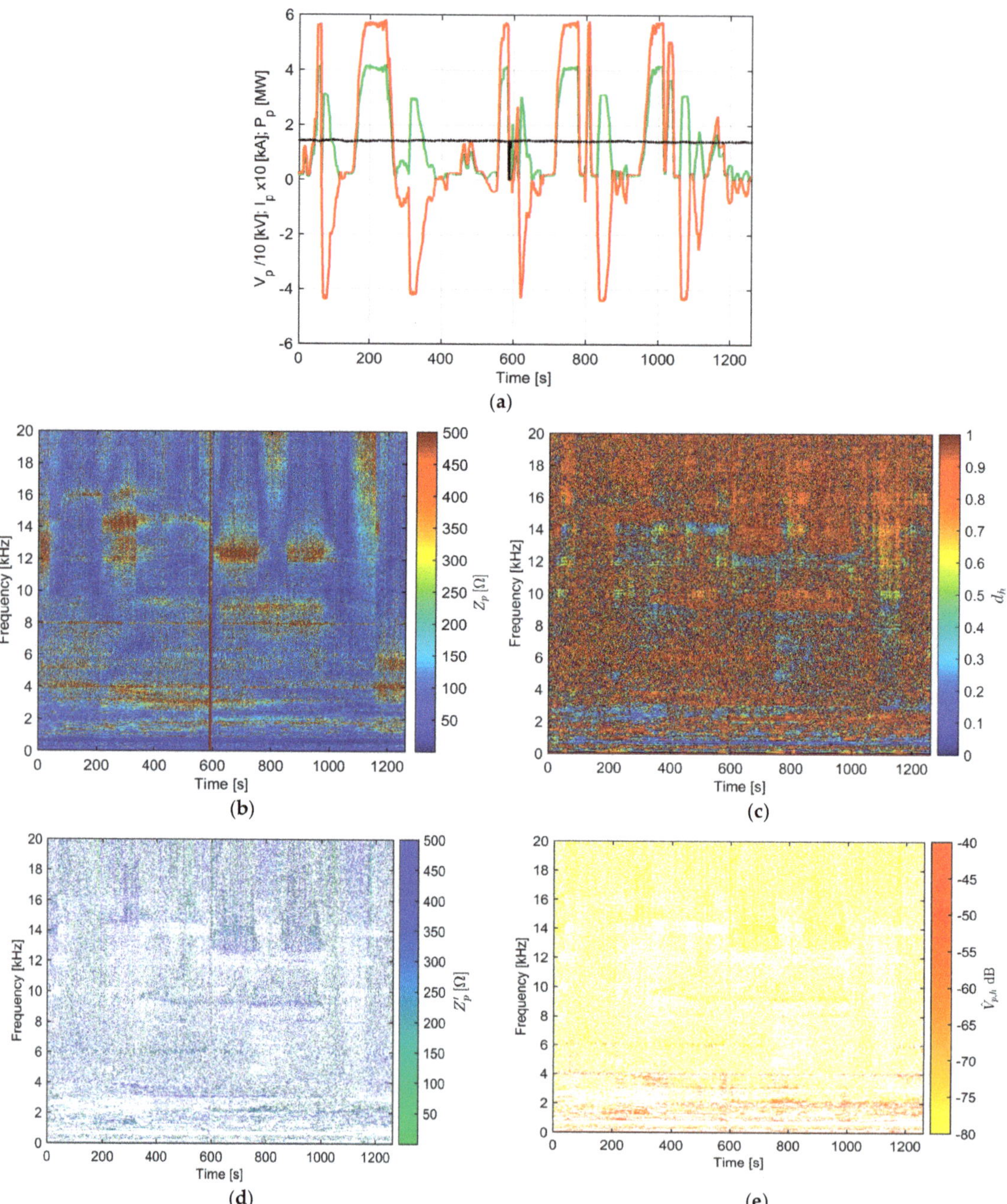

Figure 2. Switzerland example: (**a**) voltage (black), current (green), and active power (red) profiles; (**b**) harmonic impedance Z_p; (**c**) harmonic displacement factor d_h; (**d**) filtered impedance Z'_p; (**e**) normalized harmonic voltage $\hat{V}_{p,h}$.

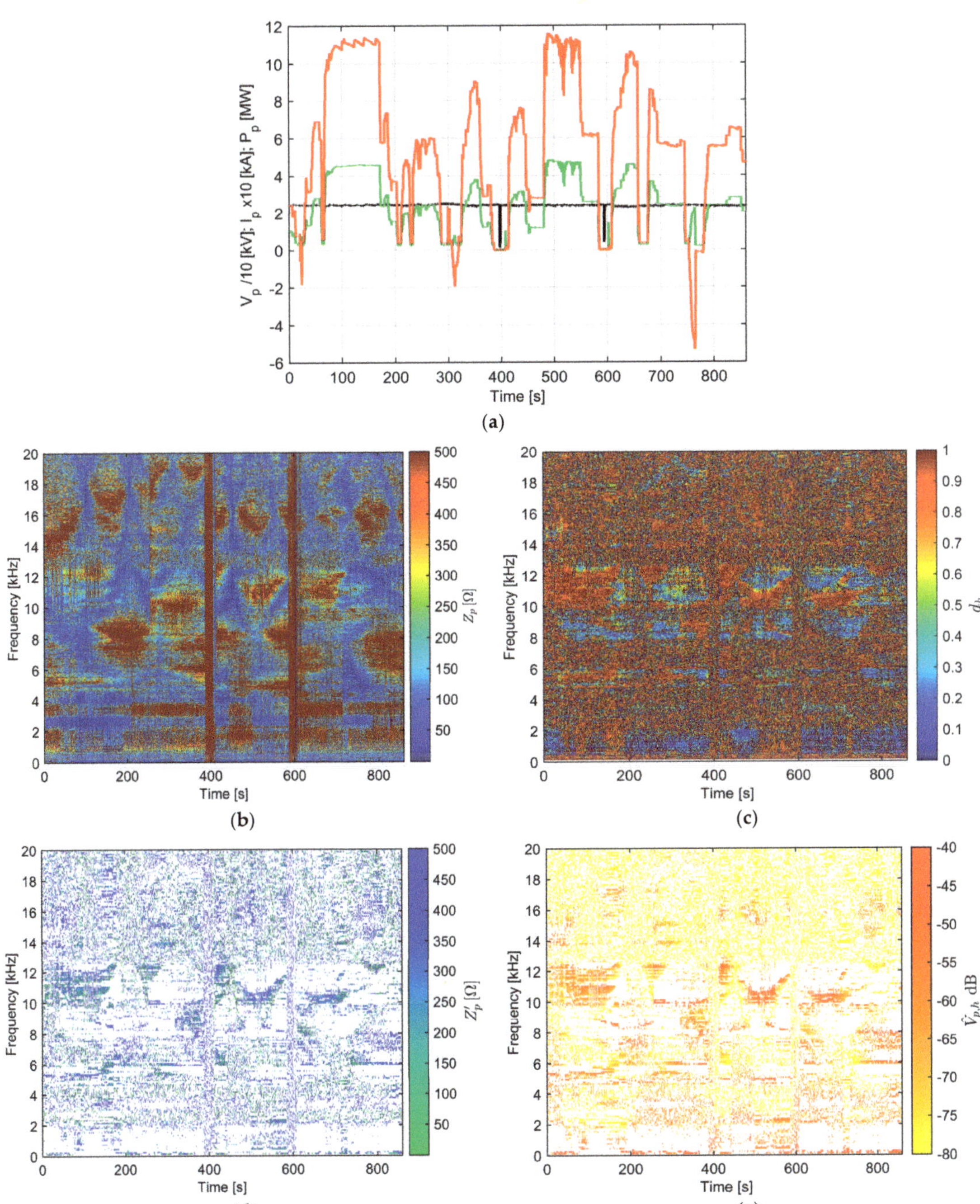

Figure 3. France example: (**a**) voltage (black), current (green), and active power (red) profiles; (**b**) harmonic impedance Z_p; (**c**) harmonic displacement factor d_h; (**d**) filtered impedance Z'_p; (**e**) normalized harmonic voltage $\hat{V}_{p,h}$.

There are several red areas of Z_p (Figure 3b), which however do satisfy *Rule 4* ($d_h \geq 0.9$), as shown by comparing with Figure 3c. The persistent horizontal lines at about 2, 4, and 6 kHz in Figure 3e are not backed up by corresponding large impedance values of Z'_p in Figure 3d, and they are in fact the harmonic emissions of a 4QC converter working at 2 kHz switching frequency.

It is interesting to observe two red areas of Z_p with a shape that is typical of a line resonance at about 8–9 kHz and 10–11 kHz, at time 200 s and 300 s, respectively, but they do not pass the *Rule 4* verification.

At some locations, this network shows a slight deviation from the theoretically grounded assumption that resonances do not depend on train position; this occurs between the two phase-separation sections, between 500 s and 650 s. Around 11–12 kHz an approximate U shape is barely visible in the Z_p graph, and it is better highlighted when combined with P_h as per *Rule 4*, as shown in Figure 3d,e. A slight dependency on train position is in reality possible when the line is not straight, but rather a joint with a third line segment at a junction.

Anti-resonances are also clearly visible as between 10 and 12 kHz, ascending first and then descending, centered on 200 s.

Observing Figure 4b, there is a triangularly shaped set of low values (from 12 kHz to about 18 kHz and then back to 14 kHz) indicating an anti-resonance, as the harmonic power P_h is also maximum (Figure 4c), and the values of Z'_p and $\hat{V}_{p,h}$ are at their minimum (greenish and yellowish, respectively). The triangular anti-resonance begins and ends with slightly larger values of Z'_p and $\hat{V}_{p,h}$, and at the vertex shows its minima. Correspondingly, there is a moderately large value of Z'_p at about 12 kHz, making the base of this triangular area and showing distortion values $\hat{V}_{p,h}$ that are moderately large (around –60 dB, that is 0.1%). In the original Z_p map (Figure 4b), there is a larger red area of high impedance values that do not pass the confirmation d_h test of *Rule 4* and disappear when considering Z'_p in Figure 4d.

Starting from the previous tests' cases, the behavior of the band limited THD was evaluated, as shown in Figure 5 for Switzerland and Figure 6 for France. Bands for calculation of THD_i were selected depending on the characteristics of the $\hat{V}_{p,h}$ spectrum, starting from a frequency of 500 Hz and separating the low-frequency interval from the successive bands H_i = 1 kHz located above it. THD_i curves were smoothed before plotting using a moving average filter of order 11.

The two systems confirm their major differences, with the France AC network featuring higher distortion, and not only in the first low-frequency interval.

Resonance occurrence corresponds to time intervals with large THD_i, although situations of increased rolling stock emissions also fall into this category and a clear distinction without Z_p information is not possible. In addition, the prevalence of active harmonic power is an indicator that the distortion corresponds to a resonance situation, whereas without it intervals of current pulling might also be included.

The *THD* for Switzerland is all caused by network components in THD_0, so below 500 Hz. Considering the high-frequency *THD*, it is evident that the *THD* channels of 1 kHz are not very selective in tracking resonances, as even THD_{14} can follow the two resonances at 13–14.5 kHz occurring in the intervals 650–800 s and 850–1000 s. The THD_{14} profile loses its dynamic due to other voltage components for which there is no prevalence of harmonic active power and where $\hat{V}_{p,h}$ would have discarded. Similarly, THD_9 misses the resonance at 9.5 kHz before 600 s and tracks it only between 600 and 950 s.

Observing France in Figure 6, THD_7 captures two resonances at about 200 and 350 s, also visible in Figure 3e. At higher frequency, THD_{11} and THD_{12} track the two resonances at about 10.5 kHz, occurring in particular at 500 and 650 s. THD_5 and THD_6 should have tracked the discontinuous resonances occurring between 4.5 and 6.5 kHz, and they are partially successful, at the beginning and around 500–550 s.

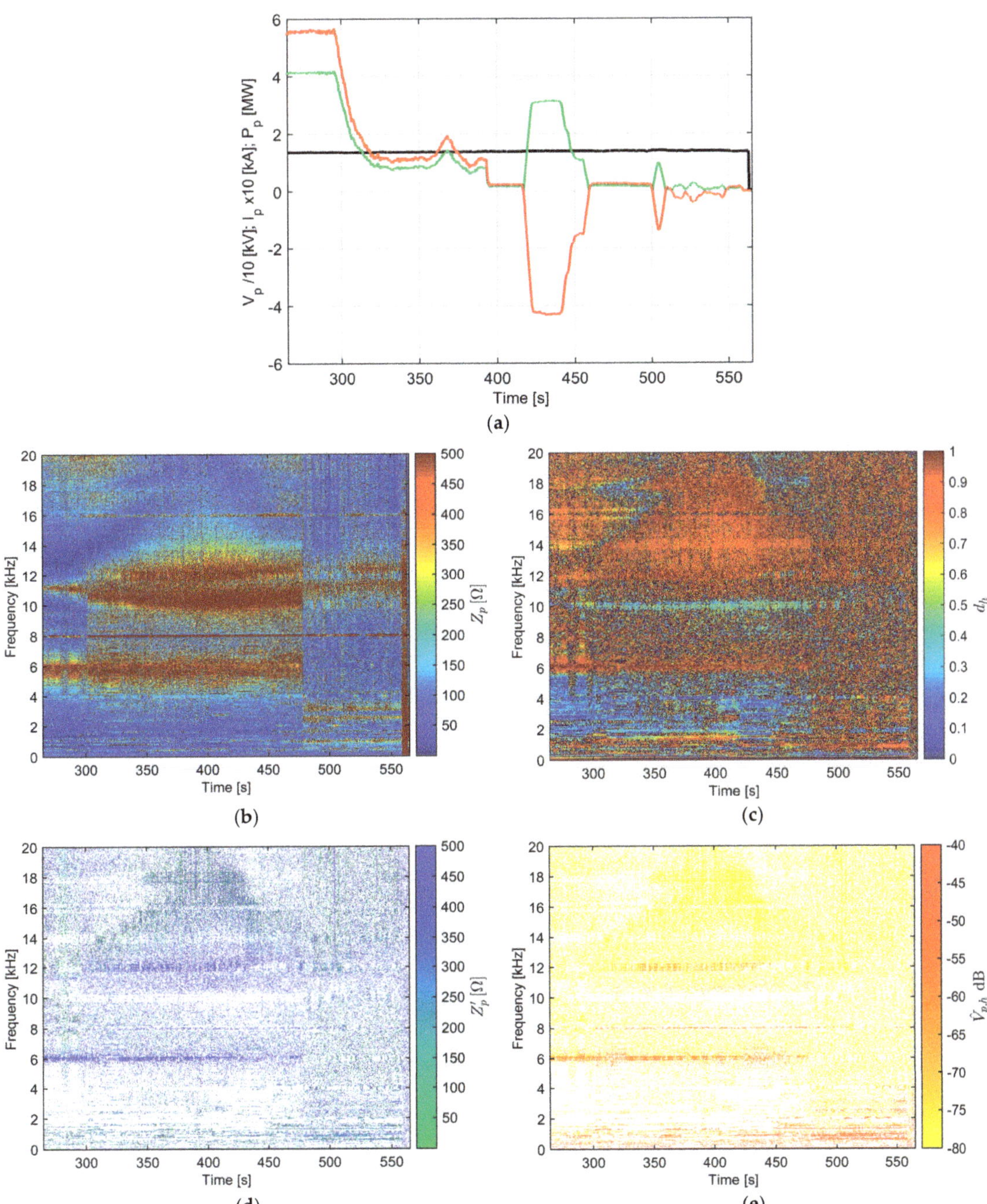

Figure 4. Switzerland example with zoom of time-varying anti-resonance: (**a**) voltage (black), current (green), and active power (red) profiles; (**b**) harmonic impedance Z_p; (**c**) harmonic active power P_h; (**d**) filtered impedance Z'_p; (**e**) normalized harmonic voltage $\hat{V}_{p,h}$.

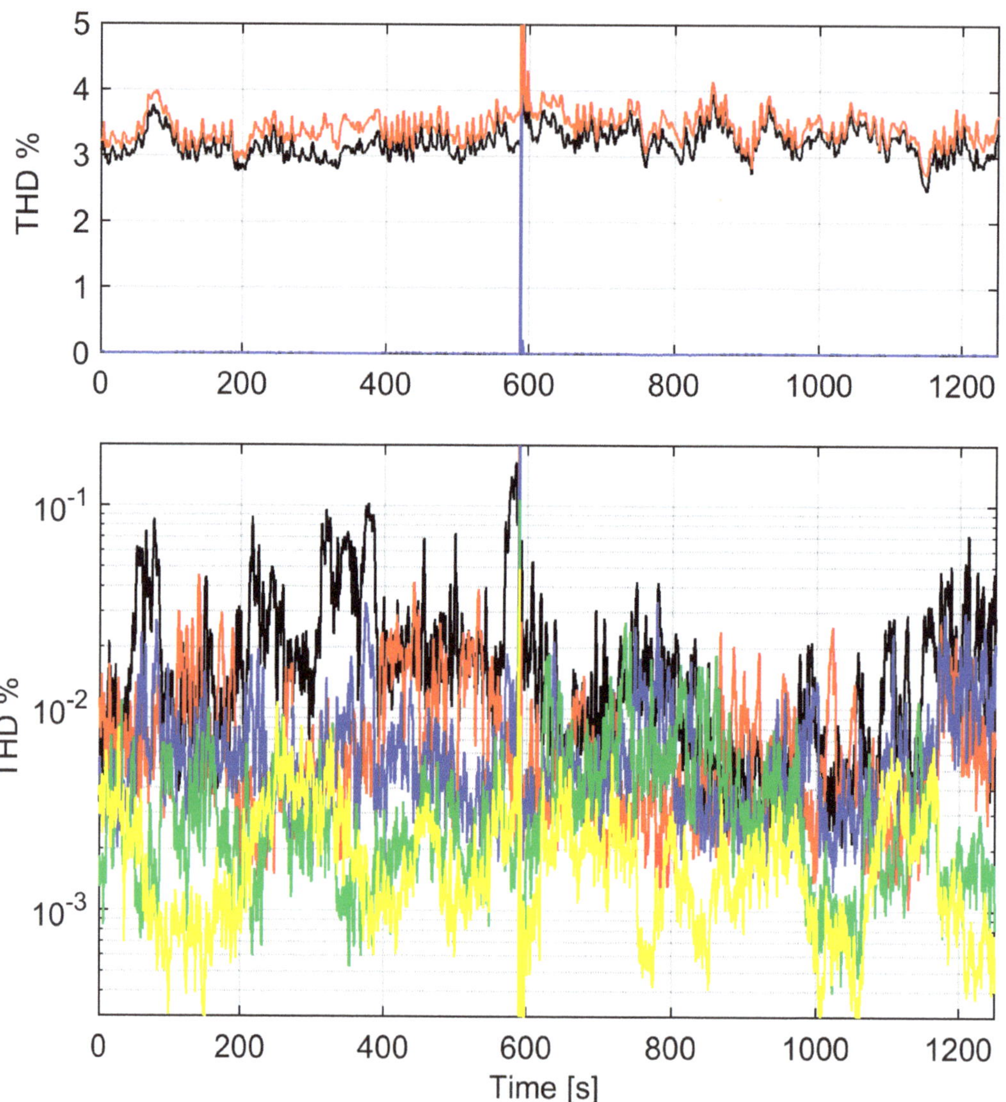

Figure 5. Switzerland case of Figure 2 (16.7 Hz): harmonic distortion profiles THD_i calculated at some frequency bands H_i = 1 kHz together with an indication of low-frequency distortion THD_0 calculated over the first 500 Hz (H_0 = 500 Hz). *THD*, THD_0, THD_1 (above); THD_4, THD_5, THD_6, THD_9, THD_{14} (below).

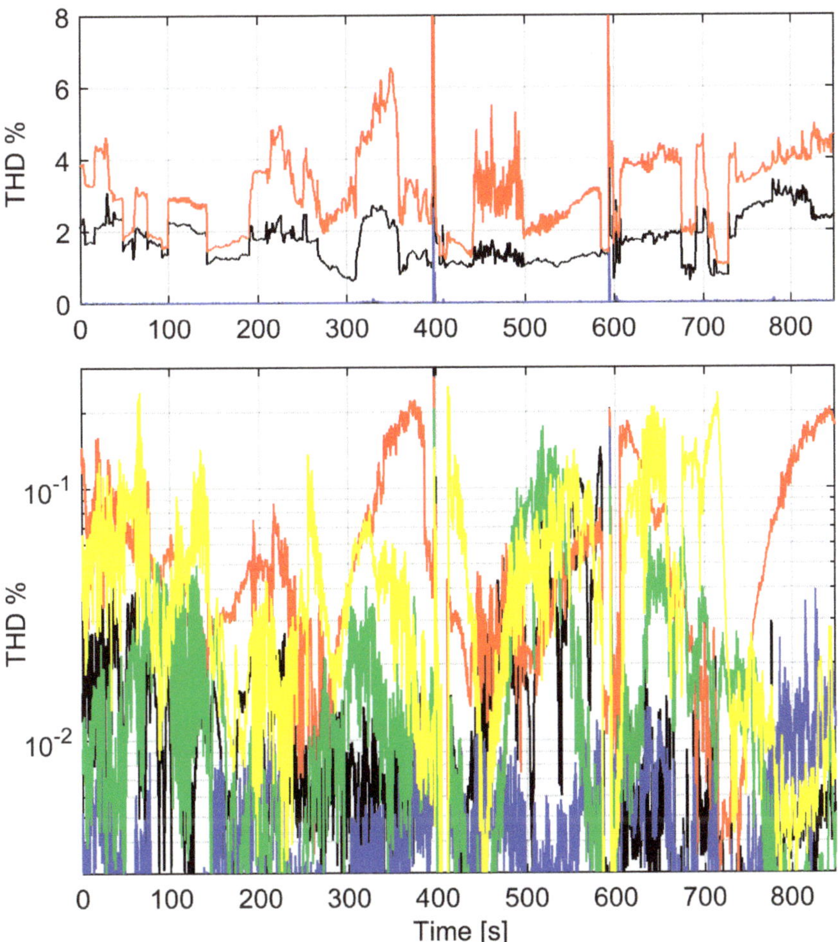

Figure 6. France case of Figure 3 (50 Hz): harmonic distortion profiles THD_i calculated at some frequency bands H_i = 1 kHz together with an indication of low-frequency distortion THD_0 calculated over the first 500 Hz (H_0 = 500 Hz). THD, THD_0, THD_1 (above); THD_5, THD_6, THD_7, THD_{11}, THD_{12} (below).

5. Outline of Real-Time Implementation

The discussed method is based on electrical quantities available onboard (V_p and I_p) and processed by means of STFT, on which the ratio and the product are calculated to derive Z_p and P_h, that, once normalized, gives d_h. The involved operations are thus basic operations available in a wide range of microcontrollers and microprocessors. The processing to produce the figures in Section 4 was purposely performed using a large number of bins on the frequency axis with a 8 Hz frequency resolution for graphical reasons; what is discussed in the following is the optimization of time and frequency resolution, as well as the size of matrices to fit available memory and process data in real time. Real time is defined as a time frame suitable to feedback on the monitored system following the detection of events that trigger a control action. In the present case, the identification of an incipient major resonance condition with increasing overvoltage caused by distortion should trigger a change of the rolling stock operating conditions, besides tagging the chainage position for later inspection. An estimate of the real-time requirement can be based on experience and confirmed by plots of resonance conditions vs. time that are available in the literature; Figure 3 in [9] shows an almost linear increase in line voltage between 28 kV at 9:35 am and more than 30 kV at 9:36 am. A detection and action time of some seconds is thus suitable for real-time reaction to these types of phenomena.

The discussion begins with a flowchart and description of data structures (Figure 7).

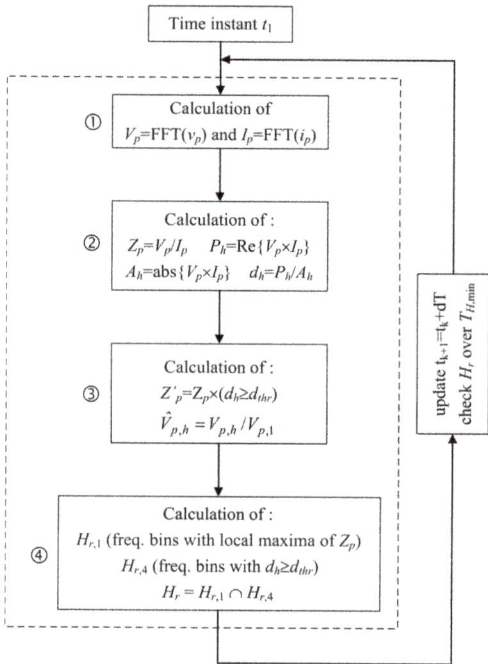

Figure 7. Flowchart of the harmonic resonance detection method and data structures used to support estimate of computational and storage complexity.

The number of points N is determined by the sampling frequency f_s and the frequency resolution df: $N = f_s/df$ (of course, N is an integer and should be rounded if needed). Then, the sampling frequency must be set to a minimum of twice the maximum frequency that is to be evaluated. In our case, we also reasoned on components that are slightly above 10 kHz, which implies $f_s \geq 25$ kSa/s (in Section 4 the value of f_s was 50 kSa/s).

Keeping df = 16.7 Hz (T = 60 ms), the resulting N ranges between 1500 and 3000 for f_s ranging between 25 and 50 kSa/s. Let's assume the same f_s of Section 4 and N = 3000.

Such calculation can be accommodated in any DSP or microcontroller with a math core. For the ongoing discussion, a basic 10 MFlops floating-point calculation speed may be considered, as reserved for running the algorithm ("Flops" stands for floating point operations per second); it is a small fraction of the whole DSP calculation performance in excess of several hundred MFlops (exemplified by [48,49] covering almost 20 years of DSP production). For complex operations, a factor of four should be considered for multiplication or division, and a factor of two for summation. A reserved 10 MWords memory area may be also assumed, having for simplicity indicated with "words" a unit suitable for a single-precision floating point number; although, in case of 32-bit words a double-precision floating point number requires two words. Similarly, this reference memory budget is a small fraction of the total system memory, that can range up to several hundred MB.

At every time instant t_k the vectors V_p and I_p are calculated with a fast Fourier transform (FFT) of N points and the complexity, as known, is in the order of $1.5N\log_2 N$.

This amounts to 52k Flops for the two N values, to be multiplied by four assuming all operations are complex multiplications, resulting in 2.1% of the 10 MFlops basic calculation speed. The occupied memory is $2N$ words for each complex vector, so 12 kWords, or 0.12%.

Calculation of Z_p, then, requires $(2 + 2 + 1)N$ for the absolute value of the numerator and denominator, separately, with two square roots (single CPU instruction with a math coprocessor) and a ratio of real vectors, so another $3N$, which is in total $8N$ Flops (24 kFlops, or 0.24%). The necessary memory is $2N$ words as a complex vector.

Calculation of P_h requires the calculation of apparent power A_n (with the same complexity of Z_p) and the extraction of the real part that has no complexity. The, for d_h, reusing the pre-calculated A_n (that must be stored somewhere in memory, requiring $2N$ words, or 6 kWords), there is only the ratio of the absolute values of P_n and A_n, so N Flops.

Calculation of data structures Z'_p and $\hat{V}_{p,h}$ is accomplished by a comparison and flagging for the former ($2N$ operations) and by a scalar division for the latter, in total $3N$ Flops, and the storage of two real vectors ($2N$ words).

The check of Z_p local peaks and large positive values of d_h, with creation of sets $H_{r,1}$ and $H_{r,4}$, needs some amount of code, possibly implementing a local peak search with comparison with neighbors, so some number of operations that may be collectively estimated to about $10N$. The storage is that of the new sets $H_{r,1}$ and $H_{r,4}$, and then the resulting H_r as intersection, in total up to $3N$ Words.

Calculations of the frequency vectors are repeated every dT seconds and the resulting vectors may be stored in adjacent memory areas to ease the comparison over the time axis. Time resolution is not demanding and dT may be chosen as a minimum equal to 300 ms for uniformity, between 16.7, 50 and 60 Hz systems, corresponding to 25 m of train movement at 300 km/h (equal to the length of one coach). Longer time intervals may be selected as well, easing computational and storage necessities and providing an adequate space resolution. Confirmation of persistence of a resonance, as commented in Section 4, is achieved if the conditions shown in *Rules* 1 to 4 persist, e.g., for tens of seconds, corresponding to about 1 km of traveled line at 300 km/h. Such an interval of observation T_{tot} represents the number of vectors stored in memory in a circular fashion, beyond which the oldest ones are replaced by the newest ones. The depth of the circular buffer will thus be $B = T_{tot}/dT$, corresponding to about 33 for the exemplified choices of dT and T_{tot}. Correspondingly, the number of repeated calculations over 1 s, M, is given by $M = 1/dT$, equal to three in the present case.

The check of persistency of membership of indexes to the set H_r (to last longer than $T_{H,\min}$) can be achieved with some logic operations, that collectively will not take more than N Flops at each time instant (in fact, $H_{r,1}$, $H_{r,4}$ and H_r are always quite sparse, possibly filled up to some % maximum).

By showing figures below 100% in the last row of Table 2, it is demonstrated that the allocated 10 MFlops and 10 MWords are sufficient for the implementation of the proposed method in real time. In case of increased requirements, e.g., a more complete representation or additional verifications, the computational and storage requirements should be scaled correspondingly.

Table 2. Summary of estimated computational and memory budget (f_s = 50 kSa/s, N = 3000). Algorithm steps are numbered as in the flowchart of Figure 7.

Algorithm Step	Float. Point Ops. (% of 10 MFlops)	Memory (% of 10 MWords)
(1) calc. V_p, I_p	2.1	0.12
(2) calc. Z_p	0.24	0.06
(2) calc. P_h with A_n	0.24	0.12
(2) calc. d_h	0.12	0.06
(3) calc. Z'_p and $\hat{V}_{p,h}$	0.09	0.06
(4) calc. $H_{r,1}$, $H_{r,4}$ and H_r	0.3	0.18
(5) check $T_{H,\min}$	0.03	0.0
Total computational ($\times M$) and storage ($\times B$) requirements	9.36	19.8

6. Conclusions

This work has discussed the problem of detection of resonances in AC railways in a rolling stock perspective, starting from the measured pantograph quantities (voltage and current) and using the derived quantities of pantograph impedance and harmonic active power. Suitable conditions are identified for the detection of resonance, focusing on parallel resonances, and a set of rules has been formulated. Local maxima of the pantograph impedance are detected and confirmed by a prevalent flow of harmonic active power, indicating the mutual cancellation of the harmonic reactive power terms. Such a condition for the harmonic active power also persists in the case of series resonance, for which minima of the pantograph impedance should be considered (an example is given in Figure 4). In both cases, further confirmation is obtained by the amplification of the harmonic components of voltage and current, respectively.

Rules are synthesized for the maxima of pantograph impedance Z_p and for the normalized harmonic active power (called harmonic displacement factor d_h) to be larger than a convenient threshold (which for the demonstration was set to 0.9). These criteria have then been validated by means of extensive experimental data measured in two different AC railway networks, one operated at 16.7 Hz (Switzerland) and one at 50 Hz (France).

The straightforward monitoring of voltage harmonic distortion was also included and compared to results in [18]. Although it is a useful indicator in general, it has issues of selectivity, due to the mix of voltage harmonic components with opposite or incoherent behavior not being able to reject those without a large active power fraction. It also always necessitates some degree of smoothing during post-processing to improve readability of obtained distortion profiles versus time. It has, however, a simple implementation, especially if implemented with a filter bank.

The comparison of voltage harmonic distortion over selected frequency bands with the results previously obtained with the pantograph impedance, combined with the harmonic displacement factor, has shown the superiority of the latter in terms of ability to locate and track resonance phenomena, avoiding the interference of spectral components of a reactive nature.

Author Contributions: Investigation, A.M. and L.S.; Methodology, A.M. and L.S.; Software, A.M. and L.S.; Writing—original draft, A.M.; Writing—review & editing, A.M. and L.S. All authors have read and agreed to the published version of the manuscript.

Funding: This research received no external funding.

Institutional Review Board Statement: Not applicable.

Conflicts of Interest: The authors declare no conflict of interest.

References

1. Kaleybar, H.J.; Brenna, M.; Foiadelli, F.; Fazel, S.S.; Zaninelli, D. Power Quality Phenomena in Electric Railway Power Supply Systems: An Exhaustive Framework and Classification. *Energies* **2020**, *13*, 6662. [CrossRef]
2. EN. *Railway Applications—Power Supply and Rolling Stock—Technical Criteria for the Coordination between Power Supply (Substation) and Rolling Stock to Achieve Interoperability*; EN 50388; CENELEC: Brussels, Belgium, 2013.
3. EN. *Railway Applications—Fixed Installations and Rolling Stock—Technical Criteria for the Coordination between Traction Power Supply and Rolling Stock to Achieve Interoperability—Part 1: General*; EN 50388-1; CENELEC: Brussels, Belgium, 2017.
4. Gazafrudi, S.M.M.; Tabakhpour, L.A.; Fuchs, E.F.; Al-Haddad, K. Power quality issues in railway electrification: A comprehensive perspective. *IEEE Trans. Ind. Electron.* **2015**, *62*, 3081–3090. [CrossRef]
5. Roudsari, H.M.; Jamali, S.; Jalilian, A. Dynamic modeling, control design and stability analysis of railway active power quality conditioner. *Electr. Power Syst. Res.* **2018**, *160*, 71–88. [CrossRef]
6. Panpean, C.; Areerak, K.; Santiprapan, P.; Areerak, K.; Shen Yeoh, S. Harmonic Mitigation in Electric Railway Systems Using Improved Model Predictive Control. *Energies* **2021**, *14*, 2012. [CrossRef]
7. Morais, V.A.; Afonso, J.L.; Carvalho, A.S.; Martins, A.P. New Reactive Power Compensation Strategies for Railway Infrastructure Capacity Increasing. *Energies* **2020**, *13*, 4379. [CrossRef]
8. Tanta, M.; Cunha, J.; Barros, L.A.M.; Monteiro, V.; Pinto, J.G.O.; Martins, A.P.; Afonso, J.L. Experimental Validation of a Reduced-Scale Rail Power Conditioner Based on Modular Multilevel Converter for AC Railway Power Grids. *Energies* **2021**, *14*, 484. [CrossRef]

9. Song, K.; Wu, M.; Yang, S.; Liu, Q.; Agelidis, V.G.; Konstantinou, G. High-Order Harmonic Resonances in Traction Power Supplies: A Review Based on Railway Operational Data, Measurements, and Experience. *IEEE Trans. Power Electron.* **2020**, *35*, 2501–2518. [CrossRef]
10. Jiang, X.; He, Z.; Hu, H.; Zhang, Y. Analysis of the Electric Locomotives Neutral-section Passing Harmonic Resonance. *Energy Power Eng.* **2013**, *5*, 546–551. [CrossRef]
11. Liu, Q.; Sun, B.; Yang, Q.; Wu, M.; He, T. Harmonic Overvoltage Analysis of Electric Railways in a Wide Frequency Range Based on Relative Frequency Relationships of the Vehicle–Grid Coupling System. *Energies* **2020**, *13*, 6672. [CrossRef]
12. Chu, X.; Lin, F.; Yang, Z. The Analysis of Time-Varying Resonances in the Power Supply Line of High Speed Trains. In Proceedings of the Internet Power Electronics Conference, Hiroshima, Japan, 18–21 May 2014. [CrossRef]
13. Lutrakulwattana, B.; Konghirun, M.; Sangswang, A. Harmonic resonance assessment of 1 × 25kV, 50Hz traction power supply system for Suvarnabhumi airport rail link. In Proceedings of the 18th International Conference on Electrical Machines and Systems (ICEMS), Pattaya, Thailand, 25–28 October 2015.
14. Holtz, J.; Kelin, H.J. The propagation of harmonic currents generated by inverter-fed locomotives in the distributed overhead supply system. *IEEE Trans. Power Electron.* **1989**, *4*, 168–174. [CrossRef]
15. Sainz, L.; Monjo, L.; Riera, S.; Pedra, J. Study of the Steinmetz Circuit Influence on AC Traction System Resonance. *IEEE Trans. Power Deliv.* **2012**, *27*, 2295–2303. [CrossRef]
16. Brenna, M.; Capasso, A.; Falvo, M.C.; Foiadelli, F.; Lamedica, R.; Zaninelli, D. Investigation of resonance phenomena in high speed railway supply systems: Theoretical and experimental analysis. *Electr. Power Syst. Res.* **2011**, *81*, 1915–1923. [CrossRef]
17. Kolar, V.; Palecek, J.; Kocman, S.; Trung, T.Y.; Orsag, P.; Styskala, V.; Hrbac, R. Interference between Electric Traction Supply Network and Distribution Power Network Resonance Phenomenon. In Proceedings of the 14th International Conference on Harmonics and Quality of Power (ICHQP), Bergamo, Italy, 26–29 September 2010. [CrossRef]
18. Lee, H.; Lee, C.; Jang, G.; Kwon, S. Harmonic analysis of the Korean high-speed railway using the eight-port representation model. *IEEE Trans. Power Deliv.* **2006**, *21*, 979–986. [CrossRef]
19. Li, J.; Wu, M.; Molinas, M.; Song, K.; Liu, Q. Assessing High-Order Harmonic Resonance in Locomotive-Network Based on the Impedance Method. *IEEE Access* **2019**, *7*, 68119–68131. [CrossRef]
20. Hu, H.; Shao, Y.; Tang, L.; Ma, J.; He, Z.; Gao, S. Overview of Harmonic and Resonance in Railway Electrification Systems. *IEEE Trans. Ind. Appl.* **2018**, *54*, 5227–5245. [CrossRef]
21. Gao, S.; Li, X.; Ma, X.; Hu, H.; He, Z.; Yang, J. Measurement-Based Compartmental Modeling of Harmonic Sources in Traction Power-Supply System. *IEEE Trans. Power Deliv.* **2017**, *32*, 900–909. [CrossRef]
22. Hu, H.; Tao, H.; Blaabjerg, F.; Wang, X.; He, Z.; Gao, S. Train–Network Interactions and Stability Evaluation in High-Speed Railways–Part I: Phenomena and Modeling. *IEEE Trans. Power Electron.* **2018**, *33*, 4627–4642. [CrossRef]
23. Zhang, G.; Liu, Z.; Yao, S.; Liao, Y.; Xiang, C. Suppression of Low-Frequency Oscillation in Traction Network of High-Speed Railway Based on Auto-Disturbance Rejection Control. *IEEE Trans. Transp. Electrif.* **2016**, *2*, 244–255. [CrossRef]
24. Hemmer, B.; Mariscotti, A.; Wuergler, D. Recommendations for the calculation of the total disturbing return current from electric traction vehicles. *IEEE Trans. Power Deliv.* **2004**, *19*, 1190–1197. [CrossRef]
25. Meyer, M.; Schöning, J. Netzstabilität in grossen Bahnnetzen. *Schweiz. Eisenb.-Rev. Eisenb.-Rev. Int.* **1999**, *7*, 312–317.
26. Pröls, M.; Strobl, B. Stabiltätskriterien für Wechselwirkungen mit Umrichteranlagen in Bahnsystemen. *Elektr. Bahnen* **2006**, *104*, 542–552.
27. Mollerstedt, E.; Bernhardsson, B. Out of control because of harmonics an analysis of the harmonic response of an inverter locomotive. *IEEE Control Syst. Mag.* **2000**, *20*, 70–81. [CrossRef]
28. Li, T.; Wu, G.; Zhou, L.; Gao, G.; Wang, W.; Wang, B.; Liu, D.; Li, D. Pantograph Arcing's Impact on Locomotive Equipments. In Proceedings of the IEEE 57th Holm Conference on Electrical Contacts (Holm), Minneapolis, MN, USA, 11–14 September 2011. [CrossRef]
29. Bongiorno, J.; Mariscotti, A. Experimental validation of the electric network model of the Italian 2 × 25 kV 50 Hz railway. In Proceedings of the 20th IMEKO TC4 Symposium on Measurements of Electrical Quantities, Benevento, Italy, 15–17 September 2014.
30. Liu, Y.; Xu, J.; Shuai, Z.; Li, Y.; Peng, Y.; Liang, C.; Cui, G.; Hu, S.; Zhang, M.; Xie, B. A Novel Harmonic Suppression Traction Transformer with Integrated Filtering Inductors for Railway Systems. *Energies* **2020**, *13*, 473. [CrossRef]
31. Zhang, R.; Lin, F.; Yang, Z.; Cao, H.; Liu, Y. A Harmonic Resonance Suppression Strategy for a High-Speed Railway Traction Power Supply System with a SHE-PWM Four-Quadrant Converter Based on Active-Set Secondary Optimization. *Energies* **2017**, *10*, 1567. [CrossRef]
32. Liu, Q.; Li, J.; Wu, M. Field Tests for Evaluating the Inherent High-Order Harmonic Resonance of Traction Power Supply Systems up to 5000 Hz. *IEEE Access* **2020**, *8*, 52395–52403. [CrossRef]
33. Mariscotti, A.; Pozzobon, P.; Vanti, M. Simplified modelling of 2 × 25 kV AT Railway System for the solution of low frequency and large scale problems. *IEEE Trans. Power Deliv.* **2007**, *22*, 296–301. [CrossRef]
34. Pilo, E.; Rouco, L.; Fernández, A.; Abrahamsson, L. A Monovoltage Equivalent Model of Bi-Voltage Autotransformer-Based Electrical Systems in Railways. *IEEE Trans. Power Deliv.* **2012**, *27*, 699–708. [CrossRef]
35. Bottenberg, A.; Debruyne, C.; Peterson, B.; Rens, J.; Knockaert, J.; Desmet, J. Network resonance detection using harmonic active power. In Proceedings of the 18th International Conference on Harmonics and Quality of Power (ICHQP), Ljubljana, Slovenia, 13–16 May 2018. [CrossRef]

36. Mariscotti, A. Impact of Harmonic Power Terms on the Energy Measurement in AC Railways. *IEEE Trans. Instrum. Meas.* **2020**, *69*, 6731–6738. [CrossRef]
37. Filippone, F.; Mariscotti, A.; Pozzobon, P. The Internal Impedance of Traction Rails for DC Railways in the 1-100 kHz Frequency Range. *IEEE Trans. Instrum. Meas.* **2006**, *55*, 1616–1619. [CrossRef]
38. Mariscotti, A.; Pozzobon, P. Measurement of the Internal Impedance of Traction Rails at Audiofrequency. *IEEE Trans. Instrum. Meas.* **2004**, *53*, 792–797. [CrossRef]
39. Robert, A.; Deflandre, T.; Gunther, E.; Bergeron, R.; Emanuel, A.; Ferrante, A.; Finlay, G.S.; Gretsch, R.; Guarini, A.; Iglesias, J.G.; et al. Guide for assessing the network harmonic impedance. In Proceedings of the 14th International Conference and Exhibition on Electricity Distribution (Part 1), Birmingham, UK, 2–5 June 1997. [CrossRef]
40. Kannan, S.; Meyer, J. Recent Developments in Harmonic Resonance Detection in Low Voltage Networks using Impedance Measurement Techniques. In Proceedings of the 8th International Conference on Power Systems (ICPS), Jaipur, India, 20–22 December 2019. [CrossRef]
41. Ritzmann, D.; Wright, P.S.; Holderbaum, W.; Potter, B. A method for accurate transmission line impedance parameter estimation. *IEEE Trans. Instrum. Meas.* **2016**, *65*, 2204–2213. [CrossRef]
42. Mariscotti, A. Experimental characterization of active and nonactive harmonic power flow of AC rolling stock and interaction with the supply network. *IET Electr. Syst. Transp.* **2021**, *11*, 109–120. [CrossRef]
43. Mariscotti, A. Behavior of Single-Point Harmonic Producer Indicators in Electrified AC Railways. *Metrol. Meas. Syst.* **2020**, *27*, 641–657. [CrossRef]
44. Lodetti, S.; Bruna, J.; Melero, J.J.; Khokhlov, V.; Meyer, J. A Robust Wavelet-Based Hybrid Method for the Simultaneous Measurement of Harmonic and Supraharmonic Distortion. *IEEE Trans. Instrum. Meas.* **2020**, *69*, 6704–6712. [CrossRef]
45. Mariscotti, A. Methods for Ripple Index evaluation in DC Low Voltage Distribution Networks. In Proceedings of the IEEE International Measurement Technology Conference IMTC, Warsaw, Poland, 1–3 May 2007. [CrossRef]
46. Mariscotti, A. Data sets of measured pantograph voltage and current of European AC railways. *Data Brief* **2020**, *30*, 105477. [CrossRef] [PubMed]
47. Mariscotti, A. Direct Measurement of Power Quality over Railway Networks with Results of a 16.7 Hz Network. *IEEE Trans. Instrum. Meas.* **2011**, *60*, 1604–1612. [CrossRef]
48. Texas Instruments, SM320C6712 Floating Point Digital Signal Processors. September 2004. Available online: https://www.ti.com/lit/gpn/SM320C6712D-EP (accessed on 19 August 2021).
49. Texas Instruments, TMS320C6652 and TMS320C6654 Fixed and Floating-Point Digital Signal Processor. October 2019. Available online: https://www.ti.com/lit/gpn/tms320c6652 (accessed on 19 August 2021).

Article

A Novel Arc Detection Method for DC Railway Systems

Yljon Seferi [1,*], Steven M. Blair [1,2], Christian Mester [3] and Brian G. Stewart [1]

1 Department of Electronic and Electrical Engineering, University of Strathclyde, Glasgow G1 1XQ, UK; steven.m.blair@strath.ac.uk or steven.blair@synapt.ec (S.M.B.); brian.stewart.100@strath.ac.uk (B.G.S.)
2 Synaptec Ltd., Glasgow G1 1XW, UK
3 Federal Institute of Metrology METAS, Lindenweg 50, 3003 Bern-Wabern, Switzerland; christian.mester@metas.ch
* Correspondence: yljon.seferi@strath.ac.uk

Abstract: Electric arcing due to contact interruption between the pantograph and the overhead contact line in electrified railway networks is an important and unwanted phenomenon. Arcing events are short-term power quality disturbances that produce significant electromagnetic disturbances both conducted and radiated as well as increased degradation on contact wire and contact strip of the pantograph. Early-stage detection can prevent further deterioration of the current collection quality, reduce excessive wear in the pantograph-catenary system, and mitigate failure of the pantograph contact strip. This paper presents a novel arc detection method for DC railway networks. The method quantifies the rate-of-change of the instantaneous phase of the oscillating pantograph current signal during an arc occurrence through the Hilbert transform. Application of the method to practical pantograph current data measurements, demonstrates that phase derivative is a useful parameter for detecting and localizing significant power quality disturbances due to electric arcs during both coasting and regenerative braking phases of a running locomotive. The detected number of arcs may be used to calculate the distribution of the arcs per kilometre as an alternative estimation of the current collection quality index and consequently used to assess the pantograph-catenary system performance. The detected arc number may also contribute to lowering predictive maintenance costs of pantograph-catenary inspections works as these can be performed only at determined sections of the line extracted by using arcing time locations and speed profiles of the locomotive.

Keywords: pantograph-catenary system; current collection quality; arc detection; predictive maintenance; railway electrical networks; Hilbert transform; rail transportation; power quality disturbance

Citation: Seferi, Y.; Blair, S.M.; Mester, C.; Stewart, B.G. A Novel Arc Detection Method for DC Railway Systems. *Energies* **2021**, *14*, 444. https://doi.org/10.3390/en14020444

Received: 7 December 2020
Accepted: 12 January 2021
Published: 15 January 2021

1. Introduction

Smooth dynamic interaction between the contact strip of the pantograph and the overhead contact line (OCL) is important for the safe and efficient performance of electrical train transportation services. A reliable contact contributes to minimising short-term power quality events and establishing good quality current collection to power the train [1,2]. Indeed, a continuous mechanical sliding contact distinguishes high and poor quality current collection performance [3–6]. However, it is unlikely that continuous contact can be maintained for an entire train journey due to mechanical oscillations of the train [7], horizontal zigzag movement of the pantograph [8], and faults or incorrect adjustment in the pantograph operating system [9]. Therefore, contact interruption between pantograph and OCL will occur leading to arcing phenomena which in turn causes a number of unwanted issues such as: conducted short-term power-quality events; current flow disturbances; undesirable radiated electromagnetic emissions [7,8,10–12]; increased temperature at contact points [13]; increased wear on contact wire and contact strip of the pantograph [3,8,12], and electric micro-welding phenomena [6]. It is worth mentioning that continuous and steady development of wear on the contact wire and contact strip of a pantograph can damage the pantograph frame and the contact strip, deteriorate the quality of current collection, and result in serious consequences leading to service interruption [8].

Standard EN 50367 [2] considers either the measured vertical contact force exercised by the pantograph to OCL or the percentage of arcing (ratio between the total duration of all arcs to the total run time) as an assessor of pantograph current collection quality. Therefore, arc detection in railway electrical networks, including their geographical localisation and characterization in terms of time duration, are fundamental for the assessment of the contact wire lifetime, pantograph lifetime and consequently for the entire safety and performance of the train transportation service. Furthermore, arc detection and characterization has a direct impact on lowering the cost of periodical maintenance which can often be expensive [14,15]. Due to the relevance of arcs and the impact on general power quality in electrified transportation services, much effort has been devoted to arc identification techniques by the research community.

In general, arc detection techniques can be grouped into three main categories: those based on image processing of recorded videos; those based on emission evaluation of physical quantities such as light, temperature and radiated electromagnetic field; and those based on the processing of electrical quantities such as voltage and current signals. In this context, [14] proposed the usage of image processing algorithms to detect arcs occurring in a pantograph-catenary system. This method requires a video camera to be installed close to the pantograph. Video frame images are processed by an algorithm in order to detect pantograph movement with changes in the surrounding background of the processed images attributed to arc occurrence. Reference [16] proposed the application of threshold values to binary converted frame images, where the evaluated ratio of white pixels to black pixels is considered as an index of arc occurrence.

Barmada et al. [15], initially proposed the use of a Support Vector Machine (SVM) based classification algorithm to detect the presence of arcs. Extracted features of pantograph recorded voltage, pantograph current, and signals from phototube sensors were used to support the proposed algorithm. The method has been shown to achieve an arc detection accuracy of 80%. Later on, Bermada et al. [17], proposed the use of clustering techniques to detect arcs by considering only the current signals, this avoiding the deployment of phototube sensors. Current signals were processed and grouped in 4 classes which indicate both the presence and magnitudes of the detected arcs.

In reference [18], the authors developed a new arc detection technique by analysing the recorded locomotive current. It was observed that arc occurrence causes significant injection of a DC component into the AC current signal, due to increased disturbances in the transformer magnetizing circuit. This proposed method requires accurate knowledge of the network infrastructure and the speed of the running train in order correlate and distinguish DC components caused by other factors, for example by the presence of phase separation sections. As a consequence, the method cannot be successfully applied for arc detection in cases when trains cross multiple power separation sections [19], and for example, when crossing borders from one country to another within Europe [20,21].

Light emission (ultraviolet emission) produced by arcs and measured by phototube sensors has been proposed in [4] as an appropriate measurement system for arc detection. Similarly, in Standard EN 50317 [1], a light detector for measuring predefined light wavelengths emitted by copper material under arcing is proposed. Despite the fact that an additional measurement sensor is needed, the proposed technique requires different sensors or sensors with appropriate wavelength tuning due to different material-to-material emission characteristics. For example when trains need to travel from one particular section characterised by copper contact wire material, to another section characterised by aluminium alloys or aluminium conductor steel reinforced contact wires [3], or for example when trains enter tunnels, and an aluminium overhead conductor rail substitutes a traditional copper contact wire [18].

Significant temperature changes developed at the contact point between pantograph strip and OCL occur during arc occurrence has been exploited by the proposed method in [13] for monitoring the current collection quality of a pantograph-catenary system. Images from a thermal camera installed close to the pantograph continuously recorded and fed a number of arc detection image processing algorithms. The temperature of the contact point was identified by using an edge detection algorithm (Canny algorithm), and the Hough transform for continuous detection of OCL movement. The main issue with this particular method appears to be the high computational burden.

Reference [22] proposes a new method for arc detection by analyzing the spectral content of the electromagnetic field radiation caused by the arc occurrence. The radiated signals were captured by an antenna and processed for the determination of possible characteristic radiated frequencies. A wide range of measured frequencies were observed in relation to arc occurrence with the main peak located around 18 MHz. A possible limitation of method is related to the fact that all analyzed arcs were created in a laboratory environment. Arcs occurring in practical rail systems may have different radiated characteristic frequencies, and be influenced by factors such as the length of OCL, number of trains, respective train positions, and network topology of the employed converters, etc.

The quality of the current collection has also been evaluated by the use of wavelet multi-resolution analyses [5]. Recorded current signals were decomposed to a 4th level, and then a comparative analysis by considering the energy of the signal quantified through Parseval's theorem employed to discriminate arc occurrences from electric welding phenomena [6]. Although the comparative analysis shows good results, it limits itself only to detecting the electric welding phenomenon. Hence, the effectiveness of the method to discriminate against other possible common phenomena in electric railway systems, for example, the presence of current spikes is still an open question.

Another interesting approach for arc detection has been proposed by measuring the vertical displacement and lateral accelerations of the OCL when the pantograph passes [9]. The proposed method clearly does not directly detect an arc event, but its consequences are evaluated. Large deviations of the displacement and acceleration have been associated to pantographs with contact strip degradation or defects, and hence the method supports maintenance activities.

Clearly, early-stage defect detection in contact wire and pantograph strip can prevent excessive wear in the entire pantograph-catenary system, and help maintenance service improve scheduling of inspection work. To this end, this paper proposes a new method for arc detection in DC railway systems. The proposed technique does not require any external equipment to be installed on the train (such as sensors, cameras, or antennas), because it is based on signal processing of measured pantograph current only. Furthermore, it does not employ extensive processing techniques, which results in a low processing overhead, and therefore it can be implemented relatively easy in real-time. The method exploits the low-frequency oscillation characteristic triggered by an arc occurrence in DC railway systems. Through application of the Hilbert transform (HT) the instantaneous phase of the generated signal associated with an arc event is quantified. It is shown that the instantaneous phase derivative can be employed to detect and localize in time the presence of arcs. The number of arcs per kilometre can be also calculated to allow a reliable estimation of the current collection quality index, and consequently form a valuable assessment of the entire pantograph-catenary system. HT is useful to analyse non-stationary time series [23], and to detect short-time disturbances. In contrast Fourier transform and the short-time version assumes signal periodicity and are optimised for stationary signals analysis [24–26].

The rest of the paper is structured as follows: Section 2 presents an overview of the arc occurrence mechanism and its impact on the railway network. In Section 3, the characteristics of the recorded current and voltage signals, containing real arcs are presented. The proposed detection method is described in Section 4, whereas the simulation results are presented in Section 5. Section 6 provides a comprehensive sensitivity analysis of the proposed method. Conclusions are summarised in Section 7.

2. Definition Mechanism of Occurrence and Impact of Electric Arcs in Railway Network

Electric arcs in railway transportation networks are physically defined as electric discharges occurring in the conductive ionizing gas, known as plasma, between the pantograph contact strip and the OCL [27]. This air gap forms during pantograph detachment and consists of three regions. The anode region and the cathode region, which are characterized by a nonlinear voltage drop, and the plasma column region characterized by a linear voltage drop. The latter is a function of physical distance between the two electrodes and the physical properties of the plasma [27].

Arcs occur because of the inability of the pantograph to continually stay attached to the contact wire. One of the main reasons causing this contact interruption is the mechanical oscillation of the train caused by the irregularity of the track geometry [7], and abrupt variation of the height of the contact wire mainly when trains enter tunnels [4].

In order to avoid rapid consumption of the pantograph contact strip, the contact wire is distributed in a zigzag manner throughout the track. Consequently, the pantograph follows this zigzag movement, and when the contact wire approaches the extreme ends of strip an increased air gap is developed between the pantograph and OCL, which causes an arc. The latter has been simulated and confirmed experimentally in a laboratory environment [8,28].

Another relevant cause of electric arcs is a fault in the pantograph mechanism or an incorrect adjustment of its operating system [9]. These cause either a low contact force to be exercised by the pantograph to the OCL, which in turn has a negative effect because it weakens the contact point and so the air gap increases, or it produces a high contact force that progresses the wear of the contact wire [29]. This wear deteriorates the contact quality and as a consequence electric micro-welding phenomena occur [6], which makes the occurrence of arcing more frequent. The arc occurrence mechanism is also negatively affected by factors such as increased train speed, the collected traction level current, as well as poor weather conditions including strong winds, snow, and ice [2,8,9,28].

Electric arcs are notorious for producing electromagnetic phenomena which propagate to the entire railway network. A wide range of injected electromagnetic frequencies have been observed during arc occurrence where some of them can excite resonant frequencies of the employed network components (such as filters and contact lines) [10,30], together with DC components induced in AC signals as a result of current interruption due to arcing [18]. Other conducted electromagnetic disturbances resulting from arcs are voltage transients and oscillations [8,30,31]. Low and high frequency oscillations spanning up to hundreds of MHz due to arcing phenomenon [7,22] have been electromagnetically radiated to nearby circuits, and potentially interfering with signaling and radio communications systems [7,8,10–12,32].

3. Characteristics of Voltage and Current Recorded Signals

This section presents the time domain characteristics of voltage and current signals measured at the pantograph of a locomotive operating on a 3 kV DC railway network in Italy. The signals used for this analysis are part of 16ENG04 MyRailS project [33,34] and can be accessed online [35]. The signals are recorded using a data acquisition system (DAQ) sampling at 50 kS/s installed on-board the train.

Figure 1 presents two examples of the measured pantograph voltages V_p and currents I_p of two different arc occurrences during the coasting phase of a running train. It can be observed that initially the pantograph voltage and current signals are relatively steady, indicating good and continuous sliding mechanical contact between the pantograph and OCL. When an arc occurs there is an immediate pantograph voltage V_p drop (V.drop 1) consistent with the research outcomes presented in [27]. Due to contact interruption, pantograph current I_p decreases because of the inability of the arc channel to conduct the level of current required by the traction drive of the locomotive. This current decrease in turn causes the stored energy in the magnetic field of the system inductance to be released instantly [8,10], and consequently causing the voltage spike (V.spike) shown in Figure 1a,b.

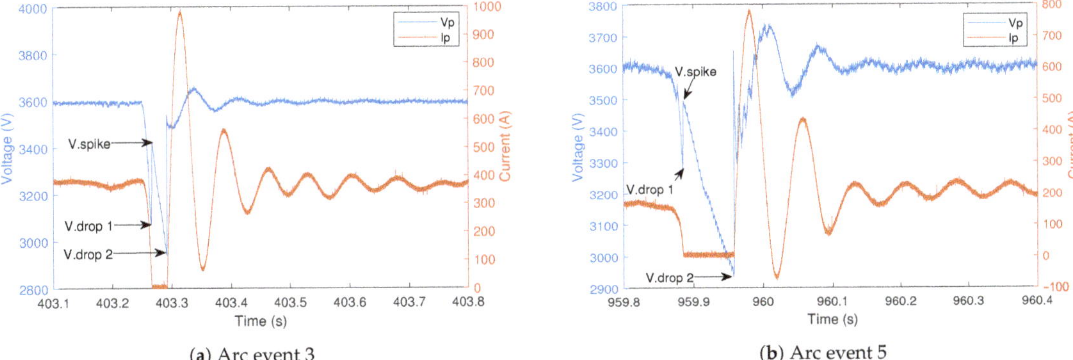

(**a**) Arc event 3 (**b**) Arc event 5

Figure 1. Time domain representation of pantograph voltage V_p and pantograph current I_p during arc event 3 and 5 occurred in a coasting phase.

After the first voltage spike, V_p continues to decrease almost linearly up to the point denoted as V.drop 2. This further decrease can be explained due to the increased gap between pantograph contact strip and the OCL, and this is again consistent with previous research studies [27,28]. After this voltage drop (V.drop 2), the pantograph restores its contact with the OCL, producing a voltage transient followed by oscillations (see the blue line—V_p). This behavior is also confirmed by previous work [7,8] and simulation [10] studies, and is a typical system response due to an applied transient/impulse. The oscillatory behavior of V_p is reflected in the current signal, causing I_p to oscillate.

Figure 2 shows recorded signals of two arc occurrences during two separate regenerative braking phases of the train. It is noted that the pantograph currents have negative values indicating current leaving the train. In Figure 2a, the arc is associated with a pantograph current magnitude decrease (from −280 A to −200 A approximately), causing a voltage spike (V.spike) due to the immediate release of stored energy in the magnetic field of the locomotive inductances. As soon as the contact restores, an immediate pantograph voltage drop (V.drop) occurs most likely caused by the loading effect of the locomotive filter, and then followed by a low-frequency oscillation observed both in V_p and I_p. A more peculiar behavior of V_p and I_p due to an arc is presented in Figure 2b, where the first pantograph detachment (characterized by an I_p magnitude decrease (I.drop 1) and V_p increase (V.spike 1) is followed by a second detachment (characterized by I.drop 2 and V.spike 2) before the final restoration of the mechanical sliding contact.

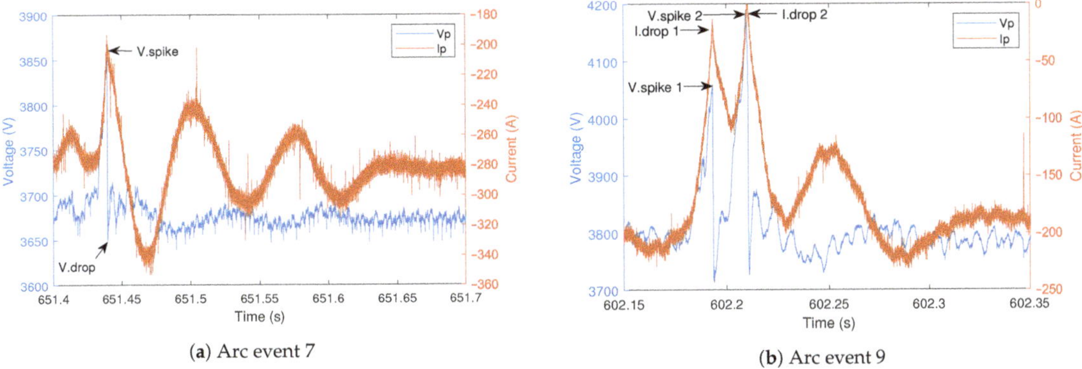

(**a**) Arc event 7 (**b**) Arc event 9

Figure 2. Time domain representation of pantograph voltage V_p and pantograph current I_p during arc event 7 and 9 occurred in a regenerative braking phase.

Many other recorded arcs are considered for the analysis in this paper. These arc events trigger similar oscillatory patterns experienced by the recorded quantities. Based on these types of wave forms, the low-frequency oscillation characteristics have been exploited by the proposed arc detection method presented in the following section.

4. Proposed Arc Detection Method

The proposed detection method is based on the instantaneous phase evaluation of the low-frequency oscillations triggered by the arc occurrence. As the raw data recorded at the pantograph level are real valued numbers, a mathematical operation is required to create an analytic complex valued signal, having both real and imaginary terms, thus enabling extraction of suitable phase information. In this analysis, the HT [36–38] is employed to quantify the instantaneous magnitude and phase of the considered time series signals. The HT is widely used in digital communication systems [39], mechanical vibration analysis [40], and recently proposed for Power Quality (PQ) disturbance monitoring [26,41] and Phasor Measurement Unit (PMU) parameter estimation [42]. Equation (1) depicts the general form of an analytic signal:

$$y(t) = y_{\mathrm{r}}(t) + \mathrm{j} \cdot y_{\mathrm{i}}(t) = y_{\mathrm{r}}(t) + \mathrm{j} \cdot HT\{y_{\mathrm{r}}(t)\} \quad (1)$$

where $y_{\mathrm{r}}(t)$ is the continuous time real signal, and $y_{\mathrm{i}}(t)$ is the imaginary terms of the analytic signal that is formed by the application of HT on $y_{\mathrm{r}}(t)$ [36–39]. In the time domain the $y_{\mathrm{i}}(t)$ is formed by the convolution operation of $y_{\mathrm{r}}(t)$ with the function $1/\pi t$ [36], as shown in Equation (2),

$$y_{\mathrm{i}}(t) = y_{\mathrm{r}}(t) * \frac{1}{\pi \cdot t} \quad (2)$$

whereas, in the frequency domain the HT operation introduces a constant phase shift of -90 degrees for every positive frequency component, and a 90 degrees for the negative frequency components present in the signal. This is achieved by using either a digital Finite Impulse Response (FIR) filter [37,39] or by using the Fourier transform approach [36]. In this paper the Fourier transform technique is employed to obtain the HT. This is normally implemented by the following steps:

- Perform the Fourier transform on the real input sequence;
- Set the DC and the Nyquist component to zero;
- Multiply the positive frequency components of the spectrum by $e^{-\mathrm{j}(\pi/2)}$, and the negative frequency components of the spectrum by $e^{\mathrm{j}(\pi/2)}$;
- Perform the inverse Fourier transform on the modified sequence to obtain the imaginary terms of the analytic signal;

Once the analytic signal is created, the instantaneous amplitude $A(t)$ (also known as envelope function of the signal), instantaneous phase $\phi(t)$, and the rate of change of the instantaneous phase (ROCOP) are computed by Equations (3)–(5), respectively.

$$A(t) = \sqrt{y_{\mathrm{r}}{}^{2}(t) + (HT\{y_{\mathrm{r}}(t)\})^{2}} \quad (3)$$

$$\phi(t) = \tan^{-1} \frac{HT\{y_{\mathrm{r}}(t)\}}{y_{\mathrm{r}}(t)} \quad (4)$$

$$ROCOP(t) = \frac{\mathrm{d}\phi(t)}{\mathrm{d}t} \quad (5)$$

Considering the discrete sampled form $y_{\mathrm{r}}[n]$ of the real time continuous signal $y_{\mathrm{r}}(t)$, and its respectively calculated analytic signal $y[n]{=}y_{\mathrm{r}}[n]{+}j{\cdot}y_{\mathrm{i}}[n]$, Equation (5) takes the form of Equation (6)

$$ROCOP[n] = \frac{\triangle\phi[n]}{\triangle t} = \frac{\phi_{\text{n}} - \phi_{\text{n-1}}}{\triangle t} \tag{6}$$

where n is the number of samples acquired in $y_r(t)$; for a time sampled signal $\triangle t$ is the sampling time which is 20 us, and $\triangle\phi$ is the difference in phase between successive.

A flow chart of the proposed algorithm is presented in Figure 3, where pre-filtering and post-filtering stages are used to attenuate the external noise of I_p signal and the ROCOP noise, respectively. These stages are explained in detail in Section 6 of the paper.

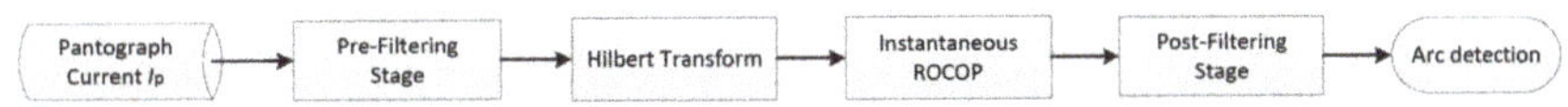

Figure 3. Flow chart of the proposed algorithm.

From the calculated parameters, ROCOP was found to be a good indicator for the arc detection in DC railway system as it is presented in following section.

5. Results of the Proposed Method

Signals (pantograph voltages and pantograph currents) have been recorded on-board the Trenitalia locomotive E464, and are categorized in two main groups [35]: arc signals detected during the traction/coasting phase that for convenience in the following analysis are identified as arc events 1 to 6, and arc signals detected during the regenerative braking phase identified as arc events 7 to 13.

Due to larger magnitudes and a longer time durations experienced by the arc triggered oscillations the following analysis considers the pantograph current as the quantity processed. Furthermore, by considering the current quantity, the method will be more immune to voltage disturbances such as voltage transients that may potentially compromise the accuracy of the method if V_p is considered.

Figures 4 and 5 present the instantaneous pantograph currents, together with instantaneous phase angles and instantaneous ROCOP values for all of the coasting and regenerative braking stage arc signals detected, respectively. To filter the incoming I_p signals from the external noise, a digital FIR filter of order 200 with cutoff frequency set at 100 Hz has been employed. In the next section an explanation for choosing the said order and cutoff frequency of the filter is provided. All the required calculations were performed in Matlab and LabVIEW programming using the HT toolsets provided in these programming environments.

The instantaneous phase angle calculations for each presented arc events in Figures 4 and 5 reflect the phase angle behavior during low-frequency oscillations triggered by the arc events. To avoid ROCOP spikes due to phase angle jumps (passing from π to -π and vice-versa), the phase angle is unwrapped. On the other hand, the ROCOP calculations provide clear evidence of the presence of an arcing event in all the considered signals, despite the apparent noise levels seen in the signal of Figure 4f. Noise may affect any arc detection triggering mechanism, but noise removal may be dealt with through appropriate filtering techniques. Arcs captured during the regenerative braking stage of the locomotive also are identified using ROCOP calculations as presented in Figure 5. Further to the arc detection, the method also can precisely localise in time the occurrence of the arc. This additional information can be useful to support, for example, future geographic arc localization algorithms.

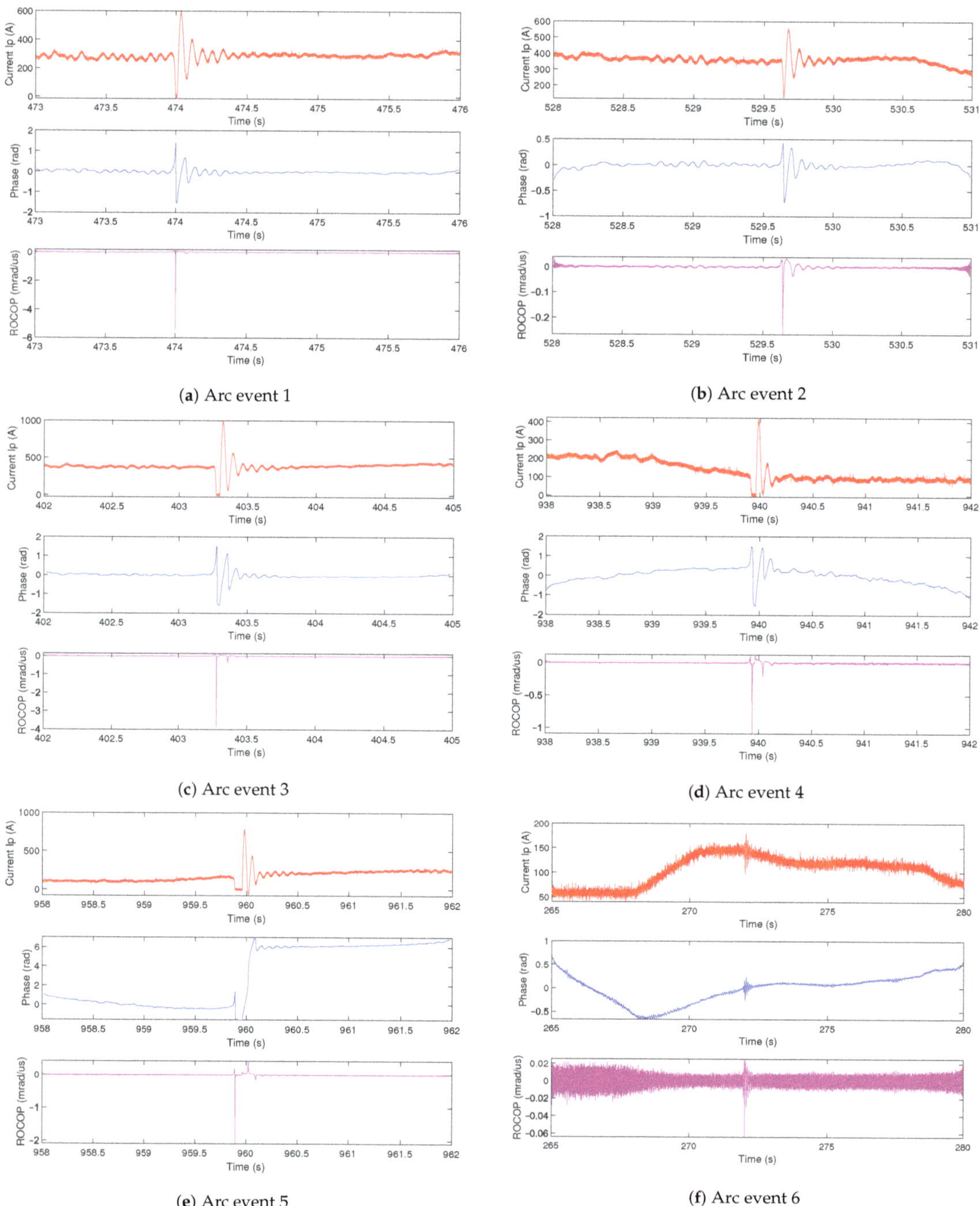

Figure 4. Instantaneous recorded I_p, instantaneous calculated phase angle, and instantaneous ROCOP for all of the arcs detected during the coasting phase.

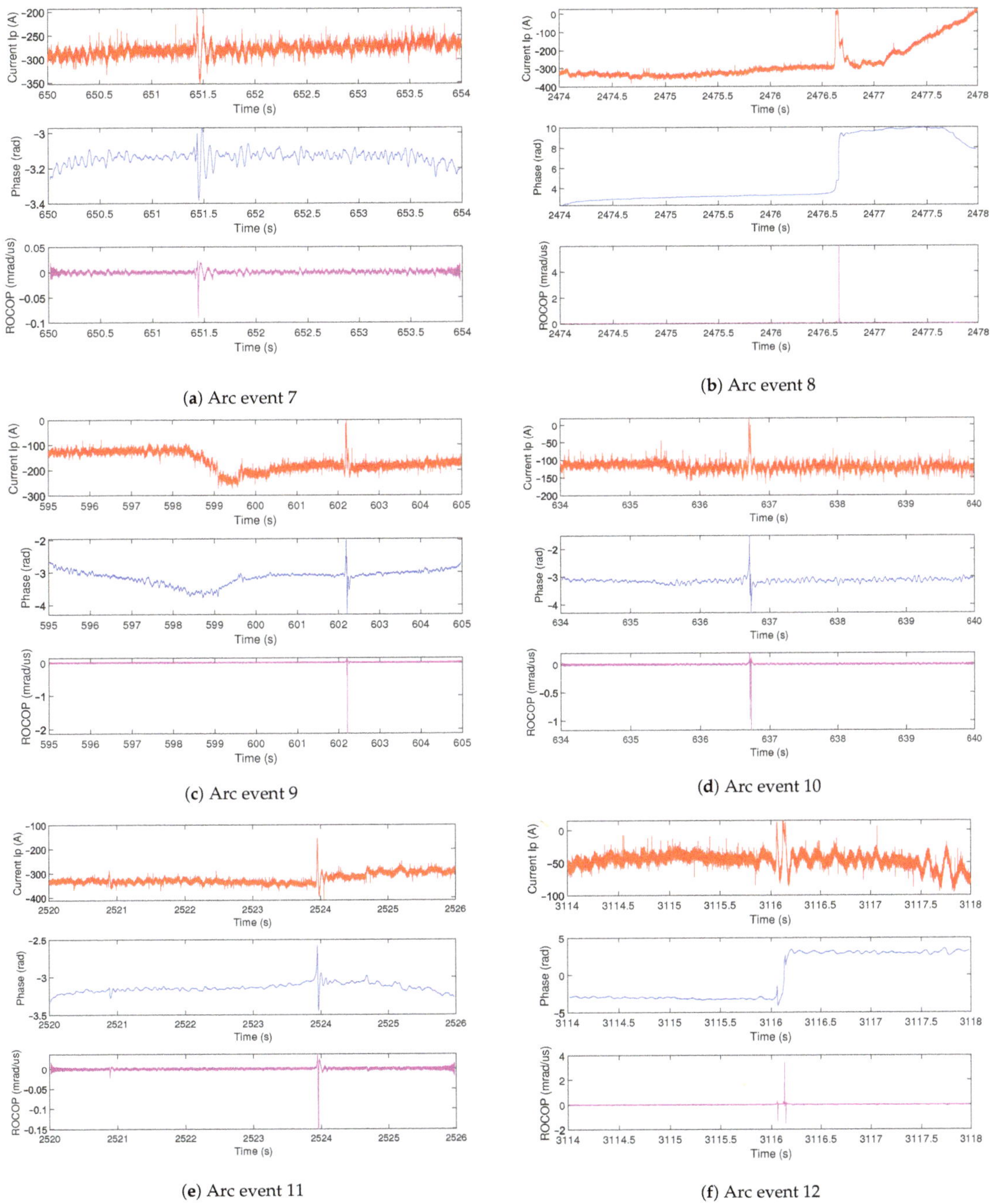

(**a**) Arc event 7 (**b**) Arc event 8

(**c**) Arc event 9 (**d**) Arc event 10

(**e**) Arc event 11 (**f**) Arc event 12

Figure 5. *Cont.*

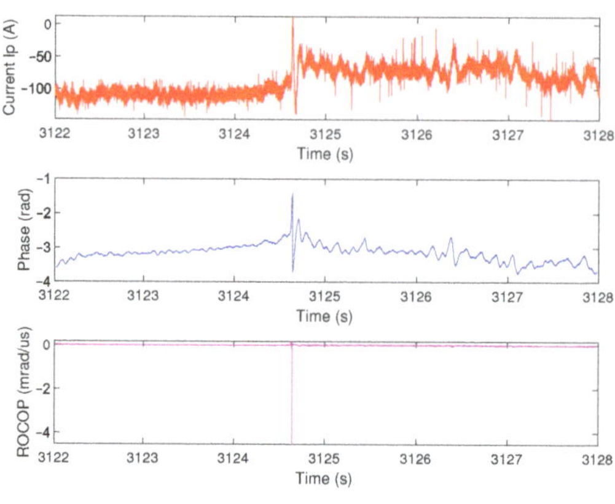

(**g**) Arc event 13

Figure 5. Instantaneous recorded I_p, instantaneous calculated phase angle, and instantaneous ROCOP for all of the arcs detected during the regenerative braking phase.

To appreciate the ROCOP noise level observed in Figure 4f, magnified versions of V_p and I_p are presented in Figure 6. Two possible factors that influence the ROCOP noise level can be: the increased noise level on top of the pantograph current signal, and the small variations in the pantograph voltage level (approximately 30 to 40 volt observed in Figure 6), that is of course reflected on the current oscillation having low magnitude variation (approximately 40 A to 45 A) during arcing events. Furthermore, indeed all the severe arcing events, characterised by increased air gap between the contact strip of the pantograph and OCL which either force the measured pantograph current to zero (observe Figures 4a,c–e and 5b,c,d,f,g, or produce significant variations of the current magnitude as in Figures 4b and 5a,e are not surrounded by significant noise level.

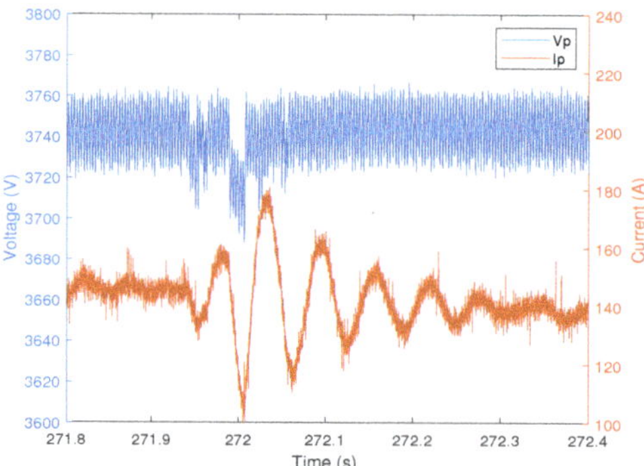

Figure 6. Time domain representation of V_p and I_p during arc event 6.

A possible reason for having such small variations in the magnitude of I_p (see Figure 4f) produced by non-significant arcs, can be explained with the locomotive position along the track and the level of current collected. By looking at Figure 4f, it can be observed that for the first 3 s (265 to 268 s) the train has been stationary (probably waiting on a station), characterized by low traction current with significant distortion. A high level of distortion when trains are near stations has previously been observed in [43]. After the stationary phase, the train has started and accelerated for approximately 2 to 3 s (268 s to 270 s), leaving the station (confirmed by the very low train speed of 27 km/h reported in [35]). While leaving stations, trains at low speed often change track position which forces the pantograph to change the OCL and the train wheels to experience mechanical oscillations causing some minor arcs to occur.

The preliminary results of the proposed method indicate the need for a fine-tuning of the algorithm to eliminate for example the noise level of calculated ROCOP (in Figure 4f) and to understand possible factors that may influence the effectiveness of the algorithm. These issues are considered in the next section of this paper.

6. Sensitivity Analysis

The effect of signal magnitude, noise and the length of the considered signal on the calculated ROCOP parameter are evaluated in this section. Pre- and post-filtering techniques are employed and presented for noise and harmonic attenuation of I_p, and smoothing of the phase derivative, respectively.

6.1. Pre-Filter Selection

To appreciate the harmonic presence and other frequency components of a DC rail network signal, a typical frequency spectrum of I_p signal during and after the arc event 1 is presented in Figure 7. The spectrum produced by the discrete Fourier transform (DFT) algorithm, has a frequency resolution of 1 Hz and is limited to 1000 Hz to clearly show the band of frequencies (5 Hz to approximately 250 Hz) excited by the arcing phenomenon. Within the frequency band, a particular frequency component (14 Hz) corresponding to the natural resonance frequency of the locomotive input filter [30,44] is also excited, causing as a consequence a significant rise in its respective magnitude.

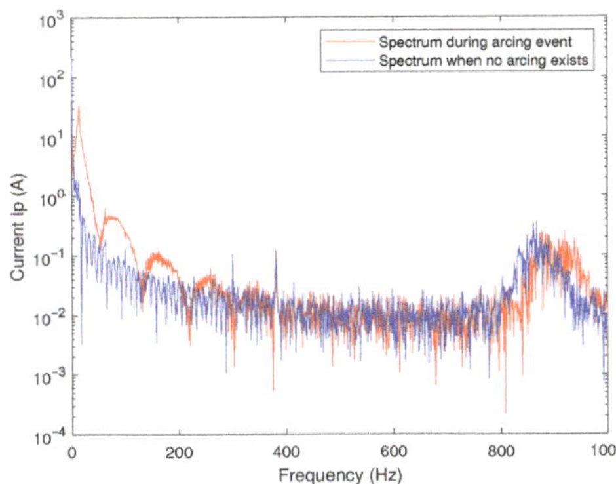

Figure 7. Frequency spectrum of I_p during and after arc occurrence.

To filter the incoming I_p signal from high frequency components and external noise, a digital Finite Impulse Response (FIR) filter is considered. ROCOP signal to noise ratio (SNR) parameter was calculated for one of the arc events (arc event 1), by considering different filter orders (ranging from 10 to 300) to filter the incoming I_p signal. The latter allows ROCOP SNR to filter order response curves to be constructed for several considered cutoff frequencies (50 Hz, 100 Hz, 200 Hz, 300 Hz, 500 Hz and 1000 Hz) as presented in Figure 8.

The analysis shows that low cutoff frequencies are needed to avoid frequency components and the noise infiltration into the ROCOP parameter. Cutoff frequency curves of 50 Hz to 300 Hz follow the same pattern and have negligible differences among them. This indicates that potentially any cutoff frequency (ranging from 50 Hz to 300 Hz) can be employed, which provide similar reproduced results. Differences begin to appear when considering the cutoff frequency of 500 Hz and are more obvious for the 1000 Hz curve, reaching differences of more than 10 dB. This large difference is caused by the presence of a band of frequencies ranging from 850 Hz to 950 Hz approximately, as shown in Figure 7.

The analysis also shows that higher filter orders provide larger ROCOP SNR, and the behaviour of the curve response is less fluctuating beyond the 150 filter order.

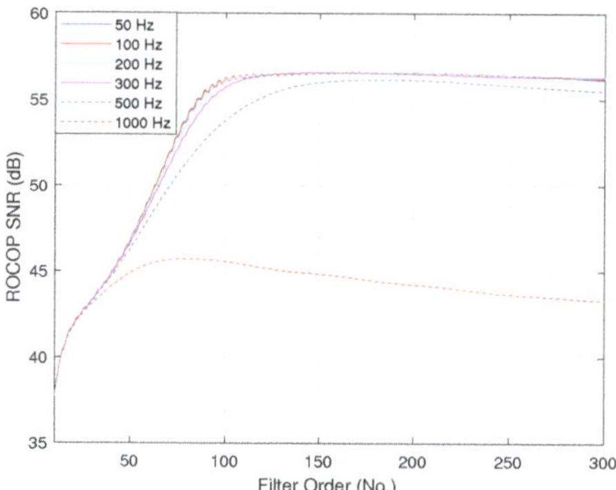

Figure 8. ROCOP SNR to filter order response curves constructed for filter cutoff frequencies of 50 Hz, 100 Hz, 200 Hz, 300 Hz, 500 Hz and 1000 Hz.

In this study, the filter order was selected to be 200, whereas the cutoff frequency 100 Hz. This cutoff frequency can also contribute to attenuate the 300 Hz component (as a result of the six-pulse rectifier installed in the electric substation) which is more distinctive when trains are near stations. The 300 Hz frequency component and other relevant components in I_p signal of arc event 6, are calculated as a function of time by the short-time Fourier transform (STFT) algorithm and are presented in Figure 9 by the spectrogram plot. The plot estimates and localizes in time the frequency content of the considered signal. It clearly indicates a 300 Hz frequency component, and a band of frequencies of 750 Hz to 850 Hz that are present continuously throughout the recording interval, and a temporary 14 Hz frequency component, excited by the arcing event. It is clear that the 300 Hz frequency component and the band of frequencies ranging from 750 Hz to 850 Hz approximately are not affected by the arc occurrence (visible after second 6).

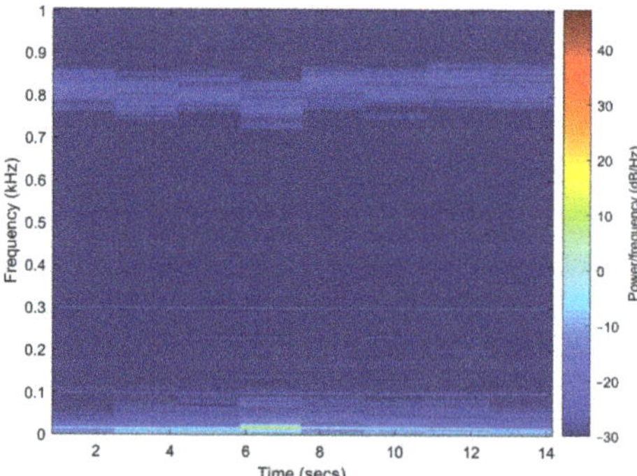

Figure 9. Spectrogram plot of I_p for arc event 6.

The selected filter parameters also have shown to remove more surrounding noise level in the ROCOP of Figure 4f.

6.2. Noise Sensitivity Analysis

To understand how noise affects the calculated ROCOP, a noise sensitivity analysis has been performed on I_p signal of arc event 1. This analysis considers the white Gaussian noise because it is the most common type of signal noise found in electric system signals [37].

Furthermore, after pre-filtering of I_p it is considered that all that remains within the passband can be reasonably approximated by additive white Gaussian noise.

While different levels of noise have been added to the recorded I_p, the maximum absolute value of the calculated ROCOP has been recorded to allow the SNR-ROCOP response curve to be constructed. Figure 10, presents the SNR-ROCOP response evaluated for different noise variances corresponding to a SNR ranging from 100 to 35 dB. An almost constant ROCOP behavior can be observed in the interval of 100 to 70 dB. Below 70 dB, the ROCOP measured parameter experiences an increased variability leading to increased uncertainty of the measured parameter.

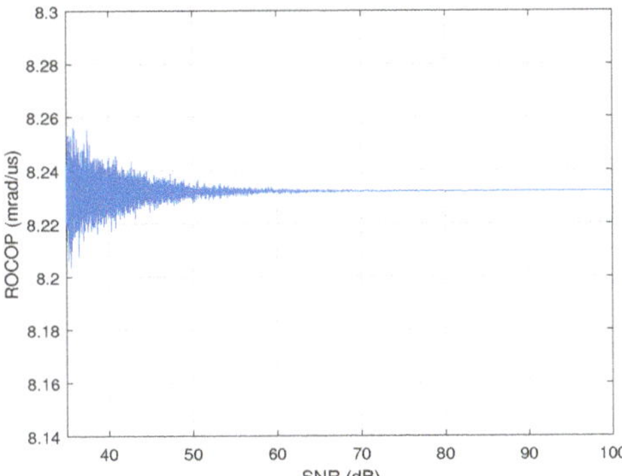

Figure 10. SNR-ROCOP response of I_p for arc event 1.

Although the analysis presented in Figure 10 is a good representation of the behavior of ROCOP magnitude at different noise levels, it does not represent the strength of the ROCOP magnitude itself to the surrounding ROCOP noise level. Therefore, further analysis is required to represent the impact of the noise floor level (developed in relation to the noise level contained in the analyzed current signal) on the measured parameter. Different levels of noise have been added to the I_p and the ROCOP SNR has been evaluated and presented in Figure 11 versus SNR corresponding to I_p.

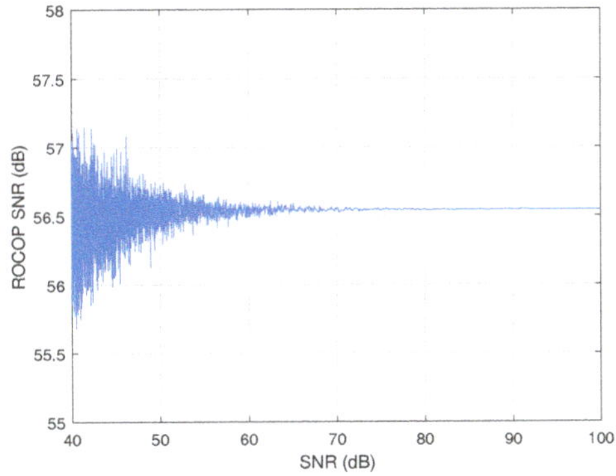

Figure 11. ROCOP SNR versus SNR of I_p for arc event 1.

As can be seen, the strength of the ROCOP magnitude to the ROCOP noise level is influenced by the increasing noise level in the I_p signal. However, this influence is almost negligible in the interval of 100 dB to 70 dB, and small (approximately only ± 1 dB) from 70 dB down to 40 dB. This is because of the efficient attenuation effect the FIR filter has on the input signal. Similarly, as observed in Figure 10, ROCOP SNR experiences an increased variability due to increased noise level starting below 70 dB SNR.

6.3. Post Filtering-ROCOP Noise Attenuation and Smoothing

The noise sensitivity analysis indicated an increase in the ROCOP noise floor level when externally applied noise magnitude increases. Because these noise levels can compromise any arc detection triggering mechanism, it is necessary to attenuate it as much as possible. Therefore, this subsection presents the use of a mitigation technique to further attenuate and smooth the noise level and fluctuations of the ROCOP calculated parameter observed in Figures 4f and 5a, allowing the algorithm to better interpret the processed data.

A moving average (MA) filter was considered to filter the ROCOP noise and spikes. As any type of filter, a MA filter offers various output responses for different filter order selection. To identify the behavior of the filter under different noise levels present on the arc event 1 signal, and to provide assistance on the selection of the filter order for the required application, Figure 12 presents the relationship between the ROCOP SNR and the MA filter order for several filter response curves.

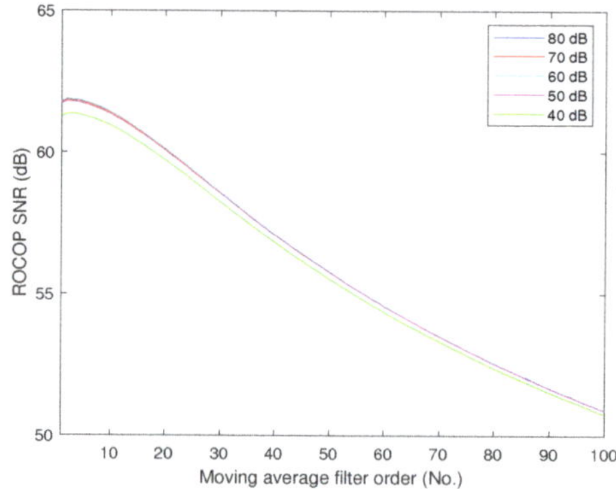

Figure 12. ROCOP SNR versus MA filter order for different SNR levels.

As can be seen, all the responses (corresponding to different noise levels present on I_p) follow the same decay and provide good attenuation (approximately 6 dB) when considering the 50th filter order. Increasing the filter order from 50 to 100 provides an additional 4.9 dB attenuation respectively for signals having SNR noise levels of 80 dB to 40 dB. Several numerical simulations have been performed for the identification of the best MA filter order for all ROCOP curves presented in Figures 4 and 5. It was found that the 50th filter order provides good attenuation of the ROCOP noise.

The absolute ROCOP calculated results filtered by the MA filter, for all the arcing captured events are presented respectively in Figures 13 and 14. These Figures clearly indicate the attenuation effect the MA filter has on the ROCOP noise. For example, the ROCOP noise level of Figure 4f has been considerably eliminated, and as a consequence the ROCOP parameter during arcing is more clearly differentiated from the noisy environment.

Another important feature of the MA filter is the smoothing effect it has on the ROCOP fluctuations. For example, the ROCOP spikes resulting from current spikes in Figure 5a,d,f,g, are all smoothed by both FIR and the MA filter. Consequently, ROCOP spikes will not be counted as arcing events with a proper triggering level in place. These features of the MA filter provide significant advantages to an arc detection triggering mechanism allowing the rest of the proposed method to accurately identify the input data.

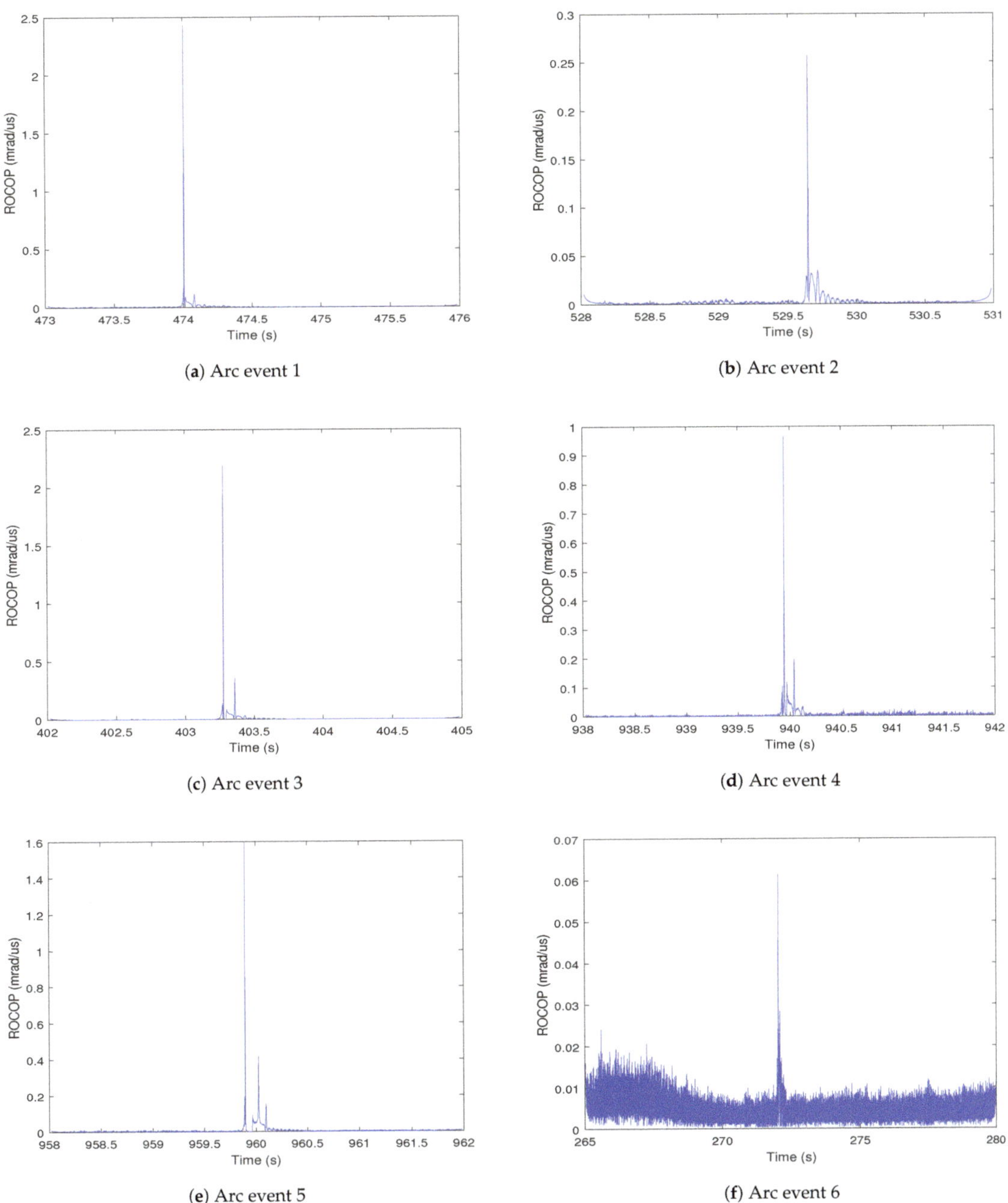

(a) Arc event 1

(b) Arc event 2

(c) Arc event 3

(d) Arc event 4

(e) Arc event 5

(f) Arc event 6

Figure 13. Absolute instantaneous ROCOP after being filtered by the MA filter for the arc events 1 to 6.

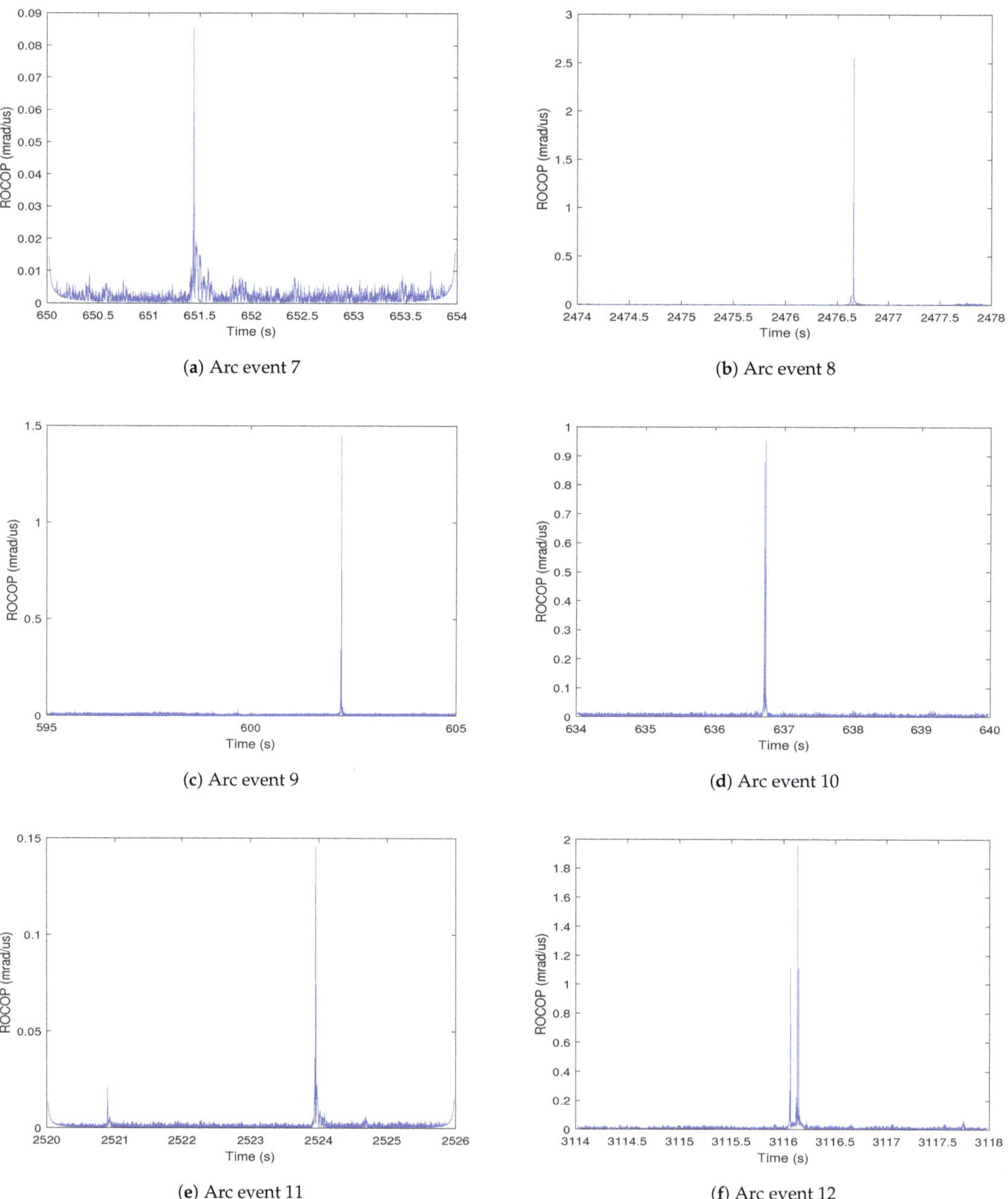

(**a**) Arc event 7 (**b**) Arc event 8

(**c**) Arc event 9 (**d**) Arc event 10

(**e**) Arc event 11 (**f**) Arc event 12

Figure 14. *Cont.*

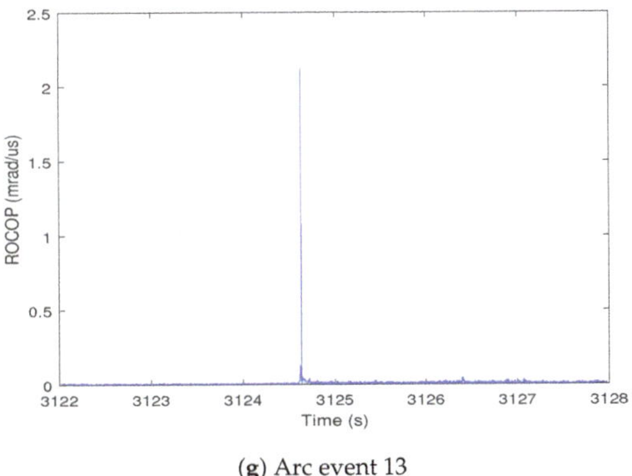

(g) Arc event 13

Figure 14. Absolute instantaneous ROCOP after being filtered by the MA filter for the arc events 7 to 13.

6.4. Magnitude Sensitivity Analysis

The following subsection presents the influence of pantograph current oscillation magnitude on the arc detection capability of the proposed method. In this analysis, I_p of arc event 11 (recorded during the regenerative braking phase) has been considered because the current was not forced to zero during the arcing event, allowing a proper scaling of the oscillation magnitude.

In such a context, the current oscillation instant during the arcing event was extracted, linearly scaled by a factor ranging between 0.9 to 0.1 in 0.1 steps, and then synthesized within the real data recordings to be used for testing the proposed method. The extracted portion of the current oscillation I_{p1} is presented in Figure 15, along with all the scaled versions denoted as I_{p2} to I_{p10}, corresponding to scaling factors 0.9 to 0.1, respectively.

The method was tested with all the 10 pantograph current versions, and the relationship between the calculated ROCOP SNR and the scaling factors is presented in Figure 16. ROCOP SNR values (for each scale factor) indicate the level of differentiation of the arc from the environment noise level. It is normally expected a decrease in ROCOP SNR with decreasing scale factors because the magnitude of the arc oscillation reduces significantly, for example to 36 A and 18 A (absolute values), respectively for scale factors 0.2 and 0.1 (corresponding to I_{p9} and I_{p10} of Figure 15.

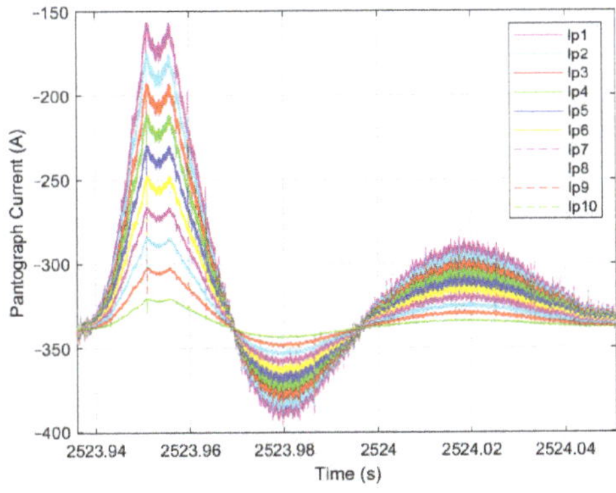

Figure 15. Pantograph current I_{p1} of arc event 11, along with scaled current versions I_{p2} to I_{p10}, respectively, for scaling factors 0.9 to 0.1.

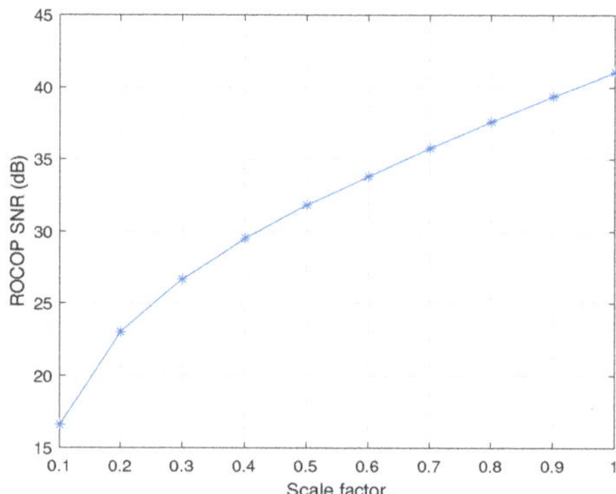

Figure 16. ROCOP SNR values versus the scaling factors for the considered arc event 11.

This magnitude sensitivity analysis has shown that the proposed method is capable of identifying all arcing events having reduced oscillation magnitudes down to 18A. However, arcing events of such a small oscillation magnitude can be considered non-significant when compared to oscillation magnitudes of the real captured arcing events (arc events 1 to 13).

6.5. Signal Window Length Impact

The effect of considered signal length on the ROCOP parameter is studied in this subsection. Pantograph current of arc event 1 has been considered for this analysis. Initially, the maximum ROCOP parameter observed within the time window processed signal is evaluated for the total length of the recorded signal (150,000 samples equals to 3 s). The signal was divided into equal lengths through different consecutive time windows and processed by the proposed method.

The maximum ROCOP values for different fixed window signal lengths are presented in Figure 17 (blue curve), where text annotations indicate the number of samples within each time window.

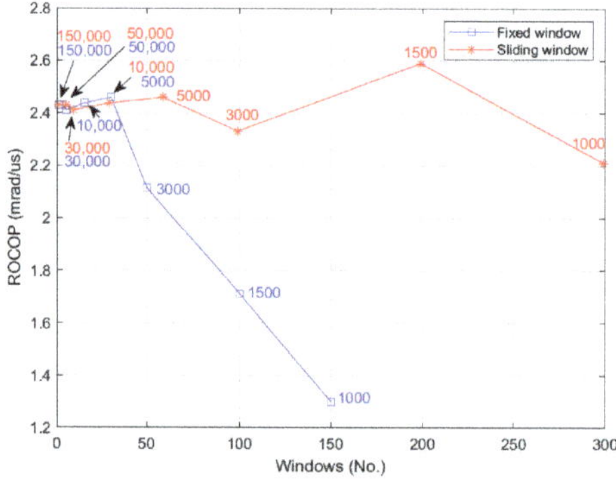

Figure 17. Maximum ROCOP values for different signal lengths.

A small difference in ROCOP magnitude exists between a long considered signal of 3 s and shorter signals of 1 s, 0.6 s, 0.2 s and 0.1 s corresponding to (50,000, 30,000, 10,000 and 5000) samples as presented by the blue curve in Figure 17. Decreasing further the length of the time window to 3000 samples, 1500 samples or 1000 samples produces a decrease in ROCOP magnitude. This behaviour is caused by short windows that fail to accommodate a compete arcing event. This is demonstrated when sliding windows of 50% overlap are used (see the red curve in Figure 17), which better accommodate the arcing event for processing, and consequently produce ROCOP magnitudes of small differences for short windows.

The analysis presented in this section has indicated that the proposed arc detection method is capable of detecting arcs of different magnitude at different noise levels in the DC rail network.

In addition, standard EN 50317 [1] requires the current collection quality of the pantograph-catenary system to be evaluated. For this purpose the percentage of arcing is proposed in [1], to be calculated as the ratio of total duration of all arcs to the total run time for pantograph current values above the 30% of the nominal current per pantograph. The proposed arc detection method clearly cannot evaluate the time duration of the arcs, but the detected arcs can be used to compute the number of arcs per kilometre, as an alternative criterion (also acceptable by standard EN 50317 [1]) for evaluating the current collection quality.

As presented, the method does not employ an extensive processing technique, and therefore it can be implemented in real-time. One option to accomplish this could be to embed the algorithm within the actual train computational and monitoring units. Alternatively, the on-board instrumentation of the train can be considered for the possibility of sharing the measured pantograph current with external acquisition and computational units, such as microcontrollers having real-time computational capabilities.

7. Conclusions

This paper has presented a novel arc detection technique for DC Railway Systems. The method quantifies the instantaneous ROCOP of the pantograph current oscillating signal during the arc occurrence with the help of the Hilbert transform. Real current signals measured at a 3 kV DC pantograph level of a running locomotive have been applied to test the performance of the algorithm, and it was demonstrated that the phase derivative can detect and localize in time all the electric arcs occurring at both the coasting and regenerative braking phase.

The sensitivity of the algorithm to external applied white Gaussian noise, various synthesized oscillating magnitudes representing different arcing events, and the length of the considered signal is also evaluated. It was demonstrated that the noise level in the pantograph current has little influence in ROCOP parameter due to efficient noise attenuation provided by the FIR filter. ROCOP parameter measured for signals having SNR between 100 dB to 70 dB was high repeatable, whereas, below the 70 dB of SNR, ROCOP becomes variable, leading to increased uncertainty.

A moving average filter was employed to attenuate and smooth the ROCOP noise level and spikes, and it was demonstrated that the ROCOP parameters after the filtering stage were clearly differentiated from the rest of the noisy environment.

The magnitude sensitivity analysis has demonstrated that the proposed method is also capable to identify minor arcing events having oscillation magnitudes down to 18 A. These arcs are typical in very low train speeds and are probably caused when trains change the track position.

A small difference in ROCOP magnitude for different considered signal lengths (3 s and 1 s, 0.6 s, 0.2 s and 0.1 s lengths) has also been demonstrated. For very short considered time windows (1000 to 3000) samples, it was shown that sliding windows outperforms fixed length windows because complete arcing events are better accommodated, providing so consistently results.

The DC arc detection method in this paper provides a new tool for accurate, real-time condition monitoring of critical electrified rail infrastructure. This has the potential to enable predictive maintenance, thereby reducing operational costs and improving safety.

Author Contributions: Y.S. undertook the literature review and main writing of the paper; Y.S. and B.G.S. developed the arcing algorithms, analysed the signals and processed the results; C.M. and S.M.B. reviewed, commented and edited the whole document including the figures; B.G.S. edited the manuscript and guided the research work. All authors read and agreed to the published version of the manuscript.

Funding: The University of Strathclyde and METAS received funding under 16ENG04 MyRailS project, which has received funding from the EMPIR programme co-financed by the Participating States and from the European Union's Horizon 2020 research and innovation programme. This work also received funding from the European Union's Horizon 2020 research and innovation programme MEAN4SG under Marie Skłodowska-Curie Grant Agreement 676042.

Acknowledgments: The authors would like to thank the partners of MyRailS project for making available the data for the analysis presented in this paper.

Conflicts of Interest: The authors declare no conflict of interest.

References

1. *CENELEC EN 50317:2012 Railway Applications–Current Collection Systems—Requirements for and Validation of Measurements of the Dynamic Interaction between Pantograph and Overhead Contact Line*; CENELEC: Brussels, Belgium, 2012.
2. *CENELEC EN 50367:2012 Railway Applications–Current Collection Systems—Technical Criteria for the Interaction between Pantograph and Overhead Line (to Achieve Free Access)*; CENELEC: Brussels, Belgium, 2012.
3. *CENELEC EN 50119:2020 Railway Applications–Fixed Installations—Electric Traction Overhead Contact Lines*; CENELEC: Brussels, Belgium, 2020.
4. Bruno, O.; Landi, A.; Papi, M.; Sani, L. Phototube sensor for monitoring the quality of current collection on overhead electrified railways. *Proc. Inst. Mech. Eng. Part F J. Rail Rapid Transit* **2001**, *215*, 231–241. [CrossRef]
5. Barmada, S.; Landi, A.; Papi, M.; Sani, L. Wavelet multiresolution analysis for monitoring the occurrence of arcing on overhead electrified railways. *Proc. Inst. Mech. Eng. Part F J. Rail Rapid Transit* **2003**, *217*, 177–187. [CrossRef]
6. Balestrino, A.; Bruno, O.; Giorgi, P.; Landi, A.; PAPI, M.; Sani, L.; Giuseppe, A.; Elettriche, S.P. Electric welding effect: Detection via phototube sensor and maintenance activities. In Proceedings of the CDROM of the World Congress on Railway Research, Cologne, Germany, 25–29 November 2001; pp. 19–23.
7. Tellini, B.; Macucci, M.; Giannetti, R.; Antonacci, G.A. Conducted and radiated interference measurements in the line-pantograph system. *IEEE Trans. Instrum. Meas.* **2001**, *50*, 1661–1664. [CrossRef]
8. Midya, S.; Bormann, D.; Schutte, T.; Thottappillil, R. Pantograph arcing in electrified railways—Mechanism and influence of various parameters—Part I: With DC traction power supply. *IEEE Trans. Power Deliv.* **2009**, *24*, 1931–1939. [CrossRef]
9. Betts, A.; Hall, J.; Keen, P. Condition monitoring of pantographs. In Proceedings of the International Conference on Main Line Railway Electrification 1989, York, UK, 25–28 September 1989; pp. 129–133.
10. Crotti, G.; Giordano, D.; Roccato, P.; Delle Femine, A.; Gallo, D.; Landi, C.; Luiso, M.; Mariscotti, A. Pantograph-to-OHL Arc: Conducted Effects in DC Railway Supply System. In Proceedings of the 2018 IEEE 9th International Workshop on Applied Measurements for Power Systems (AMPS), Bologna, Italy, 26–28 September 2018; pp. 1–6.
11. Crotti, G.; Delle Femine, A.; Gallo, D.; Giordano, D.; Landi, C.; Luiso, M.; Mariscotti, A.; Roccato, P.E. Pantograph-to-OHL arc: Conducted effects in DC railway supply system. *IEEE Trans. Instrum. Meas.* **2019**, *68*, 3861–3870. [CrossRef]
12. Bormann, D.; Midya, S.; Thottappillil, R. DC components in pantograph arcing: Mechanisms and influence of various parameters. In Proceedings of the 2007 18th International Zurich Symposium on Electromagnetic Compatibility, Munich, Germany, 24–28 September 2007; pp. 369–372.
13. Landi, A.; Menconi, L.; Sani, L. Hough transform and thermo-vision for monitoring pantograph-catenary system. *Proc. Inst. Mech. Eng. Part F J. Rail Rapid Transit* **2006**, *220*, 435–447. [CrossRef]
14. Aydin, I. A new approach based on firefly algorithm for vision-based railway overhead inspection system. *Measurement* **2015**, *74*, 43–55. [CrossRef]
15. Barmada, S.; Raugi, M.; Tucci, M.; Romano, F. Arc detection in pantograph-catenary systems by the use of support vector machines-based classification. *IET Electr. Syst. Transp.* **2013**, *4*, 45–52. [CrossRef]
16. Aydın, İ.; Yaman, O.; Karaköse, M.; Çelebi, S.B. Particle swarm based arc detection on time series in pantograph-catenary system. In Proceedings of the 2014 IEEE International Symposium on Innovations in Intelligent Systems and Applications (INISTA) Proceedings, Alberobello, Italy, 23–25 June 2014; pp. 344–349.
17. Barmada, S.; Tucci, M.; Menci, M.; Romano, F. Clustering techniques applied to a high-speed train pantograph–catenary subsystem for electric arc detection and classification. *Proc. Inst. Mech. Eng. Part F J. Rail Rapid Transit* **2016**, *230*, 85–96. [CrossRef]
18. Huang, H.H.; Chen, T.H. Development of method for assessing the current collection performance of the overhead conductor rail systems used in electric railways. *Proc. Inst. Mech. Eng. Part F J. Rail Rapid Transit* **2008**, *222*, 159–168. [CrossRef]

19. Seferi, Y.; Clarkson, P.; Blair, S.M.; Mariscotti, A.; Stewart, B.G. Power Quality Event Analysis in 25 kV 50 Hz AC Railway System Networks. In Proceedings of the 2019 IEEE 10th International Workshop on Applied Measurements for Power Systems (AMPS), Aachen, Germany, 25–27 September 2019; pp. 1–6. [CrossRef]
20. Mariscotti, A. Results on the power quality of French and Italian 2× 25 kV 50 Hz railways. In Proceedings of the 2012 IEEE International Instrumentation and Measurement Technology Conference Proceedings, Graz, Austria, 13–16 May 2012; pp. 1400–1405.
21. Seferi, Y.; Blair, S.M.; Mester, C.; Stewart, B.G. Power Quality Measurement and Active Harmonic Power in 25 kV 50 Hz AC Railway Systems. *Energies* **2020**, *13*, 5698. [CrossRef]
22. Gao, G.; Yan, X.; Yang, Z.; Wei, W.; Hu, Y.; Wu, G. Pantograph–Catenary arcing detection based on electromagnetic radiation. *IEEE Trans. Electromagn. Compat.* **2018**, *61*, 983–989. [CrossRef]
23. Huang, N.E.; Shen, Z.; Long, S.R.; Wu, M.C.; Shih, H.H.; Zheng, Q.; Yen, N.C.; Tung, C.C.; Liu, H.H. The empirical mode decomposition and the Hilbert spectrum for nonlinear and non-stationary time series analysis. *Proc. R. Soc. Lond. Ser. A Math. Phys. Eng. Sci.* **1998**, *454*, 903–995. [CrossRef]
24. Mahela, O.P.; Shaik, A.G.; Gupta, N. A critical review of detection and classification of power quality events. *Renew. Sustain. Energy Rev.* **2015**, *41*, 495–505. [CrossRef]
25. Ozgonenel, O.; Yalcin, T.; Guney, I.; Kurt, U. A new classification for power quality events in distribution systems. *Electr. Power Syst. Res.* **2013**, *95*, 192–199. [CrossRef]
26. Jayasree, T.; Devaraj, D.; Sukanesh, R. Power quality disturbance classification using Hilbert transform and RBF networks. *Neurocomputing* **2010**, *73*, 1451–1456. [CrossRef]
27. Ammerman, R.F.; Sen, P. Modeling high-current electrical arcs: A volt-ampere characteristic perspective for AC and DC systems. In Proceedings of the 2007 39th North American Power Symposium, Las Cruces, NM, USA, 30 September–2 October 2007; pp. 58–62.
28. Midya, S.; Bormann, D.; Larsson, A.; Schutte, T.; Thottappillil, R. Understanding pantograph arcing in electrified railways-influence of various parameters. In Proceedings of the 2008 IEEE International Symposium on Electromagnetic Compatibility, Detroit, MI, USA, 18–22 August 2008; pp. 1–6.
29. Usuda, T.; Ikeda, M.; Yamashita, Y. Prediction of contact wire wear in high speed railways. In Proceedings of the 9th World Congress on Railway Research, Paris, France, 21 September 2011; pp. 1–10.
30. Mariscotti, A.; Giordano, D. Experimental characterization of pantograph arcs and transient conducted phenomena in DC railways. *Acta Imeko* **2020**, *9*, 10–17. [CrossRef]
31. Li, T.; Wu, G.; Zhou, L.; Gao, G.; Wang, W.; Wang, B.; Liu, D.; Li, D. Pantograph arcing's impact on locomotive equipments. In Proceedings of the 2011 IEEE 57th Holm Conference on Electrical Contacts (Holm), Minneapolis, MN, USA, 11–14 September 2011; pp. 1–5.
32. Mariscotti, A.; Marrese, A.; Pasquino, N.; Moriello, R.S.L. Time and frequency characterization of radiated disturbance in telecommunication bands due to pantograph arcing. *Measurement* **2013**, *46*, 4342–4352. [CrossRef]
33. MyRailS Website. Available online: https://myrails.it/ (accessed on 14 January 2021).
34. Giordano, D.; Clarkson, P.; Gamacho, F.; van den Brom, H.E.; Donadio, L.; Fernandez-Cardador, A.; Spalvieri, C.; Gallo, D.; Istrate, D.; De Santiago Laporte, A.; et al. Accurate Measurements of Energy, Efficiency and Power Quality in the Electric Railway System. In Proceedings of the 2018 Conference on Precision Electromagnetic Measurements (CPEM 2018), Paris, France, 8–13 July 2018; pp. 1–2.
35. Giordano, D.; Signorino, D.; Mariscotti, A.; Cipolletta, G.; Delle Femine, A.; Gallo, D.; Biancucci, A.; Donadio, L.; Balic, F.; Quintana, J. Pantograph Arcing in DC Railway Systems: Transient Behavior of Voltage and Current Recorded at Pantograph. *Mendeley Data* **2020**, *V1*. [CrossRef]
36. Marple, L. Computing the discrete-time "analytic" signal via FFT. *IEEE Trans. Signal Process.* **1999**, *47*, 2600–2603. [CrossRef]
37. Hussain, Z.M.; Sadik, A.Z.; O'Shea, P. *Digital Signal Processing: An Introduction with MATLAB and Applications*; Springer Science & Business Media: Berlin, Germany, 2009.
38. Van Drongelen, W. *Signal Processing for Neuroscientists*; Academic Press: Cambridge, MA, USA, 2018.
39. Romero, D.T.; Jovanovic, G. Digital FIR Hilbert transformers: Fundamentals and efficient design methods. *MATLAB-A Fundam. Tool Sci. Comput. Eng. Appl.* **2012**, *1*, 445–482.
40. Feldman, M. Hilbert transform in vibration analysis. *Mech. Syst. Signal Process.* **2011**, *25*, 735–802. [CrossRef]
41. Santos-Hernandez, J.A.; Valtierra-Rodriguez, M.; Amezquita-Sanchez, J.P.; Romero-Troncoso, R.D.J.; Camarena-Martinez, D. Hilbert filter based FPGA architecture for power quality monitoring. *Measurement* **2019**, *147*, 106819. [CrossRef]
42. Razo-Hernandez, J.R.; Valtierra-Rodriguez, M.; Granados-Lieberman, D.; Tapia-Tinoco, G.; Rodriguez-Rodriguez, J.R. A phasor estimation algorithm based on Hilbert transform for P-class PMUs. *Adv. Electr. Comput. Eng.* **2018**, *18*, 97–105. [CrossRef]
43. Crotti, G.; Giordano, D.; Signorino, D.; Delle Femine, A.; Gallo, D.; Landi, C.; Luiso, M.; Biancucci, A.; Donadio, L. Monitoring Energy and Power Quality On Board Train. In Proceedings of the 2019 IEEE 10th International Workshop on Applied Measurements for Power Systems (AMPS), Aachen, Germany, 25–27 September 2019; pp. 1–6.
44. Hill, R.J.; Fracchia, M.; Pozzobon, P.; Sciutto, G. A frequency domain model for 3 kV DC traction DC-side resonance identification. *IEEE Trans. Power Syst.* **1995**, *10*, 1369–1375. [CrossRef]

Article

Decentralized Control Strategy for an AC Co-Phase Traction Microgrid

Lan Ma [1], Yuhua Du [2,*], Leilei Zhu [1], Fan Yang [1], Shibiao Xiang [1] and Zeliang Shu [1]

[1] School of Electrical Engineering, Southwest Jiaotong University, Chengdu 611756, China; mlan@swjtu.edu.cn (L.M.); zhuleilei@my.swjtu.edu.cn (L.Z.); jalorfy@my.swjtu.edu.cn (F.Y.); xsb@swjtu.edu.cn (S.X.); shuzeliang@swjtu.edu.cn (Z.S.)

[2] College of Engineering, Temple University, Philadelphia, PA 19122, USA

* Correspondence: yuhua.du@temple.edu

Abstract: High speed and heavy loads have become more prevalent in the traction power supply system recently. To ensure system operating stability, better power quality, and sufficient power capacity, improvements are needed over the conventional traction system. Inspired by the concept of a microgrid (MG), an AC co-phase traction MG system was proposed. Substations were connected to the traction grid as distributed generators operate in islanded mode. Droop control was adopted as the primary control to stabilize the system's operating frequency and voltage. Considering the operating features of the substation and locomotive load, a de-centralized secondary control strategy was proposed for AC co-phase traction MG system operation with enhanced resiliency. The proposed control strategy could increase system stability and prevent circulation currents between substations. Moreover, the proposed de-centralized coordination between substations does not rely on communication, which promotes the system's "plug-and-play" functionality. Stability analysis was undertaken and the proposed controller was proved to be exponentially stable. The dynamic response of the proposed controller was validated using comprehensive case studies in MATLAB/Simulink.

Keywords: circulation current; co-phased traction system; secondary control; microgrid

Citation: Ma, L.; Du, Y.; Zhu, L.; Yang, F.; Xiang, S.; Shu, Z. Decentralized Control Strategy for an AC Co-Phase Traction Microgrid. *Energies* **2021**, *14*, 7. https://dx.doi.org/10.3390/en14010007

Received: 6 November 2020
Accepted: 14 December 2020
Published: 22 December 2020

Publisher's Note: MDPI stays neutral with regard to jurisdictional claims in published maps and institutional affiliations.

1. Introduction

With the development of high-speed rail, the requirements for traction power quality and capacity have increased [1]. A conventional railway traction power system suffers from critical power quality problems, such as voltage unbalance, poor power factor, and harmonic distortion. To compensate for the reactive power and harmonics of the catenary, a co-phase traction power supply system was proposed based on the design of an active power compensator [2,3]. A benefit of the co-phase system is that half of the neutral sections can be canceled. To link all the catenaries of the substations without any neutral sections, an advanced co-phase system was proposed in [4]. Instead of a traction transformer, a three-phase to single-phase converter was adopted by the substation [5]. Therefore, the traction voltage can be controlled and there's only one catenary in the system. However, the system could be destabilized with varying line impedances or loading conditions.

The substations could be controlled as distributed generators (DGs), and the advanced co-phase traction power grid could be treated as a special type of microgrid (MG). Different from a conventional MG [6,7], there are several unique characteristics of a traction MG. First, the locomotive load moves on the traction grid, which means that the system impedance distribution changes regularly. Second, due to the unique features of the traction power infrastructure, the distance between two substations is fixed to dozens of kilometers, which results in a non-ignorable line impedance effect. Third, unlike the DGs with various capacities, converter-based substations with an identical power capacity are connected to the traction grid. Automatic power-sharing in response to the line impedance variation should be realized [8], which is different from the proportional power-sharing in a conventional

MG [9,10]. Lastly, since the traction power system is an autonomous AC system, the AC traction MG operates constantly under islanded operation mode [11,12].

Coordinated operation between multiple DGs in the context of autonomous MG operation has been frequently discussed in the literature. Conventional islanded MGs contain multiple inverter-based DGs that are installed in parallel [13]. Differing in their control objectives, there are three typical control modes, namely, P/Q control, V/F control, and droop control [14]. Under P/Q control, a DG inverter operates as a current source whose power output is as commanded. In contrast, a fixed voltage and frequency reference is adopted by the inverter in V/F control. Due to the differences in inverter impedance and capacity, an overcurrent problem would be frequently generated when multiple V/F-controlled inverters operate in parallel [15]. Moreover, a V/F-controlled DG normally operates as the only slack node in an islanded MG, which requires its installed power capacity to be sufficiently large. To automatically stabilize the system frequency and voltage while achieving DG power-sharing, droop control is proposed [16,17]. However, due to system line impedance variation, mismatched DG reactive power-sharing is inevitable in conventional droop control [18]. To ensure system operation synchronization, voltage regulation, power balance, and load sharing, many advanced control strategies are carried out. Improved droop control [19,20], virtual impedance-based control [21,22], and improved hierarchical control strategies [23] are often adopted to eliminate the frequency and voltage deviations caused by droop control and achieve accurate power-sharing [24,25].

Similar challenges to voltage regulation and power-sharing exist in an advanced co-phase traction system. Inspired by the concept of an MG, an AC co-phase traction MG was proposed in this study.

- Considering the characteristics of a traction system, a decentralized control strategy was proposed to enhance the traction MG operating performance. In the proposed strategy, no communication between substations was introduced, which makes it easier for the substation to realize plug-and-play functionality.
- Due to the benefit of automatic power-sharing between substations, the capacity of the high-speed railway could easily be expanded through the connection of more substations without modification, which differs from the ones in a conventional traction grid.
- As the primary control is implemented, the traction grid operating frequency and voltage can be stabilized via autonomous regulation through multiple substations coordinating together. In addition, the phase and voltage magnitude error is eliminated in the secondary control. Despite the effect of the long-distance line impedance, the traction grid voltage is maintained in an acceptable range along the catenary line. Moreover, the circulation current is eliminated due to the elimination of the substation voltage difference.
- Without neutral sections, there is only one catenary in a traction MG, which makes it more flexible for distributed renewable resources to integrate with the traction grid in the future.

The rest of this paper is constructed as follows. The proposed AC co-phase traction MG system is introduced in Section 2. In Section 3, the line-impedance-based power-sharing and circulation current are analyzed and the proposed control strategy is presented. The small-signal stability of the system is analyzed in Section 4. Simulations results based on three substations are derived in Section 5 to validate the effectiveness of the proposed strategy for the AC co-phase traction MG system.

2. AC Co-Phase Traction Microgrid

In an advanced co-phase system, only one catenary line is utilized, which does not require a neutral section. All substations are connected to the single-phase AC traction grid [4]. Inspired by the concept of an MG, an AC co-phase traction MG is proposed in this section. The system topology is presented schematically in Figure 1.

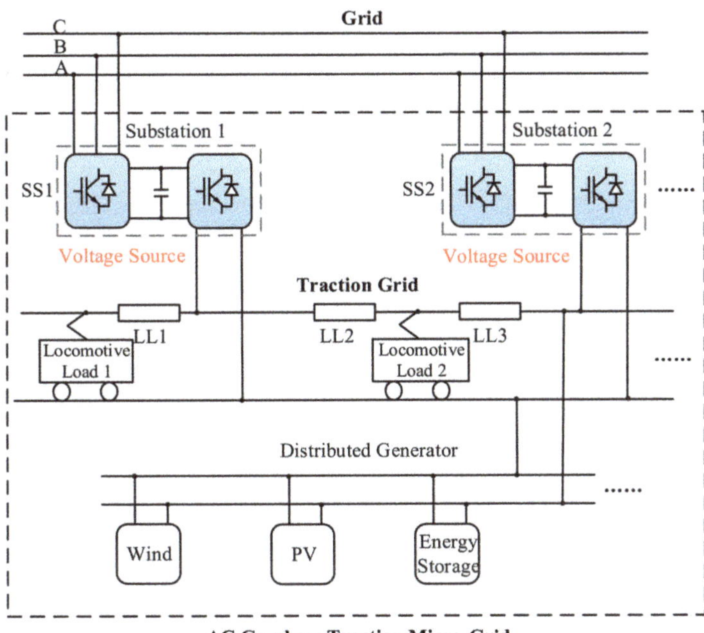

Figure 1. AC co-phase traction microgrid (PV: photovoltaics).

As shown in Figure 1, substations are utilized to stabilize the traction grid and energize the locomotive loads. The traction MG and the main grid are interconnected by substations through AC/AC converters (SS1 and SS2). Compared to a conventional transformer, a converter-based substation operates as a controllable voltage source with better controllability. Due to the characteristics of an AC co-phase traction MG, the substation converter always operates under islanded mode. The distance between the substations is normally at the kilometer (km) level and the effect from the line impedance (LL1, LL2, and LL3) is non-negligible. Additionally, the capacity of each substation is identical, which is different from the case in a typical MG where the generation capacities would vary.

Only a locomotive load is presented in the traction MG. As the trains move, the loads are not connected at fixed points. Line impendence between the source and load is constantly changing, and thus, power-sharing between substations is difficult to achieve using the conventional droop controller. Based on a line impedance change, an automatic power-sharing strategy can be adopted. The locomotive load will automatically be energized by the nearest substation. Such a method is not only easy to apply but also maintains the operation voltage of the loads within an acceptable range all the time.

Additionally, DGs can also be directly connected to the traction grid. The power generated by the DGs can be utilized to support loads or flow into the grid through the substation. Without any neutral section, it is also more flexible for DGs to connect with the traction gird. As a result, the proposed traction MG will help with renewable penetration along the railway.

3. Proposed Traction MG

In an islanded MG, at least one voltage source that acts as a slack bus needs to be present. However, in a traction grid, all the substations are distributed on one long feeder. If only one slack bus is presented, due to the line impedance effect, the system will be destabilized by the substations that are far away [5]. As the voltage difference between the substations increases, there will also be more power losses caused by the circulation current.

To achieve a robust system, all the substations are operated as a voltage source, which supports the voltage of the traction MG. A decentralized control strategy is proposed for the substations on the traction grid, as shown in Figure 2. In the traction system, the catenary line impedance changes the locomotive location. For an effective load power allocation, an automatic power-sharing function can be achieved under this control strategy. Furthermore, a two-layer-based control strategy is designed for the system stability enhancement and the output voltage synchronization. In the following section, the circulation current between the substations is first analyzed; then, primary control and secondary control strategies are introduced for the substation operation in the traction MG.

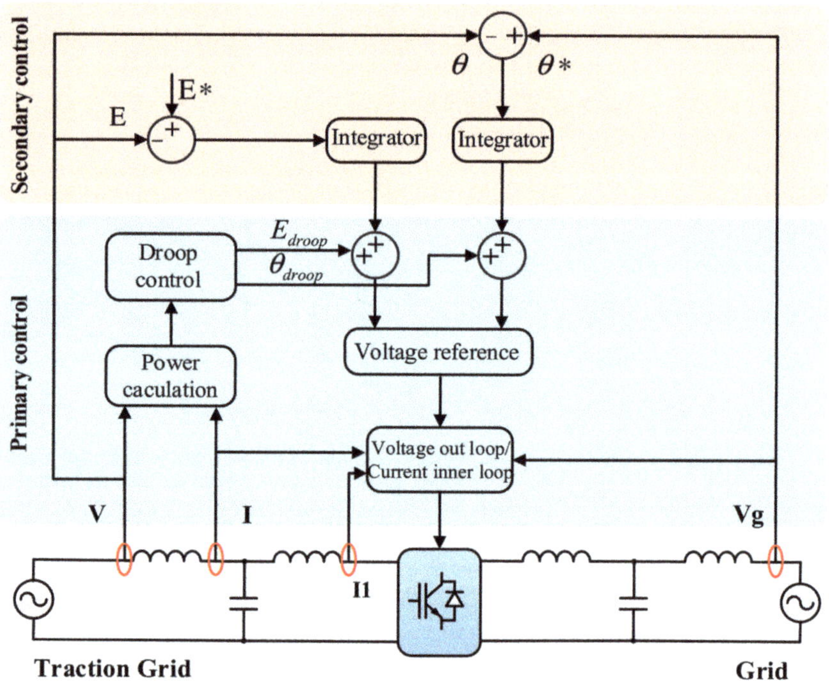

Figure 2. The proposed decentralized control strategy for the substation in the traction microgrid.

3.1. Line-Impedance-Based Power-Sharing and Circulation Current between the Substations

In a conventional traction power grid, only one substation provides power in each interval. There will be a significant voltage drop at the end of each interval due to the long-line impedance. Without a neutral section, the substations in a traction MG could operate in coordination and the voltage drop could be effectively reduced. Figure 3 presents an equivalent circuit of a traction MG with two substations.

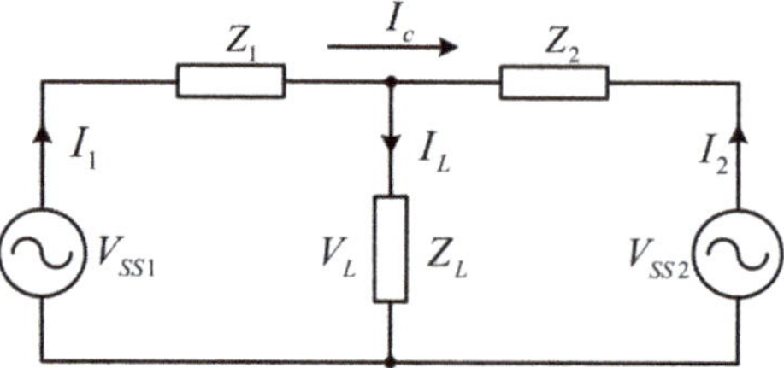

Figure 3. Equivalent circuit of two substations.

In Figure 3, Z_1 and Z_2 represent the line impedances between two substations V_{SS1} and V_{SS2}; Z_L is the reactance of the locomotive load; I_1, I_2, and I_L are the currents of V_{SS1}, V_{SS2}, and the load. When the output voltages of V_{SS1} and V_{SS2} are equal, the system current flows are described as follows:

$$I_1 = \frac{Z_2}{Z_1 + Z_2} I_L, \tag{1}$$

$$I_2 = \frac{Z_1}{Z_1 + Z_2} I_L. \tag{2}$$

The values of Z_1 and Z_2 depend on the location of the locomotive load. It can be seen that when the load moves toward the substation V_{SS1}, the line impedance Z_1 gets smaller. Referring to Equations (1) and (2), in this case, the current I_1 increases while the current I_2 decreases. This indicates the fact that the load will get more energy from the nearest substation. In conclusion, automatic line-impedance-based power-sharing is achieved in this traction power system. Compared to a conventional traction grid, the voltage drop on the line impedance would be maintained in an acceptable range, which will be discussed in detail in Section 5.

A voltage difference between two substations will result in circulation currents. Such currents will flow through the line impedance between two sources and eventually be fed back to the grid through a converter. This would not only cause extra energy loss due to the line impedance and the power electronic devices but will also have a negative effect on the converter power stress and lifetime. The circulation current is defined as:

$$I_c = \frac{V_1 - V_2}{Z_1 + Z_2}, \tag{3}$$

such that Equations (1) and (2) can be rewritten as:

$$I_1 = \frac{Z_2}{Z_1 + Z_2} I_L + I_c, \tag{4}$$

$$I_2 = \frac{Z_1}{Z_1 + Z_2} I_L - I_c. \tag{5}$$

If the output voltages of substations V_1 and V_2 are regulated as rated, there would be no voltage difference (i.e., $V_1 = V_2 = E^*$ and $V_1 - V_2 = 0$) and the circulation current can be eliminated.

3.2. Primary Control

In this work, a three-phase converter was adopted to stabilize the DC link voltage, while a single-phase converter was applied to maintain the traction grid voltage and support the locomotive load consumption. Three control loops were introduced under primary control, namely, droop control, a voltage out loop, and a current inner loop. DC bias current control was adopted for the inner current loop [26]. The D/Q voltage reference was generated by the voltage out loop. In [5], the voltage loop reference came from the sampling voltage/phase values from the connection point of each substation in the traction grid. When the initial state is changed, such as in the load condition, the system will collapse. Moreover, there is only one voltage source in the advanced co-phase traction grid. The long-distance line impedance between the substations and voltage source will cause a considerable voltage drop and become a threat to the stability of the system. Droop control is widely implemented in islanded MG to stabilize the voltage. In a high voltage traction power system where the line impedances are inductive (X >> R), droop control is suitable for this system. In order to autonomously stabilize the traction grid operating frequency and voltage using multiple substations in coordination, droop control is implemented at each substation. Based on the calculation of the power at the output connection point, the voltage amplitude and phase reference are generated according to the droop function.

3.3. Secondary Control

Due to the inherent characteristics of droop control, steady-state deviations will be introduced to both the system operating frequency and voltage magnitude [27]. In this study, secondary control was implemented for further regulation. As shown in Figure 2, E^* and θ^* represent the rated voltage magnitude and phase, while E and θ represent the substation output voltage magnitude and phase. In the AC co-phase traction MG system, the rated phase is synchronized with the grid all the time. The proposed voltage and phase regulation are defined as:

$$\dot{\theta} = \theta^* - mP + k_\omega \Delta\omega, \tag{6}$$

$$\dot{E} = E^* - nQ + k_e \Delta e, \tag{7}$$

where m and n represent the droop coefficients of the P–f and Q–V droops, respectively; k_ω and k_e are the designed control gains; $\Delta\omega$ and Δe are the designed secondary control variables:

$$\Delta\dot{\omega} = -k_w(\theta - \theta^*), \tag{8}$$

$$\Delta\dot{e} = -k_e(E - E^*). \tag{9}$$

By implementing integrators, the phase and voltage magnitude error is eliminated. Despite the effect of long-distance line impedance, the traction grid voltage is maintained in an acceptable range along the catenary line. Moreover, the circulation current is eliminated due to the elimination of the substation voltage difference.

4. Stability Analysis

4.1. Modeling of System Operation States

To derive the stability conditions for the proposed controller, a system small-signal model was developed. The following lemmas are introduced for the subsequent analysis.

Lemma 1 [28]: If M is positive definite and N is positive definite/semi-definite, then M + N is positive definite/semi-definite.

To avoid unnecessary technical complications, we ignored the delay in adjusting the output frequency and modeled the delay in adjusting the output voltage magnitude as a first-order low-pass filter [29]. The system operation states under primary control can be modeled as:

$$\Delta\dot{\delta}_i = \omega_i - \omega^* = -m_i \frac{E_i E_P}{X_i} sin(\Delta\delta_i) + \Delta\omega_i, \tag{10}$$

$$\omega_v{}^{-1}\dot{E} = -(E_i - E^*) = -n_i \frac{(E_i - E_P)E_P}{X_i} cos(\Delta\delta_i) + \Delta e_i, \tag{11}$$

where E_P presents the equivalent voltage magnitude at point of common coupling (PCC); X_i and $\Delta\delta_i = \delta_i - \delta_{PCC}$ represent the equivalent reactance and voltage phasor phase mismatch between the ith DG and the PCC and ω_v represents the cut-off frequency of the equivalent low-pass filter from the output voltage adjustment; $\Delta\omega_i$ and Δe_i are the designed secondary control variables. To derive the system's small-signal stability, it was assumed that $\sin\Delta\delta_i \approx \Delta\delta_i$ and $E_P \approx E^*$. Additionally, referring to the conditions where the AC droop control is validated [30], it was assumed that in Equation (8), $E_i \approx E_0$ was constant, and in Equation (9), $\cos\Delta\delta_i \approx \cos\delta_0$ was constant. Then, the system operation states modeling can be reduced to:

$$\Delta\dot{\delta}_i = -m_i M_i \Delta\delta_i + \Delta\omega_i, \tag{12}$$

$$\omega_v{}^{-1}\dot{E}_i = -E_i - n_i N_i E_i - (\frac{N_i}{\cos\delta'} - 1)E^* + \Delta e_i, \tag{13}$$

where $M_i = \frac{E' \bullet E^*}{X_i}$ and $N_i = \frac{\cos\delta' \bullet E^*}{X_i}$.

It is noteworthy that the time scale of the secondary control is at the millisecond level, while the variation of the system impedance, X_i can be considered to be a constant at such a time scale. For subsequent analysis, X_i, and thus M_i and N_i, are treated as constants. Denote $M = \text{diag}(M_i)$, $N = \text{diag}(N_i)$, $m = \text{diag}(m_i)$, and $n = \text{diag}(n_i)$. The proposed secondary controller is modeled as:

$$\Delta\dot{\omega}_i = -k_\omega(\Delta\delta_i + \delta_{ref}), \tag{14}$$

$$\Delta\dot{e}_i = -k_e(E_i + E_{ref}), \tag{15}$$

where k_ω and k_e are the designed control gains, while δ_{ref} and E_{ref} represent the reference voltage and phase reference for each substation, respectively.

4.2. Phase Regulation Stability

Denote $mM = \text{diag}(m_iM_i)$ and let I represent the unit matrix and 0 represent the zero matrix. The linearized system model can be expressed in matrix form as:

$$\dot{X} = WX + U, \tag{16}$$

where X = $(\Delta\delta/\Delta\omega)$ represents system states and $W = \begin{pmatrix} -mM & I \\ -k_\omega I & 0 \end{pmatrix} = \begin{pmatrix} -W_1 & I \\ -W_2 & 0 \end{pmatrix}$.

U represents variables that are independent from X. The characteristic polynomial of W can be simplified using a Schur complement and is derived as: $\det(sI - W) = \det(sI + W_1)\det(2sI + sW_1 + W_2) = 0$. It can be concluded that W_1 is positive definite. The roots of $\det(sI - W) = 0$ satisfies $\text{Re}(s) < 0$ if and only if the following conditions are satisfied:

$$\lambda_{\min}(W_1 + {W_1}^T) > 0, \tag{17}$$

$$\lambda_{\min}(W_2 + {W_2}^T) > 0. \tag{18}$$

Conditions (17) and (18) are satisfied as both W_1 and W_2 are positive definite. Therefore, it can be concluded that the system is exponentially stable with the proposed controller.

4.3. Voltage Regulation Stability

Denote $nN = \text{diag}(n_iN_i)$. The linearized system model can be expressed in matrix form as:

$$\dot{X} = \omega_v WX + \omega_v U, \tag{19}$$

where X = $(\Delta E/\Delta e)$ represents the system states $W = \begin{pmatrix} -1 + nN & I \\ -k_e I & 0 \end{pmatrix} = \begin{pmatrix} -W_1 & I \\ -W_2 & 0 \end{pmatrix}$.

U represents variables that are independent from X. The characteristic polynomial of W can be simplified using a Schur complement and is derived as: $\det(sI - W) = \det(sI + W_1)\det(2sI + sW_1 + W_2) = 0$. It can be concluded that W_1 is positive definite, referring to Lemma 1. The roots of $\det(sI - W) = 0$ satisfies $\text{Re}(s) < 0$ if and only if the following conditions are satisfied:

$$\lambda_{\min}(W_1 + W_1^T) > 0, \tag{20}$$

$$\lambda_{\min}(W_2 + W_2^T) > 0. \tag{21}$$

Conditions (20) and (21) are satisfied as both W_1 and W_2 are positive definite. Therefore, it can be concluded that the system is exponentially stable with the proposed controller.

5. Results

To verify the proposed decentralized control strategy for the AC co-phase traction MG power system, simulations using MATLAB/Simulink were studied. The single-phase AC traction voltage RMS value was rated as 400 V, the unite line impedance of the catenary was defined as 0.076 + j0.176 Ω/km. The locomotive load was modeled as a 10 Ω resistor parallel with a 50 mH inductor. The structure of the traction power system under simulation is presented in Figure 4. Three substations and two locomotive loads were involved. L1, L2, L3, and L4 represent the line impedances between the load and the substation. Comparison results regarding the system's operational stability, circulation current elimination, and load voltage drop regulation are carried out in the following sections.

5.1. Voltage/Frequency Regulation

As previously discussed, compared to the conventional large-scale power grid, the operational stability of an AC co-phase traction grid can be easily affected by the line impedance and load condition and could be improved by the proposed control strategy. In this section, compared to the traditional control strategy in [4], the simulation results for the system stability under different conditions are discussed. In Figure 4, the proposed control strategy was adopted by all the substations SS1, SS2, and SS3. All the substations operated as a voltage source. With the same system structure, there was only one voltage source SS1 that supported the traction grid voltage in the former research, while SS2 and SS3 operated as power sources. The output voltage waveforms are given when the substation that is connected with the traction grid has a different load condition.

5.1.1. No-Load Condition

In this simulation, SS1 was connected to the traction grid at 0 s, SS2 was connected to the traction grid at 3 s, SS3 was connected to the traction grid at 6 s, load 1 is connected to the traction grid at 9 s, load 2 is connected to the traction grid at 12 s. The lengths of L1 = L2 = L3 = L4 = 3, which means the load was connected at the middle point between the two substations

As shown in Figure 5a, as the only voltage source, SS1 kept a constant output voltage V1 all the time. At time T1, SS2 was connected with the traction grid. The voltage reference for the voltage loop was only affected by line impedances L1 and L2. The output voltage of SS2 V2 went to a steady state after nearly 1 s of dynamic response. At T2, SS3 was connected with the traction grid. The voltage reference for the voltage loop was affected by line impedances L1, L2, L3, and L4. The output voltage of SS3 V3 went to a steady state after nearly 1 s of dynamic response. It can be seen that V3 was getting smaller than V2, because of the voltage droop on the bigger line impedance. Load 1 and load 2 were connected with the grid at T3 and T4, respectively. After a quick dynamic response, the entire system maintained a steady state.

The simulation results for the proposed control strategy are shown in Figure 5b. In this AC co-phase traction MG, all the substations were operating as voltage sources. There was a dynamic process when SS1 was connected with the traction. Due to the additional droop and secondary control, the speed of the system dynamic response was affected. However, same as SS2 and SS3, all the substations had a steady output voltage after 1 s. The voltage reference for the voltage loop was no longer just the sampling value on the grid. It depended on the power consumption according to the control strategy. All the output voltages were steady after load 1 and load 2 were connected with the grid.

5.1.2. On-Load Condition

To validate the system stability under different load conditions, the substation was connected to the grid with a load. SS1 was connected to the traction grid at 0 s, load 1 was connected to the traction grid at 3 s, SS2 was connected to the traction grid at 6 s, load 2 was connected to the traction grid at 9 s, and SS3 was connected to the traction grid at 12 s. All the loads were still connected at the middle point of two substations. The simulation results are shown in Figure 6.

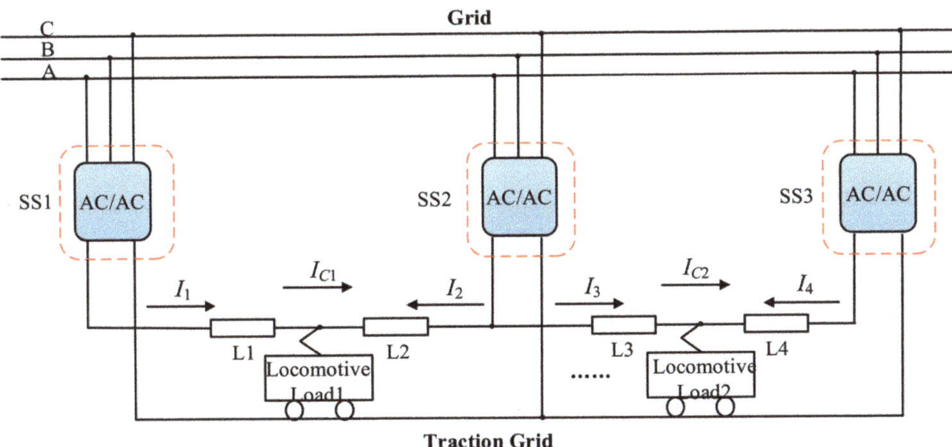

Figure 4. The structure of the traction microgrid system for the simulation model.

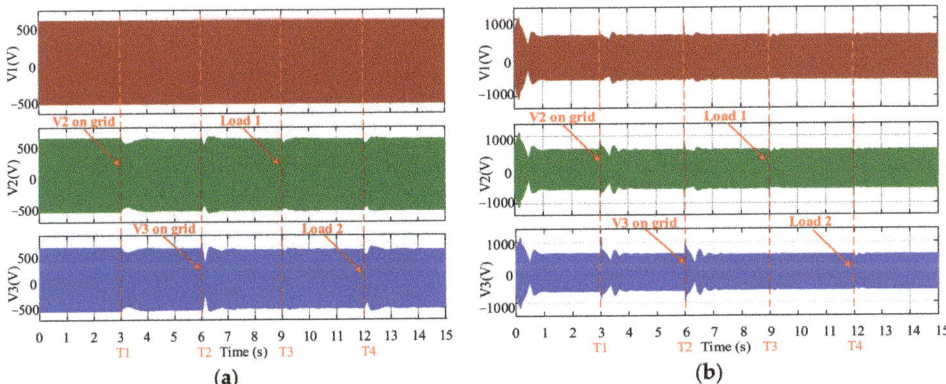

Figure 5. Three substation output voltages of different control strategies under a no-load condition: (**a**) the traditional control strategy; (**b**) the proposed strategy.

As shown in Figure 6a, SS1 kept a constant output voltage V1 all the time. The system was stable until SS3 was connected to the traction grid. Before T3, there were already two loads on the grid. The voltage reference error of SS3 had a great effect on the system's stability. In contrast, the system remained stable under on-load conditions. Based on the decentralized control strategy, the inverter of the substation not only provided the power consumption but also stabilized the traction grid voltage. In addition, when the line impendence was getting bigger, the stability of the proposed system was guaranteed, while the system became unstable in [4].

5.2. Circulation Current

A circulation current between substations is one of the key issues for a system's efficiency and lifetime. Same as the section before, two load conditions will be discussed. The voltage difference and circulation current comparison results are given in the following.

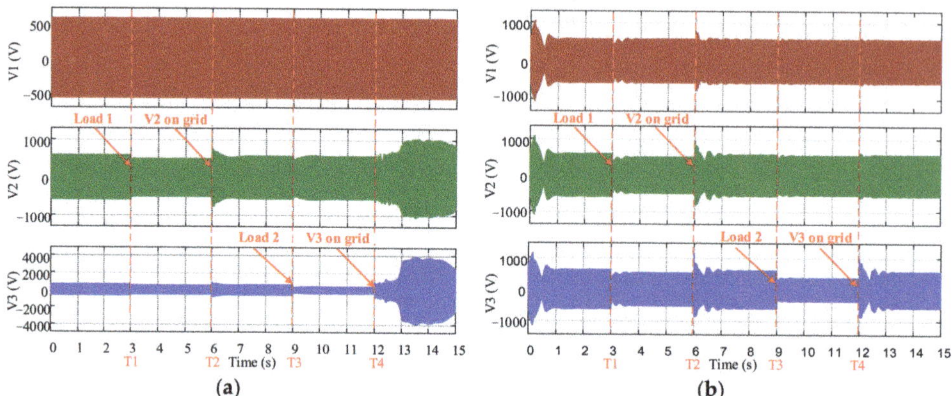

Figure 6. Three substation output voltages of different control strategies under on-load conditions: (**a**) the traditional control strategy; (**b**) the proposed strategy.

5.2.1. No-Load Condition

With the same no-load condition as before, the loads were connected to the grid after all the substations were on-grid. As it can be observed, Figure 7 gives the voltage difference V1 − V2 and V2 – V3 results of different control strategies under no-load conditions. In Figure 7a, when the loads connected to the grid, V1 − V2 and V2 − V3 had dynamic responses every time, which indicates that the system was sensitive to the load condition. In contrast, the voltage differences with the proposed strategy displayed better dynamic performance.

For the steady-state part, the zoomed-in results at the last second are given in Figure 8. It can be seen that nearly 50% of the voltage difference was cut down by the application of the proposed control strategy for the traction MG.

With the decrease of the voltage difference, the circulation current could be eliminated. The circulation currents' results with different strategies are depicted in Figure 9. I_{C1} is the circulation current between SS1 and SS2, while I_{C2} is the circulation current between SS2 and SS3. In this simulation, the line impedances of L1, L2, L3, and L4 were same. According to (3), $I_{C1} = 0.5(I1 - I2)$, while $I_{C2} = 0.5(I3 - I4)$. It can be seen that the circulation current was relative to the voltage difference in Figure 9.

In the zoomed-in results in the steady state for the last second, I_{C1} = 15 A and I_{C2} = 8 A with the traditional strategy, while I_{C1} = 5 A and I_{C2} = 1 A with the proposed strategy. The circulation current was reduced by more than 66%.

5.2.2. On-Load Condition

As seen in Figure 6, the system was unstable after SS3 connected to the grid with the traditional strategy. As a result, the voltage differences increased, which caused a large circulation current, as shown in Figure 10a. However, the voltage differences remained in an acceptable range with the proposed control strategy in Figure 10b.

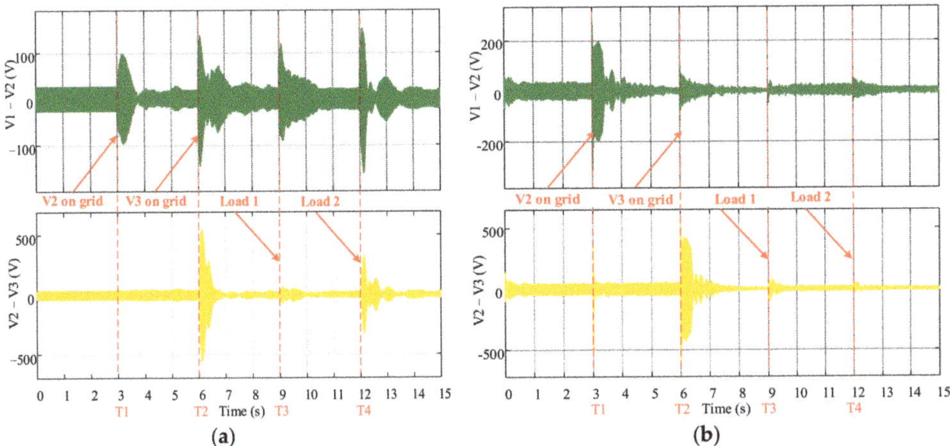

Figure 7. The voltage differences different control strategies under no-load conditions: (**a**) the traditional control strategy; (**b**) the proposed strategy.

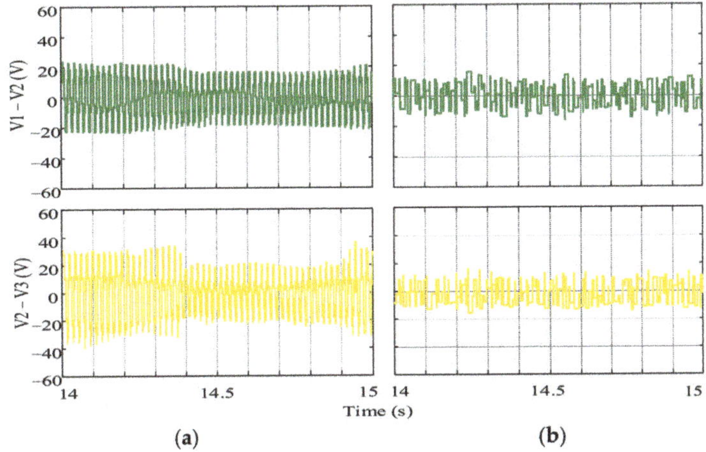

Figure 8. The zoomed-in voltage differences in the steady state under a no-load condition: (**a**) the traditional control strategy; (**b**) the proposed strategy.

The zoomed-in results of the voltage differences and circulation currents of the proposed strategy at the last second are also given in Figure 11. Both V1 − V2 and V2 − V3 were less than 20 V. I_{C1} and I_{C1} were smaller than 5 A. The circulation currents between substations were eliminated under the on-load conditions.

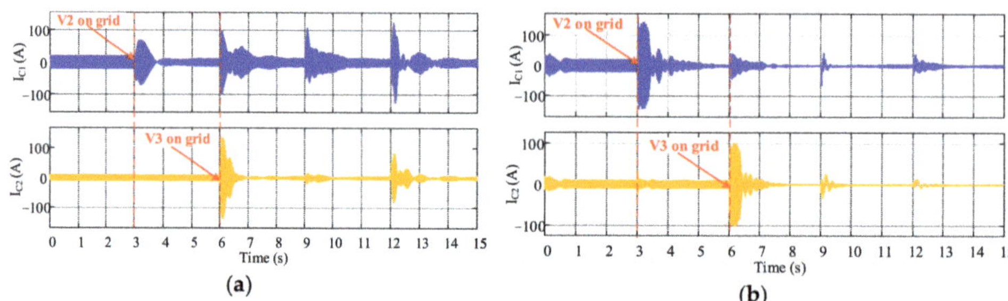

Figure 9. The circulation currents when they were connected with the traction grid without a load: (**a**) the traditional control strategy; (**b**) the proposed strategy.

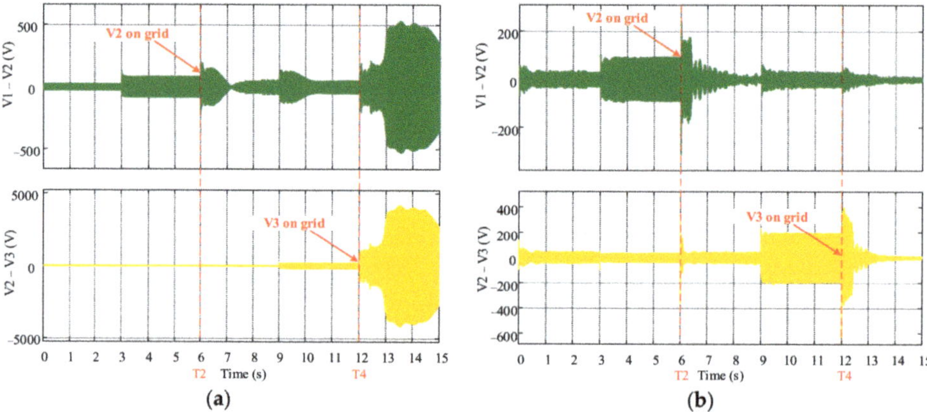

Figure 10. The voltage differences when they were connected with the traction grid with a load: (**a**) the traditional control strategy; (**b**) the proposed strategy.

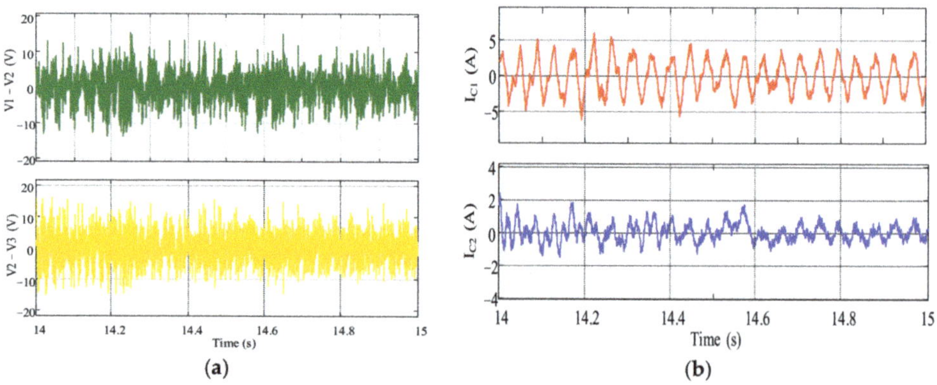

Figure 11. The zoomed-in voltage differences (**a**) and circulation currents; (**b**) in the steady state when they were connected with the traction grid without a load with the proposed strategy.

5.3. Power Sharing/Voltage Drop

In the existing traction grid, only one substation provides the power for one interval due to the neutral section.

There was an obvious voltage drop at the end of the catenary. The situation was different with the proposed traction MG system. The simulation was modified based on the structure in Figure 4. As shown in Figure 12, six same-value line impedances L were connected in series. SS1, SS2, SS3, and load 1 were already connected to the grid before 8 s. Load 2 was connected to the traction grid at 8 s, 10 s, 12 s, 14 s, and 16 s in sequence. The length of L was 1 km, which is a reasonable number considering the speed of the locomotive and the system response. For the traditional traction grid simulation, only SS2 supported the load consumption.

The current and the voltage of the load comparison results with different traction grids is given in Figure 13. As shown in Figure 13a, the output current of SS2, namely I_{SS2}, was getting smaller when load 2 moved away from the station. With the line impedance increase, the load voltage was getting smaller too. As a result, it was hard to meet the voltage requirement when the load was far away from the station. Compared to the traditional traction grid, there was only a small voltage drop on the load in the proposed AC co-phase traction grid. As can be seen in Figure 13a, there were two output currents from SS2 and SS3, which were I_{SS2} and I_{SS3}, respectively. When the load moved from SS2 to SS3, I_{SS2} became smaller, while I_{SS2} got bigger. The largest voltage drop happened when the load was moving at the middle point. As seen in the analysis in Section 3, the load got more power from the near substation. The line-impedance-based power-sharing required no additional control and could maintain the load voltage in an acceptable range all the time.

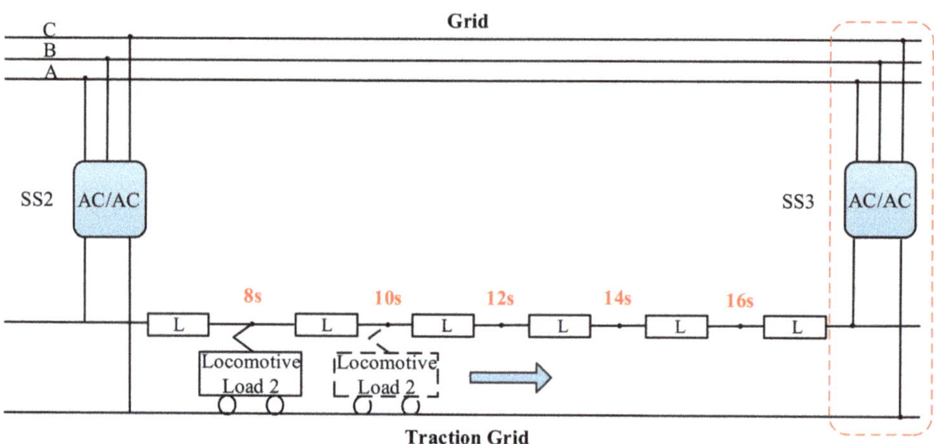

Figure 12. The system structure when the load is moving.

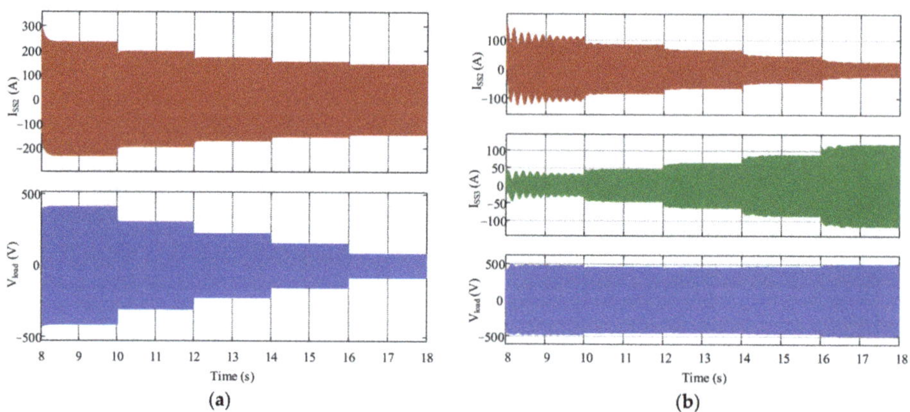

Figure 13. Output current and load voltage of different traction grids: (**a**) the traditional traction grid (**b**) the proposed AC co-phase traction grid.

6. Conclusions

In this study, the AC co-phase traction MG system structure was introduced. To improve the system's operating performance, a decentralized control strategy was designed for the converter-based substation. A small-signal stability analysis was presented for the proposed controller. Finally, the proposed work was validated through extensive simulations. With the proposed decentralized control strategy, the system's stability was improved under various load and line impedance conditions. The circulation current was reduced by more than 66%, which indicates the higher efficiency of the system. In addition, the quality of the load voltage was guaranteed despite the significant effect of the line impedance. With the simplified control and power-sharing strategy, additional substations can be easily integrated for further increases of the system's capacity for future high-speed trains. Moreover, without communications between the substations or to the grid, "plug-and-play" functionality is realized. In the following research, optimization of the control parameters for the system's robustness and the application of distributed renewable energy in the traction MG will be further studied.

Author Contributions: Conceptualization, L.M. and Z.S.; methodology, Y.D.; software, L.M.; validation, Y.D., L.M. and L.Z.; formal analysis, Y.D.; investigation, F.Y.; resources, S.X.; data curation, L.Z.; writing—original draft preparation, L.M.; writing—review and editing, Y.D.; visualization, Z.S.; supervision, Z.S.; project administration, School of Electrical Engineering Southwest Jiaotong University; funding acquisition, Southwest Jiaotong University. All authors have read and agreed to the published version of the manuscript.

Funding: This research was funded by [Basic scientific research operating expenses of central universities] grant number [2682020CX24].

Conflicts of Interest: The authors declare no conflict of interest.

References

1. Xie, B.; Li, Y.; Zhang, Z.; Hu, S.; Zhang, Z.; Luo, L.; Cao, Y.; Zhou, F.; Luo, R.; Long, L. A Compensation System for Cophase High-Speed Electric Railways by Reactive Power Generation of SHC&SAC. *IEEE Trans. Ind. Electron.* **2018**, *65*, 2956–2966.
2. Shu, Z.; Xie, S.; Lu, K.; Zhao, Y.; Nan, X.; Qiu, D.; Zhou, F.; Gao, S.; Li, Q. Digital Detection, Control, and Distribution System for Co-Phase Traction Power Supply Application. *IEEE Trans. Ind. Electron.* **2012**, *60*, 1831–1839. [CrossRef]
3. Li, Q.; Liu, W.; Shu, Z.; Xie, S.; Zhou, F. Co-phase power supply system for HSR. In Proceedings of the 2014 International Power Electronics Conference (IPEC-Hiroshima 2014-ECCE ASIA), Hiroshima, Japan, 18–21 May 2014; pp. 1050–1053.
4. He, X.; Shu, Z.; Peng, X.; Zhou, Q.; Zhou, Y.; Zhou, Q.; Gao, S. Advanced Cophase Traction Power Supply System Based on Three-Phase to Single-Phase Converter. *IEEE Trans. Power Electron.* **2014**, *29*, 5323–5333. [CrossRef]
5. Shu, Z.; He, X.; Wang, Z.; Qiu, D.; Jing, Y. Voltage Balancing Approaches for Diode-Clamped Multilevel Converters Using Auxiliary Capacitor-Based Circuits. *IEEE Trans. Power Electron.* **2013**, *28*, 2111–2124. [CrossRef]
6. Mahmood, H.; Michaelson, D.; Jiang, J. Reactive Power Sharing in Islanded Microgrids Using Adaptive Voltage Droop Control. *IEEE Trans. Smart Grid* **2015**, *6*, 3052–3060. [CrossRef]
7. Mao, M.Q.; Dong, Z.; Ding, Y.; Chang, L. A unified controller for a microgrid based on adaptive virtual impedance and conductance. In Proceedings of the 2014 IEEE Energy Conversion Congress and Exposition (ECCE), Pittsburgh, PA, USA, 15–18 September 2014; pp. 695–701.
8. Zhong, Q.-C. Robust Droop Controller for Accurate Proportional Load Sharing Among Inverters Operated in Parallel. *IEEE Trans. Ind. Electron.* **2013**, *60*, 1281–1290. [CrossRef]
9. Du, Y.; Lu, X.; Wang, J.; Lukic, S. Distributed Secondary Control Strategy for Microgrid Operation with Dynamic Boundaries. *IEEE Trans. Smart Grid* **2019**, *10*, 5269–5282. [CrossRef]
10. Lee, T.-Z.; Cheng, P.-T. Design of a New Cooperative Harmonic Filtering Strategy for Distributed Generation Interface Converters in an Islanding Network. *IEEE Trans. Power Electron.* **2007**, *22*, 1919–1927. [CrossRef]
11. He, J.; Li, Y.; Guerrero, J.M.; Blaabjerg, F.; Vasquez, J.C. An Islanding Microgrid Power Sharing Approach Using Enhanced Virtual Impedance Control Scheme. *IEEE Trans. Power Electron.* **2013**, *28*, 5272–5282. [CrossRef]
12. Blaabjerg, F.; Chen, Z.; Kjaer, S. Power Electronics as Efficient Interface in Dispersed Power Generation Systems. *IEEE Trans. Power Electron.* **2004**, *19*, 1184–1194. [CrossRef]
13. Rocabert, J.; Luna, A.; Blaabjerg, F.; Rodríguez, P. Control of Power Converters in AC Microgrids. *IEEE Trans. Power Electron.* **2012**, *27*, 4734–4749. [CrossRef]
14. He, J.; Li, Y.W. Analysis, Design, and Implementation of Virtual Impedance for Power Electronics Interfaced Distributed Generation. *IEEE Trans. Ind. Appl.* **2011**, *47*, 2525–2538. [CrossRef]
15. Brabandere, K.D.; Bolsens, B.; Keybus, J.V.D.; Woyte, A.; Driesen, J.; Belmans, R.; Leuven, K. A voltage and frequency droop control method for parallel inverters. *IEEE Trans. Power Electron.* **2007**, *22*, 1107–1115. [CrossRef]
16. Yao, W.; Chen, M.; Matas, J.; Guerrero, J.M.; Qian, Z. Design and Analysis of the Droop Control Method for Parallel Inverters Considering the Impact of the Complex Impedance on the Power Sharing. *IEEE Trans. Ind. Electron.* **2011**, *58*, 576–588. [CrossRef]
17. Han, Y.; Li, H.; Shen, P.; Coelho, E.A.A.; Guerrero, J.M. Review of Active and Reactive Power Sharing Strategies in Hierarchical Controlled Microgrids. *IEEE Trans. Power Electron.* **2016**, *32*, 2427–2451. [CrossRef]
18. Li, Y.W.; Kao, C.-N. An Accurate Power Control Strategy for Power-Electronics-Interfaced Distributed Generation Units Operating in a Low-Voltage Multibus Microgrid. *IEEE Trans. Power Electron.* **2009**, *24*, 2977–2988. [CrossRef]
19. Chen, Y.; Luo, A.; Zhou, J.; Bai, L.; Tu, C. Rapid reactive power control method for parallel inverters using resistive-capacitive output impedance. In Proceedings of the 2013 1st International Future Energy Electronics Conference (IFEEC), Tainan, Taiwan, 3–6 November 2013; pp. 98–102.
20. Gu, Y.; Li, W.; He, X. Frequency-Coordinating Virtual Impedance for Autonomous Power Management of DC Microgrid. *IEEE Trans. Power Electron.* **2015**, *30*, 2328–2337. [CrossRef]
21. Sreekumar, P.; Khadkikar, V. A new virtual harmonic impedance scheme for harmonic power sharing in an islanded microgrid. *IEEE Trans. Power Del.* **2016**, *31*, 936–945. [CrossRef]
22. Milczarek, A.; Malinowski, M.; Guerrero, J.M. Reactive Power Management in Islanded Microgrid—Proportional Power Sharing in Hierarchical Droop Control. *IEEE Trans. Smart Grid* **2015**, *6*, 1631–1638. [CrossRef]

23. Du, Y.; Tu, H.; Lukic, S. Distributed Control Strategy to Achieve Synchronized Operation of an Islanded MG. *IEEE Trans. Smart Grid* **2018**, *10*, 4487–4496. [CrossRef]
24. Liu, S.; Wang, X.; Liu, P.X. Impact of Communication Delays on Secondary Frequency Control in an Islanded Microgrid. *IEEE Trans. Ind. Electron.* **2015**, *62*, 2021–2031. [CrossRef]
25. Li, Q.; Chen, F.; Chen, M.; Guerrero, J.M.; Abbott, D. Agent-Based Decentralized Control Method for Islanded Microgrids. *IEEE Trans. Smart Grid* **2016**, *7*, 637–649. [CrossRef]
26. Liu, X.; Wang, P.; Loh, P.C. A Hybrid AC/DC Microgrid and Its Coordination Control. *IEEE Trans. Smart Grid* **2011**, *2*, 278–286. [CrossRef]
27. Ahumada, C.; Cardenas, R.; Saez, D.; Guerrero, J.M. Secondary Control Strategies for Frequency Restoration in Islanded Microgrids With Consideration of Communication Delays. *IEEE Trans. Smart Grid* **2016**, *7*, 1430–1441. [CrossRef]
28. Roger, A.H.; Charles, R.J. *Matrix Analysis*; Cambridge University Press: Cambridge, UK, 1990.
29. Simpson-Porco, J.W.; Shafiee, Q.; Dörfler, F.; Vasquez, J.C.; Guerrero, J.M.; Bullo, F. Secondary frequency and voltage control of islanded microgrids via distributed averaging. *IEEE Trans. Ind. Electron.* **2015**, *62*, 7025–7038. [CrossRef]
30. Guerrero, J.M.; Vasquez, J.C.; Matas, J.; De Vicuña, L.G.; Castilla, M. Hierarchical control of droop-controlled ac and dc microgrids—A general approach toward standardization. *IEEE Trans. Ind. Electron.* **2011**, *58*, 158–172. [CrossRef]

Article

Four-Quadrant Operations of Bidirectional Chargers for Electric Vehicles in Smart Car Parks: G2V, V2G, and V4G

Tingting He [1], Dylan Dah-Chuan Lu [2], Mingli Wu [1], Qinyao Yang [1], Teng Li [1,*] and Qiujiang Liu [1]

[1] School of Electrical Engineering, Beijing Jiaotong University, Beijing 100044, China; hetingting@bjtu.edu.cn (T.H.); mlwu@bjtu.edu.cn (M.W.); 19126189@bjtu.edu.cn (Q.Y.); qjliu@bjtu.edu.cn (Q.L.)
[2] Faculty of Engineering and Information Technology, University of Technology Sydney, Sydney, NSW 2007, Australia; Dylan.Lu@uts.edu.au
* Correspondence: liteng@bjtu.edu.cn

Abstract: This paper presents the four-quadrant operation modes of bidirectional chargers for electric vehicles (EVs) framed in smart car parks. A cascaded model predictive control (MPC) scheme for the bidirectional two-stage off-board chargers is proposed. The controller is constructed in two stages. The model predictive direct power control for the grid side is applied to track the active/reactive power references. The model predictive direct current control is proposed to achieve constant current charging/discharging for the EV load side. With this MPC strategy, EV chargers are able to transmit the active and reactive powers between the EV batteries and the power grid. Apart from exchanging the active power, the vehicle-for-grid (V4G) mode is proposed, where the chargers are used to deliver the reactive power to support the grid, simultaneously combined with grid-to-vehicle or vehicle-to-grid operation modes. In the V4G mode, the EV battery functions as the static var compensator. According to the simulation results, the system can operate effectively in the full control regions of the active and reactive power (PQ) plane under the aforementioned operation modes. Fast dynamic response and great steady-state system performances can be verified through various simulation and experimental results.

Keywords: model predictive control; bidirectional two-stage charger; electric vehicle

Citation: He, T.; Lu, D.D.-C.; Wu, M.; Yang, Q.; Li, T.; Liu, Q. Four-Quadrant Operations of Bidirectional Chargers for Electric Vehicles in Smart Car Parks: G2V, V2G, and V4G. *Energies* **2021**, *14*, 181. https://doi.org/10.3390/en14010181

Received: 20 November 2020
Accepted: 24 December 2020
Published: 31 December 2020

Publisher's Note: MDPI stays neutral with regard to jurisdictional claims in published maps and institutional affiliations.

1. Introduction

Electric vehicles (EVs), including railway, public service, and personal vehicles, play an important role in building a clean and efficient environment in the transportation market [1–5]. They reduces the pollution, gas emissions, and the reliability of fuel oil. With the permeation of EVs in recent decades, they are considered a new major load in the main grid and distributed systems [6]. When connected to the grid, EV batteries can deliver power to the grid as energy storage devices. When the EVs consume power, it is denoted as a grid-to-vehicle (G2V) operation. On the other hand, it is known as a vehicle-to-grid (V2G) operation when the energy is provided from the EVs to the grid [7,8]. Therefore, the installation of bidirectional EV charging stations is needed to fulfill the demands mentioned above.

As a new paradigm of transport devices, EV batteries have the ability to compensate the grid with active power via the bidirectional chargers. The current chargers can be classified into several types in terms of the location (on-/off-board) and function (unidirectional/bidirectional) [9]. In a recent work, Kwon et al. proposed an electrolytic capacitor bidirectional EV charger to exchange the active power between the grid, EVs, and smart homes [10]. In [11], Zahid designed a bidirectional DC/DC resonant converter to reduce the cost and size of the chargers for V2G operation. As control methods for converters have developed in recent years, it is now possible for the EVs to exchange the reactive power with the main grid, in addition to the active power. A system in which EVs can participate in the microgrids as the reactive power provider was presented in [12]. The economic efficiency from different aspects, including the energy market, reactive market, and lost opportunities, was analyzed to show the effectiveness of EV participation. The reactive power compensation capabilities of EVs in two different bidirectional battery converter topologies were investigated in [13]. EV batteries were used as static var compensators to support the main grid; this was named the vehicle-for-grid (V4G) mode in this paper. Under this working condition, the reactive power is transmitted between the EVs and the power grid. In [14], the EV batteries could operate only in the V2G mode or with reactive power operation. It is impossible for the controller to meet the active and reactive power requirements at the same time. In [15,16], reactive power could be controlled to be provided from/to the EVs. According to the types of the operation modes, the current references of the EV and grid sides were generated and sent to be used according to the pulse width modulation (PWM) strategy. Then, the bidirectional chargers were able to operate in the V2G/G2V/V4G modes in smart grids or homes.

Compared with the PWM strategies for current-tracking targets used in [14–16], an advanced model predictive control (MPC) algorithm was proposed with no need for modulation. Due to its easy implementation and its combination of multiple objectives, the MPC scheme has recently attracted researchers [17,18]. The adoption of the MPC method strongly depends on the calculation speed of the components and system. With the application of microprocessors, it is now possible to apply it in the MPC method to control powered electronic devices [19]. A cost function was designed to select an optimal switching state for the next sampling period [20]. The model predictive direct power control with duty cycle optimization was proposed to control the PWM rectifier [21]. A nonzero vector and a zero vector operating with optimal control times were selected to achieve better system performance. In [22], the model predictive current control was applied on the grid side to track the grid current reference. Currently, the predictive control strategy is used for unidirectional power flow in general. The active power is delivered from the grid side to the load side.

To achieve bidirectional active/reactive power flow (four-quadrant operation), a cascaded model predictive control is proposed in this paper. The controller is divided into two parts. One is for the active and reactive power control used in the grid side, denoted as the model predictive direct power control (MPDPC). The combination of the active and reactive powers' errors is designed as the cost function. It is used to track the active/reactive power demands for the power grid. Another is the current control for the EV side to achieve a constant current charging target, which is named the model predictive direct current control (MPDCC). The key novelties of the paper can be summarized as follows: (1) The active and reactive powers can be controlled independently with the cascaded predictive control; (2) EV batteries regarded as the reactive power generator (V4G mode) are proposed to support the main grid; (3) the initial constant current control for charging/discharging operation is available on the load side.

In this paper, a three-phase laboratory converter and a DC/DC half-bridge converter were developed as the bidirectional EV charger. Simulation and experimental tests were implemented to verify the proposed method in the four-quadrant operation. The detailed working scenarios were: (1) the pure G2V operation mode; (2) the pure V2G operation mode; (3) the pure inductive V4G operation mode; (4) the pure capacitive V4G operation mode; (5) the G2V along with the inductive V4G operation mode; (6) the V2G along with the inductive V4G operation mode; (7) the V2G along with the capacitive V4G operation mode; (8) the G2V along with the capacitive V4G operation mode. It should be noted that the inductive/capacitive operation refers to positive/negative reactive power during the transmission.

The rest of this paper is presented as follows. Section 2 discusses the topologies, application fields, and pros and cons of the unidirectional and bidirectional chargers. In Section 3, the selected off-board EV battery charger prototype is introduced and the proposed MPC methods for both the first stage and second stage are explained in detail.

Sections 4 and 5 present the simulation and experimental results of the bidirectional charger under the G2V, V2G, and V4G operation modes. Finally, the conclusions are shown in Section 6.

2. Electric Vehicle Battery Chargers for Smart Car Park Integration

According to the power flow direction, the types of EV chargers can be summarized as: unidirectional/bidirectional with on-/off-board chargers [23]. The active and reactive power (PQ) coordinate in Figure 1 shows the full control region for EV battery chargers. The positive directions of P and Q represent the power being delivered from the grid to the EVs. According to the direction of active and reactive power transmission, the PQ plane can be divided into eight operation modes. The xy axes of the PQ frame, defined as Modes I–IV, represent the pure G2V, inductive V4G, V2G, and capacitive V4G operations, respectively. The other four quadrants show the V2G or G2V operation along with the capacitive or inductive V4G operation.

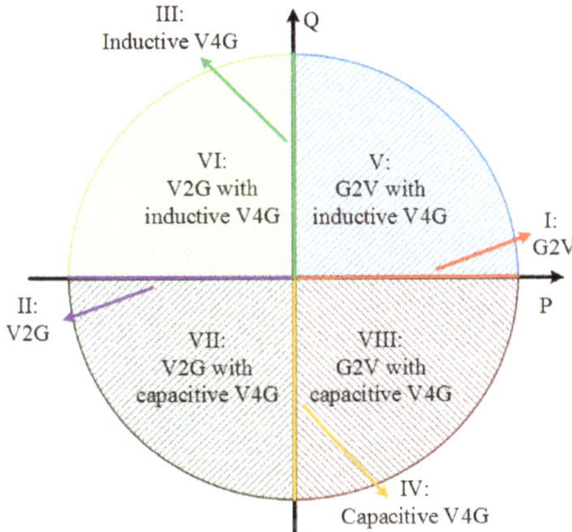

Figure 1. Four-quadrant operation for chargers.

2.1. Unidirectional Chargers

Unidirectional chargers can only operate on the positive x-axis (i.e., $P \geq 0$) of the P-Q coordinates (Modes I, V, and VIII), as shown in Figure 1. EVs can only be charged. On the other hand, reactive power is able to be transmitted during the battery charging process. However, low-frequency current harmonics will be introduced in the utility, which is inadvisable [24]. Therefore, these chargers are used to achieve a unity power factor operation, as presented in Mode I.

Various topologies have been studied for both single- and three-phase unidirectional chargers with half-bridge, full-bridge, or multilevel topologies. Considering the cost, size, and component stresses of the chargers, different topology circuits are applied in diverse application fields. The power factor correction boost converter, a conventional topology, is used in the low-power-level application [24]. To increase the application power level, some advanced topologies were proposed and discussed, such as the interleaved AC/DC boost converter, inverting or positive buck/boost PFC converter, and the multilevel converter [23,25].

2.2. Bidirectional Chargers

To make full use of EV batteries, smart chargers with V2G technology have been proposed and extensively studied. With the G2V and V2G technologies, the active power can be transferred to the EV batteries and returned back to the grid. They are controlled for operation in all regions of the PQ plane. This kind of bidirectional charger takes into account the grid's and EV drivers' energy demands. Along with optimal charging/discharging schemes, potential commercial benefits stemming from EV chargers can be achieved, as discussed in our previous work [26]. Apart from transmitting the active power in V2G/G2V operations, reactive power (inductive or capacitive) can be provided by the batteries to improve the power quality. During the reactive power operation process, the EV batteries can be regarded as capacitor banks, static compensators, etc. Note that it is impossible for the state of charge (SOC) and lifetime of the batteries to be affected when operating as a reactive power provider [16]. However, due to the increasing charge–discharge cycles, the DC-link filter capacitor is affected under this reactive power support condition.

Figure 2 shows the typical single-/three-phase bidirectional battery chargers, respectively. There are three power level types for EV chargers, namely, Level 1, Level 2, and Level 3. Most EV chargers are able to be plugged into a home or office outlet for Level 1 (slow charging) or Level 2 [23]. They are usually designed for on-board chargers for private or public facilities with a single-phase topology. For commercial applications, such as shopping malls and official car parking spots, three-phase structures are normally applied with Level 2 or 3 (fast-charging) chargers installed for off-board charging. The EVs can be controlled to fully charge within two or three hours. Therefore, the three-phase circuits are selected and studied for the smart car park in this paper.

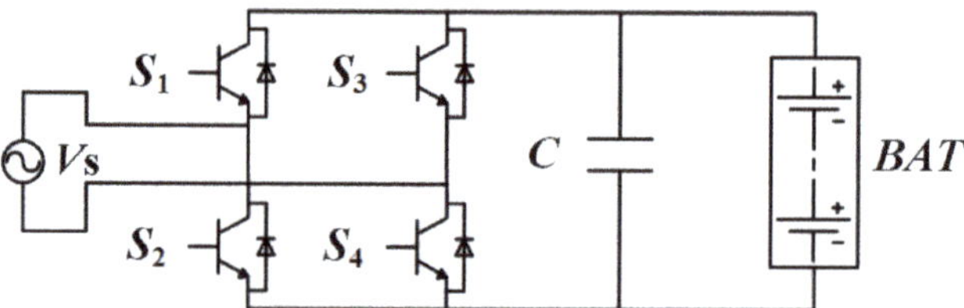

(**a**) Circuit topology of a single-phase bidirectional charger.

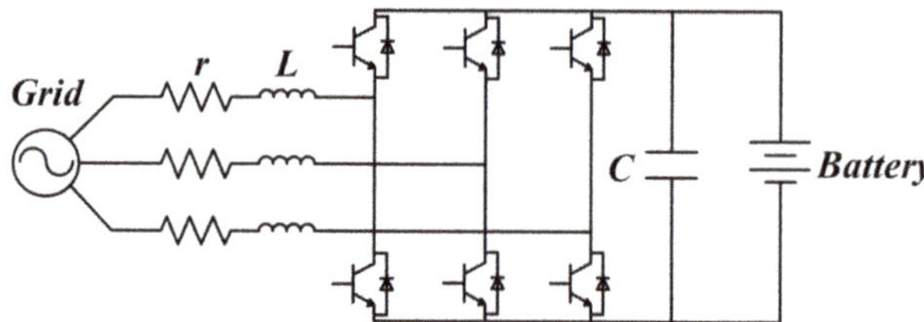

(**b**) Circuit topology of a three-phase bidirectional charger.

Figure 2. Bidirectional charger structures.

3. Proposed System and Control Scheme for Bidirectional Power Flow

With the integration of EVs, car parks located in workplaces or shopping malls show enormous potential for getting EVs involved in the grid. To exploit this commercial potential, a smart car park where large populations of EVs would be parked was proposed in a previous work [26]. The proposed smart car park system consists of bidirectional charging stations, a control center, and energy storage devices, as presented in Figure 3. The active/reactive power management system can be regarded as a central controller. The power demands from the EVs and storage system (such as a supercapacitor/flywheel) are sent to this central controller. Optimal control strategies are applied to generate the control signals. The switching states are transferred to drive the relevant converters or devices.

The bidirectional chargers should be controlled to track the active/reactive power and charging/discharging current references provided from the grid and EVs, respectively. Due to the easy implementation and ability to fulfill multiple objectives, the proposed MPC method is designed to track these targets. The classical topology in Figure 2b is used for energy storage systems with the bidirectional power flow [27]. However, as a boost converter, the range of the battery voltage is limited, varying from the grid peak voltage to the battery maximum voltage [28], which will be proved later in this section.

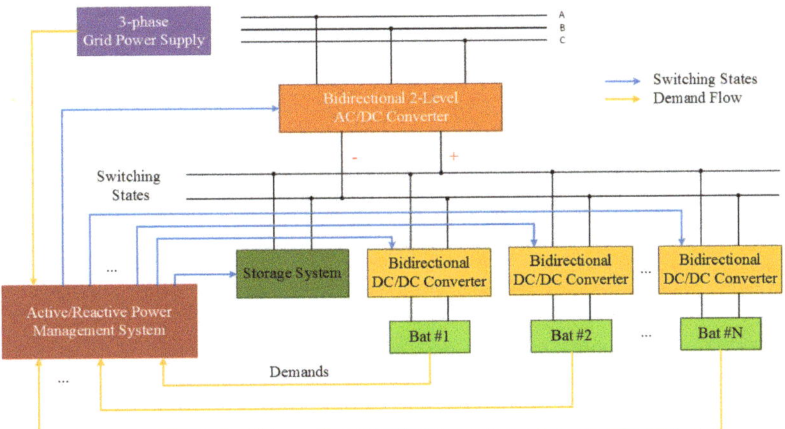

Figure 3. Framework of the smart car park system.

Considering that the common nominal battery voltage is between 300 and 400 V, an AC/DC boost converter, the typical topology, is preferred in the rectification stage with a 120/240 V grid connection [25]. A DC/DC half-bridge bidirectional converter is used to interface with the EV batteries. With this DC/DC converter, the low boundary of the charging/discharging range can be enlarged from the peak value of the grid voltage to 0 V.

Based on the above analysis, a three-phase two-stage converter is used as the off-board bidirectional charger, as depicted in Figure 4. The two-level AC/DC converter is connected with the power grid via a line resistance (R) and a filter inductor (L). A filter capacitor (C) is connected on the DC bus to smooth the DC voltage. The half-bridge DC/DC converter is applied as the second stage. An output inductor (L_{dc}) connecting the battery is used on the DC side.

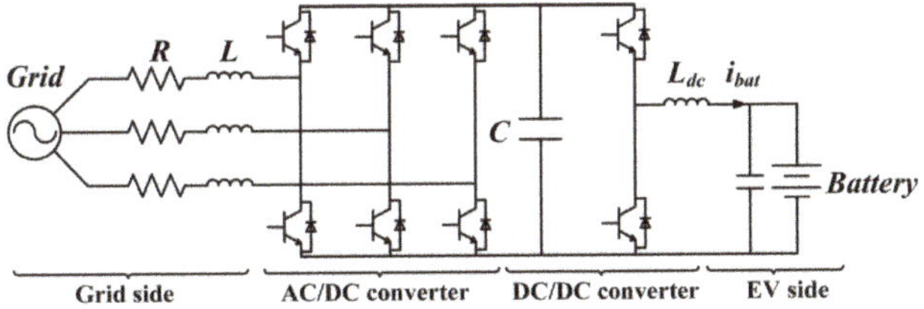

Figure 4. Three-phase off-board charger.

The switching signals of the three-phase AC/DC converter are defined as

$$S_k = \begin{cases} 1 & S_1 \text{ or } S_3, \text{ or } S_5 \text{ is on} \\ 0 & S_2 \text{ or } S_4, \text{ or } S_6 \text{ is on} \end{cases}, \tag{1}$$

where $k = a, b, c$ phase.

The switching state is transformed into the stationary $\alpha\beta$ coordinate system, written as

$$\mathbf{S}_{\alpha\beta} = \frac{2}{3}\begin{bmatrix} 1 & -\frac{1}{2} & -\frac{1}{2} \\ 0 & \frac{\sqrt{3}}{2} & -\frac{\sqrt{3}}{2} \end{bmatrix}\mathbf{S}_{\mathbf{abc}}, \tag{2}$$

where

$$\mathbf{S}_{\alpha\beta} = \begin{bmatrix} S_\alpha \\ S_\beta \end{bmatrix}, \quad \mathbf{S}_{\mathbf{abc}} = \begin{bmatrix} S_a \\ S_b \\ S_c \end{bmatrix}.$$

Similarly, the input voltage of the AC/DC converter in the $\alpha\beta$ coordinates can be expressed as:

$$\mathbf{V}_{\alpha\beta} = \begin{bmatrix} V_\alpha \\ V_\beta \end{bmatrix} = \mathbf{S}_{\alpha\beta} V_{dc} = \begin{bmatrix} S_\alpha \\ S_\beta \end{bmatrix} V_{dc}. \tag{3}$$

The eight possible switching states and the corresponding input voltage vectors are summarized in Table 1.

Table 1. Input voltage vectors of the AC/DC converter in the $\alpha\beta$ coordinates.

No.	S_a	S_b	S_c	S_α	S_β	V_α	V_β	$\|\|V_\alpha+jV_\beta\|\|$
0	0	0	0	0	0	0	0	0
1	1	0	0	$\frac{2}{3}$	0	$\frac{2V_{dc}}{3}$	0	$\frac{2V_{dc}}{3}$
2	1	1	0	$\frac{1}{3}$	$\frac{1}{\sqrt{3}}$	$\frac{V_{dc}}{3}$	$\frac{V_{dc}}{\sqrt{3}}$	$\frac{2V_{dc}}{3}$
3	0	1	0	$-\frac{1}{3}$	$\frac{1}{\sqrt{3}}$	$-\frac{V_{dc}}{3}$	$\frac{V_{dc}}{\sqrt{3}}$	$\frac{2V_{dc}}{3}$
4	0	1	1	$-\frac{2}{3}$	0	$-\frac{2V_{dc}}{3}$	0	$\frac{2V_{dc}}{3}$
5	0	0	1	$-\frac{1}{3}$	$-\frac{1}{\sqrt{3}}$	$-\frac{V_{dc}}{3}$	$-\frac{V_{dc}}{\sqrt{3}}$	$\frac{2V_{dc}}{3}$
6	1	0	1	$\frac{1}{3}$	$-\frac{1}{\sqrt{3}}$	$\frac{V_{dc}}{3}$	$-\frac{V_{dc}}{\sqrt{3}}$	$\frac{2V_{dc}}{3}$
7	1	1	1	0	0	0	0	0

3.1. The Charging Rate Analysis

Assuming that the reference voltage V_{ref} is located in sector 1 of the vector diagram presented in Figure 5, based on the parallelogram law, it can be obtained from the two closest voltage vectors (V_1 and V_2), expressed as

$$\begin{cases} \frac{T_1}{T_s} V_1 + \frac{T_2}{T_s} V_2 = V_{ref} \\ T_1 + T_2 + T_0 = T_s \end{cases}, \tag{4}$$

where T_1, T_2, and T_0 are the action periods for the three basic voltage vectors, V_1, V_2, and $V_{0,7}$, respectively.

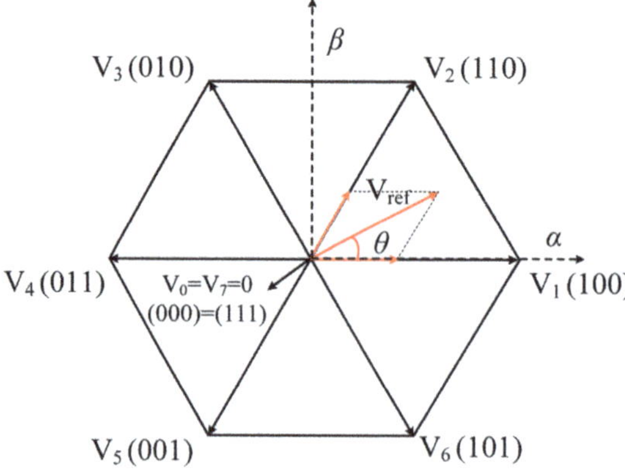

Figure 5. Eight possible input voltage vectors.

In the $\alpha\beta$ coordinate system, the included angle between the reference value (V_{ref}) and the voltage vector (V_1) is denoted as θ. Based on the law of sines, the relationship among V_1, V_2, and V_{ref} can be recorded as

$$\frac{|V_{ref}|}{sin(\frac{2\pi}{3})} = \frac{|\frac{T_1}{T_s}V_1|}{sin(\frac{\pi}{3}-\theta)} = \frac{|\frac{T_2}{T_s}V_2|}{sin\theta}. \tag{5}$$

Because V_1 and V_2 have the same amplitude, which is equal to $\frac{2V_{dc}}{3}$, (4) and (5) can be rewritten as

$$\begin{cases} T_1 = mT_s sin(\frac{\pi}{3}-\theta) \\ T_2 = mT_s sin\theta \\ T_0 = T_s - T_1 + T_2 \end{cases}, \tag{6}$$

where m is the modulation in index, which can be calculated by

$$m = \frac{\sqrt{3}|V_{ref}|}{V_{dc}}. \tag{7}$$

In the space vector modulation, a system constraint condition should be met:

$$T_1 + T_2 \leq T_s. \tag{8}$$

Substituting (6)–(8), the system constraint can be represented as

$$|V_{ref}| \leq \frac{V_{dc}}{\sqrt{3}sin(\frac{\pi}{3}+\theta)}. \tag{9}$$

In order to satisfy (9) for any included angle θ, it can be deduced that

$$V_m = |V_{ref}| \leq \frac{V_{dc}}{\sqrt{3}}, \tag{10}$$

which can be rewritten as

$$V_{dc} \geq \sqrt{3}V_m = V_{ph-ph}, \tag{11}$$

where V_m is amplitude of the phase voltage and V_{ph-ph} is the phase-to-phase voltage, known as the line voltage.

Therefore, the output DC-link voltage V_{dc} of the three-phase AC/DC converter has a minimum value. With the topology in Figure 2b, the DC-side output voltage is limited to between the voltage-line voltage and the fully charged voltage of the battery. In order to achieve the full range of charging/discharging from the minimum to the maximum battery voltage, a DC/DC bidirectional half-bridge converter is connected with the battery, as described in Figure 4.

3.2. Model Predictive Direct Power Control for the Grid Side

The MPC strategy for tracking the grid active/reactive power—named MPDPC—is proposed in this section. A cost function is designed as the combination of the error between the predicted and expected values of the active and reactive powers. The appropriate switching state that minimizes the cost function is chosen for the next sampling period.

A balanced source power is considered on the AC side. The system model in the standard $\alpha\beta$ frame transformation can be written as

$$L\frac{d\mathbf{i}_{g,\alpha\beta}}{dt} = \mathbf{V}_{g,\alpha\beta} - R\mathbf{i}_{g,\alpha\beta} - \mathbf{V}_{\alpha\beta}, \tag{12}$$

where $\mathbf{V}_{g,\alpha\beta}$ and $\mathbf{i}_{g,\alpha\beta}$ represent the grid voltage and phase current vectors, respectively, and are expressed by

$$\mathbf{V}_{g,\alpha\beta} = \begin{bmatrix} V_{g,\alpha} \\ V_{g,\beta} \end{bmatrix} = \frac{2}{3}\begin{bmatrix} 1 & -\frac{1}{2} & -\frac{1}{2} \\ 0 & \frac{\sqrt{3}}{2} & -\frac{\sqrt{3}}{2} \end{bmatrix}\begin{bmatrix} V_{g,a} \\ V_{g,b} \\ V_{g,c} \end{bmatrix},$$

$$\mathbf{i}_{g,\alpha\beta} = \begin{bmatrix} i_{g,\alpha} \\ i_{g,\beta} \end{bmatrix} = \frac{2}{3}\begin{bmatrix} 1 & -\frac{1}{2} & -\frac{1}{2} \\ 0 & \frac{\sqrt{3}}{2} & -\frac{\sqrt{3}}{2} \end{bmatrix}\begin{bmatrix} i_{g,a} \\ i_{g,b} \\ i_{g,c} \end{bmatrix}.$$

Based on the forward Euler approximation, the derivative of the grid current can be assumed as

$$\frac{d\mathbf{i}_{g,\alpha\beta}}{dt} = \frac{\mathbf{i_g}(k+1) - \mathbf{i_g}(k)}{T_s}. \tag{13}$$

The grid current (12) in the discrete time domain can be written as

$$\mathbf{i_g}(k+1) = \frac{T_s}{L}(\mathbf{V}_{g,\alpha\beta}(k) - R\mathbf{i}_{g,\alpha\beta}(k) - \mathbf{V}_{\alpha\beta}(k)) + \mathbf{i}_{g,\alpha\beta}(k). \tag{14}$$

Then, the active/reactive powers for the next time instant $(k+1)$ are predicted as

$$\begin{aligned} P(k+1) &= \frac{3}{2}Re\left\{\mathbf{V_g}\mathbf{i_g^*}\right\} \\ &= \frac{3}{2}(V_{g,\alpha}(k+1)i_{g,\alpha}(k+1) + V_{g,\beta}(k+1)i_{g,\beta}(k+1)) \end{aligned} \tag{15}$$

$$\begin{aligned} Q(k+1) &= \frac{3}{2}Im\left\{\mathbf{V_g}\mathbf{i_g^*}\right\} \\ &= \frac{3}{2}(V_{g,\beta}(k+1)i_{g,\alpha}(k+1) - V_{g,\alpha}(k+1)i_{g,\beta}(k+1)), \end{aligned} \tag{16}$$

where $V_{g,\alpha}(k+1)$, $V_{g,\beta}(k+1)$, $i_{g,\alpha}(k+1)$, and $i_{g,\beta}(k+1)$ are the predicted values. $i_{g,\alpha}(k+1)$ and $i_{g,\beta}(k+1)$ are obtained from (14). Due to the large value of the rate between the grid frequency and sampling frequency, $V_{g,\alpha}(k+1)$ and $V_{g,\beta}(k+1)$ can be assumed to be same as the measured values, $V_{g,\alpha}(k)$ and $V_{g,\beta}(k)$, respectively.

The cost function is designed as

$$g = \sqrt{(P_{ref} - P(k+1))^2 + (Q_{ref} - Q(k+1))^2}. \tag{17}$$

The commands of the active/reactive powers (P_{ref} and Q_{ref}) are determined by the center controller to satisfy the requirements of the grid and the EV customers.

3.3. Model Predictive Direct Current Control for the EV Side

For the half-bridge DC/DC converter, the switching state G is defined as

$$G = \begin{cases} 1 & \text{upper switch } G_1 \text{ is on} \\ 0 & \text{lower switch } G_2 \text{ is on} \end{cases}. \tag{18}$$

This means that the switching state G is set to 1 when the upper switch is on and the lower one is off, or vice versa.

Based on Kirchhoff's voltage law (KVL), when G is controlled to be 1, the DC-link voltage can be obtained as

$$V_{dc} = L\frac{di_{bat}}{dt} + V_{bat}, \tag{19}$$

where V_{dc} is the voltage on the DC bus, V_{bat} is the battery voltage, and i_{bat} is the battery charging/discharging current.

Similarly, when the switching state G is 0, the KVL value of the DC/DC converter can be expressed as

$$0 = L\frac{di_{bat}}{dt} + V_{bat}. \quad (20)$$

Based on (19) and (20), the mathematical model of the bidirectional DC/DC converter can be written as

$$GV_{dc} = L\frac{di_{bat}}{dt} + V_{bat}. \quad (21)$$

Then, the predicted battery current at the $k+1$ time instant is obtained as

$$i_{bat}(k+1) = i_{bat} + \frac{T_s}{L}(GV_{dc} - V_{bat}). \quad (22)$$

For constant current charging, the cost function g_{cc} is defined as

$$g_{cc} = (i_{bat} - i^*_{bat})^2, \quad (23)$$

where i^*_{bat} is the reference value of the battery current.

Neglecting the power loss in the transmission, the grid active power reference (P_{ref}) can be calculated as

$$P_{ref} = \sum_{n=1}^{N} P^*_{bat\#n} + P_{stor} = \sum_{n=1}^{N} V_{bat\#n} i^*_{bat\#n} + P_{stor}, \quad (24)$$

where $P^*_{bat\#n}$, $V_{bat\#n}$ and $i^*_{bat\#n}$ are the charging power reference, voltage, and expected charging current of the nth EV battery, respectively, N is the total number of the parked EVs, and P_{stor} is the power delivered to the storage system. In this paper, as an example, N is set to 1 without the storage system, which leads to $P_{ref} = P^*_{bat} = V_{bat} i^*_{bat}$. The proposed control algorithm is shown in Figure 6.

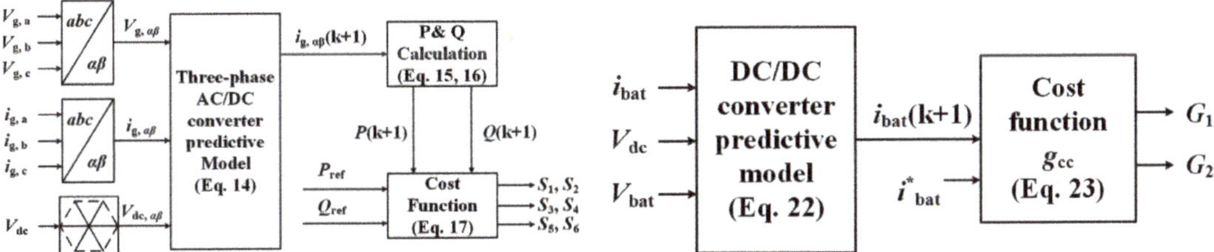

(a) Model predictive direct power control (MPDPC) for the AC/DC converter.

(b) Model predictive direct current control (MPDCC) for the DC/DC converter.

Figure 6. Proposed control algorithm.

4. Simulation Results

In [14], Metin et al. used a proportional-integral controller (PI) controller for V2G reactive power operation with an off-board charger. With the proposed system controller, the reactive power reference can be tracked effectively while charging the EV battery. However, it took around three grid cycles (60 ms) to respond to a new command.

By using the proposed MPC method for the system in [14] under the same transient simulation operation, the response time can be reduced dramatically. From the zoomed version of the active and reactive power in Figure 7a, it can be seen that the response time for the system changing from a unity power factor operation to a 0.4 pf (leading) operation is less than 2 ms. Compared with the three-cycle (60 ms) response time in [14,15], the response speed is improved significantly. The grid current in Figure 7b can also reach the new command operation within 2 ms.

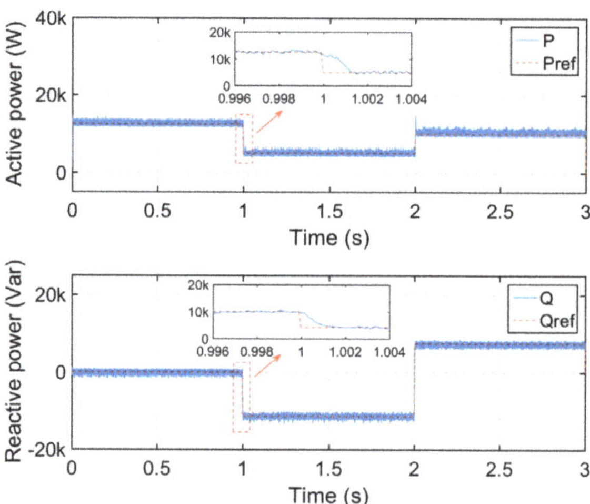

(a) Active and reactive power performance.

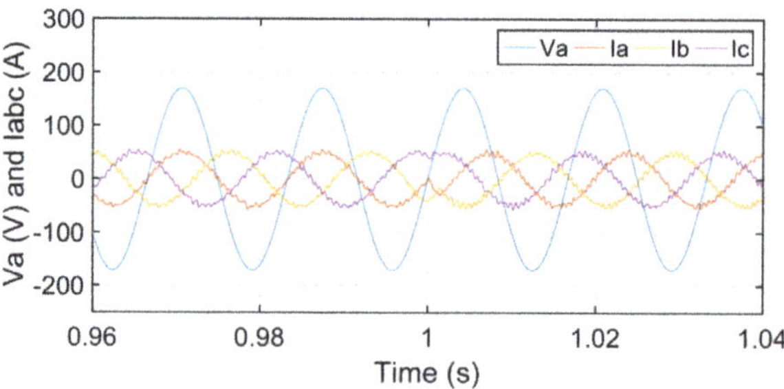

(b) Grid voltage and current performance.

Figure 7. Performance by using the model predictive control (MPC) for the system in Ref. [14].

This section provides the simulation results of the proposed control method for the four-quadrant P-Q operation in Matlab/Simulink. The parameters of the bidirectional charger were chosen as shown in Table 2. The simulation scenarios are designed as follows:

(1) The charging current i^*_{bat} declines from 6.67 to -6.67 A at t = 1.04 s, which leads to the active power reference P_{ref} dropping from 2 to -2 kW. It steps up to 0 A at t = 1.08 s and stays constant for the next four cycles. Meanwhile, the reactive power reference stays at 0 kvar during the first four-cycle period. Then, it rises to 1 kvar at t = 1.08 s and drops to -1 kvar at t = 1.12 s.

(2) The charging current is decreased from 6.67 to 3.33 A at t = 1.04 s, which means that the active power reference P_{ref} is reduced from 2 to -1 kW, and after two cycles, it is restored to 6.67 A. At t = 1.12 s, it steps down to -3.33 A with P_{ref} equal to -1 kW. During this period, the reactive power reference is reduced from 1 to -1 kvar at t = 1.08 s.

Figure 8 presents the results of the first operation condition, which was used to describe the system performance in the independent operation of G2V, V2G, and (capacitive/inductive) V4G. The active/reactive power demands can be tracked effectively by using the proposed MPC strategy. In Modes I and II, no reactive power is transferred in this operation. Based on the active power reference, the bidirectional charger works effectively in the G2V and V2G modes. The grid current is controlled to be in phase with the grid voltage, as in the operation in the G2V mode. The EV batteries charge the energy from the source. The total harmonic distortion (THD) of the grid current is around 4.23%. Conversely, the active power is delivered from the EV to the grid in the V2G mode. It can be seen that the current is in phase opposition with the grid voltage. In Modes III and IV, the converter supports the grid with the reactive power compensation (V4G operation). With the reactive power being delivered from the grid to the load side, the grid current is 90° and leading the grid voltage; otherwise, it is 90° and lagging behind, as shown in Figure 8c.

Table 2. System parameters for the simulation test.

Parameter	Value
Line resistance per phase, R	0.25 Ω
Filter inductor per phase, L	10 mH
DC capacitor, C	470 µF
DC inductor, L_{dc}	100 mH
Output capacitor, C_{dc}	470 µF
Grid r.m.s voltage, V_{LL}	100 V/50 Hz
Sampling frequency, f_s	20 kHz
Battery voltage, V_{bat}	300 V

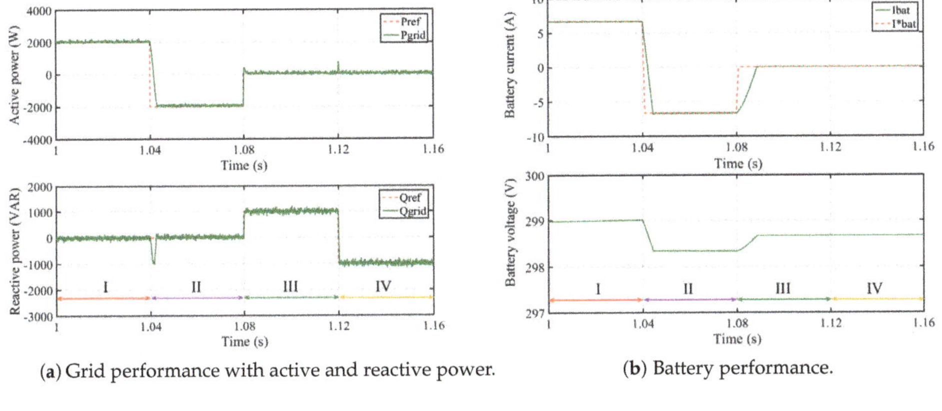

(**a**) Grid performance with active and reactive power. (**b**) Battery performance.

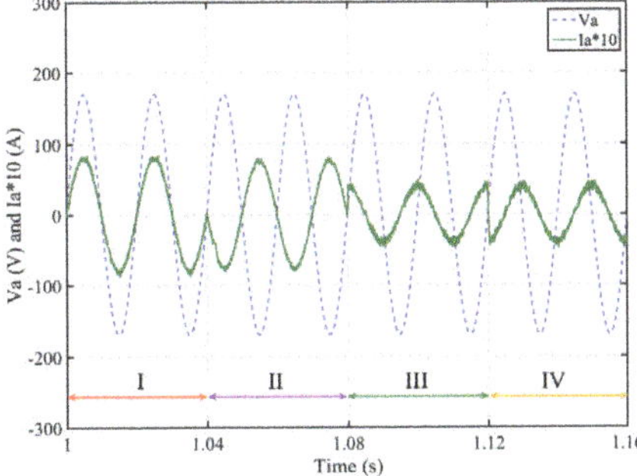

(**c**) Power grid voltage and current.

Figure 8. Simulation results in Quadrants I, II, III, and IV.

The second simulation was performed to show the system dynamic and steady-state performance in the four-quadrant regions of the PQ plane, as described in Figure 9.

The bidirectional charger works in G2V or V2G with the capacitive or inductive V4G operation modes. Similarly, it can be observed clearly in Figure 9a that the system can track the references of the active and reactive powers at the same time. Figure 9b,c shows the battery operation details and the grid-side performance in the a-phase, respectively. In Mode V, where the charger operates in G2V with inductive V4G operation, the main power grid provides the positive active/reactive power to the load. The grid current leads its voltage, as depicted inFigure 9c. The system operates as V2G with the inductive V4G mode in Mode VI, delivering the active power from the EV battery to the power grid. The system serves as a static var generator to compensate the reactive power in V2G or G2V with capacitive V4G operation (Modes VII and VIII). The grid current lags behind the same-phase voltage in both operation modes. The battery is charged and discharged according to the transmission direction of the active power.

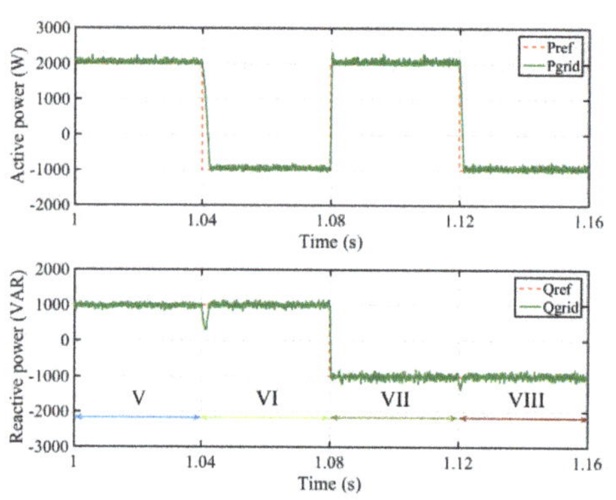

(**a**) Performance of the grid with active and reactive power.

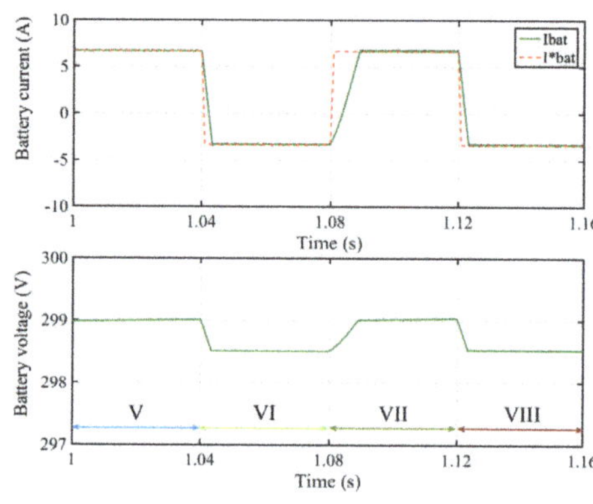

(**b**) Battery performance.

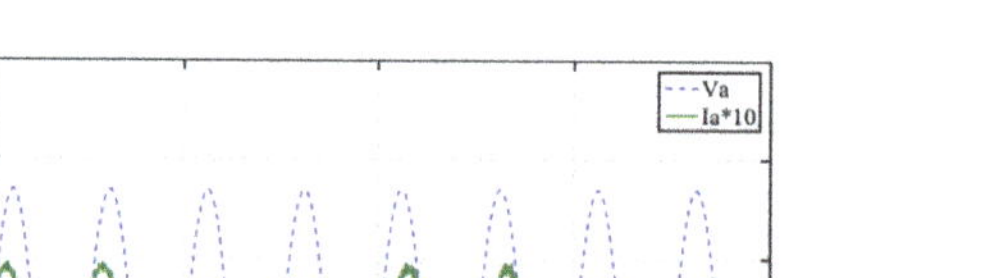

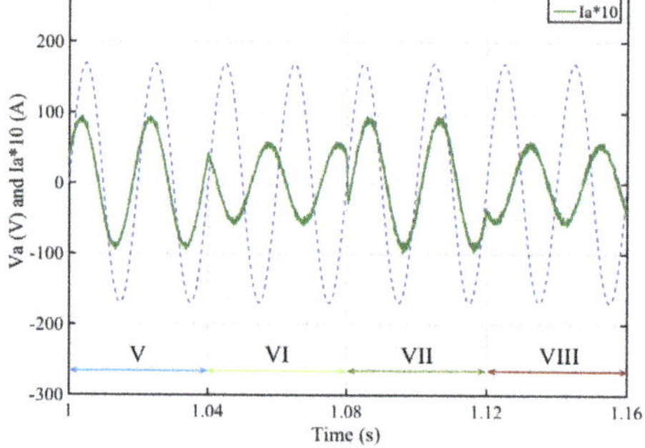

(**c**) Power grid voltage and current.

Figure 9. Simulation results in Quadrants V, VI, VII, and VIII.

5. Experimental Results

The proposed control strategy was further verified by a scaled-down experiment that used a laboratory setup, as shown in Figure 10. The setup was composed of the following devices: an insulated-gate bipolar transistor-based three-phase two-stage bidirectional converter, three AC filter inductors, and DC power sources. Twelve 12 V 24 Ah sealed lead-acid batteries connected in series were used to simulate an EV battery. Therefore, the nominal voltage and the capacity of this sealed battery were 144 V and 24 Ah, respectively. The experimental parameters are listed in Table 3. The control algorithm was implemented on a dSPACE DS1104 processor board. The operations were performed with a sampling frequency equal to 10 kHz. All of the measured parameters were viewed through the dSPACE control desk. Note that during the experimental test, the battery contactor should be closed prior to turning on the main power on the AC side. Otherwise, a current spike occurs due to the voltage change between the DC bus (V_{dc}) and the battery (V_{bat}). Since there is a direct connection between the converter output capacitor and the battery, this current spike cannot be controlled. It might even destroy the converter output contactor and the battery contactors, as well as other semiconductor devices.

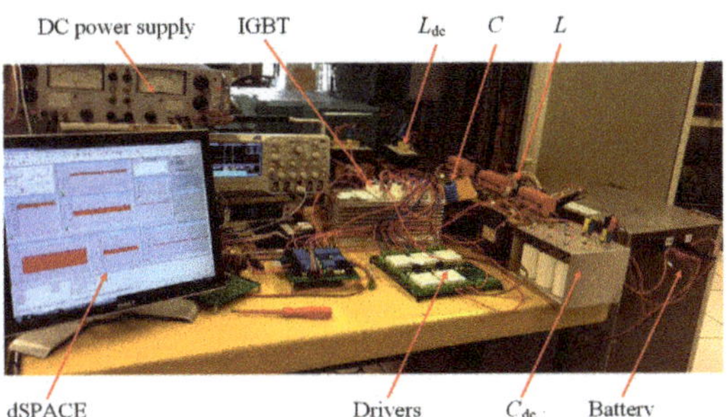

Figure 10. Experimental setup.

Table 3. System parameters for the experimental test.

Parameter	Value
Filter inductor per phase, L	20 mH
DC capacitor, C	680 µF
DC inductor, L_{dc}	35 mH
Output capacitor, C_{dc}	68 µF
Grid r.m.s. voltage, V_{LL}	100 V/50 Hz
Sampling frequency, f_s	20 kHz
Battery voltage, V_{bat}	144 V

In the experimental test, the charging and discharging current references were set to be 2 and −2 A, respectively. The battery voltage was about 154 V. The active power demand delivered from the grid was around 308 and −308 W for the charging and discharging processes, separately. For inductive and capacitive operations, the reactive power references were 200 and −200 var, respectively.

Figure 11 shows that the system operates properly in Quadrants I, II, III, and IV. The battery can be charged and discharged with the proposed MPDPC and MPDCC strategies. The grid active and reactive power is well maintained according to the reference values presented in Figure 11a. The battery current can track the reference value (2 and −2 A) effectively, as shown in Figure 11b. Figure 11c depicts the experimental current and voltage waveforms in the a-phase. The input current has a nearly sinusoidal waveform. When the reactive power is 0 var, it is in or out of phase with the grid voltage depending on the direction of the active power.

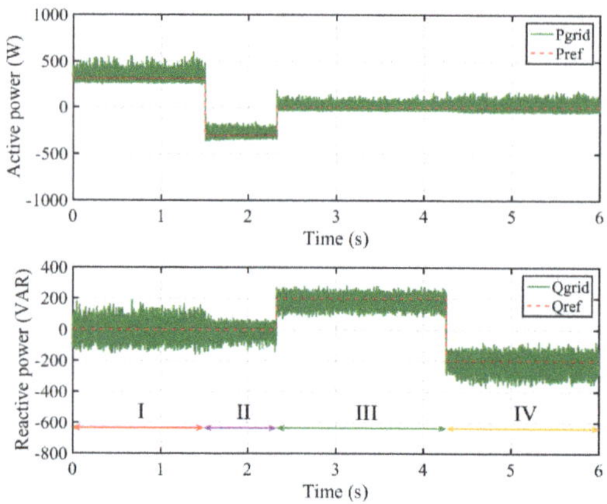

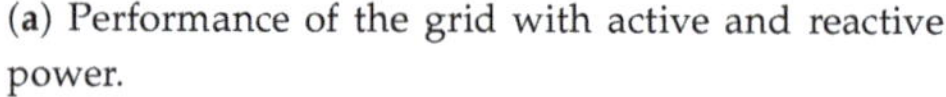

(**a**) Performance of the grid with active and reactive power.

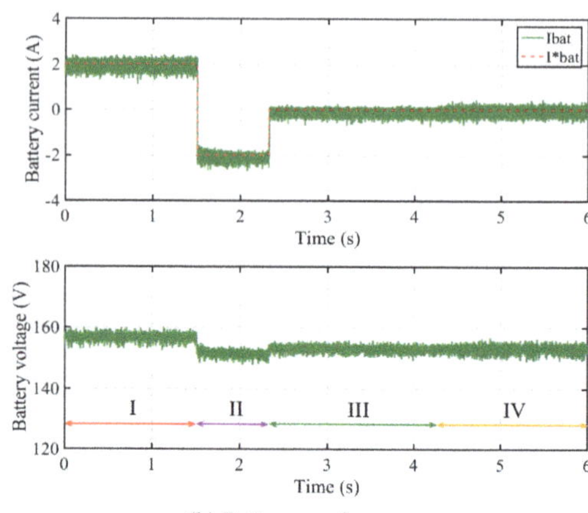

(**b**) Battery performance.

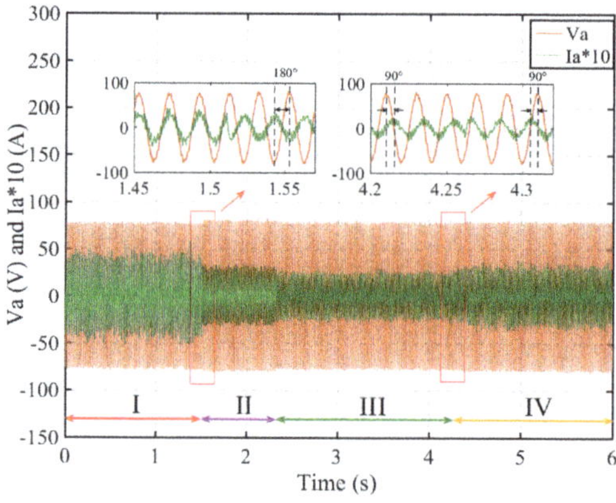

(c) Power grid voltage and current.

Figure 11. Experimental results in Quadrants I, II, III, and IV.

In Figure 12, the reactive power is varied from 200 to −200 var, and the active power is controlled to change between 308 and −308 W. It can be seen clearly that during this process, not only can the active power be delivered between the EVs and the grid, but the reactive power can also be provided or absorbed by the EV battery. Figure 12c presents the grid voltage and current waveforms. The response time is only approximately 2 ms for the system to reach the next demand condition.

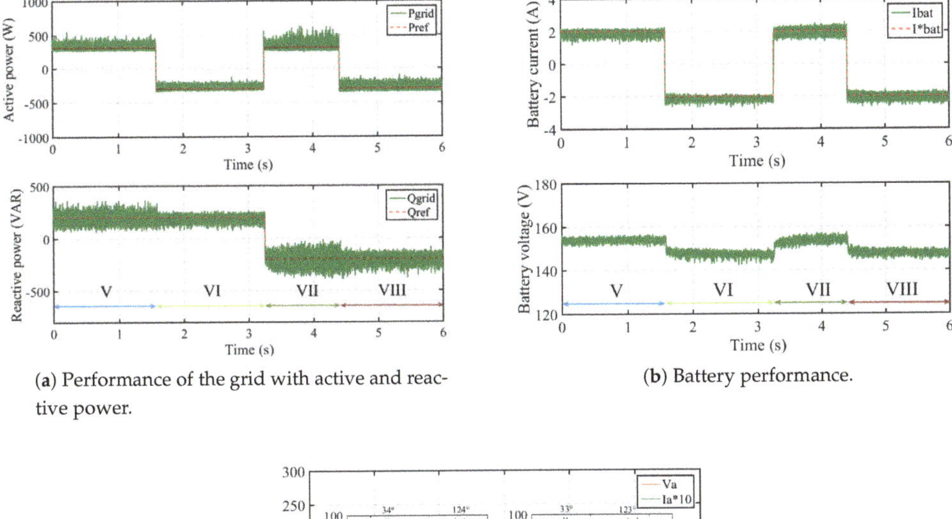

(**a**) Performance of the grid with active and reactive power.

(**b**) Battery performance.

(**c**) Power grid voltage and current.

Figure 12. Experimental results in Quadrants V, VI, VII, and VIII.

6. Conclusions

This paper proposed a model predictive control composed of model predictive direct power control and a model predictive direct current control schemes for three-phase two-stage off-board bidirectional EV chargers. The MPDPC method was designed to track the grid active/reactive power references for the AC/DC converter. The MPDCC scheme was applied in a second-stage DC/DC converter for constant current charging. The proposed system controller receives the discharging and charging currents from the EV charging stations. With the proposed method, the bidirectional charger can operate in G2V or V2G with/without the capacitive/inductive V4G modes. The EV battery can not only exchange active power with the grid, but can also function as a static var compensator to improve the power quality based on the grid requirements. It can be seen from the simulation and experimental results that the designed MPC scheme has a fast dynamic response and good steady-state performance. The demands from both the grid and the EV batteries can be met effectively. However, there are some gaps between the laboratory and real-life off-board EV chargers, including the power level, charging conditions, and battery types. (a) In the real-life EV chargers, the power level is much higher than in the laboratory ones. In the US Society of Automotive Engineers (SAE) J1772 standard, DC Level 1 and DC Level 2 are the two levels of fast DC charging, with a rated power equal to 80 and 400 kW, respectively. (b) Only the initial constant current charging operation is considered in this paper. Generally, the constant voltage charging mode in the final process is required in most real cases. (3) In the current EV market, due to their long lifetime and high energy density, lithium batteries are used in the real vehicles, rather than sealed lead-acid batteries. (4) An isolated DC/DC converter is used in real-life chargers to protect EV batteries and the grid. To narrow the gap between the scaled-down laboratory setup and the real-life deployment, a high-power-level topology with an isolated DC/DC converter interfaced with lithium batteries will be considered in future work.

Author Contributions: All authors contributed to the research in the paper. Conceptualization, T.H.; formal analysis, T.H. and D.D.-C.L.; methodology, M.W. and T.H.; project administration, M.W.; supervision, D.D.-C.L.; validation, T.H. and Q.L.; writing-original draft, T.H.; writing-review and editing, T.L. and Q.Y.; all the authors have read and approved the final manuscript. All authors have read and agreed to the published version of the manuscript.

Funding: This work was supported in part by Fundamental Research Funds for the Major Projects: KEJB0220001536.

Institutional Review Board Statement: Not applicable.

Informed Consent Statement: Not applicable.

Data Availability Statement: Not applicable.

Conflicts of Interest: The authors declare no conflict of interest.

References

1. Anamika, D.; Santoso, S. Electric Vehicle Charging on Residential Distribution Systems: Impacts and Mitigations. *IEEE Access* **2015**, *3*, 1871–1893.
2. Choi, S.Y.; Gu, B.W.; Jeong, S.Y. Advances in wireless power transfer systems for roadway-powered electric vehicles. *IEEE J. Emerg. Sel. Top. Power Electron.* **2015**, *3*, 18–36. [CrossRef]
3. Mi, C.C.; Buja, G.; Choi, S.Y. Modern Advances in Wireless Power Transfer Systems for Roadway Powered Electric Vehicles. *IEEE Trans. Ind. Electron.* **2016**, *63*, 6533–6546. [CrossRef]
4. Tian, Z.; Wang, Y.; Zhang, F. Real-time charging station recommendation system for electric-vehicle taxis. *IEEE Trans. Intell. Transp. Syst.* **2016**, *17*, 3098–3109. [CrossRef]
5. Sjafie-khah, M.; heydarian-Forushani, E.; Osório, G.J. Optimal Behavior of Electric Vehicle Parking Lots as Demand Response Aggregation Agents. *IEEE Trans. Smart Grid* **2016**, *7*, 2654–2665. [CrossRef]
6. Farzin, H.; Aghtaie, M.M.; Firuzabad, M.F. Reliability studies of distribution systems integrated with electric vehicles under battery-exchange mode. *IEEE Trans. Power Deliv.* **2016**, *31*, 2473–2482. [CrossRef]
7. Karfopoulos, E.L.; Panourgias, K.A.; Hatziargyriou, N.D. Distributed Coordination of Electric Vehicles providing V2G Regulation Services. *IEEE Trans. Power Syst.* **2016**, *31*, 2834–2846. [CrossRef]
8. Alam, M.J.E.; Muttaqi, M.; Sutanto, D. Effective utilization of available pev battery capacity for mitigation of solar PV impact and grid support with integrated V2G functionality. *IEEE Trans. Smart Grid* **2016**, *7*, 1562–1571. [CrossRef]
9. Yilmaz, M.; Krein, P.T. Review of the impact of vehicle-to-grid technologies on distribution systems and utility interfaces. *IEEE Trans. Power Electron.* **2013**, *28*, 5673–5689. [CrossRef]
10. Kwon, M.; Choi, S. An Electrolytic Capacitorless Bidirectional EV Charger for V2G and V2H Applications. *IEEE Trans. Power Electron.* **2017**, *32*, 6792–6799. [CrossRef]
11. Zahid, Z.U.; Dalala, Z.M.; Chen, R. Design of Bidirectional DC–DC Resonant Converter for Vehicle-to-Grid (V2G) Applications. *IEEE Trans. Transport. Electrific.* **2015**, *1*, 232–244. [CrossRef]
12. Rabiee, A.; Farahani, H.F.; Khalili, M.; Muttaqi, J.A.K.M. Integration of plug-in electric vehicles into microgrids as energy and reactive power providers in market environment. *IEEE Trans. Ind. Inform.* **2016**, *12*, 1312–1320. [CrossRef]
13. Buja, G.; Bertoluzzo, M.; Fontana, C. Reactive power compensation capabilities of V2G-enabled electric vehicles. *IEEE Trans. Power Electron.* **2017**, *32*, 9447–9459. [CrossRef]
14. Kesler, M.; Kisacikoglu, M.C.; Tolbert, L.M. Vehicle-to-grid reactive power operation using plug-in electric vehicle bidirectional off-board charger. *IEEE Trans. Ind. Electron.* **2014**, *61*, 6778–6784. [CrossRef]
15. Monteiro, V.; Pinto, J.G.; Afonso, J.L. Operation modes for electric vehicle in smart grids and smart homes: Present and proposed modes. *IEEE Trans. Veh. Technol.* **2016**, *65*, 1007–1020. [CrossRef]
16. Kisacikoglu, M.C. Single-phase on-board bidirectional PEV charger for V2G reactive power operation. *IEEE Trans. Smart Grid* **2015**, *6*, 767–775. [CrossRef]
17. Khan, S.A.; Guo, Y.; Zhu, J. Model predictive observer based control for single-phase asymmetrical T-type AC/DC power converter. *IEEE Trans. Ind. Appl.* **2015**, *55*, 2033–2044. [CrossRef]

18. Kouro, S.; Cortes, P.; Vargas, R.; Ammann, U.; Rodriguez, J. Model predictive control—A simple and powerful method to control power converters. *IEEE Trans. Ind. Electron.* **2009**, *56*, 1826–1838. [CrossRef]
19. Rodriguez, P.; Cortes, J. (Eds.) Predictive control of power converters and electrical drives. In *Predictive Control of Power Converters and Electrical Drives.*; John Wiley & Sons, Ltd.: West Sussex, UK, 2012.
20. Lezana, P.; Aguilera, R.; Quevedo, D.E. Model predictive control of an asymmetric flying capacitor converter. *IEEE Trans. Ind. Electron.* **2009**, *56*, 1839–1846. [CrossRef]
21. Zhang, Y.; Xie, W.; Li, Z.; Zhang, Y. Model predictive direct power control of a PWM rectifier with duty cycle optimization. *IEEE Trans. Power Electron.* **2013**, *28*, 5343–5350. [CrossRef]
22. Calle-Prado, A.; Alepuz, S.; Bordonau, J.; Nicolas-Apruzzese, J.; Cortés, P.; Rodriguez, J. Model predictive current control of grid-connected neutral-point-clamped converters to meet low-voltage ride-through requirements. *IEEE Trans. Ind. Electron.* **2015**, *628*, 1503–1514. [CrossRef]
23. Yilmaz, M.; Krein, P.T. Review of battery charger topologies, charging power levels, and infrastructure for plug-in electric and hybrid vehicles. *IEEE Trans. Power Electron.* **2013**, *28*, 2151–2169. [CrossRef]
24. Kisacikoglu, M.C. EV/PHEV bidirectional charger assessment for V2G reactive power operation. *IEEE Trans. Power Electron.* **2013**, *28*, 5717–5727. [CrossRef]
25. Dusmez, S.; Khaligh, A. A compact and integrated multifunctional power electronic interface for plug-in electric vehicles. *IEEE Trans. Power Electron.* **2013**, *28*, 5690–5701. [CrossRef]
26. He, T.; Bai, Y.; Zhu, J. Optimal charging strategy of electric vehicles customers in a smart electrical car park. In Proceedings of the 8th IET International Conference on Power Electronics, Machines and Drives (PEMD 2016), Glasgow, UK, 19–21 April 2016.
27. Akter, M.P.; Mekhilef, S.; Tan, N.M.L. Modified model predictive control bidirectional AC/DC converter based on lyapunov function for energy storage systems. *IEEE Trans. Ind. Electron.* **2013**, *63*, 704–715. [CrossRef]
28. Singh, B.; Singh, B.N.; Chandra, A. A review of three-phase improved power quality AC/DC converters. *IEEE Trans. Ind. Electron.* **2004**, *51*, 641–660. [CrossRef]

 energies MDPI

Article

The Problem of Train Scheduling in the Context of the Load on the Power Supply Infrastructure. A Case Study

Szymon Haładyn

Faculty of Mechanical Engineering, Wroclaw University of Science and Technology, 50-370 Wroclaw, Poland; szymon.haladyn@pwr.edu.pl

Abstract: This article deals with the new challenges facing modernising railways in Poland. We look at the problem of the efficiency of the power supply system (3 kV DC) used in the context of the increasing use of electric vehicles, which have a higher demand for electricity than the old type. We present and characterise the power supply system in use, pointing out its weaknesses. We consider a case study. The load of the power supply network generated by the rolling stock used in Poland was examined using a microsimulation. A real train timetable was taken into account on a fragment of one of the most important railway line sections in one of the urban agglomerations. Then the results were compared with the results of a microsimulation in which old units were replaced by new trains. These tests were carried out in several variants. We found critical points in the scheduling of railway system use. Our results indicate that it is becoming increasingly necessary to take into account the permissible load capacity of the supply network in certain traffic situations in the process of timetable construction.

Keywords: railway DC power supply system; railway case study; quality of rail power supply

Citation: Haładyn, S. The Problem of Train Scheduling in the Context of the Load on the Power Supply Infrastructure. A Case Study. *Energies* **2021**, *14*, 4781. https://doi.org/10.3390/en14164781

Academic Editors: Andrea Mariscotti, Leonardo Sandrolini and Tseng King Jet

Received: 25 May 2021
Accepted: 3 August 2021
Published: 6 August 2021

1. Introduction

More trains are running than ever before. New units are replacing trains which are several decades old. Railway undertakings, thanks to multi-million-euro subsidies from the European Union, have record amounts of money and can afford to carry out bold, large-scale rolling stock projects. An increase in the pace of such investments has been noticeable in Europe in recent years.

The economic crisis caused in the late 1970s and early 1980s by the inefficiency of the centrally planned economy was exacerbated by economic transformations in the early 1990s. At that time, the state did not invest in railways. There was a lack of funding. The neoliberal approach to the state-owned mass railways maintained by public funds did not give any chance for the development of this mode of transport. Many sections of the railway line, totalling several thousand kilometres in length in Poland alone, were closed. Individual transport was promoted very strongly. The motorisation rate increased significantly. At the time, the railways were carrying a record low number of passengers. The deteriorating state of the infrastructure discouraged new passengers. Worn-out trains were not replaced by younger units. In recent years, we have been struggling with the consequences of the lack of funding at that time. Many of the trains running on the tracks are reaching the end of their useful lives and so-called technical death. This has necessitated the replacement, or at least modernisation, of the rolling stock.

Thanks to external funding, new trains can be purchased. Railway lines are also being repaired and modernised. Due to the measures taken, the number of rail passengers is gradually increasing and new transport needs are appearing, both of which generate demand for rolling stock. Modern rolling stock is being rolled out on tracks in ever greater numbers. In some parts of Poland, usually in urban agglomerations, a record number of connections is starting to be recorded.

Modern trains increase the comfort of travel. They provide passengers with a completely different quality of rail travel. They are designed according to modern standards. They are also better adapted to the type of connections they serve. They are equipped with more powerful drive units. Electric units designed for regional and agglomeration connections have several times more installed power in comparison to trains manufactured several decades ago. This allows more dynamic driving and greater acceleration [1,2]. This greater acceleration leads to shorter journey times because of stops. This is particularly noticeable on railway lines with large numbers of intermediate stops. On the other hand, long-distance passenger trains are designed to achieve higher maximum speeds. Due to the exponential nature of the train's resistance to motion (in regard to size), increasing the permissible speed must also involve increasing the power of the power unit [3,4]. On agglomeration and regional lines, units with several hundred kilowatts of power are being replaced by electric multiple units with 2–3 MW of power. Long-distance trains are increasingly being driven by 3–6 MW motive power units instead of locomotives with an installed power of 2–3 MW [5]. Besides, systems that were not used in the old trains, such as air-conditioning, also generate additional load on the electricity systems associated with the railway lines. The standard nowadays is also not equipping every seat with an electric socket with 230 V AC voltage.

In the case of the 3 kV DC power supply system used on the Polish railway network, such an increase in energy consumption by trains and the installation of additional equipment onboard railway vehicles may lead to overloading the power supply infrastructure [6]. The relatively low voltages (in comparison to the 15 kV AC and 25 kV AC supply systems) and the accumulation of higher-powered units mean very high current values [7]. These are too high for the overhead contact line cross-sections used—hence the necessity of introducing restrictions on the maximum loads of individual traction substations. An overload may result in a temporary voltage drop in the overhead contact line due to the tripping of fast circuit breakers. Such an undesirable event generates delays in rail traffic, which are particularly acute on lines with high train loads.

Electrified with 3 kV DC, Poland's railway network faces the difficult challenge of raising the speed of train traffic and allowing traction units to run at far higher power levels than those originally used. The DC power system allows for speeds of up to 230–250 km/h [5,7,8]. The powers of traction units exceed by several times the values known even a few years ago. More and more often, it is necessary to take into account not only the capacity of railway lines but also the permissible load of the power supply network when designing the timetable. The passage of units with high installed capacity disrupts traffic, especially in conurbations, or significantly reduces the capacity of railway lines [6]. In the absence of plans to change the railway supply system, it is reasonable to check the extent to which the supply network generates critical points, which must be taken into account in the process of scheduling the use of the railway system.

The problem of power quality in the DC rail power system is particularly significant in situations where there are scheduling disruptions. The overlap of several trains with significant power consumption can cause an excessive load on the traction substation. This can consequently exacerbate delays and make the railway system unable to regain punctuality. Additionally, the problem of the resilience of the railway system to adverse events and their consequences has been described [9,10].

In this article, the power supply system used in Poland is described. Potential critical points in the railway infrastructure in the context of the power supply system are pointed out. Next, a model of traction vehicle movement is discussed. The built model is then adapted for the case study. Next, calculations of the power supply system load are illustrated based on the train timetable for a representative railway line running through one of the major Polish agglomerations and starting from a junction station. The loads are investigated in several variants, taking into account different types of rolling stock operating the connections. Operating situations are identified which—due to significant current loads—may lead to overloading of the power network. This, in turn, can result in a temporary stop of traffic due to a voltage drop in the traction network. The results are discussed and commented on.

2. Literature Review

Several works have been produced on the subject of powering trains. They deal with problems related to power supply systems for electric railways, which are different for DC and AC systems. The justification for the development of this type of power supply was emphasised in [1], by stating that electric railways are characterised by high traffic speeds, reliability, and the possibility of using large amounts of power. It was pointed out in [11] that railway power supply systems are among the largest end-users of electricity networks, generating high power demands. The choice of the power system is determined by historical, geographical, and economic considerations [12]. DC technology is used in systems with lower requirements and lower capacities, pointing mainly to urban railways, metros, tramway networks, and light rail systems [11]. However, there are several DC power systems in place for full-scale railways [13] in Italy, Poland [5], Spain (3 kV voltage), and France (1.5 kV voltage). In these countries, however, in the construction of high-speed railway lines with the highest energy demands, separate lines are being built (or in the case of Poland—are planned to be built) that are supplied with AC at higher voltages.

The main problems associated with DC power supply include, in addition to the limitation of the transmitted power to about 10 MW [7,8], primarily: waveform transients, system asymmetry, voltage and current harmonics, low power factor [11], and the occurrence of stray currents [2,12,14]. Another problem identified for electric traction is that of substation harmonics [15–18]. The occurrence of stray currents is related to the fact that the currents—instead of running on the rail—flow through the ground and re-enter the rails. This phenomenon can lead to electrochemical corrosion of metallic equipment, and to the occurrence of step voltages near the railway line [19–23].

The problem of traction substation loading in a DC power system has been addressed [24]. The loading of an electrical substation feeding a metro system was studied. A simulation model mimicking a 51 km long line with 27 traction substations supplying the line with 750 V DC was built and calculations were carried out for a so-called minimum-time run. The Matlab tool was used for this. However, only one type of rolling stock was modelled; the trains ran cyclically. Previous work constructed a simulation model representing the power supply of a tramway network [25]. The model was implemented in Matlab. Relating the conditions for the tram system to the railway system is difficult, due to the specifics of the two systems—different voltages, powers, and higher schedule rigidity for the railway. In the case of a DC rail system, the problem of electricity demand was addressed [26]. The energy consumption of trains moving on a long gradient (in the uphill and downhill variants) was studied. However, the total energy demand was not considered, but focused on the possibility of recovering braking energy from the downhill train and using it to accelerate the uphill train. The energy demand and other parameters characterising train running were also addressed in [27], where, however, the profile of the railway line and the specifics of the train formation were not taken into account. Another study [13] looked at how to improve the energy efficiency of DC railway power supply systems.

Many microsimulation models, which are based on solving the equation of motion, focus on minimising the energy demands of a vehicle (among other works: [28–34]) or a small grid—mainly by using braking energy recovery and supplying it through the power system to the accelerating vehicle, or using an energy storage tank [35–37]. However, these models ignore issues related to the permissible load on the power supply network, focusing instead on its ability to accommodate the electricity returned by the train.

3. Specifics of the Supply System

Electric multiple units EN57 produced in the years 1961–1993 had continuous power of 580 kW [38]. Contemporary units, however, have a power requirement several times higher. For example, the Newag Impuls 31WE MEU, designed to operate regional trains, was equipped with engines of 2 MW [39]. The Impuls has a similar capacity to the EN57 unit and is often used as a replacement for the EN57 when replacing rolling stock. The ET41 locomotive, built in the 1980s and designed to handle the heaviest goods trains, and regarded in the past as one of the most powerful of those in service in Poland, has engines with a total power of 2.6 MW less than the modern locomotive EU44 "Siemens EuroSprinter". Increased engine power allows trains to develop higher maximum speeds (a desirable postulate for long-distance trains with a small number of stops) and ensures greater acceleration of start-up, which in turn significantly shortens the travel time of agglomeration and local trains [40].

The current drawn from the overhead line by the electric train is proportional to the power and inversely proportional to the voltage:

$$I_n = \frac{P}{U_n} \tag{1}$$

The 3 kV DC [41–43] power supply system used in Poland is the reason for the occurrence of significant current loads on the network, an order of magnitude higher than in Western countries. Low supply voltage means very high current consumption and requires—especially in the case of using heavy networks (such networks are mainly present on the PKP PLK SA network)—appropriately frequent sectioning [44].

Enclosure 2.12 of the Network Statement of PKP PLK SA [45] contains a tabular list of catenary network parameters; information regarding catenary network type, e.g., YC150–2CS150, C120–2C, or YwsC120–2C; maximal speed a train on a part of the railway line with the specified type of network; maximal current-carrying capacity—that is, the maximal current which can be drawn during train passage; and the minimal distance between operating current collectors. These relations result directly from the overhead contact line design, especially from [46]:

- The cross-sectional area of contact wires, suspension cables and hangers [46,47];
- The working temperature of the overhead contact line [48,49];
- Vibrations occurring in the overhead contact line and dynamic interaction of the current collector with the overhead contact line [50–52].

In the national rail infrastructure manager's network, there are traction networks in which the admissible train current capacity is between 1010 and 2730 A—most often 1725 A. This means that in practice, trains with acceptable power ratings for traction (P = 8 MW) can run on Polish railway tracks, which translates into the fact that in the case of Polish power system it is possible to operate trains with maximal speeds of up to 250 km/h [7,8]. Modern power units of high-speed trains, used in Western Europe and Asia, are characterized by traction power, even reaching 10–15 MW [7,8,44]. In the case of the Western European power supply system, this means that the current drawn by such a train does not exceed the value of 500 A, whereas in Poland such a train would draw the current of 4 kA. However, national technical standards [43] foresee running trains—after [50]—with current up to 2.5 kA and even—in the case of newly built lines—4 kA. The above was confirmed by tests carried out on the national network [44]. The electrification of the Warsaw junction carried out before the war [51,52] and the extension after the war of the voltage applied there to the whole country, together with the failure of the communist authorities to decide on a change in the power supply system, is now a considerable obstacle to the introduction of a high-speed railway in the country.

The heavy current network type commonly used in Poland (C120–2C) may be easily overloaded. Significant current loading may expose the contact wire to high temperatures and even to reductions in mechanical parameters [44].

Power supply distances of many kilometres determine the occurrence of voltage drops in contact wires. These are directly proportional to the rated current and the distance between the power source and the load, and inversely proportional to the conductivity and contact wire area:

$$\frac{\Delta U}{U_n} = \frac{2 \cdot I_n \cdot l}{\sigma \cdot U_n \cdot s} \quad (2)$$

The voltage on the overhead contact line, therefore, varies linearly with increasing distance between the train and the substation (Figure 1):

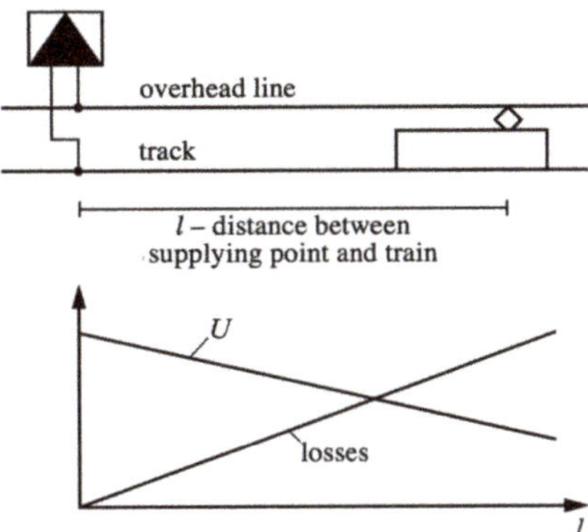

Figure 1. Changing overhead contact line voltage as a function of distance from the power supply point.

As previously communicated, the railway line is divided into shorter sections, the so-called traction sections [44,46]. Each section is supplied by an electrical substation, the task of which is to change the voltage from—usually—15 kV AC (where the national power grid operates at medium voltages) to 3 kV DC (flowing in the contact wires). The length of one section does not exceed 30 km [43] but is usually between 7 and 20 km [44].

Each of the tracks is powered independently of the others, simultaneously from two sides, i.e., from two different power substations. The use of such a solution allows one to limit the voltage drops occurring on the network and makes it possible to connect neighbouring sections in emergencies.

4. Computational Model

The characteristics of the driving parameters of a railway vehicle can be obtained by solving the equation of motion of a train [3,4,53–55]. The equation of motion of a train is based on Newton's second law of dynamics. It is described by a second-order differential equation [3,4,53,54]:

$$F(v,x) = m \cdot k \cdot \frac{d^2x}{dt^2} \quad (3)$$

where F is the resultant force, m the mass and k the coefficient of swirling mass.

Equation of motion can be noted as:

$$x = m \cdot k \int_{v_1}^{v_2} \frac{v dv}{F} \quad (4)$$

or:

$$t = m \cdot k \int_{v_1}^{v_2} \frac{dv}{F} \quad (5)$$

We also know that:

$$P = F \cdot \frac{dx}{dt} \quad (6)$$

After the transformation of Equation (1) and subrogation of Equations (3) and (6), the formula for train run amperage can be obtained:

$$I = \frac{m \cdot k \cdot \frac{d^2x}{dt^2} \cdot \frac{dx}{dt}}{U} [A] \tag{7}$$

The magnitude of current consumption recorded in an electrical substation is higher than that derived from the solved equation of train movement. This is due to current transmission losses and the inefficiency of technical equipment, especially traction motors, and the presence of other consumers on the train—air conditioning, heating, lighting, passenger information systems, brake compressor drive, propulsion control system, etc. It is, therefore, necessary to account for these factors when determining the actual load on an electrical substation. For this purpose, the value determined by Equation (7) has to be corrected:

$$I_{ent} = I \cdot \frac{1}{\eta_1} \cdot \frac{1}{\eta_2} \cdot \frac{1}{\eta_3} \cdot \frac{1}{\eta_4} \cdot \frac{1}{\eta_5} [A] \tag{8}$$

where: η_1 is the efficiency, considering other energy consumers (0.96); η_2 is the efficiency counting train's interior heating (0.93); η_3 is the efficiency of contact line (0.91);.η_4 is the efficiency, counting aberrant movement work (0.98); η_5 is the efficiency of the electricity substation (0.94) [3,54]. The values given above are averages taken globally for the entire electric traction at 3 kV DC. Railway administrations adopt—from experience—values similar to those given above. In practice, electric traction efficiency values are given in ranges of a few percent [3].

The increasing speed phase consists of switching on the traction motors and thus overcoming the forces resisting movement. Maintaining a constant speed consists of balancing, through the traction force, the sum of the resistance to motion of the vehicle and the local resistances determined by the course of the railway line. In this case, the acceleration force is zero. Cruising is moving with the engines not taking up any energy (driving force Z = 0)—the accelerating force F takes negative values equal to the opposite of the sum of the resistances to motion and the railway line resistance. During breaking, the accelerating force is the opposite of the sum of the resistance of motion, the resistance of the railway line, and the braking forces generated by applying the brakes. The existing resistance to motion contributes de facto to faster deceleration of the trainset. The value of the force accelerating the train depending on the phase of motion is described by the following Equation [4]:

$$F(v,x) = Z(v) - W(v) - I(x) \tag{9}$$

$$F(v,x) = 0,\ Z(v,x) = W(v) + I(x) \tag{10}$$

$$F(v,x) = -W(v) - I(x) \tag{11}$$

$$F(v,x) = B(v) + W(v) + I(x) \tag{12}$$

where Z is the tractive effort, B is the braking force, W is the movement resistance force, and I is the sum of the resistance forces, which are dependent on the tenor of the railway line.

Equation (9) is assumed for the mode in which the train increases its speed (so-called pull mode). Equation (10) is true for constant speed operation. Equation (11) shows running with the engine off. The value of the acceleration force takes the form of Equation (12) for braking. Running modes are shown in Figure 2.

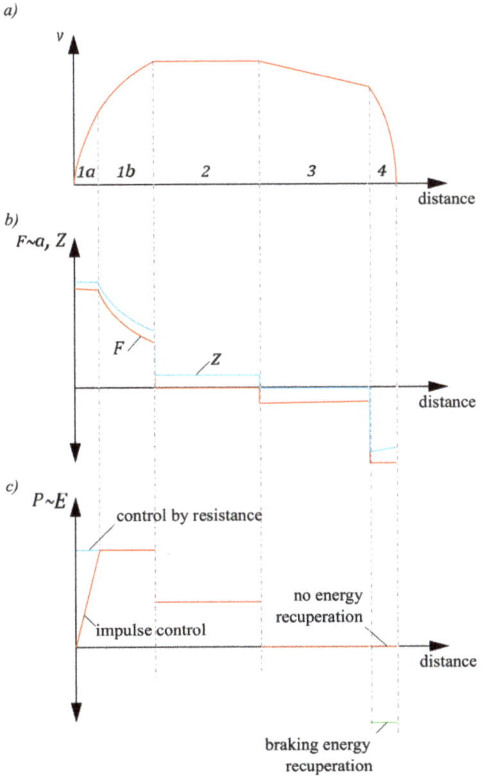

Figure 2. Four train running modes: 1a—starting (according to the curve of constant acceleration), 1b—acceleration (according to the course of the hyperbola of constant power), 2—running at a constant speed, 3—running from coasting, 4—braking. (**a**) Velocity versus distance graph; (**b**) acceleration and acceleration force (red line) and tractive effort (blue line) versus distance graph; (**c**) power and energy of running train with impulse (red on first movement phase) and resistance (blue) control or (in phase 4) braking energy with recuperation (green) and without recuperation (red).

The value of the driving force is determined by the traction characteristics of the vehicle. It is a curve of the dependence of the tractive force on the wheels of the traction vehicle as a function of the running velocity.

The magnitude of the resistance to train movement is determined using the so-called Strahl's formula. It is a quadratic equation with three coefficients: A, B, and C. It approximates the value of resistance to motion acting on a train as a function of its speed [3,4]:

$$W(v) = A + Bv + Cv^2 \tag{13}$$

This formula is also called the "Davis equation", or "Leitzmann formula", or "Barbier function", or "von Borries formula".

Railway research centres all over the world commonly use the quadratic equation to approximate the resistance to motion. They empirically determine the values of coefficients A, B, and C for different types of trains. The determination of coefficient values for the Strahl equation was also handled by Armstrong and Swift [56]. In Poland the values for this formula were determined by the Railway Institute.

The publications [57,58] contain sets of formulas for the specific resistance of train movement designated by centres from around the world, including German, French, Spanish, Japanese, and Chinese centres. The total resistance to motion is the sum of the resistance to motion of the traction vehicle and the resistance to motion of all types of cars in the train [59]:

$$W(v) = W_L(v) + \sum_{i=1}^{n} W_{w_i}(v) \tag{14}$$

The total resistance of the locomotive movement is determined by the equation [3,59]:

$$W_L(v) = j_l \cdot \left(\left(0.9 + \frac{0.015}{3.6} \right) \cdot Q_L + 147 \cdot n + \frac{23}{241.2} \cdot v^2 \right) \tag{15}$$

where j_l is the number of locomotives, Q_L is the locomotive weight [kN], and n is the number of axles on the locomotive [–].

The resistances of wagons of the i-th type are [58]:

$$W_{W_i}(v) = \left(0.65 + 0.15 \cdot \frac{v}{36} \right) Q_{W_i} + 150\chi + \beta(2.5 + j) \cdot \frac{v^2}{36} \tag{16}$$

where Q_{Wi} is the weight of wagons of i-th type [kN], χ is the number of axles of wagon group i-th type [–], β is the braking weight [kN], and j is the number of wagons of type i [–].

Equation (14) (and therefore, also Equations (15) and (16)) is applicable for trains consisting of locomotives and cars. However, if the service is operated by multiple units, Equation (17) is applicable [56]:

$$W = \left(K + 1.5 \cdot \frac{v}{10} \right) \cdot m_j + 150 \cdot n_o + 10 \cdot (2.7 + n) \cdot \left(\frac{v}{10} \right)^2 \tag{17}$$

where K is the bearing type factor, which for roller bearings is K = 6.5 [–] and for plain bearing is K = 9.0 [–]; m_j is the weight of motive power unit with passengers [44]; n_o is the number of axles in the unit [–]; n is the number of coaches in the unit [–].

There is also local resistance to train movement when the traction vehicle is moving. The local resistance is defined in [51] as additional resistance to motion occurring in specific places or sections of a route. These resistances include the resistance of the medium (air) in a tunnel, resistance caused by crossing hills, the resistance of running on curved tracks, and wind resistance (in mountainous and coastal areas).

The occurrence of bow resistance results directly from the fact that both wheels of a wheelset are seated on a common axis. An obvious consequence of this design solution is that both wheels have the same angular velocity. In a curve, when the outer wheels of a rail vehicle travel a longer distance than the inner wheels, this leads to partial slippage of one or both wheels. The resulting resistance of the curve is taken into account by adding the equivalent of the additional resistance of the train to the actual gradient of the profile.

To replicate train movement, a model was built in the MATLAB Simulink environment to determine the technical transit time of the train. In the model, the train movement parameters described above were taken into account.

Each train in the microsimulation model was mapped as a multiblock with a timetabled start time for the simulation. From each of the multiblocks, information is given about the position of the train on the railway line and the current measured at the contact between the catenary and the current collector. This allows the determination of the load of the substations supplying electricity to the train located in the area of the respective supply section. To correct for the performance of the overhead contact line, the train current consumption values recorded at the substation are added together. The result is information about the total load of the electrical substation by trains on the considered railway line.

5. Case Study

This section describes the case study under consideration. The railway infrastructure is presented together with its operational constraints. The specifics of the supply network are also defined. The vehicles running on the network are characterised and the scenarios studied are presented.

5.1. The Rolling Stock

The most important components of the system are the traction vehicles. They generate the load on the electrical substations through their movement and the resulting demand for electricity. The model was designed to investigate how much load is generated by different operating situations depending on the type of rolling stock used. Trains (power units) differ from one another. Modern trainsets are characterised by higher installed capacities, and therefore higher start-up acceleration. They, therefore, generate higher current consumption. On the other hand, the value of motive power for old locomotives and multiple units, still quite commonly used in Poland to run trains, is regulated using a reluctance control. As a result, during the start-up phase, the current consumption of such a vehicle is higher than when implementing control using an inverter.

This case study investigates the loading of feeder stations by trains operated by: EN57 electric multiple units—an old-type unit, most often used for regional services on electrified lines; old-type locomotives: EU07 carrying mainly fast trains and EP09 carrying express trains; a modern Newag Impuls 31WE electric multiple unit for regional services; ES64U4 locomotive carrying long-distance trains; an ED250 electric multiple unit carrying express trains.

The modern Newag Impuls traction unit has been adapted for regional travel thanks to its high starting acceleration value ($a = 1$ m/s^2) in the speed range from 0 to 40 km/h. The electric multiple unit weighs 172 t and is 74.4 m long. It is driven by traction engines with a total power of 2 MW. They ensure the possibility of moving at a speed of 160 km/h. These—and more detailed data on the said unit—are contained in [39] and in the vehicle's markings. The operating parameters of the unit are presented below in Figure 3a. These units commonly replace 31 WE electric multiple units.

The EN57 was produced between 1961 and 1993 and was the backbone of the regional electric traction rolling stock. The applied speed control (resistive) is characterised by maximum current consumption regardless of the acceleration phase. The excess energy is dissipated in the form of heat by resistors connected to the main circuit of the EMU. The vehicle has a mass of 126.5 t and a length of 64.97 m. The engines with a total power of 580 kW give the unit a maximum speed of 110 km/h [38]. The basic, simplified characteristics of this EMU are illustrated in Figure 3b.

Like the EN57, the EU07 locomotive controls speed using resistance. Engines with a power of 2 MW give it a maximum speed of 125 km/h. It is, therefore, widely used to run long-distance trains on routes where—due to technical parameters of the line or economic issues—trains are not run at top speeds. The locomotive weighs 80 t. Traction characteristics of the electric locomotive are shown in Figure 3c.

The EP09 locomotive was designed to run high-speed passenger trains. It has a higher installed power (2.92 MW) than the EU07 locomotive and reaches a maximum speed of 160 km/h. It was manufactured in 1986–1997, and its weight is 83.5 t. Initially used to service only express trains, it now also operates high-speed trains for which the parameters of the EU07 locomotive are insufficient. A graph of the traction characteristics of this locomotive is shown in Figure 3d.

The EU44 Siemens Eurosprinter locomotive is a unit equipped with 6 MW engines. It operates long-distance and express trains. The weight of the locomotive is 87 t [60]. It is the most powerful locomotive used to service passenger trains in Poland. The traction characteristics of the locomotive are illustrated in Figure 3e.

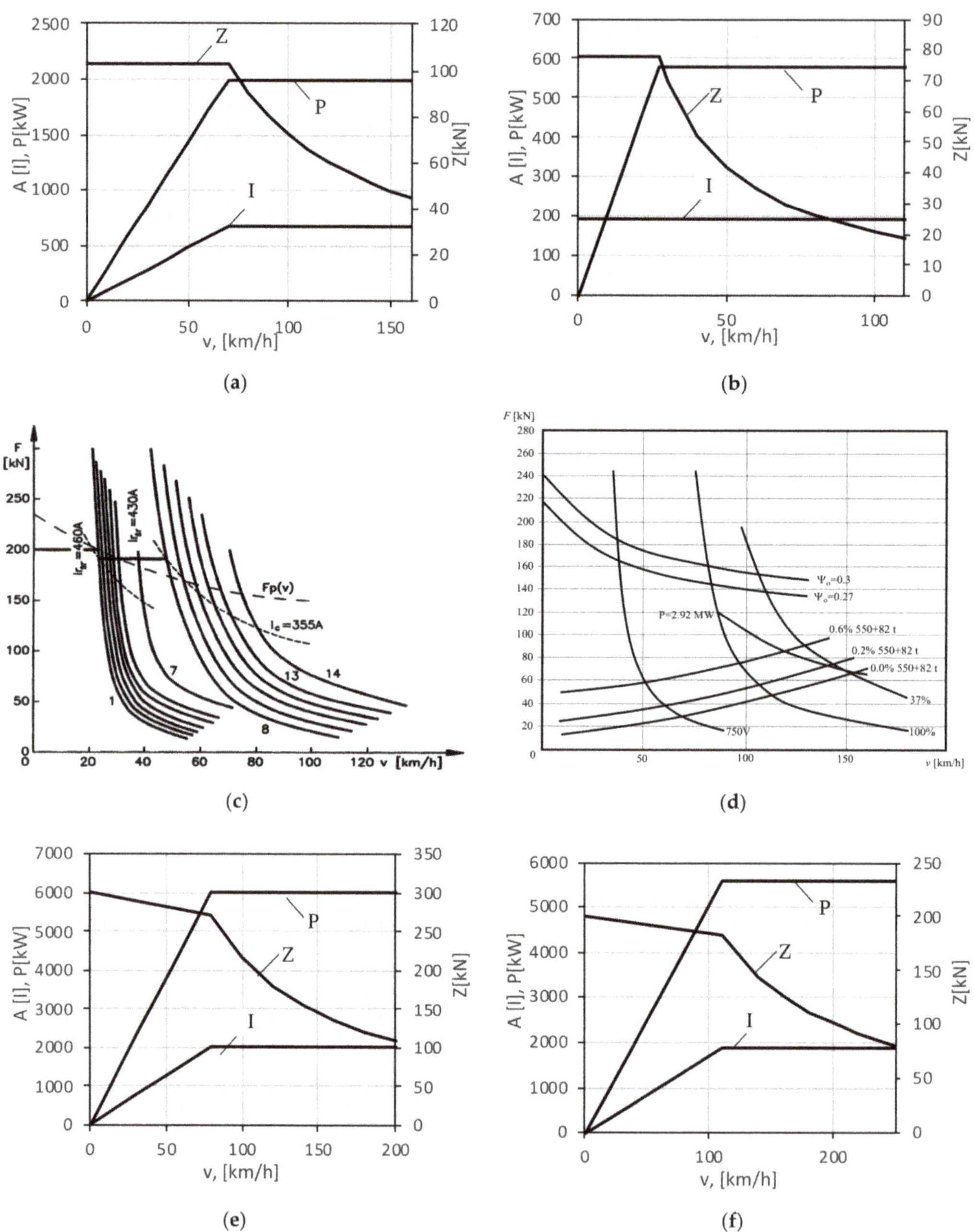

Figure 3. Traction characteristics of trains. Tractive effort (F(v) or Z(v)), power P(v), and electric current consumption I(A): (**a**) Electric multiple unit 31WE "Newag Impuls"; (**b**) electric multiple unit EN57 (simplified) (**c**) EU07 locomotive [61]; (**d**) EP09 locomotive [6]; (**e**) EU44 locomotive 'Siemens EuroSprinter' [59]; (**f**) Electric multiple unit ED250 'Pendolino' [6].

The fastest passenger trains of the highest-quality category are operated with ED250 electric multiple units produced by Alstom, Saint-Ouen-sur-Seine (France)—Pendolino

(EMU250). In Poland, they operate at speeds of up to 200 km/h. The trains have eight asynchronous engines, each with a power of 708 kW. The traction characteristics are shown in Figure 3f.

5.2. Railway Line

The rail line framework considered in the case study starts with a large passenger interchange station, serving a total of several hundred regional, long-distance, and express trains each day. Freight traffic operates on the initial section of the railway line in question very rarely and only in emergencies—in terms of the timetable it is a strict passenger railway line. The line departs in an easterly direction from the start station and continues in a south-easterly direction. The first boundary between supply sections is located approximately 2.8 km downstream of the junction station. In this section the railway line is multi-tracked; much of it runs alongside the extended eastern head of the station. Moreover, it runs alongside several groups of tracks. On these tracks, trains wait to be substituted before starting their service routes [45].

The railway line forks at approximately a km 3.4 from the junction station. A double-track line with regional and long-distance passenger traffic branches off to the south. It consists of 21 pairs of regional connections and one pair of long-distance connections (of which 18 pairs of trains are operated by electric multiple units, and the rest by diesel multiple units). The railway line considered in this case study leads towards a passenger stop less than 2 km away [45].

After leaving the passenger station, where only regional trains stop, the railway line leads towards a branch station. At the branch station, the railway line is connected to the railway switch, which is generally freight-oriented. In addition, a mixed passenger/freight line departs from the railway line under consideration in this case study. The freight railway line mentioned earlier is the freight bypass of a large nodal station from which the railway line under consideration in this case study starts. At least several dozen pairs of trains pass through the freight ring every day. The passenger-freight line diverging from this line operates 21 pairs of regional trains per day, and freight (several dozen pairs daily) departs from the line in the case study. The railway line continues for c.a. 2.1 km to the next station. Only the regional ones stop at this station. Directly behind the eastern head of the station is another boundary between supply sections [45].

The next feeder section is an unbranched line. It has two passenger stops. On this feeder section of the line, 27 pairs of passenger trains, 25 pairs of long-distance trains, and 2 pairs of fast trains run daily in each direction. Several pairs of freight trains also run on the line. The low load of freight traffic on this railway line results from the fact that it is a railway line dedicated to passenger transport. There is a priority line for freight traffic running parallel to this line at a distance of approximately 10 km.

Another section of power supply (about 12 km long) covers only one railway station, where regional trains and most of the fast (long-distance) trains stop.

The last supply section under consideration has one passenger stop and one station where some long-distance trains stop. This section has a total length of approximately 13 km. The occupancy of passenger electric trains of this section is the same as the occupancy of the previous supply section.

The railway line under consideration in this case study is part of a European railway trunk line. Government documents that plan connection offers indicate a target of increasing the load on the line with additional long-distance connections, including international connections, in the next few years. The railway line is double tracked, fully electrified, and suitable for a maximum speed of 160 km/h [45]. A diagram of the railway line under consideration is shown in Figure 4.

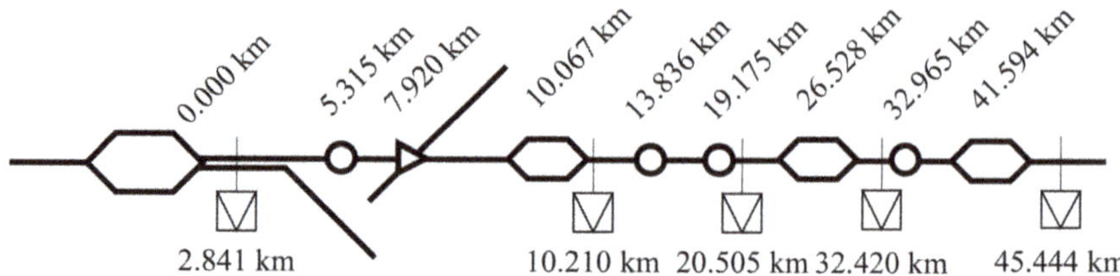

Figure 4. The schematic layout of the railway line we analysed.

The characteristics of the different sections of the railway line supply are shown in Table 1.

Table 1. Supply sections of the line under consideration.

Supply Section	Start of Section [km]	End of Section [km]	Section Length [km]	Remarks
A	2.841	10.210	7.369 + 3420 + 0.040 + 4.111	Mainline and its branches
B	10.210	20.505	10.295	-
C	20.505	32.420	11.915	-
D	32.420	45.444	13.024	-

The maximum current consumption of trains moving along the railway line is also conditioned by the speed of movement of the trains under consideration. The train accelerates until it reaches its target speed, which is conditioned either by the technical possibilities of the rolling stock or by the speed restriction of a section of the railway line.

According to Equation (11), a train that is moving at a constant speed will normally use only part of its available motive power. Such a train has to overcome only specific resistance (mainly rolling resistance and aerodynamic resistance) [3] and resistance of place, which results from overcoming hills and horizontal curves [4]. For this reason, the sections where many trains are accelerating but have already reached speeds that allow the use of constant power hyperbolas are valuable in terms of considering the amount of current consumption (for modern trainsets whose speed is not controlled by resistance).

In the case of the railway line under consideration, traversing the head of the junction station (located at the beginning of the line) is carried out at maximum speeds of 80 to 100 km/h [45]. In practice, however, due to the numerous switches, when traversing the eastern head of the junction station, trains usually travel at a maximum speed of 40 km/h.

In the case of trains moving through switches according to the main direction, permissible traffic speeds on the considered railway line are summarized in the table below (Table 2).

As can be seen from the table above, the permitted maximum traffic speeds on the route in question are relatively high. Passenger trains can reach speeds of up to 160 km/h; goods trains—up to 120 km/h. This means that the railway line under consideration is of M160 standard (according to the nomenclature PKP PLK S.A.).

The line is over flat terrain, with maximum gradients ranging from −3‰ to +3‰. Such gradients are almost negligible from the perspective of traction calculations.

5.3. Tested Scenarios

The railway line considered in the case study was tested for several different options for assigning units to timetabled connections. The simulations were intended to verify how the load on the network changes depending on the types of trainsets serving particular connections. It was assumed that the offer to passengers will not change. The variants were intended to reflect the different stages at which passenger carriers replace the old-type rolling stock with new-type rolling stock.

Table 2. Maximum permissible speeds for train traffic on the line under consideration [45].

Beginning of the Section [km]	End of the Section [km]	Speed Limit Track 1 [km/h]			Speed Limit Track 2 [km/h]		
		For Multiple Units	For Passenger Trainsets Consisting of Cars	For Freight Trains	For Multiple Units	For Passenger Trainsets Consisting of Cars	For Freight Trains
0.000	2496	100	100	80	100	100	80
2496	2846	100	100	80	90	90	80
2846	6246	110	110	80	110	110	80
6246	8746	120	120	80	120	120	80
8746	26,946	160	160	120	160	160	120
26,946	27,196	130	130	120	130	130	120
27,196	41,541	160	160	120	160	160	120
41,541	42,576	120	120	80	120	120	80
42,576	45,444	160	160	120	160	160	120

5.3.1. Tested Scenarios

The baseline option analysed the operation of old-type vehicles only. Regional services were simulated using EN57 multiple units. In this variant, the long-distance trains were set up with the EU07 locomotive and wagons (from 4 to 16). Express trains were set up with an EP09 locomotive and six cars.

Subsequent variants assumed an increasing share of new trains in passenger service until the complete replacement of vehicles several decades old by modern rolling stock.

The first option assumed replacing every third regional train with a modern train and express trains with ED250 units. The second option assumed also replacing 1/3 of regional trains with modern electric multiple units and replacing some long-distance trains with modern trains. The third option assumed replacing all regional trains with new multiple units without introducing additional changes into the long-distance train service concerning the second option. The last option assumed the replacement of all trainsets with modern trainsets.

5.3.2. Tested Variants

Simulations were conducted for traffic situations occurring on the considered railway line that can potentially generate significant loads for the power infrastructure, i.e., situations where increased train traffic takes place. For this reason, traffic situations were selected in which the highest number of trains in motion and travelling at non-start-up speeds accompanied by start-ups were noted. The non-starting speeds were considered to be those at which the motive power unit uses a traction characteristic section with the shape of a constant power hyperbola.

Five traffic situations have been identified on the section of railway line under consideration, which may be characterised by significant loads on the power network:

- Situation I
    - In section A: 1 express train and 3 local trains;
    - In section B: 1 accelerating long-distance train and 1 local train;
    - In section C: 2 accelerating long-distance trains and 1 accelerating local train;
    - In section D: 1 express train and 1 local train.
- Situation II
    - In section A: 1 long-distance train and 2 local trains;

    - In section B: 1 accelerating express train, 1 accelerating long-distance train, and 1 long-distance train;
    - In section C: 1 accelerating long-distance train and 2 local trains;
    - In section D: 1 express train and 1 accelerating long-distance train.
- Situation III
    - In section A: 2 long-distance trains and 2 local trains;
    - In section B: 1 long-distance train and 2 local trains;
    - In section C: 1 express train and 2 accelerating long-distance trains;
    - In section D: 1 accelerating long-distance train and 2 local trains.
- Situation IV
    - In section A: 1 express train, 1 long-distance train and 3 local trains;
    - In section B: 1 accelerating long-distance train and 3 local trains;
    - In section C: 1 accelerating long-distance train and 2 local train;
    - In section D: 1 express train and 2 accelerating local trains.
- Situation V
    - In section A: 1 long-distance train and 1 accelerating local train, 2 local trains;
    - In section B: 1 long-distance train, 1 accelerating local train, and 1 local train;
    - In section C: 2 accelerating long-distance trains and 1 accelerating local train and 1 local train;
    - In section D: 1 accelerating express train (long train set) and 1 accelerating local train.

5.4. Results of the Simulation

The results of the loading of the individual supply sections by trains were obtained for four different variants and four operating situations and for four supply sections to which electricity is supplied from five traction substations. The simulation results are summarised in the following tables (Table 3 for one-section-load and Table 4 for multi-section-load).

An electric traction rail vehicle draws current simultaneously from two traction substations, thereby reducing the load on the substation, except that the rated primary busbar current in the traction substation shall not exceed [43]:

- 2 kA for P80 lines;
- 4 kA for new and upgraded lines T40, M80, T80, P120, M120, M120, T120, P160, and M160; and for upgraded lines P200, M200, and P250;
- 6 kA for the newly constructed P200, M200, and P250 lines.

These markings—used by PKP Polskie Linie Kolejowe S.A.—the Polish national railway infrastructure manager—indicate the purpose of the railway line and the maximum speed of traffic. The letter P indicates the passenger character of the line, T—the freight character, and M—mixed. The number indicates the permissible speed expressed in kilometres per hour.

Considering the above guidelines, the current load on substations on the considered railway line must not exceed 4 kA. At the same time—due to the two-sided supply of sections—the current load generated by trains in one supply section shall not exceed 8 kA. It is then necessary that adjacent (and subsequent) feeder sections are not loaded to a greater extent than by trains drawing 4 kA in total (measured on the feeder rails of electrical substations).

Table 3. Results of a microsimulation—one-section-load.

Situation	Variant	Section Current Load [A]			
		A	B	C	D
I	Baseline	1520.5	733.5	1301.4	1060.9
	V1	2550.7	733.5	1890.6	2851.3
	V2	2550.7	1086.3	3747.9	3440.5
	V3	3145.3	3043.7	3747.9	3440.5
	V4	3145.3	3043.7	5043.9	3440.5
II	Baseline	1133.0	1716.9	761.6	1436.3
	V1	1722.3	3447.2	975.4	3166.6
	V2	1722.3	5037.2	982.1	3166.6
	V3	2030.9	5037.2	2652.2	3166.6
	V4	2244.7	5050.6	2652.2	4436.0
III	Baseline	1440.4	825.7	1850.5	1133.0
	V1	2009.5	825.7	2224.7	1555.3
	V2	3506.0	1401.6	3146.6	1656.8
	V3	4095.3	1389.6	3146.6	1656.8
	V4	4429.3	1394.9	5344.5	2084.4
IV	Baseline	2214.0	1058.2	992.7	1380.2
	V1	1921.4	1614.1	992.7	3546.1
	V2	2517.3	1614.1	3011.6	3546.1
	V3	2540.6	2196.6	3580.8	3928.2
	V4	3572.8	2343.6	3580.8	3928,2
V	Baseline	1251.9	758.9	1560.6	1121.0
	V1	1841.2	758.9	2264.7	6032.7
	V2	2665.6	2570.7	2264.7	6032.7
	V3	2665.6	2688.3	4616.3	6032.7
	V4	4863.5	2688.4	4616.3	6032.7

The simulation carried out indicated that the current load on the sections of the railway line under consideration did not exceed the limit value. However, a significant (even fourfold) increase in the value of the currents necessary to supply the trains is noticeable. It is also noted that the power supply network in some variants of the experiment (e.g., traffic situation II, variant 4) was loaded to more than 90%, assuming that in the next supply section, which was not covered by traction calculations, the current consumption from the traction substation common to it and section D was 2 kA (and in total did not exceed 6 kA). In such a situation, the occurrence of traffic disturbances or additional trains may result in exceeding the current limit values. Traction calculations showed, however, that exceeding the permissible current consumption in the D supply section from the substation occurred in the variant V (when the express connection was performed by two electric multiple units). Such a situation is unacceptable in train operation design, as it results in a planned overload of the traction substation and consequently in disconnection of the voltage. National experience confirms that in the case of a 3 kV DC power supply, such a train must be the only one running on a given power supply section.

Table 4. Results of a microsimulation—multi-section-load.

Situation	Variant	Section Current Load [A]				
		A + B	A + B + C	B + C	B + C + D	C + D
I	Baseline	2254.0	3555.4	2034.9	3095.8	2362.3
	V1	3284.2	5174.8	2624.1	5475.4	4741.9
	V2	3637.0	7384.9	4834.2	8274.7	7188.4
	V3	6189.0	9936.9	6791.6	10,232.1	7188.4
	V4	6189.0	11,232.9	8087.6	11,528.1	8484.4
II	Baseline	2849.9	3611.5	2478.5	3914.8	2197.9
	V1	5169.5	6144.9	4422.6	7589.2	4142.0
	V2	6759.5	7741.6	6019.3	9185.9	4148.7
	V3	7068.1	9720.3	7689.4	10,856.0	5818.8
	V4	7295.3	9947.5	7702.8	12,138.8	7088.2
III	Baseline	2266.1	4116.6	2676.2	3809.2	2983.5
	V1	2835.2	5059.9	3050.4	4605.7	3780.0
	V2	4907.6	8054.2	4548.2	6205.0	4803.4
	V3	5484.9	8631.5	4536.2	6193.0	4803.4
	V4	5824.2	11,168.7	6739.4	8823.8	7428.9
IV	Baseline	3272.2	4264.9	2050.9	3431.1	2372.9
	V1	3535.5	4528.2	2606.8	6152.9	4538.8
	V2	4131.4	7143.0	4625.7	8171.8	6557.7
	V3	4737.2	8318.0	5777.4	9705.6	7509.0
	V4	5916.4	9497.2	5924.4	9852.6	7509.0
V	Baseline	2010.8	3571.4	2319.5	3440.5	2681.6
	V1	2600.1	4864.8	3023.6	9056.3	8297.4
	V2	5236.3	7501.0	4835.4	10,868.1	8297.4
	V3	5353.9	9970.2	7304.6	13,337.3	10,649.0
	V4	7551.9	12,168.2	7304.7	13,337.4	10,649.0

These results indicate that in situations of disturbing traffic or when an additional (unscheduled) transport operation is carried out, overloads can occur, leading to tripping of the fast circuit breakers in the substation and a voltage drop in the catenary.

Due to the above simulation results, it should be stated that in the case of the process of replacing the rolling stock with new, higher-powered rolling stock, it is necessary to take into account the power consumption of trains on the network at the stage of timetable construction. Modern trains—thanks to the possibility of accelerating faster through a higher power—cover a fragment of a railway line between two stops faster. This in turn means that they occupy a given railway line for less time—i.e., the capacity of the line is increased. However, trains with the highest installed power have to move on their own due to very high current values recorded at individual power sections. This in turn reduces the capacity of the railway line.

6. Conclusions

This article discussed the problem of powering trainsets with the use of a 3 kV DC power supply network, which is among those used in Poland. The influence of the modernisation of the rolling stock of the railway on the loads on the power infrastructure was described. To check the influence of replacing trains with speed-controlled using impulse control, a microsimulation model was made to solve many equations of trains' movements. It takes into account power parameters and current intensities generated by trains in motion. A case study was performed which reflected the traffic on a section of one of the most important railway lines in the southwest of Poland. The results show that as carriers upgrade their rolling stock, the nature of the loads on the network feeding the railway line changes significantly. The start-up phase of modern vehicles loads traction substations less than vehicles that are several decades old. However, as the speed of traffic continues to increase, trains with impulse control will place a greater load on the power supply infrastructure. Once the preset scheduled speed is reached, the loads on the rail power supply network are at similar levels.

As a result of the higher-power engines installed on today's trains, the durations of the largest loads on the network are different. In the case of lower-powered vehicles, it takes longer to reach the target (scheduled) velocity, so the train switches to the constant-speed phase later. On the other hand, a high-powered train reaches its scheduled speed more quickly and thus switches to constant speed mode more quickly. However, the greater acceleration of starting, resulting from the higher power engines installed in the rail vehicles, determines a greater demand for electricity, which in turn results in larger currents being recorded in the relevant electrical substation.

It should be recognised that as operators continue to replace rolling stock, the rail supply network will be increasingly stressed and could become a critical element.

Funding: This research received no external funding.

Institutional Review Board Statement: Not applicable.

Informed Consent Statement: Not applicable.

Data Availability Statement: Not applicable.

Conflicts of Interest: The author declare no conflict of interest.

References

1. Brenna, M.; Foiadelli, F.; Zaninelli, D. *Electrical Railway Transportation Systems*; John Wiley & Sons: Hoboken, NJ, USA, 2018.
2. Frey, S. *Railway Electrification Systems & Engineering*; White Word Publications: Delhi, India, 2012.
3. Madej, J. *Teoria Ruchu Pojazdów Szynowych*; Oficyna Wydawnicza Politechniki Warszawskiej: Warsaw, Poland, 2012.
4. Kwaśnikowski, J. *Elementy Teorii Ruchu i Racjonalizacja Prowadzenia Pociągu*; Wydawnictwo Naukowe Instytutu Technologii Eksploatacji—PIB: Radom, Poland, 2013.
5. Szeląg, A. Electrical power infrastructure for modern rolling stock concerning the railway in Poland. *Arch. Transp.* **2017**, *42*. [CrossRef]
6. Haładyn, S.; Restel, F.J.; Wolniewicz, Ł. Method for Railway Timetable Evaluation in Terms of Random Infrastructure Load. In *International Conference on Dependability and Complex Systems*; Springer: Brunów, Poland, 2019; pp. 235–244.
7. Kuznetsov, V.G.; Sablin, O.I.; Chornaya, A.V. Improvement of the regenerating energy accounting system on the direct current railways. *Arch. Transp.* **2015**, *36*, 35–42. [CrossRef]
8. Jefimowski, W. Simulation research of the influence of the train traffic situations on the rail potential in the power supply system 3 kV DC. *Pr. Nauk. Politech. Warsz.* **2016**, *111*, 203–213.
9. Restel, F. The railway operation process evaluation method in terms of resilience analysis. *Arch. Transp.* **2021**, *57*, 73–89.
10. Friedrich, J.; Restel, F. A fuzzy approach for evaluation of reconfiguration actions after unwanted events in the railway system. In *Method for Railway Timetable Evaluation in Terms of Random Infrastructure Load*; Springer: Cham, Switzerland, 2020; pp. 195–204.
11. Kaleybar, H.J.; Brenna, M.; Foiadelli, F.; Fazel, S.S.; Zaninelli, D. Power Quality Phenomena in Electric Railway Power Supply Systems: An Exhaustive Framework and Classification. *Energies* **2020**, *13*, 6662. [CrossRef]
12. Magro, M.C.; Mariscotti, A.; Pinceti, P. Definition of Power Quality Indices for DC Low Voltage Distribution Networks. In Proceedings of the IEEE Instrumentation and Measurement Technology Conference Proceedings, Sorrento, Italy, 24–27 April 2006; pp. 1885–1888.
13. Popescu, M.; Bitoleanu, A. A Review of the Energy Efficiency Improvement in DC Railway Systems. *Energies* **2019**, *12*, 1092. [CrossRef]
14. Kaleybar, H.J.; Kojabadi, H.M.; Brenna, M.; Foiadelli, F.; Fazel, S.S.; Rasi, A. An Inclusive Study and Classification of Harmonic Phenomena in Electric Railway Systems. In Proceedings of the 2019 IEEE International Conference on Environment and Electrical Engineering and 2019 IEEE Industrial and Commercial Power Systems Europe (EEEIC/I&CPS Europe), Genova, Italy, 11–14 June 2019; pp. 1–6.

15. Ying-Tung, H.; Lin, K.-C. Measurement and characterization of harmonics on the Taipei MRT DC system. *IEEE Trans. Ind. Appl.* **2004**, *40*, 1700–1704.
16. Skarpetowski, G.; Zajac, W.; Czuchra, W. Analytical Calculation of Supply Current Harmonics Generated by Train Unit. In Proceedings of the 12th International Power Electronics and Motion Control Conference, Portoroz, Slovenia, 30 August–1 September 2006; pp. 1378–1384.
17. Ogunsola, A.; Mariscotti, A.; Sandrolini, L. Measurement of AC Side Harmonics of a DC Metro Railway. In Proceedings of the 2012 Electrical Systems for Aircraft, Railway and Ship Propulsion, Bologna, Italy, 16–18 October 2012; pp. 1–5.
18. Terciyanli, A.; Acik, A.; Cetin, A.; Ermis, M.; Cadirci, I.; Ermis, C.; Demirci, T.; Bilgin, H.F. Power Quality Solutions for Light Rail Public Transportation Systems Fed by Medium-Voltage Underground Cables. *IEEE Trans. Ind. Appl.* **2012**, *48*, 1017–1029. [CrossRef]
19. Crotti, G.; Delle Femine, A.; Gallo, D.; Giordano, D.; Landi, C.; Luiso, M.; Mariscotti, A.; Roccato, P.E. Pantograph-to-OHL Arc: Conducted Effects in DC Railway Supply System. *IEEE Trans. Instrum. Meas.* **2019**, *680*, 3861–3870. [CrossRef]
20. Hill, R.J. Electric railway traction. Part 6: Electromagnetic compatibility disturbance-sources and equipment susceptibility. *Power Eng. J.* **1997**, *11*, 31–39. [CrossRef]
21. Morant, A.; Wisten, Å.; Galar, D.; Kumar, U.; Niska, S. Railway EMI Impact on Train Operation and Environment. In Proceedings of the International Symposium on Electromagnetic Compatibility-EMC EUROPE, Rome, Italy, 17–21 September 2012; pp. 1–7.
22. Ogunsola, A.; Mariscotti, A.; Sandrolini, L. Estimation of Stray Current from a DC-Electrified Railway and Impressed Potential on a Buried Pipe. *IEEE Trans. Power Deliv.* **2012**, *27*, 2238–2246. [CrossRef]
23. Cai, Y.; Irving, M.; Case, S. Modelling and numerical solution of multibranched DC rail traction power systems. *IEE Proc. Electr. Power Appl.* **1995**, *142*, 323–328. [CrossRef]
24. Tian, Z.; Zhao, N.; Hillmansen, S.; Su, S.; Wen, C. Traction Power Substation Load Analysis with Various Train Operating Styles and Substation Fault Modes. *Energies* **2020**, *13*, 2788. [CrossRef]
25. Chymera, M.Z.; Renfrew, A.C.; Barnes, M.; Holden, J. Modeling Electrified Transit Systems. *IEEE Trans. Veh. Technol.* **2010**, *59*, 2748–2756. [CrossRef]
26. Lu, Q.; He, B.; Wu, M.; Zhang, Z.; Luo, J.; Zhang, Y.; He, R.; Wang, K. Establishment and Analysis of Energy Consumption Model of Heavy-Haul Train on Large Long Slope. *Energies* **2018**, *11*, 965. [CrossRef]
27. Xue, Y.; Ma, D.; Wang, L. Calculation Method of Energy Consumption in Train Traction. *China Railw. Sci.* **2007**, *5*, 84–86.
28. Pröhl, L.; Aschemann, H.; Palacin, R. The Influence of Operating Strategies regarding an Energy Optimized Driving Style for Electrically Driven Railway Vehicles. *Energies* **2021**, *14*, 583. [CrossRef]
29. Cunillera, A.; Fernández-Rodríguez, A.; Cucala, A.P.; Fernández-Cardador, A.; Falvo, M.C. Assessment of the Worthwhileness of Efficient Driving in Railway Systems with High-Receptivity Power Supplies. *Energies* **2020**, *13*, 1836. [CrossRef]
30. Tan, Z.; Lu, S.; Bao, K.; Zhang, S.; Wu, C.; Yang, J.; Xue, F. Adaptive Partial Train Speed Trajectory Optimization. *Energies* **2018**, *11*, 3302. [CrossRef]
31. D'Acierno, L.; Botte, M. A Passenger-Oriented Optimization Model for Implementing Energy-Saving Strategies in Railway Contexts. *Energies* **2018**, *11*, 2946. [CrossRef]
32. Rocha, A.; Araújo, A.; Carvalho, A.; Sepulveda, J. A New Approach for Real-Time Train Energy Efficiency Optimization. *Energies* **2018**, *11*, 2660. [CrossRef]
33. Zhang, H.; Jia, L.; Wang, L.; Xu, X. Energy consumption optimization of train operation for railway systems: Algorithm development and real-world case study. *J. Clean. Prod.* **2019**, *214*, 1024–1037. [CrossRef]
34. Wang, P.; Goverde, R. Multi-train trajectory optimization for energy efficiency and delay recovery on single-track railway lines. *Transp. Res. Part B Methodol.* **2017**, *105*, 340–361. [CrossRef]
35. Cipolletta, G.; Delle Femine, A.; Gallo, D.; Luiso, M.; Landi, C. Design of a Stationary Energy Recovery System in Rail Transport. *Energies* **2021**, *14*, 2560. [CrossRef]
36. Lamedica, R.; Ruvio, A.; Palagi, L.; Mortelliti, N. Optimal Siting and Sizing of Wayside Energy Storage Systems in a D.C. Railway Line. *Energies* **2020**, *13*, 6271. [CrossRef]
37. Arboleya, P.; El-Sayed, I.; Mohamed, B.; Mayet, C. Modeling, Simulation and Analysis of On-Board Hybrid Energy Storage Systems for Railway Applications. *Energies* **2019**, *12*, 2199. [CrossRef]
38. PKP Szybka Kolej Miejska w Trójmieście sp. z o.o., Technical data. Maintenance System Documentation—Electrical Multiple Unit 5B+6B+5B, 5B+6B+6B+5B (EN57, EN71); Gdynia, Poland, 2010.
39. Newag, S.A. Electric Multiple Units 'Impuls'. Promotional Materials; 2014.
40. Podoski, J. *Transport w Miastach*; Wydawnictwa Komunikacji i Łączności: Warsaw, Poland, 1977.
41. Kacprzak, J. *Teoria Trakcji Elektrycznej*; Wydawnictwo Politechniki Warszawskiej: Warsaw, Poland, 1991.
42. Mierzejewski, L. *System Zasilania Trakcji Elektrycznej Prądu Stałego*; Wydawnictwo Politechniki Warszawskiej: Warsaw, Poland, 1989.
43. PKP Polskie Linie Kolejowe SA. *Technical Specification for Modernising or Construction of Railway Lines for Speeds Up to vmax ≤ 200 km/h (for Conventional Rolling Stock)/250 km/h (for Rolling Stock with Tilting Body) Volume IV—Electrical Traction/ Traction Power Equipment*; Polskie Linie Kolejowe SA: Warsaw, Poland, 2018.
44. Żurek, Z.H.; Duka, P. Obciążalność Prądowa Sieci Trakcyjnej Systemu 3kV w Świetle Zwiększania Mocy i Prędkości. *Pr. Nauk. Politech. Warsz. Transp.* **2017**, *119*, 529–539.
45. PKP Polskie Linie Kolejowe SA. *Regulamin Sieci 2020/2021*; Polskie Linie Kolejowe SA: Warsaw, Poland, 2019.

46. Molski, W. *Elektryfikacja Kolei*; Wydawnictwa Komunikacji i Łączności: Warsaw, Poland, 1994.
47. Figurzyński, Z. *Sieci Trakcyjne*; Wydawnictwa Komunikacji: Warsaw, Poland, 1954.
48. Hanasz, M. *Sieci i Podstacje Trakcji Elektrycznej*; Wydawnictwa Komunikacji i Łączności: Warsaw, Poland, 1965.
49. Korger, M.; Ruch, M. *Zulässige Oberstromgrenzwerte Elektrifizierten Streckennetz der DB*; DB Energie GmbH: Frankfurt am Main, Germany, 2014.
50. Polski Komitet Normalizacyjny. *PN-EN 50388:2008 Zastosowania Kolejowe—System Zasilania i Tabor—Warunki Techniczne Koordynacji Pomiędzy Systemem Zasilania (Podstacja) i Taborem w Celu Osiągnięcia Interoperacyjności*; Polski Komitet Normalizacyjny: Warsaw, Poland, 2008.
51. Podoski, J. *Zasady Trakcji Elektrycznej*; Wydawnictwa Komunikacji i Łączności: Warsaw, Poland, 1980.
52. Smolarz, W. *Przystosowanie Kolei do Zwiększonych Szybkości i Dużych Przewozów*; Wydawnictwa Komunikacji i Łączności: Warsaw, Poland, 1969.
53. Chwieduk, A. *Dyr. T. Projektowanie Ruchu Pociągów*; Zakład Poligraficzny Politechniki Radomskiej: Radom, Poland, 1997.
54. Wyrzykowski, W. *Ruch Kolejowy*; Wydawnictwa Komunikacji: Warsaw, Poland, 1951.
55. PKP Polskie Linie Kolejowe SA. *Instrukcja o Rozkładzie Jazdy Pociągów (Ir-11)*; PKP Polskie Linie Kolejowe SA: Warsaw, Poland, 2015.
56. Armstrong, D.S.; Swift, P.H. *Lower Energy Technology. Part A, Identification of Energy Use in Multiple Units, Report MR VS 077*; British Rail Research: Derby, UK, 1990.
57. Biliński, J. Resistance of traction vehicles movement—Empirical equations (1). *TTS Tech. Transp. Szyn.* **2019**, *26*, 34–39.
58. Biliński, J. Resistance of traction vehicles movement—Empirical equations (2). *TTS Tech. Transp. Szyn.* **2019**, *26*, 33–39.
59. Gruszczyński, J. *Eksploatacja Pojazdów Trakcyjnych*; Wydawnictwa Komunikacji i Łączności: Warsaw, Poland, 1975.
60. Siedlecki, J. Locomotive ES64U4. *TTS Tech. Transp. Szyn.* **2009**, *15*, 45–51.
61. COBiRTK. *Selection of Traction Units for Running Trains at Higher Speeds*; Auxiliary Materials: Warsaw, Poland, 1973.

Article

Electrodynamics of Reactive Power in the Space of Inter-Substation Zones of AC Electrified Railway Line

Mykola Kostin [1], Anatolii Nikitenko [2,*], Tetiana Mishchenko [3] and Lyudmila Shumikhina [4]

[1] Department of Electrical Engineering and Electromechanics, Dnipro National University of Railway Transport Named after Academician V. Lazaryan, Lazaryana St. 2, Room 238, 49-010 Dnipro, Ukraine; nkostin@ukr.net
[2] Electric Traction Division, Electrical Power Engineering Institute, Warsaw University of Technology, Koszykowa St. 75, 00-662 Warsaw, Poland
[3] Department of Intelligent Power Supply Systems, Dnipro National University of Railway Transport Named after Academician V. Lazaryan, Lazaryana St. 2, Room 334, 49-010 Dnipro, Ukraine; mishchenko_tn@ukr.net
[4] Department of Philological Disciplines and Foreign Languages, Dnipro College of Railway Transport and Transport Infrastructure, Pushkin Av. 77a, Room 504, 49-006 Dnipro, Ukraine; mila.shum80@gmail.com
* Correspondence: anatolii.nikitenko@pw.edu.pl; Tel.: +48-2223-476-16

Abstract: In railway traction, the definition of "electromagnetic field" is functionally connected to the concept of the reactive power consumed by the electric rolling stock, and characterized by the running and standing electromagnetic waves in the space of the inter-substation zones from the site of the AC traction system. Such a definition is established and theoretically justified by the theory of electromagnetic fields. This article uses the methodology of this theory, in particular, a method for power balance estimation in electromagnetic fields based on Maxwell's equations, as well as methods for the analysis of running and standing electromagnetic waves based on the theory of reflection, propagation and transmission of plane harmonic waves. The research considers the regularities of standing electromagnetic waves in the space of inter-substation zones of electric traction systems, which occur due to the incomplete reflection of incident waves from the contact wire and metal parts of the roof surface and the frontal part of the body of the electric rolling stock. The flow of electricity to the roof surface and the frontal part of the body of an electric locomotive is considered. The possibility of using existing methods to reduce wave reflections and thereby to effectively compensate for reactive power in the space of inter-substation zones is discussed.

Keywords: reactive power; railway; electric rolling stock; inter-substation zone; Poynting vector; Maxwell's equations; electromagnetic field; incident waves; standing wave; film coating

Citation: Kostin, M.; Nikitenko, A.; Mishchenko, T.; Shumikhina, L. Electrodynamics of Reactive Power in the Space of Inter-Substation Zones of AC Electrified Railway Line. *Energies* **2021**, *14*, 3510. https://doi.org/10.3390/en14123510

Academic Editors: Andrea Mariscotti and Leonardo Sandrolini

Received: 20 April 2021
Accepted: 8 June 2021
Published: 13 June 2021

1. Introduction

The fundamental role of reactive power is widely recognized in the electric power industry, and the ambiguity of the concept and formulae for determining reactive power is well known. In five years from now, we will celebrate the 100th anniversary of the "reactive power" concept [1,2], which has brought about a long-lasting discussion that has continued to the present day; let us review some of them.

The works of Constantin J. Budeanu [2] and Stanisław S. Fryze [3,4] were the initial studies at the first stage of the development of the reactive power definition, which lasted until the 1980s. Comprehensive review, classification and analysis of the methods developed during this period, and the names of their authors, are given in monographs by Leszek S. Czarnecki [5] and Volodymyr E. Tonkal [6].

The beginning of the second stage in the development of the power theory was laid in [7–9], proved by Hirofumi Akagi et al. in 1982. The main direction in this work is to obtain mathematical relationships and develop algorithms for the control systems of active filters when solving problems of compensating the reactive component of the first harmonic of the current and suppressing its higher harmonics. As a result, the author proposed the so-called "generalized theory of the instantaneous reactive power in three-phase circuits", which is frequently known as the "*p-q* theory". Unfortunately, in this theory, the authors do not explain reactive power's physical meaning, considering it as a calculated value when analyzing three-phase three-wire systems without a neutral wire. The improved *p-q* theory was proposed in [10], in which the limitations of its original version were overcome.

As opposed to the *p-q* theory, based on a plane, the power theory, based on the use of a spatial Cartesian coordinate system, was proposed by Fang Z. Peng et al. in [11] for the three-phase four-wire systems. This theory is applicable for all the following conditions of operation: sinusoidal or non-sinusoidal, as well as balanced or not.

Further development of the previous theory is given in the studies [12,13]. Considering the essence and degree of generalization of the obtained version, the authors proposed calling it the "*p-q-r* theory". The supply voltage and consumer currents are converted into rotating rectangular spatial *pqr* coordinates, where the reactive powers q_q and q_r are calculated as scalar and vector products of voltage and current vectors. A comparative analysis of the considered three power theories is given in the review article [14].

In recent years, the actual publications, including [15–17], state the precise physical foundations of the fundamental theory of instantaneous active and reactive power.

Compared with previous power theories, the mathematical relationships of which are intended to determine the instantaneous active and reactive power in three-phase circuits with valve converters, the expressions of instantaneous reactive power in [18–20] are proposed together with a physical interpretation. It allowed determining the reactive power in single-phase circuits and individual phases of three-phase systems, including circuits equipped with converters.

From the history of reactive power development, it is seen that the main subject of long-lasting discussions and scientific research is the ambiguity of the concept and formulae for its definition in electrical circuits with non-sinusoidal voltages and currents. From the time of the first work of C. J. Budeanu and up to the 1980s, such ambiguity has already led to the emergence of more than ten methods for determining and calculating reactive power, proposed by many authors. Moreover, in each of these methods, various researchers proposed their approaches to assessing reactive power. For a simple comparison, reactive power was calculated for the DE1 electric locomotive (Ukraine) according to crucial theories. The calculation results are shown in [20] and Table 1. The below theories and expressions were used:

Table 1. Comparison of reactive powers calculated by different methods.

No	Active Power P [MW]	Total Power S [MVA]	Reactive Power [Var] by Different Theories				
			Budeanu $Q_B \cdot 10^4$	Fryze $Q_F \cdot 10^6$	Differential $Q_d \cdot 10^4$	Integral $Q_i \cdot 10^4$	Generalized $Q_g \cdot 10^4$
1	1.50	2.58	−1.83	1.60	−5.8	−2.22	6.79
2	2.27	3.59	−4.78	2.24	1.04	−3.67	9.33
3	1.93	3.09	−2.06	1.87	−3.7	−1.73	5.73
4	1.34	2.37	1.35	1.47	−2.7	1.50	4.92
5	1.45	2.56	3.55	1.64	−1.1	2.72	4.77
6	1.55	2.54	2.51	1.47	14.6	5.12	5.67
7	1.63	2.68	3.73	1.59	−1.6	2.86	5.18
8	1.41	2.28	1.73	1.21	4.9	1.20	5.20
9	2.76	4.10	2.88	2.46	22.3	2.66	10.41
10	1.00	1.83	3.00	1.25	4.9	4.06	4.37
11	1.90	3.08	2.84	1.90	2.4	2.39	5.29

-by C. J. Budeanu

$$Q_B = \sum_{k=1}^{n} Q^{(k)} = \sum_{k=1}^{n} U^{(k)} I^{(k)} \sin \varphi^{(k)},$$

-by S. S. Fryze

$$Q_F = \sqrt{S^2 - P^2},$$

-by differential

$$Q_d = \sum_{k=1}^{n} kQ^{(k)} = \sum_{k=1}^{n} kU^{(k)} I^{(k)} \sin \varphi^{(k)},$$

-by integral

$$Q_i = \sum_{k=1}^{n} \frac{Q^{(k)}}{k} = \sum_{k=1}^{n} \frac{1}{k} U^{(k)} I^{(k)} \sin \varphi^{(k)},$$

-by generalized

$$Q_g = \sqrt{Q_d Q_i},$$

where U—pantograph voltage, I—locomotive current, k—harmonic number, φ—phase shift, S—total (apparent) power, P—active power, S—total (apparent) power, Q—reactive (inactive) power, Q_d—the sum of the reduced reactive powers of all of the circuit elements whose voltage and current have the same harmonics, Q_i—the sum of the reduced reactive powers of all of the circuit elements whose voltage and current have different harmonics (according to O. A. Maevski).

The different values of reactive powers in Table 1, as well as the different signs, show clear evidence of the ambiguity of the reactive power defined by these theories. In our opinion, it has not been excluded because of numerous reasons and, perhaps, the most important of them is a state of the problem of "reactive power", of which the solution is based on the "circuit" approach, that is, on the theory of electric and magnetic circuits. However, the theory of circuits is known to historically originate from the nucleation and development of a more complex fundamental theory of electromagnetic fields. Many physical concepts, such as the theory of electric and magnetic circuits, like most laws, are based precisely on the formulas and postulates of the field theory. The "circuit" principles for the analysis of processes in electrical devices are widespread, but also provide numerous obligatory assumptions and simplifications of electromagnetic processes, which determine their nature and mechanisms only approximately, and in some cases give erroneous results. Moreover, electric power processes in all electrotechnical devices and systems are the processes of creation, transformation and propagation of electromagnetic fields and the interaction of fields with electric charges. Therefore, to solve electric power problems in electric transport systems, a "field" approach is needed.

This approach is based on the fact that electromagnetic energy is transmitted from traction substations to the electric rolling stock (electric locomotives, electric trains, trams, metro trains) of electric transport systems through the airspace of inter-substation zones, by means of electromagnetic waves (i.e., by the movement of the electromagnetic field) rather than through the traction network (i.e., catenary system [21] and rails [22]). In this case, the amount and direction of the transmitted energy are determined by the Poynting vector $\vec{S}$, known from the theory of electrical engineering and radio engineering [23], as follows:

$$\vec{S} = \vec{E} \times \vec{H}$$

where $\vec{E}$ and $\vec{H}$—the values of the vectors of the strengths of electric and magnetic fields at points within the space of inter-substation zones.

In this case, the incident (direct) running electromagnetic waves fall at the surface of the contact wire, rails and body of the electric rolling stock, and form reverse (reflected) waves, which play a double role in the consumption of active and reactive power. Firstly, the flow of active energy decreases, since it is equal to the difference between the energy fields of the incident and reflected waves. Secondly, when combined with the incident wave under certain conditions, the reflected waves cause the appearance of standing waves, which are a characteristic feature of reactive power.

Classical approaches [23–25] describing the processes of the complete reflection of incident waves, and thus the formation of standing waves, which currently prevail in theoretical electrical engineering, are limited to the "circuit" approach (according to the theory of electrical circuits) and mainly to the analysis of the operating modes of a long line in the limiting cases of its load, as follows: idle, short circuit, reactive load, etc. To solve the problem set in this article, a "field" approach must be applied, in which the feeder or substation zones of the electrified section are considered simple conducting systems with special properties for the propagation of electromagnetic waves from traction substations to the electric rolling stock, which is associated with currents flowing in the catenary system and rails and, finally, the charges arising on them.

A review of the existing investigations has shown that there are currently no scientific publications on this issue. Some exceptions are articles [26–29], which, using elements of the electromagnetic field theory, make a physical explanation for the occurrence of reactive power Q and distortion power D in electrical circuits with nonsinusoidal voltages and currents. They mostly use the well-known expression of the Umov–Poynting theorem in its complex form to express the balance of powers over volume V of the periodic electromagnetic field, and conclude that the real part of this expression describes active power P and the imaginary part describes reactive power Q. Finally, the authors obtain expressions for the E, D, H and B quantities of the electromagnetic field using EMF, voltage and current in an external nonlinear electric circuit. Ultimately, the authors adhere to the concept of C. J. Budeanu that the total power is determined by the following ratio:

$$S = \sqrt{P^2 + Q^2 + D^2},$$

thus, the processes of wave propagation of energy in the electromagnetic field are not considered in these works.

The need for a physical interpretation of the inactive components of the total power using the Poynting vector is also emphasized by the results of the research presented in [24]. The author correctly describes the mechanism of the flow of electricity that propagates in the dielectric space around the wires of a three-phase power line. According to this research, the electrical energy of running electromagnetic waves, dissipated in the conductors of the supply line, is distributed over the phases of the load or accumulated and returned by inductions or capacitances. The author divides these waves into two groups. The first group consist of waves with a unidirectional flow of energy from source to load. These waves, making oscillations (called "active"), have a nonzero average value of the transferred energy. The waves of the second group, making oscillations (called "inactive"), do not transfer energy. They spread in the dielectric space around the supply line conductors.

As well as in the previous research, research [30] states that a plane electromagnetic wave is characterized by two flows of energy and two, active and reactive, densities. As these energy flows are inseparable at the phase shift angle $\varphi \neq 0$. If $\varphi = 0$, the reactive components of the energy flow will disappear. The author mathematically and physically substantiated the algorithm for separating the active and reactive components of the energies of electromagnetic waves, and expressed them in the following way:

$$\mathbf{\Pi}_a = \frac{A^2}{|\rho|} e^{-2\alpha z} \cos\varphi \cdot \cos^2(\omega t - \beta z) \cdot e_z,$$

$$\mathbf{\Pi}_r = \frac{A^2}{2|\rho|} e^{-2\alpha z} \sin\varphi \cdot \sin(2(\omega t - \beta z)) \cdot e_z,$$

where $\mathbf{\Pi}_a$ and $\mathbf{\Pi}_r$ active and reactive components of Poynting vector, A—power, α—attenuation coefficient, β—phase coefficient, ω—angular frequency, ρ—wave impedance, z—coordinate in the direction of which the Pointing vector is moved. The importance of such division of the energy of electromagnetic waves is shown by the author in an example of the propagation speed of wave energy in a medium with losses.

Quite an original consideration of the components of the total power and their balance is shown in [31]. This article proves that in order to balance complex power, the following applies:

$$\underline{S} = P + jQ,$$

that is, the balance of this expression for both sides, time $t - js$ must be considered. This is due to the fact that there cannot exist a balance of reactive energy in the "active" (real) time. This balance must be considered at "reactive" time $t - js$. The authors try to show the stated idea on the example of time-dependent, i.e., parametric, RLC circuits.

The fulfilment of the power balance in the electromagnetic field is also studied in the processes of dissipation and absorption of the energy of electromagnetic waves. In particular, a finite-size coated sphere scatterer is considered in [32], which is illuminated by incident electromagnetic waves. The percentage of absorbed and scattered energies determined by the Poynting vector is analyzed. The study found that the absorbed energy, called active energy, is about 25% of the energy of the incident electromagnetic field. It is stated that the imaginary part of the complex power balance expression is the dissipated power, and its occurrence is due to the phase shift between the incident electromagnetic field and polarization currents. Finally, it concludes that the diffuser must be optimally designed to eliminate the reactive power.

From the above analysis follows that the authors of the publications, recognizing the presence of active and inactive components of the energy of electromagnetic waves, do not consider the processes of wave reflection, and do not associate the physical interpretation of reactive power with the emergence and suppression of standing electromagnetic waves in the dielectric space of the electromagnetic field.

2. Objective of the Article

The article continues and develops the research presented in [33], dedicated to the "field" approach to substantiate the nature and processes of electricity transmission from traction substations (TS) to the electric rolling stock (ERS).

The primary purpose is to establish and provide theoretical grounds for the "field" functional connection between reactive power consumed by the electric rolling stock, and the occurrence and damping of standing electromagnetic waves in the space of inter-substation zones of electric transport systems, based on the electromagnetic field theory.

3. Novelty of Scientific Results of the Article in Relation to Existing Publications

1. For the first time in the Ukraine and Europe, a "field" (i.e., based on the electromagnetic field theory) interpretation of functional connectivity of reactive power, transmitted from a traction substation to the electric rolling stock, with occurrence and dissipation processes of standing electromagnetic waves in inter-substation zones of electric transport systems, is hereby proposed and substantiated.
2. Dependencies have been found to describe the occurrence of standing waves in the space of inter-substation zones under the incomplete reflection of incident electromagnetic waves from a contact wire, rails, metal and dielectric surfaces of roofs, front and side parts of the body of the ERS.
3. For the first time, existing methods for the suppression of standing electromagnetic waves are adapted to compensate the reactive power in electric transport systems by applying single- or multi-layered film coatings, with specific electromagnetic properties, on the surfaces of feeder zone devices.

4. Reactive Power as a Measure of Asymmetry of the Rates of Change of Electric Power in Electrical and Magnetic Fields

Let us consider the starting point of a 560 km long railway line supplied by a 25 kV AC 50 Hz traction power system. Imagine that the space of the specified line assumed to be limited by a traction substation, catenary system, railway vehicle and the rails, has a volume V limited by the enclosing surface A_v, as shown in Figure 1. The trains in this section are driven by single-phase AC electric locomotives, types VL80 and VL85 (Ukraine). The locomotives have thyristor power converters supplying the traction motors. The frequency of voltage higher harmonics on the output changes are in the range of 1250 to 1950 Hz, according to [34,35].

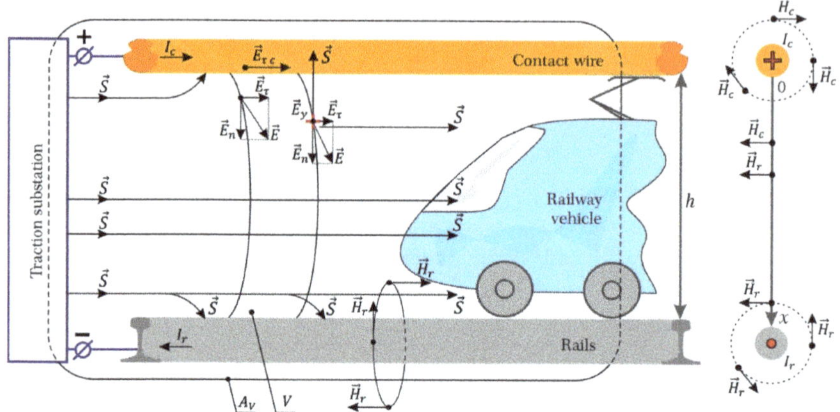

Figure 1. Representation of the Poynting vector $\vec{S}$ propagation in the process of electric power transmission from traction substation to the railway vehicle in the inter-substation zone, where V—volume, A_v—enclosing surface, $\vec{H}_c$ and $\vec{H}_r$—vectors of magnetic field tensions in the space of catenary and rails, $\vec{E}_n$ and $\vec{E}_\tau$—normal and tangential vectors of electric field tension $\vec{E}$, $\vec{E}_y$—vector of electric field tension at the point y, I_c and I_r—currents in catenary and rails, h—distance between the contact wire of catenary system and rails, x—coordinate axis for height.

Let this space be linear, homogeneous and isotropic with the following electromagnetic properties: permittivity ε, conductivity σ and permeability μ. Then, the alternating electromagnetic field in this space could be described by Maxwell's equations in its complex form according to [23], as follows:

$$rot\vec{\underline{H}} = \sigma\vec{\underline{E}} + j\omega\varepsilon\vec{\underline{E}} \tag{1}$$

$$rot\vec{\underline{E}} = -j\omega\mu\vec{\underline{H}}, \tag{2}$$

where $\vec{\underline{E}}$ and $\vec{\underline{H}}$ are the vectors of electric and magnetic field tensions in the space of inter-substation zones.

Since both tensions $\vec{\underline{E}}$ and $\vec{\underline{H}}$ are created by a single energy source, i.e., by a traction substation, electricity is transferred according to Poynting vector $\underline{S}$, and takes place in one direction from TS to ERS.

Like complex power $\underline{S}$ in the theory of circuits, we form a complex Poynting vector (Figure 1) as follows:

$$\vec{\underline{S}} = \left[\vec{\underline{E}}\vec{\underline{H}}''\right], \tag{3}$$

where

$$rot\vec{\underline{H}}'' = \sigma\vec{\underline{E}}'' + j\omega\varepsilon\vec{\underline{E}}''. \tag{4}$$

Next, let us transform system (1)–(4) as follows: multiply expression (4) by $\vec{\underline{E}}$, and (2) by $\vec{\underline{H}}''$. Then the following expressions could be written:

$$\vec{\underline{E}}rot\vec{\underline{H}}'' = \sigma\vec{\underline{E}}\vec{\underline{E}}'' + j\omega\varepsilon\vec{\underline{E}}\vec{\underline{E}}'' \tag{5}$$

$$\overrightarrow{\underline{H}}'' \, rot \overrightarrow{\underline{E}} = -j\omega\mu \overrightarrow{\underline{H}}\overrightarrow{\underline{H}}'' \quad (6)$$

Subtracting (6) from (5) and considering that the difference of the left parts of (5) and (6) is equal to $-div \overrightarrow{\underline{S}}$, we have the following:

$$-div \overrightarrow{\underline{S}} = \sigma E^2 + j\omega\left(\mu H^2 - \varepsilon E^2\right). \quad (7)$$

Then the flow of the Pointing vector through the enclosing surface A_V into the volume V of inter-substation zones (Figure 1) will be equal to the following:

$$-\int_V div \overrightarrow{\underline{S}} dV = \int_V \sigma E^2 dV + j2\omega \int_V \left(\frac{\mu H^2}{2} - \frac{\varepsilon E^2}{2}\right) dV \quad (8)$$

By analogy with the theory of circuits and according to the Joule–Lenz law in its differential form [23], the first component on the right side of Expression (8) is the average value of heat losses in volume V of inter-substation zones over the period of the supply voltage, and the second component, i.e., the imaginary part, is the average value of reactive power Q entering into volume V through the enclosing surface A_V, as follows:

$$Q = 2\omega \int_V \left(\frac{\mu H^2}{2} - \frac{\varepsilon E^2}{2}\right) dV \quad (9)$$

where E and H are the effective values of the harmonic tensions.

Given that $\overrightarrow{B} = \mu \overrightarrow{H}$ and $\overrightarrow{D} = \varepsilon \overrightarrow{E}$, and considering Formula (9), the expression of the energy of the "field" (by the electromagnetic field vectors) for the representation of reactive power can be written in the following form:

$$Q = 2\omega \int_V \left(\frac{\mu H^2}{2} - \frac{\varepsilon E^2}{2}\right) dV \approx \int_V \left(\overrightarrow{\underline{H}}\frac{\partial \overrightarrow{\underline{B}}}{\partial t} - \overrightarrow{\underline{E}}\frac{\partial \overrightarrow{\underline{D}}}{\partial t}\right) dV = \int_V \frac{\partial}{\partial t}\left(\frac{\overrightarrow{\underline{H}}\overrightarrow{\underline{B}}}{2} - \frac{\overrightarrow{\underline{E}}\overrightarrow{\underline{D}}}{2}\right) dV. \quad (10)$$

Therefore, reactive power in the space of inter-substation zones by volume V is a measure of the asymmetry of the rates of change in the electric and magnetic components of the energy of the electromagnetic field. Numerically, it is equal to the difference between average energies stored in the magnetic and electric fields in a given volume V, multiplied by 2ω.

5. Volume of Reactive Power in the Space of Inter-Substation Zones of Electrified Section

Using Expressions (9) and (10), let us consider the reactive power Q spreading in the space of an inter-substation zone consisting of seven sections with a total length of 560 km. Each inter-substation zone has a one-sided power supply via the catenary of PBSM-95+MF-100 type and the rails of R65 type, which could be specified as follows: PBSM-95 catenary wire—bimetal steel copper wire with a cross-section of 95 m^2; MF-100 contact wire—solid copper wire with a cross-section of 100 m^2, R65 rails—carbon steel rails with a specific weight of 64.88 kg/m [36]. The VL85 electric locomotive with a train is moving along the inter-substation zone at 160 km/h and taking power $P = 5369$ kW. According to the experimental research of this study, at this speed, the voltage at the pantograph is $U = 26.12$ kV, the current in the contact wire and the electric locomotive is $I_c = 208.9$ A, and in the rails of the one-side AC traction system is $I_r = 83.6$ A.

Let us define the tensions of the electric and magnetic fields in the space of the inter-substation zones.

The tension of the electric field in a dielectric (i.e., in the air) has the following two components: tangential E_τ and normal E_n parts, which are shown in Figure 1. The tangential component E_τ in the air of inter-substation zones, according to the "conductor-dielectric" boundary condition [23], is equal to the tangential component in conductor $E_{\tau\ c}$, which can be determined from the differential Ohm's law, as follows:

$$J = \sigma E_{\tau\ c}, \tag{11}$$

where J is the current density in the equivalent conductor of the catenary system, and σ is the specific electrical conductivity of the material of the contact wire.

Given that the cross-section area of an equivalent contact wire is $A = 193.3 \times 10^{-6}$ m^2, with a current of $I_c = 208.9$ A, we find that the current density is $J = 1.08 \times 10^6$ A/m^2, assuming that the conductivity of copper is $\sigma = 5.8 \times 10^7$ S/m. Then, the tangential component of tension E_τ in the air, according to (11), equals 0.186 V/m. The distance between the contact wire and the head of the rail usually varies from $h_{min} = 5.75$ m to $h_{max} = 6.8$ m. Assuming the maximum distance and the above-mentioned voltage $U = 26.12$ kV, we find the normal component of the tension, which is equal to $E_n = 3841$ V/m. Consequently, as far as $E_n \gg E_\tau$, we could find that electric field tension in the airspace of inter-substation zones is $E = E_n = 3841$ V/m.

The tension of the magnetic field at any point of the inter-substation space $\vec{H}$ is defined as the vector sum of the components created by the currents in the contact wire $\vec{H}_c$ and in the rail $\vec{H}_r$, as follows:

$$\vec{H} = \vec{H}_c + \vec{H}_r \tag{12}$$

Each of these components has to be defined using the law of total currents in catenary I_c and rails I_r (Figure 1), as below:

$$H_c(x) = \frac{I_c}{2\pi(x + r_c)} \tag{13}$$

$$H_r(x) = \frac{I_r}{2\pi(6.8 - x)} \tag{14}$$

where x is the wire-to-rail coordinate, which varies from an equivalent contact wire to an equivalent rail and is mostly equal to the distance between them $h = 6.8$ m, according to Figure 1; r_c and r_r are the radii of equivalent contact wires and rails, which are equal to $r_c = 7.85 \times 10^{-3}$ m and $r_r = 72.7 \times 10^{-3}$ m, respectively.

So, we can calculate, according to Formulae (12)–(14), the dependence of the resulting tension of the magnetic field H in the inter-substation space on the x coordinate, i.e., on height h. The results are shown in Figure 2 and Table 2.

Let us calculate the possible reactive power Q in a volume V of the space of the inter-substation zone. The volume could be found using the following data: the total length of the space of the inter-substation zone of 560 km, its width is taken to be the width of the railway track of 1.524 m, and its height of $h = 6.8$ m; then, $V = 5,803,392$ m^3.

Given that the electric field is uniformly distributed over the inter-substation space and magnetic power is distributed according to Figure 2, the reactive power spreading from TS to ERS is found by specified Formula (10), as follows:

$$Q = 2\omega \int\limits_V \left(\frac{\mu_0 H^2(h)}{2} - \frac{\varepsilon_0 E^2}{2} \right) dV \approx 4\pi f \sum_{h_i=0.1}^{6.8} \left(\frac{\mu_0 H^2(h_i)}{2} - \frac{\varepsilon_0 E^2}{2} \right) \Delta V_i \tag{15}$$

Finally, based on Expression (15) and Table 3, we can conclude that an electric locomotive in the specified mode with electrical parameters $U = 26.12$ kV, $I_r = 208.9$ A, $I_r = 83.6$ A, totally could transfer a reactive power of 2930.5 kvar from the feeder zone.

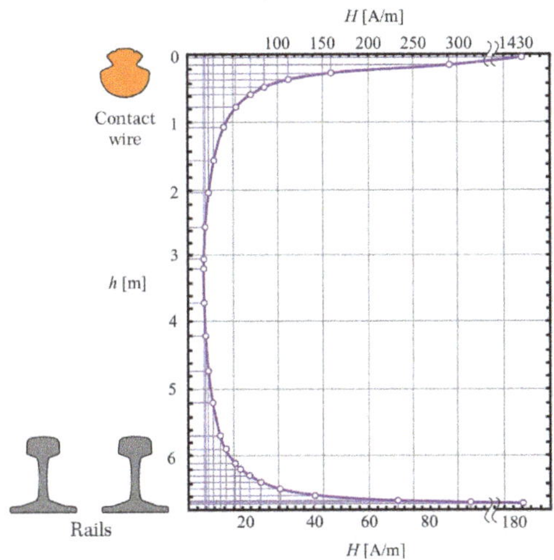

Figure 2. Distribution of magnetic field strengths in the space between the equivalent contact wire and the rails.

Table 2. Calculated physical parameters of the equivalent contact wire and rails.

Element of Feeder Zone	Parameters of Metal		
	Electrical Conductivity σ [S/M]	**Relative Magnetic Permeability μ [R.U.]**	**Radii of Rail and Wire, or Sheet Thickness [Mm]**
Equivalent contact wire	3.82×10^7	10	7.85
Equivalent rails	8×10^6	1000	72.7
Electric locomotive, steel sheet	1.8×10^6	5200	5

Table 3. Results of calculations of basic electromagnetic coefficients according to Expression (15).

x_i [m]	H_c [A/m]	H_r [A/m]	H [A/m]	$\frac{\mu_0 H^2}{2} \cdot 10^{-3}$ [r.u.]	$(\frac{\mu_0 H^2}{2} - \frac{\varepsilon_0 E^2}{2}) \cdot 10^{-3}$ [r.u.]
0	4237.5	4.9	4242.4	11.3	11.3
0.1	308.43	2.0	310.4	60.5	60.4
0.2	160.0	2.0	162	16.48	16.4
0.3	108.1	2.05	110	8.1	7.6
0.4	81.56	2.08	84	4.13	4.36
0.5	65.5	2.14	68	2.9	2.84
0.7	47.0	2.2	49.5	1.5	1.46
1.0	33	2.3	35.3	0.783	0.17
1.5	22.06	2.5	24.50	0.377	0.311
2.0	16.57	2.8	18.8	0.347	0.28
3.0	11.06	3.5	14.56	0.246	0.18
4.0	8.3	4.8	13	0.255	0.19
5.0	6.642	7.4	14	0.396	0.33
5.5	6.04	10.22	16.2	0.628	0.563
5.8	5.73	13.3	19.0	0.95	0.89
6.0	5.54	16.6	22	1.39	1.33
6.2	5.36	22.2	27.6	2.32	2.26
6.4	5.19	33.3	38.5	4.9	4.84
6.6	5.03	66.5	71.5	2.42	2.25
6.7	4.96	53	58	71.6	5.9
6.8	4.89	183	188	134	134

6. Conditions for the Formation of Standing and Mixed Electromagnetic Waves

The alternating electromagnetic field created in the airspace of inter-substation zones is characterized by the harmonic plane running waves of electric and magnetic field tensions that transmit electricity from TS to ERS [23], as follows:

$$E(y,t) = E_{m\ in}\sin(\omega t - \beta y) + E_{m\ rf}\sin(\omega t + \beta y) \tag{16}$$

$$H(y,t) = H_{m\ in}\sin(\omega t - \beta y) + H_{m\ rf}\sin(\omega t + \beta y) \tag{17}$$

Let us rewrite Expressions (16) and (17) in the following complex form:

$$\underline{E}(y) = E_{m\ in}e^{-j\beta y} + E_{m\ rf}e^{j\beta y} \tag{18}$$

$$\underline{H}(y) = H_{m\ in}e^{-j\beta y} - H_{m\ rf}e^{j\beta y} \tag{19}$$

As it is known, a reflected wave will be formed if the incident wave meets another medium (dielectric or conductor) on its way. For a medium with infinite electrical conductivity σ (a lossless medium), when $|\Gamma| = 1$, the amplitude of the reflected wave is equal to the amplitude of the incident wave and, as a result of their adding together, a standing wave is formed. In this regard, given that $E_m = E_{m\ in} = E_{m\ rf}$, $H_m = H_{m\ in} = H_{m\ rf}$, we obtain the following expressions of standing waves:

$$E(y,t) = 2E_m\cos(\beta y)\sin(\omega t) \tag{20}$$

$$H(y,t) = -2H_m\sin(\beta y)\cos(\omega t) \tag{21}$$

A graphical representation of the standing waves of electric and magnetic fields reflected from the railway vehicle is shown in Figure 3.

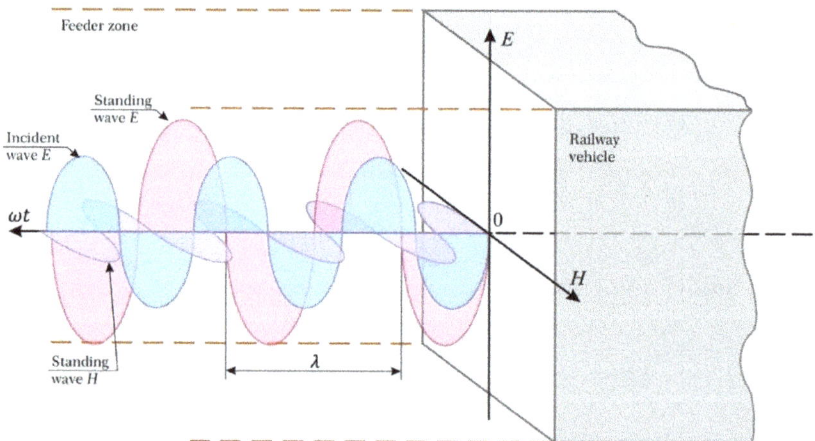

Figure 3. Representation of incident and standing electromagnetic waves reflected from the railway vehicle body.

Consequently, it is generally accepted that a necessary condition for the occurrence of standing electromagnetic waves is a complete reflection of the incident waves. Let us analyze this issue, considering [23,25,37–39], as follows.

As it is widely recognized, if a plane sinusoidally changing the electromagnetic wave moving in medium 1 (in our case, it is the air), with properties ε_1, σ_1, μ_1, encounters on its way the interface of another medium 2 (parts and devices of the railway vehicle), with properties ε_2, σ_2, μ_2, and this wave falls (the incident wave having tensions $\vec{\underline{E}}_{in}$, $\vec{\underline{H}}_{in}$)

perpendicularly to the boundary of the media, it is partially reflected $\vec{\underline{E}}_{rf}$, $\vec{\underline{H}}_{rf}$ with a reflection ratio $\underline{\Gamma}$.

$$\underline{\Gamma} = \frac{\underline{\eta}_2 - \underline{\eta}_1}{\underline{\eta}_2 + \underline{\eta}_1} = \frac{1 - \underline{T}}{1 + \underline{T}} \tag{22}$$

where $\underline{T} = \underline{\eta}_1 / \underline{\eta}_2$—running wave coefficient; $\underline{\eta}_1$ and $\underline{\eta}_2$—wave complex resistances of media 1 and 2, defined as follows:
- for dielectrics

$$\eta = \sqrt{\mu/\varepsilon} \tag{23}$$

- for the conducting medium

$$\underline{\eta} = \sqrt{j\omega\mu/\sigma} \tag{24}$$

Let us analyze the flow of electricity from the traction substation to the devices of the inter-substation zone.

7. Energy Flow to the Contact Wire

At the boundary of the contact wire, the electromagnetic wave falling from the air of the inter-substation zone is partly reflected from the wire surface and partly penetrates into it (Figure 1). Let us determine the properties of the reflected wave.

For air, medium 1, the wave resistance is $\eta_{air} = \eta_1 = 376.7\ \Omega$.

Medium 2 is a conductive medium represented by the catenary system of the PBSM-95 + MF − 100 type. The cross-section area of the equivalent contact wire is $193.3 \times 10^{-6}\ \text{m}^2$, hence the specific conductivity is $\sigma_2 = \sigma_{eq\ w} = 5 \times 10^7$ S/m and the permeability is $\mu_{r2} = 10$.

The higher harmonics of the feeder current in the normal operating mode have frequencies up to 750–1950 Hz, which will be taken into account too. The wave resistance of the material of the equivalent wire for $f = 1950$ Hz is 1.33 Ω according to (24), at the same time for $f = 50$ Hz it is equal to 0.212 Ω. Then, the reflection ratio on the boundary "air-to-wire" surface, for example, for the current harmonic with a frequency of $f = 1950$ Hz is defined as $\underline{\Gamma}_{12} \approx -1$ according to Expression (22).

Then, according to Expressions (23)–(24), the field tensions of the reflected wave are as follows:

$$\vec{\underline{E}}_{rf1} \approx -\vec{\underline{E}}_{in1} \text{ and } \vec{\underline{H}}_{rf1} \approx \vec{\underline{H}}_{in1}$$

that is, the almost completely reflected wave changes the sign of vector $\vec{E}$, and a similar result is obtained for the rails.

8. Incidence of the Poynting Vector on the ERS Roof

As defined in [33], the electromagnetic energy supplying the ERS comes from the traction network of the power supply system in two ways (Figure 4). The first way is contact wire 2—roof bushing insulator 4—appliances of the high-voltage cabinet 6—traction motors. The second way is rails 3—bottom part of the ERS—traction motors and the high-voltage cabinet 6. A significantly minor amount of the energy of the electromagnetic waves comes from the airspace 1 of the inter-substation zones to any part of the EPS, as follows: metal body 8 and to the glass of the driver's cabin 7 in (Figure 4).

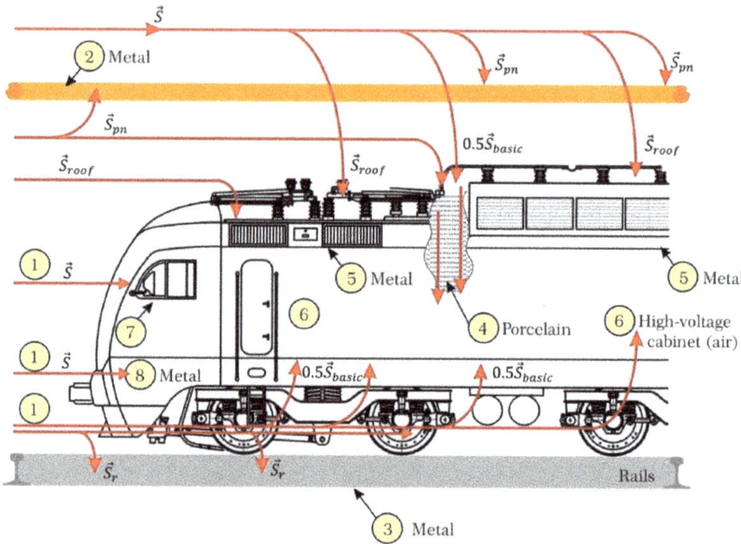

Figure 4. Representation of electric power flows using the Poynting vector $\vec{S}$ on the surface of the contact wire ($\vec{S}_{pn}$), rails ($\vec{S}_r$), roof ($\vec{S}_{roof}$), and front parts of the electric rolling stock.

At the outset, let us analyze the basic stream in which the following components of the Poynting vector carrying this stream can be identified (Figure 4):

1. Basic vector $0.5\vec{S}_{basic}$ corresponding to the density of energy flow coming into the elements of the traction power circuit of the ERS (i.e., into the high-voltage cabinet) through the roof bushing porcelain insulator 4. To analyze the electromagnetic waves carrying $\vec{S}_{basic}$, the coefficients of reflection Γ_{14} and refraction T_{14} (i.e., transmission, penetration) of the incident wave on the "air 1-to-porcelain 4" border have to be defined. For perfect dielectrics like air $\varepsilon_{r1} = 1.0$ and porcelain $\varepsilon_{r4} = 6.0$, so for Γ and T coefficients we will get the following:

$$\Gamma_{14} = \frac{\sqrt{\varepsilon_{r1}} - \sqrt{\varepsilon_{r4}}}{\sqrt{\varepsilon_{r1}} + \sqrt{\varepsilon_{r4}}} = -0.42 \text{ and } T^E_{14} = \frac{2\sqrt{\varepsilon_{r1}}}{\sqrt{\varepsilon_{r1}} + \sqrt{\varepsilon_{r4}}} = 0.58 \quad (25)$$

As we can see, some part of the energy is reflected from the insulator, as follows:

$$E_{rf14} = -0.42E_{in1} \text{ and } T^H_{14} = \frac{2\sqrt{\varepsilon_{r4}}}{\sqrt{\varepsilon_{r1}} + \sqrt{\varepsilon_{r4}}} = 1.42 \quad (26)$$

$$H_{rf14} = -(0.42H_{in1}) = 0.42H_{in1}$$

$$E_{pn14} = 0.58E_{in1}$$

$$H_{pn14} = 1.42H_{in1}$$

where rf index denotes the reflected value, in index denotes the incident value, pn index denotes the refracted (penetrated, transmitted) value.

The second stream of $\vec{S}_{basic}$ (Figure 4) penetrates into the high-voltage cabinet of the ERS from the space around rails 3 (i.e., from the bottom of ERS). The wave propagation and power transmission in this space need more in-depth study, which is not provided in this article.

2. Vector $\overrightarrow{S}_{roof}$ is normally pointed to the roof surface (Figure 4). Its material is a sheet structural steel of 2 mm thick, with the properties of $\sigma = 7 \times 10^6$ S/m and $\mu_r = 1000$. The electromagnetic harmonic waves of various frequencies, incident from air 1 on the border with the surface of metal roof 5 (Figure 4), are partially reflected and partially penetrate into it. They are gradually damped, resulting in electric power losses in the metal roof.

The wave resistance of the roof metal η_5, found by Expression (24) in the case of the current harmonic with $f = 1950$ Hz, is 1.33 Ω, and for air it is $\eta_1 = 376.7\ \Omega$. Substituting for η_1 and η_5 in (22), the coefficient of wave reflection from the roof surface being $\Gamma_{15} \approx -1$, the refraction (transmission, penetration) coefficient equals $T_{15} \approx 0.87 \times 10^{-5}$. A 1950 Hz wave penetrates to the depth of $d_5 = 0.115$ mm. In the case of basic current harmonic with a frequency of 50 Hz, the wave resistance of the roof metal η_5 is 0.212 Ω, hence the wave penetrates 0.85 mm deep. The metal sheet of the roof is "impenetrable" for the waves, since they are completely damped out within it without reaching an opposite surface. We obtain similar results for the side surfaces of the ERS body.

9. Incidence of the Waves on the Frontal Part of the Locomotive Body

Electromagnetic waves spread in the middle part of the air space between the contact wire and the rail, and fall on the frontal part of the locomotive body consisting of glass and metallic parts.

The windshield has a thickness of 15 mm and permittivity $\varepsilon_r = 5.5 \ldots 10$ (let us assume 9.0). At the "air-to-glass" boundary, the waves are partially reflected with a ratio of $\Gamma = -0.5$, according to the following:

$$\Gamma = \frac{\sqrt{\varepsilon_{r1}} - \sqrt{\varepsilon_r}}{\sqrt{\varepsilon_{r1}} + \sqrt{\varepsilon_r}} \tag{27}$$

The metal part of the lower frontal part of the ERS body is made of structural steel sheets of 7 mm thick, with parameters $\sigma = 7 \times 10^6$ S/m and $\mu_r = 1000$. Then, the wave resistance of this metal body, according to (25), is equal to $\eta_{front} = \eta_2 = 33.6 \times 10^{-5}\ \Omega$. Substituting $\eta_{air} = \eta_1$ and $\eta_{front} = \eta_2$ in (22), the reflection ratio of the waves from the metal part of the body equals $\Gamma_1 = -1$.

The performed calculations of the reflection coefficient of electromagnetic waves from electric traction devices (i.e., contact wire, glass, porcelain bushing, as well as metal roofs and sidewalls of the body) show that incomplete reflection of waves from the surfaces of specified parts and devices occurs in the process of electric locomotive operation. Simultaneously, in the space of the inter-substation zone, both active and reactive powers are transferred from TS to ERS. Consequently, there are running and standing waves that superimpose in this space and form mixed electromagnetic waves. Let us show that the necessary condition for standing waves occurrence, namely, the existence of total reflection, is not so strict. The standing waves in the inter-substation space, which characterize the reactive energy [37], are also possible with incomplete reflection, i.e., when the reflection ratio is $0 \leq \Gamma < 1$.

Given that $E_{m\ in}/H_{m\ in} = \eta_{air} = \eta_1$, where η_{air} is a wave resistance of the airspace of inter-substation zones, let us rewrite Expressions (18) and (19) in the following way:

$$\underline{E}(y) = \underline{E}_{m\ in} e^{-j\beta y}\left(1 + \Gamma e^{j2\beta y}\right) \tag{28}$$

$$\underline{H}(y) = \frac{\underline{E}_{m\ in}}{\eta_1} e^{-j\beta y}\left(1 - \Gamma e^{j2\beta y}\right) \tag{29}$$

Then, according to the theory of the well-established symbolic method for the analysis of electric circuits, the instantaneous complex values of electric and magnetic field strengths can be found after formal multiplication of Equations (28) and (29) by operator $e^{j\omega t}$, as follows:

$$\underline{E}(y) = \underline{E}_{m\ in}\left(1 + \Gamma e^{j2\beta y}\right)e^{j(\omega t - \beta y)} \tag{30}$$

$$\underline{H}(y) = \frac{\underline{E}_{m\ in}}{\eta_1}\left(1 - \Gamma e^{j2\beta y}\right)e^{j(\omega t - \beta y)} \tag{31}$$

Compared to Expressions (18) and (19), where the fields are represented as the sum of the incident and reflected waves, the form of writing of Expressions (30) and (31) corresponds to only one incident wave with an amplitude varying from the coordinate, as follows:

$$\underline{E}_m(y) = \underline{E}_{m\ in}\left|1 + \Gamma e^{j2\beta y}\right| \tag{32}$$

$$\underline{H}_m(y) = \frac{\underline{E}_{m\ in}}{\eta_1}\left|1 - \Gamma e^{j2\beta y}\right| \tag{33}$$

These functions are periodic in the inter-substation space and acquire maximum values with $(1 + \Gamma)$, which are much larger than the amplitude of the incident wave at points called "wave antinodes". Between them, there are "wave nodes", where the amplitude is minimal and equals $(1 - \Gamma)$ of the incident wave amplitude. The spatial position of these points does not change over time; therefore, for the expressions described by (30) and (31), the definition of "standing wave" is used [40]. Consequently, in a non-perfect conductor (such as the body of an electric locomotive), the reflected wave has a smaller amplitude than the incident wave. So, a small running wave perpendicular to the conductor surface is superimposed on the standing wave, which compensates for power losses in the conductor.

For more advanced analysis, let us consider the dependence of electromagnetic waves type on the coefficient of running wave T. For this, we rewrite the system of Expressions (28) and (29), considering the well-established Euler's formula and Expression (22), then we obtain the following:

$$\underline{E}(y) = \underline{E}_{in}\cdot\frac{2}{1+\mathrm{T}}\cdot(\cos(\beta y) - j\mathrm{T}\cdot\sin(\beta y)) \tag{34}$$

$$\underline{H}(y) = \frac{\underline{E}_{in}}{\eta_1}\cdot\left(2 - \frac{1}{\mathrm{T}}\right)\cdot\left(\cos(\beta y) - j\frac{1}{\mathrm{T}}\cdot\sin(\beta y)\right) \tag{35}$$

After some simple transformations of Expressions (34) and (35), and a transition into instantaneous quantities, we obtain the following system:

$$E(y,t) = E_{in}\cdot\frac{4}{1+\mathrm{T}}\cdot\left(\sin(\omega t - \beta y) + E_{in}\cdot\frac{2\cdot(1-\mathrm{T})}{1+\mathrm{T}}\cdot\cos(\beta y)\cdot\sin(\omega t)\right) \tag{36}$$

$$H(y,t) = \frac{E_{in}}{\eta_1}\cdot\frac{1}{\mathrm{T}}\cdot\sin(\omega t - \beta y) + \frac{E_{in}}{\eta_1}\cdot 2\cdot\left(1 - \frac{1}{\mathrm{T}}\right)\cdot\cos(\beta y)\cdot\sin(\omega t) \tag{37}$$

It follows from Expressions (36) and (37) that the expressions for the electric and magnetic field strengths are the sums of running and standing waves. It follows from them that if the coefficient of wave transmission is $\mathrm{T} = 1$, i.e., if the air space of the inter-substation zones is consistent with the materials of the contact wire and the body of the electric rolling stock ($\eta_1 = \eta_2$), then standing waves do not arise in the inter-substation space, since there is no wave reflection. If $\mathrm{T} \neq 1$, i.e., space and materials are not consistent ($\eta_1 \neq \eta_2$), there is an incomplete reflection and both running and standing waves arise in the inter-substation air space, as a result of which a mixed wave is formed. Incomplete reflection is also indicated by the fact that the metals of both the contact wire and the locomotive body are "real" materials, i.e., they have a finite (and not infinite) conductivity. Therefore, it follows that not all the energy of the incident wave is reflected, and some part of it passes into the metal and dissipates in it as heat.

The above suggests that active energy is transmitted from traction substations to locomotives in the inter-substation airspace by running waves. Thus, the energy exchanged between electric and magnetic fields characterizing the standing waves involves the exchange of reactive power too.

10. Reactive Power Compensation by "Suppression" of Standing Electromagnetic Waves and Discussion of Results

Reactive power consumed by the ERS of different types is quite significant, as in some modes it reaches up to 40–65% of the consumed power [41]. Therefore, the problem of its compensation is significant for electric traction systems. As it follows from the above analysis, this problem has to be solved by "suppression" of standing waves in the space of inter-substation zones. Since standing waves result from reflected waves, reflections of incident waves need to be reduced to compensate for reactive power. Therefore, it is necessary to investigate issues of the suppression of standing waves and how to damp or eliminate reflected waves. As it follows from Expression (22), it is necessary to ensure that wave resistances along both sides of their separation boundary are as close as possible to each other. There are several ways to reduce the reflection coefficient of monochromatic electromagnetic waves [40,42].

The easiest way to reduce reflections is a thin-film method [40]. For the problem solved in this article, this means that a layer of electrically conductive or dielectric coating of certain thickness d with wave resistance η_2 should be applied to the contact wire, rails and the body of the electric rolling stock. In this case, we will get a circuit model consisting of three environments (1, 2, 3), shown in Figure 5, where η_1, η_2 and η_3 are wave resistances of the environments, and K_1, K_2 and K_3 are their wave numbers defined as $K_i = 2\pi/\lambda_i$ (where $i = 1, 2, 3$).

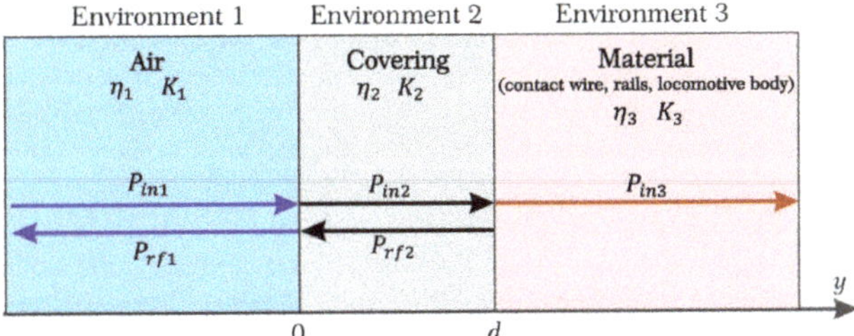

Figure 5. A circuit model of three environments with incident and reflected waves.

Air and the contact wire material are considered to be semi-infinite environments, since they are "impermeable" to electromagnetic waves. The incident waves P_{in1} and P_{in2} both meet partial reflections P_{rf1} and P_{rf2}, respectively, from the first and the second boundaries; the wave P_{in3} passes into the third environment.

The influence of the coating layer (environment 2 in Figure 5) on the formation of reflected waves is analyzed using analogies of the expressions obtained in the theory of electric circuits with distributed parameters for long lines. These analogies allow us to obtain the reflection coefficient from the first boundary of the environments, i.e., in the air of the inter-substation zone, in the form as below [40]:

$$\underline{\Gamma} = \frac{\eta_2 \cdot (\eta_3 - \eta_1) + j(\eta_2^2 - \eta_1 \cdot \eta_3) \cdot \tan(K_2 \cdot d)}{\eta_2 \cdot (\eta_3 + \eta_1) + j(\eta_2^2 - \eta_1 \cdot \eta_3) \cdot \tan(K_2 \cdot d)} \quad (38)$$

Let us consider under what conditions the reflection coefficient Γ in Expression (39) disappears, i.e., when there are no reflected and, therefore, standing waves.

According to the first variant, if $\eta_1 = \eta_2 = \eta_3$, then the real and imaginary parts of the numerator in Expression (39) are equal to zero, and hence $\Gamma = 0$. However, such conditions are practically impossible if environment 3 is a contact wire or metal parts of the locomotive body, since η_1 is the wave resistance of the air, η_1 =376.7 Ω, and η_3 is the resistance of the metal, $\eta_3 = 1.33\ \Omega$; as a result, $\eta_1 > \eta_3$. At the same time, this condition is feasible for dielectric environment 3, that is, the passage porcelain insulator of the locomotive roof or the glass of its frontal part. However, first, it is necessary to choose the dielectric permittivity of porcelain and glass, equal to the dielectric permittivity of the air. This will provide the following condition: $\eta_1 = \eta_3$. Second, the thickness of environment 2 (coating) must be, regardless of its η_2, equal to a multiplication of half the wavelength in this environment ($d = n{\cdot}\lambda_2/2;\ n = 1,2,3$). In this case, the reflection coefficient of the waves from the first boundary in Expression (39) is equal to $\tan(K_2{\cdot}d) = 0$. So, for the first boundary, the following applies:

$$\Gamma_1 = \frac{\eta_2 - \eta_1}{\eta_2 + \eta_1}$$

and for the second boundary, the following:

$$\Gamma_2 = \frac{\eta_3 - \eta_2}{\eta_3 + \eta_2}$$

Since $\eta_1 = \eta_3$, the coefficients Γ_1 and Γ_2 have equal amplitudes, but opposite signs, and therefore opposite phases in the reflection planes. However, the passage of the wave reflected from the second boundary, the distance d "back and forth", causes an additional incursion of its phase by the value of $2K_2d = 2\pi n$. As a result, the waves in the air environment 1, as equal in amplitude, but opposite in phase, cancel out each other and the resulting wave will be absent.

Another way to reduce reflection is based on using the fact that the wave resistance of a non-conductive material is defined as $\eta = \sqrt{\mu/\varepsilon}$. By selecting μ and ε, we can get η equal to the air resistance. If the hysteresis loops $\mu(H)$ and $\varepsilon(E)$ are the same, so that for any pair of tensions H and E ratio μ/ε is the same, then the layer of this absorbing material will be an empty space for the incident wave with a normal incidence.

It is somewhat more complicated to provide a complete or partial reflection of waves from metal parts of the roof and the locomotive body. The most effective way is to make the body with intermediate media with graded or multi-layered (from three to four layers) coatings made of different materials. They could be based on thin films made from metal, magnetodielectric, semiconductor and superplastic materials [40]. In this case, the required value of the reflection coefficient is obtained by a change in thickness of the layers and properties ε, μ, σ of the material. It is necessary to create such an intermediate coating, either graded or multi-layered, in which wave resistance would change exponentially as $\eta_2 = \eta_1 e^{\beta x}$ [40]. It is always possible to choose such value β as to match the impedances of this layer and the air of inter-substation zones. The advantages of multi-layered coatings over single-layered coatings are clear from researches presented in [42,43]. It follows that double-layered coatings have significantly lower reflection Γ (on 12–16%), transmission T (on 37–42%) and absorption A (on 49–58%) coefficients than single-layered coatings.

It should also be noted that the presented-above calculations show results for basic and higher harmonics created by power converters feeding the traction motors. The higher harmonics can be significantly reduced by appropriate configuration of converters and the application of harmonic filters, which exclude waves created by higher frequencies mentioned above in the article. Such a solution substantially improves power quality and reduces reactive energy transferred by higher harmonics. At the same time, the basic component stills transferer reactive power. So, the proposed multi-layered and composite materials could be applied to compensate this power by the way presented above.

These materials have beneficial reflective properties [44–46]. It is also necessary to use the fact that a rough (reflexed, corrugated) metal surface reduces the reflection coefficient [43]. The analogical result could be achieved if the same is done with micro-layered coatings. Nowadays, layered materials are the most advanced and comparatively cheap solution; their design technique is shown in [47]. We have to note that a more elaborated manufacture of the body of the electric rolling stock, by applying multi-layered reflective coatings, will pay back quickly and noticeably. This conclusion is evident due to the large amounts of reactive power exchanged in inter-substation zones and the ever-increasing cost of electricity spent on trains' electric traction.

In conclusion, it must be added that the presented idea in this article must be continued and added with the more comprehensive analysis of the described-below aspects. First, the article considered the field created by the return current flowing in rails because of the high values of magnetic fields. This current concentrates in rails, but its value equals only 40% of the current taken by locomotive. The other part, known as the stray current, enters to the ground and flows through the metal parts and pipelines. Because of the complexity of the process and difficulties to calculate the values in the nearest underground metal infrastructure [48–51], this current is not considered in the article. Second, for simplification, the influence of eddy currents in the metal body of the electric locomotive is not considered. In our opinion, this issue is very complex and requires separate special studies. Third, as it was written above, the method is in the conceptual phase and its model must be validated through field tests. To establish the adequacy of the "field" approach for reactive power estimation, one should take the following steps. On the first stage:

1. Select an extended electrified railway line with a few sections, with a length longer than 160 km;
2. Choose an AC electric locomotive, which will move along this section;
3. Perform hourly monitoring of field strengths at the surface of the electric loc motive's roof using broadband electric and magnetic field strength meter. Simultaneously, record the voltage across the current collector and locomotive current by oscilloscope and define its harmonic spectrum;
4. Calculate the volume of reactive energy Q_1 according to Formula (15).

At the second stage of the experimental tests, it is necessary to cover the surface of the roof of the electric locomotive with a single-layer or multi-layer coating film and perform the exact step-by-step procedure for determining the reactive power Q_2. Finally, after comparing obtained values of the reactive power, if $Q_2 > Q_1$, we could conclude that the "field" approach is adequate.

11. Conclusions

1. The currently existing methods for analyzing electric power processes in electric traction systems do not consider the fundamental law of the propagation of electromagnetic energy. It does not account accurately the energy spent on train traction;
2. Electromagnetic harmonic waves spreading in the space of inter-substation zones from the traction substation to electric rolling stock, incident on the surface of the contact wire, as well as on the frontal, lateral and roof surfaces of the body of electric rolling stock, partially reflected and summing up with the incident waves, form standing waves whose energy is reactive;
3. Reactive power in the space of inter-substation zones is a measure of the asymmetry of the rates of change in the electric and magnetic components of the energy of the electromagnetic field existing in that zone. This power is numerically equal to the difference between the average values of the energies stored in the magnetic and electric fields in the volume of the feeder zone, multiplied by 2ω;
4. To compensate for the reactive power existing in the space of inter-substation zones, it is necessary to "suppress" (damp) standing waves by applying single-layered or multi-layered film coatings with certain electromagnetic properties to particular parts of the surface of the body of the electric rolling stock;
5. The stated and theoretically substantiated "field" (based on the theory of the electromagnetic field) approach to the processes of formation, transmission, and "damping" of standing electromagnetic waves, as characteristics of reactive power in inter-substation zones, is applicable to any type of electric locomotives and AC electric traction systems. However, there are no theoretical limitations.

Author Contributions: Conceptualization, M.K.; methodology, M.K. and T.M.; software, A.N.; validation, M.K., T.M. and A.N.; formal analysis, M.K.; investigation, M.K., T.M. and A.N.; resources, M.K. and A.N.; data curation, M.K. and A.N.; writing—original draft preparation, M.K. and A.N.; writing—review and editing, M.K., A.N. and L.S.; visualization, A.N.; supervision, M.K. All authors have read and agreed to the published version of the manuscript.

Funding: This article was published with financial support under the Open Science Program which was launched at the Warsaw University of Technology as part of the project "Initiative of Excellence—Research University" (IDUB). It was also supported with internal funds of the Electrical Power Engineering Institute.

Institutional Review Board Statement: Not applicable.

Informed Consent Statement: Not applicable.

Data Availability Statement: Not applicable.

Conflicts of Interest: The author declares no conflict of interest.

References

1. Lyon, W.V. Reactive power and unbalanced circuits. *Electr. World* **1920**, *75*, 1417–1420.
2. Budeanu, C. Probleme de la presence der pussances reactsves dans les installstson de production et distribution d'energie electrigue. *Rap. Discuss. Sur la Puissance React.* **1929**, *3*, 117–218.
3. Fryze, S. Active, reactive and apparent power in circuits with nonsinusoidal voltage and current. *Przegląd Elektrotechniczny* **1931**, *7*, 193–203. (In Polish)
4. Fryze, S. Wirk-, Blind-und Scheinleistung in elektrischen Stromkreisen mit nichtsinusförmigem Verlauf von Strom und Spannung. *Elektrotechnische Z.* **1932**, *25*, 596–599. (In German)
5. Czarnecki, L.S. Interpretacja, identyfikacja i modyfikacja właściwości energetycznych obwodów jednofazowych z przebiegami odkształconymi. *Zeszyty Naukowe Politechniki Śląskiej. Elektryka* **1984**, *91*, 130. (In Polish)
6. Tonkal, V.E.; Novoseltsev, A.V.; Denisyuk, S.P.; Zhuikov, V.Y.; Strelkov, V.T.; Yatsenko, Y.A. *The Balance of Energies in Electrical Circuits*; Scientific Opinion: Kyiv, Ukraine, 1992; p. 312. (In Russian)
7. Akagi, H.; Kanazawa, Y.; Nabae, A. Generalized theory of the instantaneous reactive power in three-phase circuits. In Proceedings of the IPEC'83 International Power Electronics Conference, Tokyo, Japan, 1983; pp. 1375–1386.
8. Akagi, H.; Kanazawa, Y.; Fujita, K.; Nabae, A. Generalized theory of the instantaneous reactive power and its applications. *Electr. Eng. Jpn.* **1983**, *103*, 58–66. [CrossRef]
9. Akagi, H.; Kanazawa, Y.; Nabae, A. Instantaneous reactive power compensators comprising switching device without energy storage components. *IEEE Trans. Ind. Appl.* **1984**, *IA-20*, 625–630. [CrossRef]
10. Akagi, H.; Ogasawara, S.; Kim, H. The theory of instantaneous power in three-phase four-wire systems: A comprehensive approach. In Proceedings of the Conference Record of the IEEE Industry Applications Conference, Thirty-Forth IAS Annual Meeting, Phoenix, AZ, USA, 3–7 October 1999; pp. 431–439. [CrossRef]
11. Peng, F.Z.; Ott, G.W.; Adams, D.J. Harmonic and reactive power compensation based on the generalized instantaneous reactive power theory for three-phase four-wire systems. *IEEE Trans. Power Electron.* **1998**, *13*, 1174–1181. [CrossRef]
12. Kim, H.S.; Akagi, H. The instantaneous power theory on the rotating p-q-r reference frames. In Proceedings of the IEEE 1999 International Conference on Power Electronics and Drive Systems. PEDS'99, Hong Kong, China, 27–29 July 1999; pp. 422–427. [CrossRef]
13. Kim, H.; Blaabjerg, F.; Bak-Jensen, B. Spectral analysis of instantaneous powers in single-phase and three-phase systems with use of p-q-r theory. In Proceedings of the 2001 IEEE 32nd Annual Power Electronics Specialists Conference, Vancouver, Canada, 17–21 June 2001; pp. 54–61. [CrossRef]
14. Domnin, I.F.; Zhemerov, G.G.; Krylov, D.S.; Sokol, E.I. Modern power theories and their use in converter electronics systems. *Tech. Electrodyn.* **2004**, *1*, 80–91. (In Russian)
15. Czarnecki, L. On Some Misinterpretations of the Instantaneous Reactive Power p−q Theory. *IEEE Trans. Power Electron.* **2004**, *19*, 828–836. [CrossRef]
16. Akagi, H.; Watanabe, E.H.; Aredes, M. *Instantaneous Power Theory and Applications to Power Conditioning*, 2nd ed.; Wiley-IEEE Press: Hoboken, NJ, USA, 2017; p. 454. [CrossRef]
17. Zhezhelenko, I.V. Modern concept of reactive power. *Bull. Priazovsky State Tech. Univ.* **1995**, *7*, 192–197. (In Russian)
18. Nabae, A.; Tanake, T. A new definition of instantaneous active-reactive current and power based on instantaneous space vectors on polar coordinates in three-phase circuits. *IEEE Trans. Power Deliv.* **1996**, *11*, 1238–1243. [CrossRef]
19. Szelag, A.; Kostin, M.; Nikitenko, A. Instantaneous reactive power in DC electric traction systems. In Proceedings of the Konferencja Naukowa WD'2016 Warsztaty Doktoranckie, Lublin, Poland, 11–13 June 2016; pp. 130–131.

20. Nikitenko, A.; Kostin, M.; Szelag, A.; Jefimowski, W.; Mishchenko, T. Instantaneous reactive power in systems with stochastic electric power processes. In Proceedings of the IEEE 6th International Conference on Energy Smart Systems (ESS), Kyiv, Ukraine, 17–19 April 2019; pp. 52–57. [CrossRef]
21. Song, Y.; Liu, Z.; Rxnnquist, A.; Navik, P.; Liu, Z. Contact Wire Irregularity Stochastics and Effect on High-speed Railway Pantograph-Catenary Interactions. *IEEE Trans. Instrum. Meas.* **2020**, *69*, 8196–8206. [CrossRef]
22. Sahebdivani, S.; Arefi, H.; Maboudi, M. Rail Track Detection and Projection-Based 3D Modeling from UAV Point Cloud. *Sensors* **2020**, *20*, 5220. [CrossRef] [PubMed]
23. Kostin, M.O.; Sheikina, O.G. *Theoretical Foundations of Electrical Engineering*; Publ. Dep. of Dnipropetrovsk National University of Railway Transport named after Academician V. Lazaryan: Dnipropetrovsk, Ukraine, 2001; Volume 3 Pt 2, p. 352. (In Ukrainian)
24. Emanuel, A.E. Poynting Vector and the Physical Meaning of Nonactive Powers. *IEEE Trans. Instrum. Meas.* **2005**, *54*, 1457–1462. [CrossRef]
25. Ida, N. *Engineering Electromagnetics*; Springer: New York, NY, USA, 2000; p. 1233. [CrossRef]
26. Kadomskiy, D.E. Active and reactive power—Characteristics of the average values of operation and energy of the electromagnetic field in the elements of periodic circuits. *Electricity* **1987**, *7*, 39–43. (In Russian)
27. Agunov, M.V.; Agunov, A.V. Determination of reactive power on the basis of an electromagnetic field in a nonlinear environ-ment. *Electricity* **1993**, *2*, 67–72. (In Russian)
28. Agunov, M.V.; Agunov, A.V. On energy relations in electrical circuits with nonsinusoidal modes. *Electricity* **2005**, *4*, 53–56. (In Russian)
29. Agunov, M.V.; Agunov, A.V.; Globenco, I.G. Energy balance in electric circuits with nonsinusoidal voltage and current. *IEEE Trans. Pow. Syst.* **1997**, *12*, 1507–1510. [CrossRef]
30. Naidenko, V.I. Active and reactive energy of electromagnetic waves. In Proceedings of the XI International Conference on Antenna Theory and Techniques (ICATT), Kyiv, Ukraine, 24–27 May 2017; pp. 135–139. [CrossRef]
31. Jeltsema, D.; Kaiser, G. Active and reactive energy balance equations in active and reactive time. In Proceedings of the X International Conference on Antenna Theory and Techniques (ICATT), Kyiv, Ukraine, 29 June–1 July 2016; pp. 21–26. [CrossRef]
32. Liberal, I.; Ziolkowski, R.W. Analytical and Equivalent Circuit Models to Elucidate Power Balance in Scattering Problems. *IEEE Trans. Antennas Propag.* **2013**, *61*, 2714–2726. [CrossRef]
33. Kostin, M. Electrodynamics of electric power transmission and losses in devices of electric transport systems. In Proceedings of the MATEC Web of Conferences, 13th International Conference Modern Electrified Transport—MET'2017, Warsaw, Poland, 5–7 October 2017; Volume 180, pp. 1–6. [CrossRef]
34. Savoskin, A.N.; Kulinich, Y.M.; Alekseev, A.S. Mathematical modelling of electromagnetic processes in the dynamic system "contact network-electric locomotive". *Electricity* **2002**, *2*, 29–35. (In Russian)
35. Vlasyevsky, S.V. *Mathematical Modelling of Switching Processes in Rectifier-Invert Converters of Single-Phase DC Electric Locomotives. Monograph*; Publishing House of the Far Eastern State University of Railway Transport: Khabarovsk, Russia, 2001; p. 168. (In Russian)
36. DSTU 4344:2004. *Normal Rails for Full-Gauge Railways. General Technical Conditions. State Standard of Ukraine*; Ukrainian Science and Research Institute of Metals "UkrNDIMet": Kyiv, Ukraine, 2005. (In Ukrainian)
37. Kazakov, O.A. Reactive power as a characteristic of the transformation of the components of the electromagnetic field energy. *Eng. Phys.* **1999**, *1*, 50–59. (In Russian)
38. Pimenov, Y.V.; Volman, V.I.; Muravtsov, A.D. *Technical Electrodynamics*; Radio and Communication: Moscow, Russia, 2002; p. 536. (In Russian)
39. Turowski, J.; Turowski, M. *Engineering Electrodynamics: Electric Machine Transformer and Power Equipment Design*, 1st ed.; CRC Press Taylor&Francis Group: Boca Raton, FL, USA, 2014; p. 568.
40. Ivanov, V.B. *Wave Theory*; Pabl. Dep. of Irkutsk University: Irkutsk, Russia, 2006; p. 209. (In Russian)
41. Mishchenko, T.M. Prospects of circuit engineering solutions and simulation of subsystems of electric traction at high-speed trains. *Electr. Eng. Electr. Power Eng.* **2014**, *1*, 19–28. (In Ukrainian)
42. Latypova, A.F.; Kalinin, Y.Y. The analysis of perspective radio-absorbing materials. *Bull. Voronezh State Tech. Univ.* **2012**, *8*, 70–76. (In Russian)
43. Ashulevich, A.P. Reflection and Absorption of Electromagnetic Radiation by Multilayer and Composite Media. Ph.D. Thesis, Chelyabins State University, Chelyabinsk, Russia, 2009; p. 23. (In Russian)
44. Zhang, W.; Bie, S.; Chen, H.; Lu, Y.; Jiang, J. Electromagnetic and microware absorption properties of carbonyl iron/MnO_2 composites. *J. Magn. Magn. Mater.* **2014**, *358–359*, 1–4. [CrossRef]
45. Cheng, Y.L.; Dai, J.M.; Wu, D.J.; Sun, Y.P. Electromagnetic and microware absorption properties of carbonyl iron/$La_{0.6}Sr_{0.4}MnO_3$ composites. *J. Magn. Magn. Mater.* **2010**, *322*, 97–101. [CrossRef]
46. Zhuravlev, V.A.; Suslyaev, V.I.; Korovin, E.Y.; Dorozhkin, K.V. Electromagnetic Waves Absorbing Characteristics of Composite Material Containing Carbonyl Iron Particles. *Mater. Sci. Appl.* **2014**, *5*, 803–811. [CrossRef]
47. Vorotnitskiy, Y.I. Optimal design of multilayer electromagnetic wave absorbers. *Bolg. Fiz. Zhurnal* **1987**, *14*, 378–385. (In Russian)
48. Szeląg, A. Rail track as a lossy transmission line. Part I: Parameters and new measurement methods. *Arch. Electr. Eng.* **2000**, *49*, 407–423.

49. Szeląg, A. Rail track as a lossy transmission line. Part II: New method of measurements—simulation and in situ measurements. *Arch. Electr. Eng.* **2000**, *49*, 425–453.
50. Mariscotti, A. Stray Current Protection and Monitoring Systems: Characteristic Quantities, Assessment of Performance and Verification. *Sensors* **2020**, *20*, 6610. [CrossRef] [PubMed]
51. Chen, Z.; Koleva, D.; Breugel, K. A review on stray current-induced steel corrosion in infrastructure. *Corros. Rev.* **2017**, *35*, 397–423. [CrossRef]

Article

Review of Legal Aspects of Electrical Power Quality in Ship Systems in the Wake of the Novelisation and Implementation of IACS Rules and Requirement

Janusz Mindykowski *, Tomasz Tarasiuk and Piotr Gnaciński

Department of Ship Electrical Power Engineering, Faculty of Marine Electrical Engineering, Gdynia Maritime University, Morska St. 83, 81-225 Gdynia, Poland; t.tarasiuk@we.umg.edu.pl (T.T.); p.gnacinski@we.umg.edu.pl (P.G.)
* Correspondence: j.mindykowski@we.umg.edu.pl; Tel.: +48-58-5586-254

Abstract: This paper deals with new challenges regarding power quality in ship technology resulting from the novelisation and implementation of IACS (International Association of Classification Societies) rules and requirements. These rules, known as IACS E24 2016/2018, address harmonic distortion for ship electrical distribution systems, including harmonic filters. The reasons for the legislative changes based on a short overview of power quality-related accidents are discussed, after which a brief presentation of the updated IACS rules illustrated by a related DNV GL (Det Norske Veritas Germanischer Lloyd) case study is shown. A key part of this paper includes proposals concerning harmonics and interharmonics, distortion indices and transient disturbances. The aim of these proposals is to unify power quality indices and measurement procedures to maintain effective and comparable criteria for monitoring distortion and establish requirements for ship owners, designers, shipbuilders, classifiers, and crew members of marine objects.

Keywords: power quality; ship technology; novelisation and implementation of IACS rules

Citation: Mindykowski, J.; Tarasiuk, T.; Gnaciński, P. Review of Legal Aspects of Electrical Power Quality in Ship Systems in the Wake of the Novelisation and Implementation of IACS Rules and Requirement. *Energies* **2021**, *14*, 3151. https://doi.org/10.3390/en14113151

Academic Editors: Andrea Mariscotti and Leonardo Sandrolini

Received: 16 April 2021
Accepted: 24 May 2021
Published: 28 May 2021

1. Introduction

This paper focuses on the problem of electrical power quality and its influence on ships and shipping safety. New challenges regarding ship technology resulting from the novelisation and implementation of IACS rules and requirements will be discussed. Power quality includes two aspects [1]: continuity of a power supply and appropriate parameters of delivered and used electrical energy. This sequence is important, as electrical energy must first be continuously delivered in an appropriate quantity to the supplied system, and, second, its parameters should be kept within safety ranges. According to IEC Standard 61000-4-30 [2], power quality means 'characteristic of the electricity at a given point on an electrical system, evaluated against a set of reference technical parameters'. Usually, these parameters are voltage parameters. The related IEC Standard [3] and classification societies' rules [4–7] admit a voltage permanent deviation within −10% to +6%, ±5% frequency permanent deviation, THD (Total harmonic distortion) up to 8% and voltage unbalance up to 3%. However, aforementioned disturbances cannot be limited only to these parameters given the distinct features of a ship's power systems [8–12], which result from the specific technical solutions and operating conditions of these systems, and concern, for example, such issues as generating capacity, the power of singular load versus a single generating set, variable frequency and voltage, the application of large capacity power electronic devices, leading to a wide extension of power quality deterioration, or parallel operation of multiple power sources and a need for load-sharing control. Considering these factors, the electric power quality in ship systems should be understood as a set of parameters characterising a process of generation, distribution, and use of electric energy in all operation states of the ship (e.g., manoeuvring, sea voyage, remaining in port, cargo handling, etc.). These parameters contain the voltages, currents, and power characteristics. They were received basing on their measurement expressed by appropriate user-defined power quality indices, which must be determined. In shipbuilding technologies, it is recommended to monitor the load distribution between generating sets working in parallel. Many ship classification societies (e.g., DNV GL, ABS, or LR) define appropriate indices δPi and δQi as characterising a proportionality of the active Pi and reactive Qi powers distribution of the i-th generators working in parallel. Usually, this means that the active or reactive load of any generator is not to differ more than ±15% (active load—δPi) or 10% (reactive load—δQi) of the rated output of the largest generator from its proportionate share of the combined active or reactive load. More information concerning δPi and δQi indices, as well as a load sharing in ship electrical networks, may be found in [13]. The incorrect values of these indices are the most common cause of blackouts in ship systems. Therefore, the indices δP and δQ should be set during the operation of ship.

Although the challenges regarding power quality in ship systems formulated as power quality assessment and power quality-related safety improvements resulting from the development of ship technology have already been discussed in the literature [1,9,14,15], they should be considered again in the wake of the novelisation and implementation of the IACS E24 2016/2018 rules [16]. These rules address harmonic distortion for electrical distribution systems, including harmonic filters, and they have been written along with the recent trend to design more efficient and versatile maritime vessels and offshore objects, and new solutions have garnered attention for high-penetration power-electronic converters used in ship electric systems [9]. The availability of advanced solutions in this field have encouraged an improvement of manoeuvrability, efficiency and compactness, as well as a reduction in greenhouse gas emissions in marine vessels. Aforementioned power electronic-based solutions add many advantages to ship systems, but also increase risk factors associated with power quality and reliability. Consequently, a growing number of accidents have been registered, which threaten the safety of shipping. This problem was noticed by maritime regulatory bodies, such as IACS, and caused them to amend—and develop new—power quality standards.

Unfortunately, the IACS E24 document and other maritime power quality standards and regulations either do not cover all disturbances expected in ship power systems or are imprecise regarding the definitions of basic parameters. This concerns mainly voltage distortion disturbances, such as interharmonics and transient spikes, as there is a lack of proper indices for their assessment and established limit values. Moreover, many are doubtful about the definitions of basic factors, such as THD coefficient or reactive power [13].

Thus, we decided to concentrate on the main problem resulting from the development and wide usage of power electronic devices on-board. The solution of the main problem leads to authors' proposals to extend and unify the power quality indices and measurement procedures, and is limited only to the measurement aspects of the investigated issue.

Therefore, this paper is organised as follows: Section 2 includes the motivation for legislative changes, a brief overview of power quality-related accidents and commentary. In Section 3, a brief presentation of the updated IACS rules illustrated by a DNV GL case study is discussed. Section 4 includes the authors' proposals concerning the extension and unification of power quality indices and measurement procedures for maintaining effective and comparable criteria for monitoring distortion levels as well as establishes requirements for ship owners, shipbuilders, designers, classifiers, and crew members of marine objects. In Section 5, the short discussion and related conclusions are presented.

2. Reasons for Legislative Changes

The legislative changes introduced by IACS were motivated by an increasing number of well-documented power quality-related ship accidents. Power quality-related ship accidents [17] are closely connected with electric power quality in ship systems. This two-component interpretation, covering a risk of loss of supply continuity and a deterioration of supply quality understood as voltage characteristics in ship systems is important for analysing ship accidents. However, this does not explain why the legislative changes address harmonic distortion. A high number of ship accidents have been recently recorded [18–21], and these accidents were provoked by events, such as navigational groundings and collisions or technical failures [18–21]; however, in many cases these accidents were originally classified as power quality-related. Many of these accidents occurred on passenger ships [18,19], which are the most technologically advanced ships, often referred to as 'all-electric ships' [1,8,9]. The most technologically sophisticated ships also include large ferries, chemical and gas tankers, container vessels, oil rigs suppliers, and offshore platforms. In many cases, these are all electric ships, characterised by a rapid and continuous increase in power converters, which significantly disturb power quality in ship systems. All-electric ships are demanding in terms of power quality, technological solutions (e.g., electrical variable speed drives and harmonic filters), and staff competency (e.g., knowledge about protecting critical systems). Thus, the IACS introduced legislative changes regarding harmonic distortion for ship electrical distribution systems including harmonic filters.

To illustrate the threats connected with the impact of technological solutions and staff competencies on power quality-related accidents, an analysis of selected cases with the authors' research and professional experiences, as well as a brief overview of the existing case studies [22–25] was carried out. The results of this analysis are shown in Table 1.

Table 1. Analysis of select power quality accidents [17,22–24].

Case Study	Type and Name of Ship, Year of the Event	Type of Ship Power Plant	Kind of Accident	Accident Reasons	
				Direct	Indirect
1	Passenger cruise liner, RMS Queen Mary 2, 2010 	Engine room of CODLAG (combined diesel electric and gas turbine), integrated electric propulsion; all-electric ship, 11 kV network	The catastrophic failure of a capacitor in the aft harmonic filter room, and explosion in the aft main switchboard; temporary loss of vessel manoeuvrability	The initial degradation of harmonic filter capacitor construction	A lack of continuous monitoring of electric power quality; shortcomings in ship tests, and operation
2	Passenger ship, MS Statendam, 2002 	Diesel-electric-generation system cooperating with azipods/propulsion motors; All-electric ship, 6.6 kV network	Arc-flash event in a main circuit breaker, and fire accident in the main switchboard room	Dead short caused the failure of DG2 circuit breaker, causing an explosion and fire in the main switchboard room	A lack of analysis of damage symptoms; a lack of sufficient qualifications of marine engineers, including electricians on-board
3	Oil platform, Tern Alpha, 2006 	Platform electrical distribution system is based on all-electrical ship system solution. Technical data concerning power generation module are not publicly available	Explosion in the gas compression module, fire, personnel evacuation, stopping of the drilling process	Overheating of a high-voltage electrical motor	High level of distortion in the platform electrical distribution system

The categories related to capacitor failures in various circuits and systems, respectively, either to arc accidents or to malfunctions of protection relay systems can be assigned considering a brief overview of the power quality accidents. Explosion, fire, or loss of the main propulsion and manoeuvrability had occurred in the cases analysed. This resulted with at least the economic losses. Loss of the main propulsion was a factor in numerous investigations, and the probability of power loss in the ship electric propulsion system (SEPS) was evaluated [26]. The proposed probability-evaluation method to access SEPS power losses is based on Bayesian Belief Networks (BBNs). The BBN structure of power loss in an SEPS considers five main components under disruption: input power, cables, transformer, inverter and motor. Because this paper is focused on the power loss contingency, we considered only disruption contingencies leading to power loss. In the presented case study, the estimated value of the power loss probability and degree of importance of all components are shown, classifying the inverter component as one of the most sensitive components for disruption elements in the SEPS structure. Considering that the authors of the discussed paper [26] are aware of the need for further validation and adjustments of the model, the proposed method to evaluate the probability of SEPS power loss is promising.

Only hypothetical technical causes of the accidents were indicated in some cases by the authorities competent to investigate the circumstances and causes of ship accidents [22,27] (Table 1). Other cases concluded that the reasons were ambiguities regarding IMO (International Maritime Organization) meanings [28] and the KUP (knowledge, understanding, and proficiency) competencies of watchkeeping officers [22,23]. In some cases, both technical and competence-related components appeared jointly [22]. In the third case (Table 1), concerning a fire alert on the North Sea Platform ('Tern Alpha' [24]), more general conclusions were formulated. The authors of those papers state that 'electrical power quality is absolutely fundamental to the safety and operational integrity of drilling rigs, offshore platforms and installation worldwide'.

This opinion is justified by the related case studies [29,30]. In the aforementioned papers, Evans presents well-documented cases concerning typical power quality issues, high-frequency harmonics, and the operation of explosion-proof motors, large main AC propulsion drives, PWM drives with active front end (AFE) as variable frequency drives (VFD), and common-mode voltage (CMV) in VFDs. Moreover, in other works by this author, the themes of continuous power quality monitoring and power quality issues on existing ships are discussed. The first observation led to conclusions that electrical variable speed drives are fundamental to most operations in the oil and drilling industries. Consequently, harmonic voltage distortion offshore can exceed the recommended limits by a factor of 4–7. Finally, unacceptable power quality level can negatively impact safety, productivity, and profitability. Accident investigations can last for years and the results are usually not publicly available, so drawing conclusions from recent events is based solely on a deductive analysis of the causes and effects of ship accidents. Therefore, regulatory authorities, such as IACS should implement and oversee these processes.

The above is only a confirmation that the problems related to the marine power quality should be solved. This can also be added that authors has conducted research on board of fifteen ships and minor or serious power quality problems were detected on four vessels.

3. Updated IACS Rules and a Related DNV GL Case Study

The introduced changes are described in the document IACS E24 2016/2018 [15] and address voltage harmonic distortion for ship electrical distribution systems including harmonic filters. This document contains five sections. The first section defines the scope of the unified requirements (UR) as applied to these ships, where harmonic filters are installed on the main busbars of electrical distribution systems, excluding those installed for single-application frequency drives, such as pump motors. The second section ('General') defines a fundamental issue—that is, 'the total harmonic distortion (THD) of electrical distribution systems does not exceed 8%'.

Authors' note: unfortunately, this appointment is not comprehensive and should be clarified. The cited statement is imprecise, as it does not explain how to define the THD indices. Generally accepted approaches define the indices as the ratio of the RMS value of the sum of given harmonic components of h order to the fundamental component expressed in percentage. However, these approaches are often ineffective and lead to significant errors [1,8,14]. Two questions, then, must be answered: what kind of disturbances (not only harmonics in the limited frequency band) could be considered? Which frequency band is to be considered? The authors' proposals concerning these points are presented in Section 4 of this paper. The third section of the IACS UR is dedicated to monitoring harmonic distortion levels for a ship including harmonic filters. Subsection 3.1 states that 'the ships are to be fitted with facilities to continuously monitor the levels of harmonic distortion experienced on the main busbar, as well as alerting the crew should the level of harmonic distortion exceed the acceptable limits, where the engine room is provided with automation systems, this reading should be logged electronically, otherwise it is to be recorded in the engine log book for future inspection by the surveyor'. Subsection 3.2 states that 'harmonic distortion levels of the main busbar on board such existing ships are to be measured annually under seagoing conditions as close to the periodical machinery survey as possible so as to give a clear representation of the condition of the entire plant to the surveyor. Harmonic distortion readings are to be carried out when the greatest amount of distortion is indicated by the measuring equipment. An entry showing which equipment was running and/or filters in service is to be recorded in the log so this can be replicated for the next periodical survey'.

Additionally, information concerning the necessity of distortion measurements and their records following any modification to the ship's electrical distribution system or associated consumers by suitably trained ship's personnel (or from a qualified outside source) are formulated. Point 4 of the note in this section states that the UR E24 Rev.1—except for Subsection 3.2—is to be uniformly implemented by IACS 'for ships contracted for construction on or after 1 January 2020 or for ships where an application for a periodical or occasional machinery survey after the retrofit of harmonic filters is dated on or after 1 January 2020'. This point substitutes point 1 concerning the same issue but with the date changed from June 2016 to 'on or after 1 July 2017'. Point 2 of the note states that 'Subsection 3.2 is to be uniformly implemented by IACS for ships contracted for construction before 1 July 2017, at any scheduled machinery periodical survey having a due date on or after 1 July 2017'. Finally, point 3 explains the term contracted for construction in the context of the related date.

Section 4 of IACS UR 2016/2018 is referred to as the mitigation of the effects of harmonic-filter failure on a ship's operation, or 'where the electrical distribution system on board a ship includes harmonic filters, the system integrator of the distribution system is to show, by calculation, the effect of a failure of a harmonic filter on the level of harmonic distortion experienced'. Then, 'the system integrator of the distribution system is to provide the ship owner with guidance documenting permitted modes of operation of the electrical distribution system while maintaining harmonic distortion levels within acceptable limits during normal operation as well as following the failure of any combination of harmonic filters'.

Finally, there is an important additional requirement that 'The calculation results and validity of the guidance provided are to be verified by the surveyor during sea trials'. The last point of the IACS UR under discussion is devoted to protection arrangements for harmonic filters, and the requirements are as follows: 'Arrangements are to be provided to alert the crew in the event of activation of the protection of a harmonic filter circuit. A harmonic filter should be arranged as a three phase unit with individual protection of each phase. The activation of the protection arrangement in a single phase shall result in automatic disconnection of the complete filter. Additionally, there shall be installed a current unbalance detection system independent of the overcurrent protection alerting the crew in case of current unbalance'.

Information about the consideration of additional protection for the individual capacitor element is also added to this section.

Author's note: these new regulations are in line with and correspond well to safety lessons included in the Marine Accident Investigation Branch report [22]: '... Regular monitoring of electrical networks should be undertaken to provide early warning deterioration. Monitoring equipment should be capable of detecting transient voltage spikes, resonances, and excessive harmonic distortions levels (either continuously or periodically). ... Protection systems for critical equipment must "fail safe", and should be thoroughly tested at regular intervals to prove that all sub-components are functioning correctly. In particular, harmonic filters with current unbalance protection systems should be thoroughly checked by a competent person at the earliest opportunity'. The recurring reference to 'competent person' is consistent with the description 'by suitably trained ship's personnel' (Subsection 3.2), and they respond well to the need to compensate for the gap in staff qualification competencies shown in the analysis of reasons for power quality-related accidents (Table 1). This problem has been solved by the novelisation of the IMO International Convention on Standards of Training, Certification, and Watchkeeping for Seafarers. The STCW '78/2010 [28] Convention and is described, in the part concerning minimum standards of competence for Electro-Technical Officer (ETO). More information about this issue in particular reflects competence in terms of power systems in excess of 1000 V, being one of the most important competencies related to power quality [31]. The response of the marine sector represented by marine classification societies was to update their rules and implementation procedures regarding electrical installations for new ships, and survey requirements for existing ships.

An illustration of how the IACS UR E24 2016/2018 rules are introduced and implemented in a DNV GL case study is presented and discussed below. The implementation of the IACS Unified Requirements is illustrated in Table 2, based on comparative analysis of the IACS UR E24 [15] and the DNV GL Rules for Ships [6], respectively.

In addition to the changes in the marine classification societies rules shown in Table 2, which are focused mainly on improving technological solutions and their verification (Part 4, 7, and 8, DNV GL 2019), staff competency is also being improved (Subsection 3.2, updated IACS rules). Finally, the requirement to implement on or after 1 July 2017, has been postponed to on or after 1 January, 2020. This shows that implementing the rules was difficult, resulting from, among other things, a lack of unification of power quality indices and measurement procedures in accordance with maintaining effective and comparable criteria for monitoring harmonic distortion levels.

Table 2. Implementation of the IACS UR E24 2016/2018 rules [5], based on a case study of DNV GL 2019 [6].

IACS UR E24 2016/2018 Rules		DNV GL Rules for Ships, 2019	
Place of Description	**Considered Issue**	**Place of Description**	**Way of Implementation**
Section 1	Scope To what kind of ship are the requirements of this UR applied?	Part 4, Chapter 8, Section 1, point 2.2	Documentation related to system design shall be submitted when more than 20% of connected load is by semiconductor assemblies, in relation to connected generating capacity; also required where the electrical distribution system on board a ship includes harmonic filters. In such cases, the effects of a filter failure shall also be calculated. At the same time, in Part 4, Chapter 1, Section 2, point 1.2.8 (c), a related exclusion is placed. Harmonic filters integrated in frequency converters, or installed for single consumers, such as pump motors, may be excluded from these requirements.

Table 2. *Cont.*

IACS UR E24 2016/2018 Rules		DNV GL Rules for Ships, 2019	
Place of Description	**Considered Issue**	**Place of Description**	**Way of Implementation**
Section 2	General The limit of the total harmonic distortion (THD) and possibilities of its relaxation	Part 4, Chapter 8, Section 2, point 1.2.7	The DNV GL rules concerning the requirements connected with limit values and operation conditions of operating harmonic distortion as well as some recommendations concerning the relaxation conditions are included in Part 4, Chapter 1, Section 2, point 1.2.7: '1.2.7 Harmonic distortion (a) Equipment producing transient voltage, frequency and current variations shall not cause malfunction of other equipment on board, neither by conduction, induction or radiation. (b) In distribution systems the acceptance limits for voltage harmonic distortion shall correspond to IEC 61000-2-4 Class 2. In addition, no single order harmonic shall exceed 5%. Guidance note: IEC 61000-2-4 Class 2 implies that the total voltage harmonic distortion shall not exceed 8%.' Authors' note: THD is a ratio of the RMS value of the sum of all the harmonic components up to the 50th order to the RMS value of the fundamental component.' (c) The total harmonic distortion may exceed the values given in (b) under the condition that all consumers and distribution equipment subjected to the increased distortion level have been designed to withstand the actual levels. The system and components ability to withstand the actual levels shall be documented. (d) When filters are used for limitation of harmonic distortion, special precautions shall be taken so that load shedding or tripping of consumers, or phase back of converters, do not cause transient voltages in the system in excess of the requirements in [1.2.4]. Guidance note: the following effects should be considered when designing for higher harmonic distortion, refer to (c): - Additional heat losses in machines, transformers, coils of switchgear and controlgear; - Additional heat losses in capacitors for example in compensated fluorescent lighting; - Resonance effects in the network; - Functioning of instruments and control systems subjected to the distortion; - Distortion of the accuracy of measuring instruments and protective gear (relays); -Interference of electronic equipment of all kinds, for example regulators, communication and control systems, position-finding systems, radar and navigation systems. A declaration or guarantee from system responsible may be an acceptable level of documentation.'

Table 2. *Cont.*

IACS UR E24 2016/2018 Rules		DNV GL Rules for Ships, 2019	
Place of Description	**Considered Issue**	**Place of Description**	**Way of Implementation**
Section 3	Monitoring of harmonic distortion levels for ships including harmonic filters Subsection 3.1; for new ships, facilities to continuously monitor the levels of harmonic distortion experienced on the busbar as well as alerting the crew should this level exceed acceptable levels	Part 4, Chapter 8, Section 2, point 1.2.8	Implementation of the related IACS UR concerning continuous power quality monitoring with alarm procedures is directly placed in Part 4, Chapter 1, Section 2, point 1.2.8 (b): 'Where the electrical distribution system on board a ship includes harmonic filter units, the levels of harmonic distortion experienced on the main busbar shall be continuously monitored. Should the level of harmonic distortion exceed the acceptable limits, an alarm shall be given at a manned location. For ships with class notation **E0**, the alarm shall be logged.'
	Subsection 3.2; for existing ships, as minimum harmonic distortion levels of main busbar are to be measured annually under seagoing conditions as close to the periodical machinery survey.	Part 7, Chapter 1, Section 2, point 3.1.5	Implementation of the related IACS UR concerning harmonic-distortion-levels control, together with measurement and their records conditions, in reference to existing ships, is in Part 7, Chapter 1, Section 2, point 3.1.5: 'For electrical installations the survey shall include: - Examination of main source of electrical power with respect to general condition, fire hazard and personnel safety, i.e., generators, main switchboards, distribution boards, control gear, consumers, chargers, and battery/UPS systems. - For all E0, AUT, or AUT-nh vessels (built at any time) and all vessels constructed on or after 1 July, 1998 where electricity is necessary for propulsion and steering, test of automatic start and connection to the switchboard of the standby generator set, shall be carried out. - Where the electrical distribution system on board a ship includes harmonic filters (with exception of pumps' prime movers): - Harmonic distortion levels of main busbar on board such existing ships shall to be measured annually under seagoing conditions as close to the periodical machinery survey as feasible. Records of all the above measurements shall to be made available to the surveyor. Each measurement shall be taken at maximum distortion levels and identical conditions. Guidance note: This requirement applies for ships contracted for construction before 1 July 2017.'

Table 2. *Cont.*

IACS UR E24 2016/2018 Rules		DNV GL Rules for Ships, 2019	
Place of Description	**Considered Issue**	**Place of Description**	**Way of Implementation**
Section 4	Mitigation of the effects of harmonic filter failure on a ship's operation. The system integrator of the distribution system with harmonic filters is to show, by calculation, the effect of a failure of a harmonic filter on the level of harmonic distortion experienced. The system integrator of the distribution system is to provide the ship owner with guidance documenting permitted modes of electrical distribution system, while maintaining harmonic distortion within acceptable limits during different operation conditions of the system for any combination of harmonic filters. The calculation results and validity of the guidance provided are to be verified by the surveyor during sea trials.	Part 8, Chapter 8, Section 2, point 1.2.8 and point 3.1.3	Implementation for the related Section of IACS UR is in Part 4, Chapter 1, Section 2, point 1.2.8 (a): 'Passive and active harmonic filter assemblies/units: (a) Where the electrical distribution system on board a ship includes harmonic filter units, the system integrator of the distribution system shall show, by calculation, the effect of a failure of a harmonic filter unit on the level of harmonic distortion experienced. The system integrator of the distribution system shall provide the ship owner with guidance documenting permitted modes of operation of the electrical distribution system while maintaining harmonic distortion levels within acceptable limits. The system integrator shall also calculate the harmonic distortion that will be experienced in case of a failure of a harmonic filter, and provide guidance on mitigating actions as operating modes or reduced power levels' This information is also in point 3.1.3: 'Harmonic distortion (a) All equipment shall be constructed to operate at any load up to the rated load, with a supply voltage containing harmonic distortion as given in Section 2, point 1.2.7.'
Section 5	Protection arrangements for harmonic filters. Arrangements are to be provided to alert the crew in the event of activation of the protection of a harmonic-filter circuit. The constructional contains as a three-phase unit with individual protection of each phase, as well as the conditions concerning a current unbalance detection system independent of the overcurrent protection alerting the crew.		The references in the DNV GL rules to comply with the related Section 5 of the IACS UR are included in the Chapters 8 and 1, of Parts 4 and 7, respectively. Additionally, the DNV GL rules for ships in some places are cited by the guidance note 'See IACS UR E24', and this indicates planned accordance of both documents.

4. Authors' Proposals and Future Works

Considering the need for unifying power quality indices and measurement procedures in accordance with maintaining effective and comparable criteria for monitoring distortion levels, we propose to consider the unified, extended interpretation of THD indices, as well as introduce the parameters describing interharmonics and transient phenomena to unified power quality standards. The reasons for these proposals result, first, from the IACS general philosophy that '... regular monitoring of electrical networks should be undertaken to provide early warning of deterioration. Monitoring equipment should be capable of detecting transient voltage spikes, resonances, and excessive harmonic distortion levels (either continuously or periodically)'. Second, it is a pragmatic point of view, which reflects the need to design and implement new solutions for power quality-monitoring devices. In this context, the future challenges and expectations concern the implementation of a newly designed and commonly accepted power quality module, installed in the main switchboard for continuously monitoring distortion and the implementation of a portable measuring instrument for the realisation of the same task. The former, stationary option is generally dedicated to the current-detection of power quality deterioration and corresponding threats. This option should be linked with the ship alarm system. The latter, portable option is linked to routine control and power quality troubleshooting, as well as to the periodical realisation of continuous monitoring of harmonic distortion, such as the stationary version. A new design of the measuring devices will be a good opportunity to cover other power quality measurements—for example, hitherto avoided by the classification societies, a detection of transient disturbances or interharmonics issues.

4.1. Harmonics, Subharmonics, and Interharmonics

The continuous distortion of voltage and current waveforms can be modelled by the Fourier series as harmonics, interharmonics, and subharmonics (subsynchronous interharmonics). However, the current maritime power quality requirements cover only the harmonics. Most of the rules of ship classification societies require that 'no single-order harmonic shall exceed 5%' [6]. However, if the maximum harmonics order is not determined, the 50th order is considered [7]. The exception is IEEE Standard 45.1-2017, which stipulates that 'for parallel active filters or active PWM front-end type VSDs (PWM type), related harmonics often exceed the 49th order. Therefore, when this type of VSD is used, calculations and measurements should take into consideration harmonics up to the 100th order' [32]. However, this provision is insufficient, and this problem will be discussed in the next subsection.

Still, the problem of subharmonics and interharmonics (SI) may exists. In this study, these are understood as components of the frequency less than or greater than the fundamental harmonic, being not its integer multiple (although other definitions are used [33]).The problem with this most popular definition is related to the frequency resolution of measuring algorithms.

First, on-board installation contains appliances that can generate SI of significant values. SI are produced by power-electronic equipment connecting two AC systems of various frequencies through a DC link. Examples of such equipment are inverters [33,34]. The output current of the inverter causes voltage ripples in the DC link, which can be transmitted to the input side. Consequently, the supply network may produce SI of frequencies equal to the multiple of the output inverter frequency [33,34]. For instance, in a 60-Hz power system supplied from diesel-driven generators (DG) feeding high-power inverters, the SI of the following frequencies were reported [35]: 45 Hz, 135 Hz, 225 Hz, and 405 Hz of values 0.9%, 1.17%, 0.89%, and 0.931%, respectively. This SI contamination was observed in a land power network, which shows some similarities to the ship network because of the application of DG and the high-power converters. DG may produce SI if one of the cylinders works improperly, especially if the frequency of torque pulsations corresponds to the resonant frequency of the generator (the first natural frequency of the rigid-body mode) [35].

Another significant source of SI are AC motors working with the load of the pulsating anti-torque—for example, reciprocating compressors [35,36]. Even the motor of relatively low power can inject into the grid SI of comparatively high values. For instance, according to [36], the work of a 7.5-kW induction motor driving an air compressor resulted in the occurrence of SI of values up to 0.4% and frequencies equal to 37.5 Hz and 62.5 Hz. SI are also produced by cycloconverters and time-varying loads [33].

Further, SI are particularly harmful power quality disturbances. They exert noxious impacts on various components of power systems, like synchronous generators, transformers, power electronic equipment and measurement and control systems [33,37–40]. An energy receiver that is especially sensitive to SI is an induction motor. SI—notably, of the frequency less than the doubled fundamental frequency—cause local saturation of the magnetic circuit [41], an increase of power losses [42], overheating [43], excessive vibration [44,45], and pulsation of the rotational torque [45]. For an exemplary low-power induction motor [44], the voltage subharmonic greater than 0.2%may cause excessive vibration. Moreover, extraordinary vibration was observed for voltage interharmonics [45]. The particularly detrimental effect of SI is torque pulsations. For high- and medium-power motors, their frequency may correspond to the natural frequency of the first elastic mode [45]. Possible torsional resonance may amplify torque pulsations as much as 50 times [46]; consequently, excessive torsional vibration due to SI may destroy a clutch or a shaft.

At the same time, failure of an induction motor in a land system will incur high economic losses whereas in a marine system, it may also threaten the safety of ships, the environment, passengers or crew. For example, a breakdown of an auxiliary motor in the propulsion subsystem (for instance, a motor driving a fuel pump) may stop the main engine. Thus, SI contamination in a marine power system is unacceptable.

These considerations demonstrate the necessity for imposing limitations on voltage SI in marine power systems. The simplest way to do this is to adopt admissible SI levels from standards concerning land power systems. Unfortunately, the standards generally do not impose limitations on SI, as 'levels are under consideration, pending more experience' [47]. The task of determining SI levels to include them in power quality standards is complicated [33,45]. Separate limitations should be determined according to the criterion of correct work on various electrical equipment—for example, power converters, modern light sources, induction motors—and the lowest of them should be introduced into power quality standards. The limitations should also consider harmful phenomena of a different nature. For example, in the case of induction motors, the limits should consider vibration and torque pulsations; and for subharmonics, excessive winding heating [43–45]. Determining appropriate SI limits may take a long time; however, for safety reasons, temporary limits could be introduced into the rules of the ship classification societies. The limitations should first concern the most harmful SI—namely, the frequency less than the doubled fundamental frequency.

These temporary limits may be based on the incomplete research results of SI as well as on previous proposals for SI limitations. De Abreau et al. [48] and Fuchs et al. [43] proposed limitations on subharmonics or interharmonics to 0.1%. The proposals were justified by the detrimental SI impact on induction motors. Other considered levels are 0.2%, 1%, 3%, and 5% [33]. However, some of these levels do not ensure that induction motors work correctly. As mentioned previously, subharmonics exceeding merely 0.2% may cause excessive vibration [44]. Consequently, to provide effective protection for induction motors against vibrations due to SI, the temporary limit should consider wide safety margins. Further, it is not clear which values of SI are admissible regarding torsional vibration.

Considering the above advisement as well as the previous proposals [43,48], the most appropriate temporary limit is 0.1% for any SI (or their sub-group) of the frequency less than the doubled fundamental frequency.

4.2. Distortion Indices

The current ship classification rules and IACS requirements introduce distortion parameters designated as THD, without providing a proper definition of THD, except for Lloyd, who introduces the traditional definition by the following formula:

$$THD_n = \frac{\sqrt{\sum_{h=2}^{n} V_h^2}}{V_1} \cdot 100 \quad (1)$$

where: V_h is RMS value of voltage h order harmonic, V_1 is voltage fundamental component RMS value.

The calculation of THD should cover all harmonics up to the 50th order [7]. Two main drawbacks of this formula are that it covers only harmonics, and the frequency band is not determined or is insufficient for systems with active PWM front-end drives. The topology of such a drive is presented in Figure 1. The authors' original research concerning such a system is presented in Figure 2 (voltage and current waveforms) and Figure 3 (voltage spectrum). The example was registered on-board of DP ship with electrical propulsion. During the research two generators worked in parallel with rated power of 425 kVA and 200 kVA. Rated voltage was equal to 400 V and rated frequency 50 Hz. The two AFE PWM drives with the rate power of 300 kW were used for the ship propulsion but the actual load was equal to 90 kW for each drive (sea going with reduced speed—half ahead). What is important, the switching frequency of each drive was equal to 3.6 kHz. This results in the distortions in the frequency band around the switching frequency, which is clearly visible in Figure 3a. For comparison reason the voltage spectrum registered during harbour manoeuvring of ship with drive containing 6-pulse rectifier was shown (bow thruster with rated frequency of 125 kW) in Figure 3b. The generator with rated power of 376 KVA worked during the registration. For both ships the computer with DAQ board was used. The sampling frequency was 30 kHz and cut-off frequency of anti-aliasing filter was set to 10 kHz.

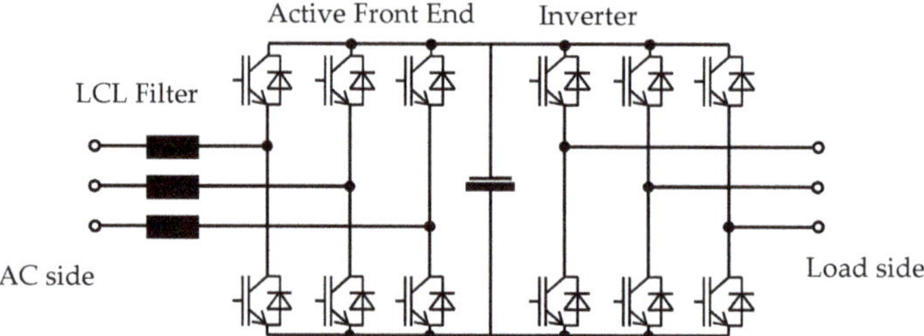

Figure 1. Application of an IGBT 'Active Front End' to an AC PWM drive.

As seen in Figure 3, the operation of the input IGBT input bridge rectifier significantly reduces the waveform distortions of conventional AC PWM drives with 6-pulse diode bridges or 12-pulse bridges. However, it can also introduce significant high-order distortions, above the 50th order. The voltage THD calculated for the example in Figures 2a and 3 is equal to 1.86%, and it does not cover the visible distortions above 50th-order harmonics. Unfortunately, most of the available harmonic-monitoring devices cover this frequency band. However, extending the typical THD definition to 100th-order harmonics, as recommended by [33], is not a solution, as the calculated THD increases up to merely 1.88%, whereas the actual level of distortions reaches 2.65% combining both: harmonics and interharmonics (Equation (2)). Therefore, the authors have proposed [49] a more traditional definition of the distortion factor as the ratio of the RMS value of the residue, after elimination of the fundamental, to the RMS value of the fundamental component, expressed as a percentage. This is directly related to the concept of distortion factor introduced in the last century [50]. Some authors proposed to designate such a distortion factor as total waveform distortion (TWD) [51]. The definition is expressed as follows:

$$TWD_n = \frac{\sqrt{V_{rms}^2 - V_1^2}}{V_1} \cdot 100 \quad (2)$$

where: V_{rms} is voltage RMS value, V_1 is voltage fundamental component RMS value.

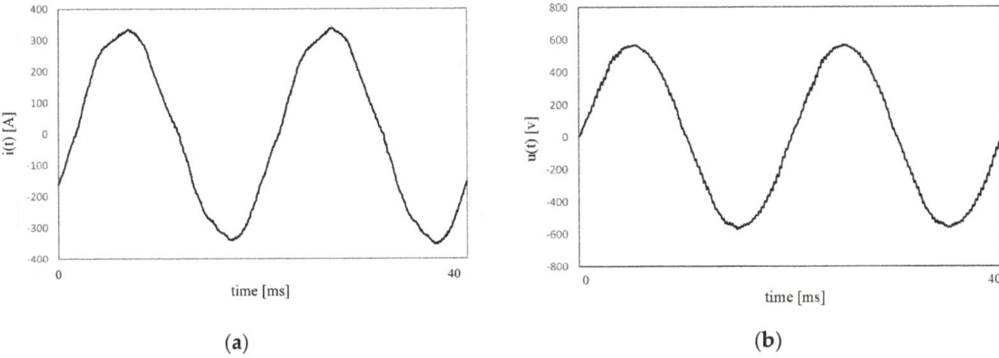

Figure 2. 'Active Front End' input current and voltage waveforms at main bus bars.

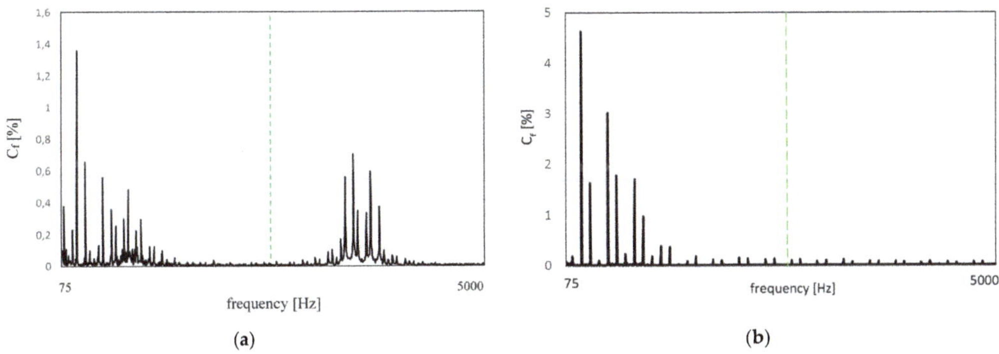

Figure 3. Voltage spectra at main bus bars of the maritime systems with AC PWM drives with an 'Active Front End' (**a**) and at main bus bars of the maritime systems with 6-pulse diode bridge (**b**). Note the distortions above the 50th order, marked by the green dashed line. C_f is the content of higher frequency components in relation to fundamental voltage, both harmonics and interharmonics.

This Equation (2) assumes values equal to or greater than THD defined by (1). Therefore, a comparison of the results of the calculation by both formulae enable a rough assessment of the interharmonics or components above the assumed harmonic frequency range. Further, detailed analysis can be carried out by the following formula (version for systems with a rated frequency equal to 50 Hz).

$$TWD_{2.5-10\text{kHz}} = \frac{V_{rms-2.5-10\text{kHz}}}{V_1} \cdot 100 \quad (3)$$

where: $V_{rms\text{-}2\cdot5-10\text{ kHz}}$ is RMS value of combined voltage component in the frequency range from 2.5 kHz up to 10 kHz, V_1 is voltage fundamental component RMS value.

In fact the factor is similar like *TWD* described by Equation (2), but narrower frequency band. This can be defined as ratio of RMS value of components in the frequency band above harmonic frequency range up to 10 kHz and RMS value of fundamental component expressed in percentage. The interpretation is similar like in the case of *THD*, which combines harmonics in harmonic frequency range. The $TWD_{2\cdot5\text{–}10\,kHz}$ combines components above the harmonic frequency range, and this can be considered as combined distortions in sub-bands of 200 Hz like proposed in IEC 61000-4-7 [52]. The new factor enables rough identification of frequency band affected by distortions and provide additional information for diagnosis and troubleshooting.

However, analysis of all the above distortion factors may be insufficient for this case analysis. This concerns the waveforms with low-level interharmonics, but which are still above the threshold values of 0.1–0.2% suggested in Subsection 4.1. As such, it is advisable to measure the total interharmonic distortion factor (TIHD), defined as follows:

$$TIHD_n = \frac{\sqrt{\sum_{ih=1.5}^{n} V_{ih}^2}}{V_1} \cdot 100 \quad (4)$$

where: V_{ih} is the RMS value of the interharmonic subgroup, V_1 is voltage fundamental component RMS value.

The interharmonic subgroup is defined in the IEC 61000-4-7 standard [52], as the square root of the sum square of the RMS values of the interharmonic frequency bins. For example, the spectrum of the original voltage registered at the navigation equipment terminal on-board a ship with AFE PWM drives is shown in Figure 4. This was registered on-board of above mentioned DP ship. The only difference is place (230 V 50 Hz system and time (before propulsion converters replacement).

In the presented case, the components above the 50th harmonic were dominant. This effect was due to poor design of the system and the choice of low-cost drives. The distortions are caused by the operation of the AFE power converters with a declared switching frequency of 3.6 kHz. Therefore, an appropriate analysis of this case requires covering the frequencies above the 50th harmonic, including interharmonics. The authors calculated the above proposed distortion indices, and the results are in Table 3.

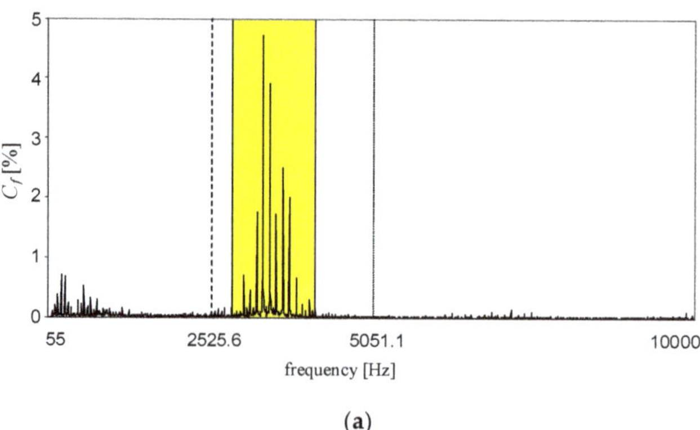

(a)

Figure 4. *Cont.*

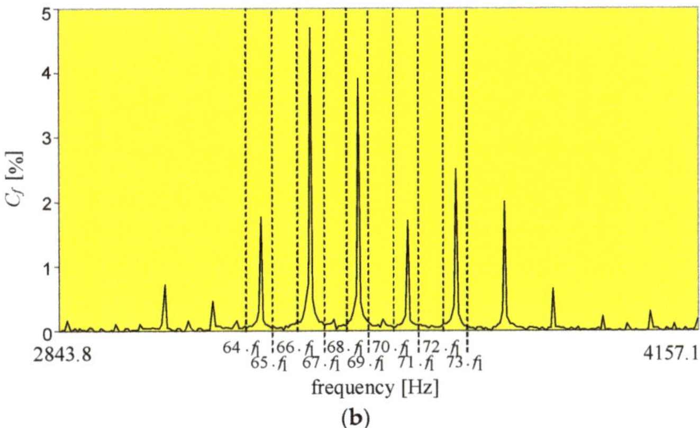

(**b**)

Figure 4. Exemplary waveform of supply voltage; (**a**) the analysed frequency sub-band is marked with yellow; the dashed line marks the frequency of the 50th harmonic whereas the dotted line marks the frequency of the 100th harmonic; (**b**) the extended frequency sub-band 2843.8–4157.1 Hz is marked with yellow; the dashed lines mark the harmonic frequencies as integer multiples of the fundamental frequency equal to 50.51 Hz [1]. C_f is the content of higher frequency components in relation to fundamental voltage, both: harmonics and interharmonics.

Table 3. Distortion indices calculated for the example from Figure 4.

Distortion Factor	Value
THD (up to the 50th harmonic)	1.39%
THD (up to the 100th harmonic)	1.61%
TWD (up to the 100th harmonic frequency range)	8.30%
TWD (up to 10 kHz)	8.37%
TWD2.5–10 kHz	8.22%
TIHD (up to the 50th harmonic frequency range)	0.68%
TIHD (up to the 100th harmonic frequency range	8.14%

Proper assessment of this case requires determining at least the following combination of the distortion factors:

- TWD up to the 100th harmonic frequency range or 10 kHz, TWD2.5–10 kHz, TIHD up to the 50th harmonic frequency range;
- THD up to 50th harmonic frequency range, TWD2.5–10 kHz, TIHD up to the 50th harmonic frequency range.

The above proposal is a necessary minimum for properly assessing waveform distortions at maritime microgrids. For an extension of the diagnosis capabilities, detailed information about harmonic subgroups and interharmonic subgroups should be provided. Moreover, the frequency range from the 50th harmonic to 10 kHz can be divided into subranges (e.g., 200-Hz wide, as proposed in the IEC 6100-4-7 standard [52]). Unfortunately, appropriate measuring devices do not exist; however, the authors have constructed a prototype device [53,54], which, after a minor software update, will be able to perform the task in real time.

4.3. Transient Disturbances

According to a Marine Accident Investigation Branch report [22], 'Monitoring equipment should be capable of detecting transient voltage spikes'. Unfortunately, this recommendation was not included in the Unified Requirements of IACS, as only the requirement for continuously monitoring the THD factor was introduced [15], despite voltage spikes also being present in maritime microgrids. The occurrence of these spikes on shipboards is related to switching generators, harmonic filters and large receivers on and off as well as malfunctioning power-electronic equipment. The consequences of these spikes can be dangerous for ships, crew and the environment. For instance, the possible reason for the harmonic filter failure aboard Queen Mary II was 'being exposed to frequent voltage transients due to increased number of switching cycles' [22]. A similar situation was observed aboard a chemical tanker; due to a malfunction in the automatic voltage regulator (AVR), the random transient was observed during idle run of the generator with rated power of 1062 kVA, an example of which is shown in Figure 5. The rated voltage was equal to 440 V and rated frequency to 60 Hz. The voltage waveform was registered with sampling frequency equal to 150,375 Hz in the frequency band up to 50 kHz (cut-off frequency of anti-aliasing filter). It must be stressed that the phenomena are usually hard to capture and determining their magnitude strongly depends on frequency bandwidth of registering device.

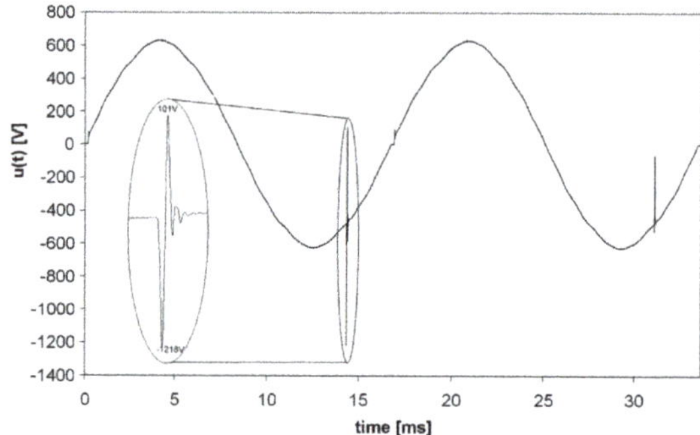

Figure 5. Example transient registered in a ship network [55].

The transient shown in Figure 5 also led to a failure of the capacitor in the generator AVR. Figure 5 also indicates the main problems with continuously monitoring such a phenomenon. These are related to their random nature, short duration, high amplitudes, and high frequencies. According to the IEEE 1159 standard [56], the transients are categorised as conducted high-frequency phenomena and can be divided into impulsive and oscillatory transients. The duration can be even below 50 ns, and frequencies above a few MHz. Typical magnitude can reach 4–8 p.u. [56]. Thus, ordinary power system performance-monitoring instruments are not appropriate for detecting and evaluating transient spikes.

Next, there are no commonly accepted indices for evaluating these phenomena. Nevertheless, some popular methods are listed in IEC Standard 61000-4-30 [2]. The most common are peak value, the rate of rise of the leading edge, frequency parameters, duration, the frequency of occurrence and energy. Hitherto, the standards related to maritime microgrids rarely deal with transient spikes. One exception is IEEE Standard 45.1-2017 [32], which states that voltage spikes should be below ±2500 V for 380–600 V systems. Unfortunately, the required minimal frequency band for transient detection was not proposed, which is vital for the results of the voltage spike magnitude measurement. For instance, the result of a magnitude measurement of the transient spike presented in Figure 5 increases linearly with an increase of the upper limit of the input frequency band of the measuring instrument [55].

The next problem concerns methods for detecting transients, for which there are many solutions. These are based on original signals or the extracted transient. The IEC 61000-4-30 lists the most common (e.g., based on absolute instantaneous value, envelope analysis after removing fundamental component, the sliding window method based on a comparison of instantaneous values with corresponding values of previous cycles or dv/dt method, etc. [2]). The authors have conducted research on the methods of transient detection in maritime microgrids. Finally, a simple solution based on a wavelet transform combined with absolute-value analysis and the dv/dt method was proposed [55]. The method consists of preliminary wavelet decomposition and partial synthesis and subsequent implementation of the following formula (5):

$$(\,|\,s_k - s_{k-1})\,|\, > tr_1 \cup \,|\,s_k\,|\,) > tr_2 \tag{5}$$

where the s_k is an instantaneous sample of the extracted transient signal, s_{k-1} is the previous sample and tr_1 and tr_2 are threshold values. If condition (5) has the logical value 'true', then an impulse occurs. The simplicity of the formula allows fast, real-time transient detection, in fact starting and ending points of spike. This is the necessary condition for determining a number of transient spike indices, e.g., duration, energy. The threshold values tr_1 and tr_2 were chosen after experimental investigation in maritime microgrids, as $tr_1 = 0.2Vn$ and $tr_2 = 0.06Vn$, where Vn is the nominal voltage amplitude [56]. The results of the application of the proposed solution for the transient depicted in Figure 5 is graphically presented in Figure 6.

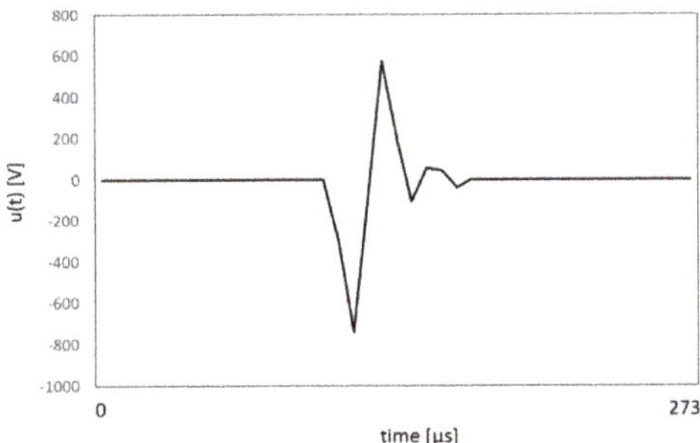

Figure 6. Extracted transient using wavelet transform and formula (5).

The proposed solution enables fast transient detection and determination of its duration. Next, the proper transient indices can be determined on the basis of instantaneous values s_k. This can be easily implemented in measuring instruments designed for power quality online monitoring. However, some precautions characteristic of the input channel of the measuring device, such as the sampling frequency and utilising an analog-to-digital converter input range, must be considered. The authors have constructed such a device, called an estimator-analyser of power quality, which is capable of measuring typical power quality indices in the frequency band up to 9 kHz, and online monitoring of voltage spikes in the frequency band up to 60 kHz [55]. For each detected transient, three indices are determined: voltage peak-to-peak value, duration, and energy. This method of detecting transients enables easy extension of the measuring device functions to add more transient parameters, including absolute peak values (with fundamental), rate of rise, and duration above predefined threshold values.

In summary, it is necessary to amend the Unified Requirements of IACS to include the requirement of the detection of transient spikes concurrently with measuring the THD factor. However, more research is yet to be done, namely, the requirements for the minimal frequency band of monitoring devices should be introduced, and threshold values for the proposed indices should be determined. This will require analysing typical voltage transient spikes in maritime microgrids as well as the susceptibility of installed equipment to detect both singular and accumulated, long-lasting spikes.

5. Discussion and Conclusions

The present regulations of ship classification societies do not provide sufficient protection of on-board equipment against malfunction due to excessive power quality disturbances. Considering numerous power quality-related ship accidents, the modification of rules of ship classification societies is proposed. The modifications concern determination of power quality indices for the assessment of voltage waveform distortions, as well as admissible levels of voltage subharmonics and interharmonics.

The presented case study and related discussion show (Section 4.2, Table 3) that implementation of IACS E24 rules encountered some difficulties, resulting from, among others, a lack of unification of power quality indices and measurement procedures in accordance with keeping the effective and comparable criteria for monitoring of harmonic distortion levels.

Additionally, the IACS E24 document and other maritime power quality standards and regulations do not cover all disturbances expected in a ship power system or are imprecise in basic parameter definitions at best. This concerns mainly the voltage distortion disturbances, such as interharmonics and transient spikes. There is a lack of proper indices for their assessment and established limit values. Moreover there are even doubts about precise definitions of such a basic factor like the THD coefficient.

Bearing in mind the aforementioned state-of-the-art, the authors decided to concentrate on the main problem resulting from technology development and wide usage of power electronic devices on the shipboard, and present their proposals concerning harmonics and interharmonics (Section 4.1), distortion indices (Section 4.2), and transient disturbances (Section 4.3). The main conclusions concerning the authors' proposals and future works in the discussed power quality-related matter may be formulated as follows: because of extraordinary harmfulness of voltage SI, limitations concerning these power quality disturbances should be introduced to the rules of ship classification societies. At the same time the investigations on the impact of SI on electrical equipment are still incomplete and for the reason determination of the target SI limitations may last for comparatively long time. Nevertheless, because of the safety when afloat, the temporary admissible limits of SI could be established, at least for the SI of the frequency less than the doubled mains frequency. Considering the considerations presented in Section 4.2, as well as the previous proposals [44,49], the most appropriate temporary limit is 0.1% for any SI (or their sub-group) of the frequency less than the doubled fundamental one.

Careful analysis of results laid in Table 3 (Section 4.2) leads to the conclusion that proper assessment of the discussed case requires determining at least following combination of the distortion factors:

- TWD up to 100th harmonic frequency range or 10 kHz, TWD2.5–10 kHz, TIHD up to 50th harmonic frequency range;
- THD up to 50th harmonic frequency range, TWD2.5–10 kHz, TIHD up to 50th harmonic frequency range.

The above presented detailed proposal can be considered as the necessary minimum for proper assessment of waveform distortions at maritime microgrids. For extension of diagnosis capabilities, more detailed information about harmonic subgroups and interharmonic subgroups should be provided. Moreover, the frequency range from the 50th harmonic to 10 kHz can be divided into subranges, e.g., 200 Hz wide as proposed in IEC 6100-4-7 standard [52]. Unfortunately, the measuring devices with required functionalities do not exist on the market. The construction of the desired power quality monitoring device for maritime microgrids is necessary and will be easy. The authors have constructed a prototype of an appropriate device [53,54], which, after a minor software update, will be able to perform the task in real time.

With regard to transient disturbances analysis (Subsection 4.3), it seems of utmost necessity to amend the Unified Requirements of IACS and include the requirement of detection of transient spikes concurrently with measurement THD factor. However some research is yet to be done. Namely, the requirements for the minimal frequency band of monitoring devices should be introduced and threshold values for the above proposed indices should be determined. This will require the analysis of typical voltage transient spike characteristics in maritime microgrids and analysis of installed equipment susceptibility for the phenomena, both singular spike as well as accumulated effect of long lasting spikes.

It is worth emphasising that all of the highlighted power quality issues can be resolved and/or prevented if the correct and comprehensive technical guidance is obtained. Specialist marine power quality consultants and experts can assist ship owners, designers, shipbuilders, surveyors, and crew members in these matters, but this must go hand-in-hand with a comprehensive upgrade of marine classification society rules.

Author Contributions: Conceptualization, J.M.; formal analysis, T.T.; investigation, J.M., T.T., and P.G.; writing—original draft preparation, J.M., T.T., and P.G.; writing—review and editing, J.M., T.T., and P.G. All authors have read and agreed to the published version of the manuscript.

Funding: This research received no external funding.

Institutional Review Board Statement: Not applicable.

Informed Consent Statement: Not applicable.

Data Availability Statement: Not applicable.

Acknowledgments: We appreciate Marcin Miotk from the classification society DNV for his consultancy in this study. We are also grateful to DNV for providing us with company procedures for implementing the updated IACS rules and requirements.

Conflicts of Interest: The authors declare no conflict of interest.

References

1. Mindykowski, J.; Tarasiuk, T. Problems of power quality in the wake of ship technology development. *Ocean. Eng.* **2015**, *107*, 108–117. [CrossRef]
2. *Electromagnetic Compatibility (EMC)—Part 4–30: Testing and Measurement Techniques—Power Quality Measurement Methods*; IEC: Geneva, Switzerland, 2015.
3. *Electrical Installations in Ships—Part 101: Definitions and General Requirements*; IEC: Geneva, Switzerland, 2018; Standard 60092-101.
4. *Requirements Concerning Electrical and Electronic Installations*; International Association of Classification Societies IACS Req.: Hamburg, Germany, 2019.
5. Part 4 Vessel systems and machinery. In *Rules for Building and Classing Steel Vessels*; American Bureau of Shipping ABS: Houston, TX, USA, 2019.
6. *Rules for Classification: Ships. Part 4 Systems and Components. Chapter 8 Electrical Installations*; July 2019 ed., Amended October 2019; DNV GL: Høvik, Norway, 2019.
7. *Rules and Regulations for the Classification of Ships*; Lloyd's Register of Shipping: London, UK, 2019.
8. Kumar, D.; Zare, F. Comprehensive review of maritime microgrids: System architectures, energy efficiency, power—Quality and regulations. *IEEE Access* **2019**, *7*, 1–11. [CrossRef]
9. Mindykowski, J. Contemporary challenges to power quality in ship systems—metrological perspective. In Proceedings of the 22nd IMEKO TC4 International Symposium 20th, International Workshop on ADC Modelling and Testing, Iasi, Romania, 14–15 September 2017; pp. 536–558.
10. Tarasiuk, T. Angular frequency variations at microgrids and its impact on measuring instrument performance. *IET Gener. Transm. Distrib.* **2016**, *10*, 3234–3240. [CrossRef]
11. Gnaciński, P.; Tarasiuk, T.; Mindykowski, J.; Pepliński, M.; Górniak, M.; Hallmann, D.; Piłat, A. Power quality and energy Efficient operation of marine induction motors. *IEEE Access* **2020**, *8*, 152193–152203. [CrossRef]
12. Tarasiuk, T.; Zunino, Y.; Bruno-Lopez, M.; Silvestro, F.; Piłat, A.; Molinas, M. Frequency fluctuations in marine microgrids: Origins and identification tools. *IEEE Electrif. Mag.* **2020**, *8*, 40–46. [CrossRef]
13. Tarasiuk, T.; Górniak, M. Load sharing in ship microgrids under nonsinusoidal conditions—Case study. *IEEE Trans. Energy Convers.* **2017**, *32*, 810–819. [CrossRef]
14. Tarasiuk, T.; Górniak, M.; Liu, W.; Savagheli, M.; Vasquez, J.C.; Chun-Lien, S.; Guerrero, J.M. Power quality assessment in shipboard microgrids under unbalanced and harmonic AC bus voltage. *IEEE Trans. Ind. Appl.* **2019**, *55*, 765–775.
15. Mindykowski, J. Case study—Based overview of some contemporary challenges to power quality in ship systems. *Inventions* **2016**, *1*, 12. [CrossRef]
16. *Requirements Concerning Electrical and Electronic Installations, Harmonic Distortions for Ship Electrical Distribution System Including Harmonic Filters*; June 2016, Rev.1 December 2018; International Association of Classification Societies, IACS UR E24: Hamburg, Germany, 2018.

17. Mindykowski, J. Impact of staff competences on power quality—Related ship accidents. In Proceedings of the 2018 International Conference and Exposition on Electrical and Power Engineering, Iasi, Romania, 1–6 October 2018.
18. Łozowicka, D.; Kaup, M. Analysis of the cause and effect of passenger ship accidents in the Baltic Sea. *Sci. J. Marit. Univ. Szczec.* **2015**, *44*, 68–73.
19. Ugurlu, O.; Yildiz, S.; Loughney, S.; Wang, J. Modified human factor analysis and classification system for passenger vessel accidents (HFACS-PV). *Ocean. Eng.* **2018**, *161*, 47–61. [CrossRef]
20. Mazaheri, A.; Montewka, J.; Nisula, J.; Kujala, P. Usability of accident and incident reports for evidence–based risk modelling—A case study on ship grounding reports. *Saf. Sci.* **2015**, *76*, 202–221. [CrossRef]
21. Chavin, C.; Lardiane, S.; Morel, G.; Clostermann, J.; Langard, B. Human and organizational factors in maritime accidents: Analysis of collisions at sea using the HFACS. *Accid. Anal. Prev.* **2013**, *59*, 26–37. [CrossRef]
22. Marine Accident Investigation Branch. *Report of the Investigation of the Catastrophic Failure of a Capacitor in the Aft Harmonic Filter Room on Board RMS Queen Mary 2 While Approaching Barcelona on 23 September 2010*; Marine Accident Investigation Branch: Southampton, UK, December 2011. Available online: http://www.maib.gov.uk/publications/investigation_reports/2011/qm2.cfm (accessed on 31 March 2021).
23. Transportation Safety Board of Canada. *Marine Investigation Report MO2W0135 Switchboard Fire, Passenger Vessel "Statendam", Strait of Georgia, British Columbia 04 August 2002*; Transportation Safety Board, Minister of Public Works Government Services: Canada, 2005. Available online: https://tsb.gc.ca/eng/rapports-reports/marine/2002/m02w0135/m02w0135.pdf (accessed on 31 March 2021).
24. *Fire alert on North Sea Platform*; Website Publication: Sumburgh, Scotland, 2006. Available online: http://news.bbc.co.uk/2/hi/uk_news/scotland/4811538.stm (accessed on 31 March 2021).
25. Evans, I.C. The perils of offshore power quality. *Offshore Eng.* **2014**, 56–59. Available online: https://www.oedigital.com/news/444795-the-perils-of-offshore-power-quality (accessed on 24 May 2021).
26. Zhang, W.; Yan, X.; Zhang, D.; Zhang, J. Evaluating the probability of power loss in ship electric propulsion system based on Bayesian belief networks. *Mar. Technol. Soc. J.* **2019**, *53*, 1–17. [CrossRef]
27. Maritime and Coastguard Agency; Draft Marine Guidance Note (MGN). *Potential Hazards of Excessive Harmonic Distortion of Current and Voltage of Onboard Electrical Systems*; Maritime and Coastguard Agency: Southampton, UK, 2010.
28. International Maritime Organization. *IMO International Convention on Standards of Training, Certification and Watchkeeping for Seafarers, including Manila Amendments, STCW Convention and STCW Code*; International Maritime Organization: London, UK, 2011.
29. Evans, I.C. High frequency harmonics and marine power quality. *Mot. Ship* **2019**, 1–5. Available online: https://www.motorship.com/news101/industry-news/high-frequency-harmonics-and-marine-power-quality (accessed on 24 May 2021).
30. Evans, I.C. The future is electric—Let us safeguard it. *Mot. Ship* **2020**, 1–5. Available online: https://www.motorship.com/news101/monitoring,-control-and-digitalization/the-future-is-electric-let-us-safeguard-it (accessed on 24 May 2021).
31. Mindykowski, J. Towards safety improvement: Implementation and assessment of new standards of competence for electro-technical officers on ships. *Marit. Policy Manag.* **2017**, *44*, 336–357. [CrossRef]
32. *IEEE Recommended Practice for Electrical Installations on Shipboard-Design*; IEEE: Piscataway, NJ, USA, 2017; p. 45.1.
33. Testa, A.; Akram, M.F.; Burch, R.; Carpinelli, G.; Chang, G.; Dinavahi, V.; Hatziadoniu, C.; Grady, W.M.; Gunther, E.; Halpin, M.; et al. Interharmonics: Theory and modeling. *IEEE Trans. Power Deliv.* **2007**, *22*, 2335–2348. [CrossRef]
34. Nassif, A.B. Assessing the impact of harmonics and interharmonics of top and mudpump variable frequency drives in drilling rigs. *IEEE Trans. Ind. Appl.* **2019**, *55*, 5574–5583. [CrossRef]
35. Arkkio, A.; Cederström, S.; Awan, H.A.A.; Saarakkala, S.E.; Holopainen, T.P. Additional losses of electrical machines under torsional vibration. *IEEE Trans. Energy Convers.* **2018**, *33*, 245–251. [CrossRef]
36. Zhiyuan, M.; Xiong, M.W.; Le, L.; Zhong, X. Interharmonics analysis of a 7.5 kW air compressor motor. *CIRED Open Access Proc. J.* **2017**, *2017*, 738–741. [CrossRef]
37. Guillen-Garcia, E.; Zorita-Lamadrid, A.L.; Duque-Perez, O.; Morales-Velazquez, L.; Osornio-Rios, R.A.; Romero-Troncoso, R.D.J. Power consumption analysis of electrical installations at healthcare facility. *Energies* **2017**, *10*, 64. [CrossRef]
38. Hou, R.; Wu, J.; Song, H.; Qu, Y.; Xu, D. Applying directly modified RDFT method in active power filter for the power quality improvement of the weak power grid. *Energies* **2020**, *13*, 4884. [CrossRef]
39. Feola, L.; Langella, R.; Testa, A. On the effects of unbalances, harmonics and interharmonics on PLL systems. *IEEE Trans. Instrum. Meas.* **2013**, *62*, 2399–2409. [CrossRef]
40. Testa, A.; Langella, R. Power system subharmonics. In Proceedings of the IEEE Power Engineering Society General Meeting, San Francisco, CA, USA, 16 June 2005; Volume 3, pp. 2237–2242.
41. Ghaseminezhad, M.; Doroudi, A.; Hosseinian, S.H.; Jalilian, A. An investigation of induction motor saturation under voltage fluctuation conditions. *J. Magn.* **2017**, *22*, 306–314. [CrossRef]
42. Ghaseminezhad, M.; Doroudi, A.; Hosseinian, S.H.; Jalilian, A. Analysis of voltage fluctuation impact on induction motors by an innovative equivalent circuit considering the speed changes. *IET Gener. Transm. Distrib.* **2017**, *11*, 512–519. [CrossRef]
43. Fuchs, E.F.; Roesler, D.J.; Masoum, M.A.S. Are harmonics recommendations according to IEEE and IEC too restrictive? *IEEE Trans. Power Deliv.* **2004**, *19*, 1775–1786. [CrossRef]
44. Gnaciński, P.; Pepliński, M.; Murawski, L.; Szeleziński, A. Vibration of induction machine supplied with voltage containing subharmonics and interharmonics. *IEEE Trans. Energy Convers.* **2019**, *34*, 1928–1937. [CrossRef]

45. Gnaciński, P.; Hallmann, D.; Klimczak, P.; Muc, A.; Pepliński, M. Effects of voltage interharmonics on cage induction motors. *Energies* **2021**, *14*, 1218. [CrossRef]
46. Feese, T.; Ryan, M. Torsional vibration problem with motor/ID fan system due to PWM variable frequency drive. In Proceedings of the 37th Turbomachinery Symposium 2008, Texas A&M University, Turbomachinery Laboratories, TX, USA, 8–11 September 2008.
47. *Voltage Characteristics of Electricity Supplied by Public Distribution Network*; EN Standard 50160: Brussels, Belgium, 2010.
48. De Abreu, J.P.G.; Emanuel, A.E. The need to limit subharmonics injection. In Proceedings of the 9th International Conference on Harmonics and Quality of Power, Orlando, FL, USA, 1–4 October 2000; Volume 1, pp. 251–253.
49. Tarasiuk, T.; Mindykowski, J. An extended interpretation of THD concept in the wake of ship electric power systems research. *Measurement* **2012**, *45*, 207–212. [CrossRef]
50. Instruments and Measurements. Annual Report of Committee on Instruments and Measurements. *Trans. Am. Inst. Electr. Eng.* **1929**, *48*, 1385–1390. [CrossRef]
51. Bollen, M.; Gu, I. *Signal processing of power quality disturbances*; Wiley-Interscience: Hoboken, NJ, USA, 2006.
52. *General Guide on Harmonics and Interharmonics Measurements for Power Supply Systems and Equipment Connected Thereto*; IEC, Std.: Geneva, Switzerland, 2008; IEC 61000-4-7.
53. Tarasiuk, T. Estimator-analyser of power quality: Part I—Methods and algorithms. *Measurement* **2011**, *44*, 238–247. [CrossRef]
54. Tarasiuk, T.; Szweda, M.; Tarasiuk, M. Estimator-analyser of power quality: Part II—Hardware and research results. *Measurement* **2011**, *44*, 248–258. [CrossRef]
55. Tarasiuk, T.; Szweda, M. DSP instrument for transient monitoring. *Comput. Stand. Interfaces* **2011**, *33*, 182–190. [CrossRef]
56. *Recommended Practice for Monitoring Electric Power Quality*; IEEE 1159–2019 IEEE: Piscataway, NJ, USA, 1994.

MDPI
St. Alban-Anlage 66
4052 Basel
Switzerland
Tel. +41 61 683 77 34
Fax +41 61 302 89 18
www.mdpi.com

Energies Editorial Office
E-mail: energies@mdpi.com
www.mdpi.com/journal/energies